SCHAUM'S OUTLINE OF

THEORY AND PROBLEMS

OF

OPERATIONS MANAGEMENT

Second Edition

•

JOSEPH G. MONKS, Ph.D.

Professor of Operations Management
Gonzaga University

SCHAUM'S OUTLINE SERIES

McGRAW-HILL

New York   San Francisco   Washington, D.C. Auckland Bogotá
Caracas Lisbon London Madrid Mexico City Milan
Montreal New Delhi San Juan Singapore
Sydney Tokyo Toronto

D1324129

Dedicated to
Mary, The Queen of Peace

JOSEPH G. MONKS is a professional mechanical and industrial engineer with a Ph.D. in Business Administration from the University of Washington. His experience includes positions with Westinghouse, General Electric, and the U.S. government and some consulting work. Dr. Monks taught at Oregon State University and in Europe prior to joining Gonzaga University. He holds professional certification in Production and Inventory Management (CPIM) and has written articles and other books in the areas of statistics and operations management.

Schaum's Outline of Theory and Problems of
OPERATIONS MANAGEMENT

Copyright © 1996, 1985 by The McGraw-Hill Companies, Inc. All rights reserved. Printed in the United States of America. Except as permitted under the Copyright Act of 1976, no part of this publication may be reproduced or distributed in any form or by any means, or stored in a data base or retrieval system, without the prior written permission of the publisher.

8 9 10 11 12 13 14 15 16 17 18 19 20 CUS CUS 09 08 07

ISBN 0-07-042764-X

Sponsoring Editor, Jeanne Flagg
Production Supervisor, Leroy A. Young
Editing Supervisor, Patty Andrews

Library of Congress Cataloging-in-Publication Data

Monks, Joseph G.
    Schaum's outline of theory and problems of operations management /
Joseph G. Monks. ‑‑ 2nd ed.
        p.   cm. ‑‑ (Schaum's outline series)
    Includes index.
    ISBN 0-07-042764-X
        1. Production management.   2. Industrial engineering.   I. Title.
    II. Title: Theory and problems of operations management.
    III. Title: Operations management.
    TS155.M674   1995
    658.5 ‑‑ dc20   94-41845
                                                          CIP

McGraw-Hill
A Division of The McGraw·Hill Companies

Preface

This Outline has three major uses: (1) as a supplement to current texts in production and operations management, (2) as a text in its own right, (3) as a study guide for practicing professionals, including those who are preparing for the American Production and Inventory Control Society (APICS) certification examinations. Key features of the Outline are its comprehensive coverage, condensed theory, and multitude of questions, examples, and solved problems.

As a supplement and/or text for production and operations management courses, the Outline presents materials normally taught at the upper college or introductory graduate (MBA) level. Although it covers all topics generally included in AACSB accredited courses, the interesting but nonessential descriptive materials are minimized. Instead, theory is straightforward, precise, and somewhat concentrated. Emphasis is on applications and examples, which are worked out in a systematic fashion.

As a reference guide for practicing managers, the Outline has independent chapters which offer relatively complete coverage of a given topic area. In addition, Chapters 2, 4, and 11 through 15 cover production and inventory control topics of special importance to production and inventory control managers. These materials can be used as one (of many) references for preparation to take the APICS certification examinations.

The Outline makes extensive use of quantitative techniques. However, a background in college algebra is sufficient for general understanding throughout. Statistical, calculus, and quantitative techniques are explained largely by example as the needs for them arise. The quantitative topic shown in smaller type after a chapter title indicates that a more detailed explanation of that quantitative method is contained in the chapter. Designed for quantitative support, the Outline tends to favor simplicity and clarity of presentation over mathematical sophistication.

In addition to a substantial updating throughout, this second edition of the Outline incorporates significant advances in both theory and application. Theory has been enhanced by the early addition of quality concepts, a recognition of the impact of instantaneous information (e.g., via internet), and the need for responsible management of competitive operations on a global basis. Applications have been enhanced by the addition of numerous answered questions in each chapter. These questions synthesize key nonquantitative concepts much the same way examples and solved problems have traditionally been used to help clarify quantitative methodologies. This second edition should thus provide students with a basic understanding of the issues facing today's operations managers.

My thanks to all the staff at Gonzaga University and at McGraw-Hill who helped bring this second edition to completion.

JOSEPH G. MONKS

Contents

Determinants and Objectives. Types of Layouts. Process (Functional) Layout Methods. Line Balancing in Product Layouts. Computerized and Mixed-Model Line Balancing.

Sampling Plans. Operating Characteristic Curves. Sampling Plans for Attributes. Sampling Plans for Variables. Average Outgoing Quality Levels. Control Charts, Control Limits, and Tolerances. Process Control for Variables via Control Charts. Process Control for Attributes via Control Charts.

Chapter 1

Operations Management:
Definition, Mission, and Productivity Concepts

HISTORY

Productive activities are the foundation of a nation's economic system. They transform human, material, capital, informational, and other resources into higher-valued goods and services. Figure 1-1 identifies some key individuals and events in the development of productive systems over the past 250 years. Although the contributors mentioned in the figure are associated primarily with the United States, significant contributions have come from other countries as well (i.e., European countries, Canada, Australia, Asia, and elsewhere).

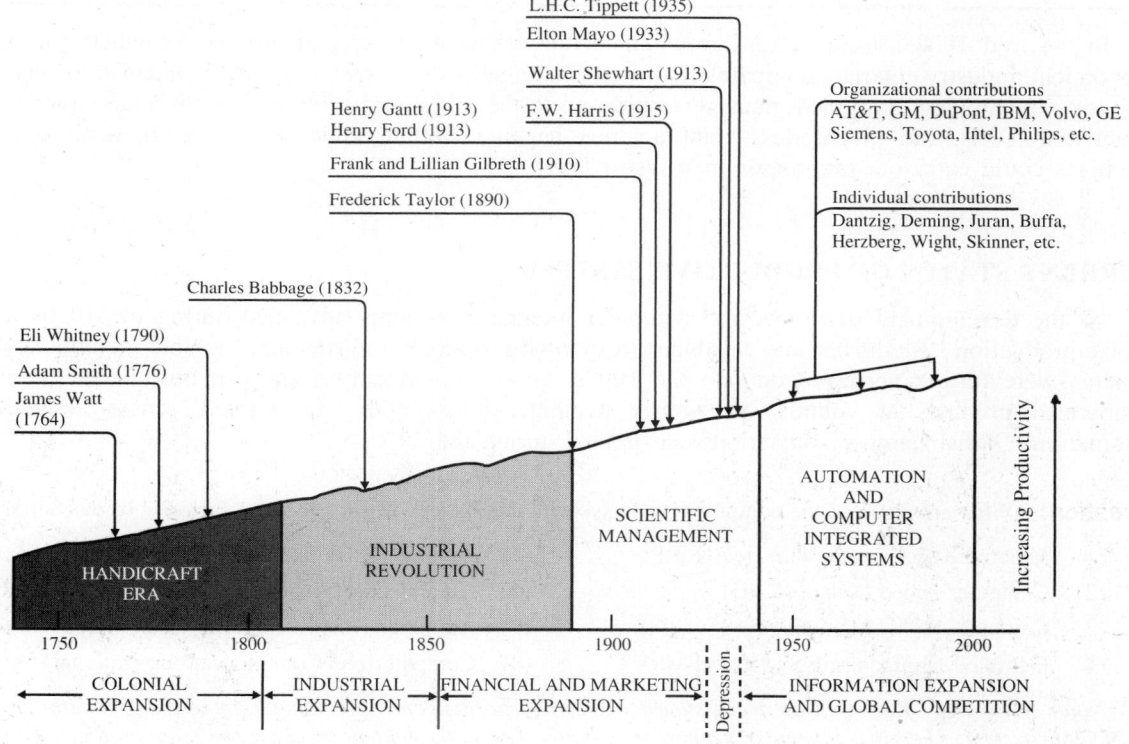

Fig. 1-1 Key individuals and events in the development of production systems

James Watt's steam engine (1764) advanced the use of mechanical power, and Adam Smith (1776) publicized the advantages of the division of labor. In the United States, the Constitution (1789) encouraged capital investment and trade, and the Civil War, along with the expanding railroad system, spurred early industrial development. Eli Whitney advanced the concept of "interchangeable parts," and Charles Babbage promoted an economic analysis of work and pay on the basis of skill requirements.

Growth of the factory system was rapid, for there was no well-established production system to supplant and unskilled labor was available. In the early 1900s Frederick Taylor's work ushered in the

1

scientific management era and endowed him with the title of "Father of Scientific Management." As better machinery and automatic controls were developed, much of the heavy manufacturing effort in the United States and Europe was directed toward mass production of similar products (e.g., "black" Ford automobiles).

Question: Identify the major contribution of key individuals associated with the scientific management era.

Contributor	Major Contribution
Frederick Taylor	Philosophy of scientific management, use of training, time study, and standards
Frank and Lillian Gilbreth	Motion economy and human factors
Henry Gantt	Use of scheduling systems
Henry Ford	Assembly-line mass production
F. W. Harris	Economic order quantity (EOQ) model
Walter Shewhart	Statistical quality control
Elton Mayo	Attention to behavioral factors
L. H. C. Tippett	Work sampling

In the mid 1900s, as operations research techniques were developed and as computers became economical, industry entered an unparalleled age of automation—which was quickly extended to service activities. Computers provided managers with up-to-the-minute information about markets, costs, production levels, and inventories. Manufacturers began installing logic units in equipment so that machines could carry out preprogrammed instructions.

CURRENT STATUS OF PRODUCTIVE SYSTEMS

As the development of robots and computer integrated systems advanced during the 1970s and 1980s, production systems became capable of responding to online information. Flexible manufacturing systems were further perfected on into the 1990s, giving manufacturers the capability of delivering customized products at volumes previously available only under "hard-wired" mass-production automation. Globalization of operations accelerated during the 1990s.

Question: What are some of the computer-based systems that have/will support the "factories of the future"?

(1) Material Requirements Planning (MRP)
(2) Computer-Aided Design (CAD)
(3) Computer-Aided Manufacturing (CAM)
(4) Flexible Manufacturing Systems (FMS)

(5) Automated Guided Vehicle (AGV)
(6) Automated Storage/Retrieval Systems (AS/RS)
(7) Automatic Identification Systems (AIS)
(8) Computerized Communications (Internet)

Note: Many of these systems are assumed to be incorporated into a computer integrated manufacturing (CIM) system, wherein a computer network essentially directs the flow of production—from customer orders back through all processing to contacts with suppliers.

Today, computers link suppliers, producers, and consumers around the world; and firms everywhere are forced to compete on a global basis. Much of this competition is for services, which now employ about three times as many workers as manufacturing activities. Manufacturing plants are even more highly computerized, and corporations are continuously "restructured" to reduce overhead and enhance their responsiveness to customers. This has resulted in a shift of some manufacturing to lesser-developed countries, and of some nonmanufacturing activities (e.g., clerical and data analysis) to homes as well as to foreign countries (via phone, FAX, the Internet, and wireless communicators, etc.).

Question: What are some of the major ways firms are meeting global competition?

(1) By utilizing *higher-technology* capital equipment (often in a foreign location)

(2) By employing part-time or *lower-paid workers* (often in a foreign location)

(3) By *improving operations* (e.g., total quality management) to world-class levels

Question: What are some of the more significant macro (economic/social) effects arising from the changes that are taking place in production systems today?

(1) Instant *communication* (voice and data) virtually anywhere in the world

(2) Extensive *computer integration* of vendors, producers, and customers

(3) *Shifting of high-tech facilities* from more-developed to less-developed countries

(4) *Oversupply of low-skilled workers* and shortage of technically trained personnel

(5) Continued *shift of employment* from manufacturing to service (and informational) activities, and a shift of these jobs to the home and to foreign locations

MANAGEMENT: MISSION, STRATEGY, AND TACTICS

The individuals most directly responsible for making an organization's resources productive, by skillfully guiding the operation of productive systems, are managers. Three prominent theories developed to explain the role of managers are the (1) functional, (2) behavioral, and (3) decision-making (systems) approaches.

Question: Briefly describe the three approaches to management.

(1) *Functional* is the traditional (classical) approach that holds that managers plan, organize, direct, and control the activities of an organization.

(2) *Behavioral* is a human relations approach that emphasizes interpersonal relationships and organizational behavior. Under it, managers work through other people to lead the activities of an organization.

(3) *Decision-making* (*systems*) is an approach that focuses upon the use of data and quantitative techniques for making decisions that facilitate system goals. Managers are primarily decision makers within an operating system.

Managers must, of course, have a blend of (1) functional (and technical) capability, (2) behavioral competence to work with people individually and in groups, and (3) analytical skills.

Question: Define management.

Management is the process of developing decisions and taking actions to direct the activities of people within an organization toward common objectives.

The objectives of organizations differ, but most have multiple goals that include the concerns of stockholders, employees, customers, and society in general. Many organizations find it useful to characterize themselves in terms of their mission, strategies, and tactics.

Question: Define (*a*) mission, (*b*) strategy, and (*c*) tactic.

(*a*) A *mission* is a general statement of the organization's purpose, or reason for existence.

(*b*) A *strategy* is a plan for using the organization's resources to accomplish an objective.

(*c*) A *tactic* is an action or means used to carry out a strategy.

The focus of attention narrows as one moves from mission to strategy to tactic. Figure 1-2 uses an upside-down pyramid concept to illustrate how broad-based objectives (e.g., good customer service) are eventually operationalized via specific rules (e.g., maintain a safety stock of 50 units). It also shows how the organization's database forms a foundation for decision-making at all levels. Note that managerial values influence objectives, policies, plans, procedures, and rules by filtering down through the organizational structure—perhaps in a subtle way. Values can influence the database, and factual data can modify values. Healthy organizations exhibit values such as honesty, respect, and justice.

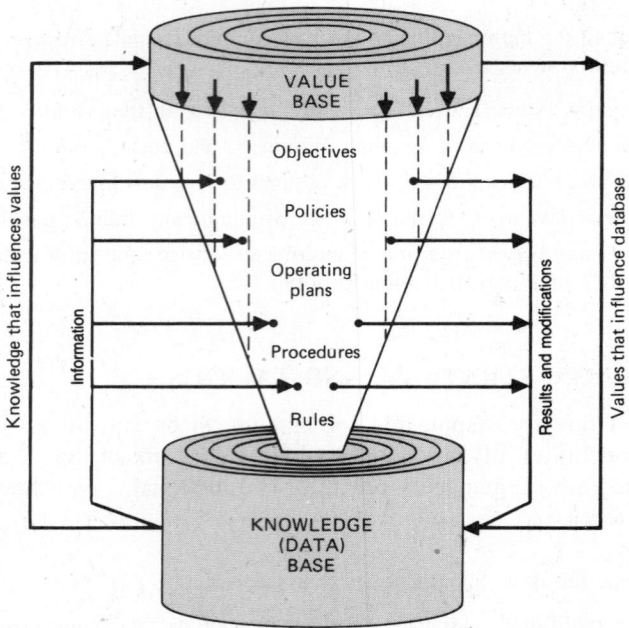

Fig. 1-2   Information flows for decision-making

OPERATIONS MANAGEMENT

Question: Define operations management.

   Operations management is that activity whereby resources, flowing within a defined system, are combined and transformed in a controlled manner to add value in accordance with policies communicated by management. Figure 1-3 depicts this process.

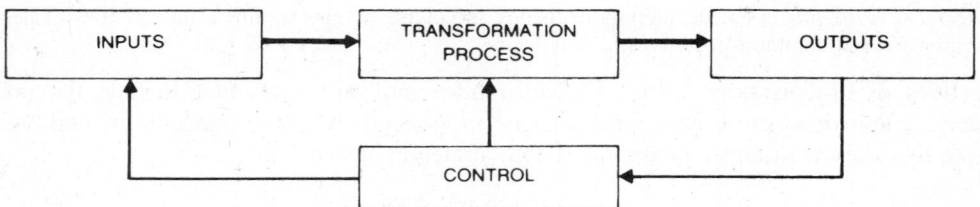

Fig. 1-3   A simplified production system

   Operations managers hold various titles (vice-president of operations, general manager, production manager, plant superintendent) and work in widely diversified industries (manufacturing, food service, medical care, government). Responsibilities generally involve bringing the inputs together under an acceptable production plan that effectively uses the materials, capacity, and knowledge available in the production facility. Given a demand on the system, the operations managers must then ensure that the work is scheduled and carried out to produce the goods and services represented by their customer demand. This requires careful control over inventory, quality, and costs; and the facility itself must be maintained. Figure 1-4 illustrates the operations management responsibilities in the form of a schematic model of a production system.

Question: Explain the meaning of the terms (*a*) resources and (*b*) system, as used in the definition of operations management.

Resources are the human, material, capital, and informational inputs. *Human* inputs (both physical and intellectual) are often the key asset. *Material* inputs include plant, equipment, inventories, and supplies such as energy. *Capital* inputs—in the form of equity, debt, taxes, and contributions—are a store of values that regulate the flow of other resources. Information is the knowledge base that guides all flows.

Systems are arrangements of components designed to achieve objectives according to plans. Our social and economic environment contains many levels of systems and subsystems, which are in turn components of larger systems. We have a free-enterprise economic system. Business firms, which are components of that system, contain personnel, engineering, finance, operations, and marketing functions, all of which are subsystems within the individual firms.

A *systems approach* emphasizes the integrative nature of all system activities and stresses the relationship and cooperation that should exist within the total system. A consistent and integrative approach can lead to *optimization* of the overall (macro) system goals. If the subsystem goals are pursued independently, *suboptimization* may result.

The ability of a system to accomplish its objectives depends upon its design and control. *Systems design* is a predetermined arrangement of components. The more structured the design, the less decision-making is involved in its operation. *Systems control* is the conformance of activities to plans or goals.

Question: Identify the four essential elements of control.

1. *Measurement* by an accurate sensory device	3. *Comparison* with standards, such as time
2. *Feedback* of information in a timely manner	4. *Corrective action* by someone with authority

Question: How do transformation activities "add value"?

Transformation activities add value by combining, or changing the resources, using some form of technology (mechanical, chemical, medical, electronic, etc.). This transformation creates new goods and services that have a higher value to consumers than the acquisition and processing costs of the inputs.

Question: What is productivity (single-factor and multi-factor)?

Productivity is a measure of the effectiveness of the use of resources to produce goods and services. Most (international) measures of productivity are *single-factor* indexes based upon an output relative to unit labor costs over a specified time period. However, the U.S. is moving toward the use of *multi-factor productivity*, which also includes capital and other nonlabor inputs, making it a better measure of overall efficiency. The ratio of the value of outputs to the cost of inputs should be greater than one.

$$\text{Productivity} = \frac{\text{value of outputs}}{\text{cost of inputs}} \qquad (1.1)$$

Example 1.1   An accounting firm generates services valued at $80,000 per day and has total costs of $50,000 per day. What is a simple measure of its productivity?

$$\text{Productivity} = \frac{\text{value of outputs}}{\text{cost of inputs}} = \frac{\$80,000}{\$50,000} = 1.6$$

Note that the value of the outputs is established by consumers in the marketplace, and the cost of inputs is dictated largely by what the firm must pay its suppliers (accountants, in this example).

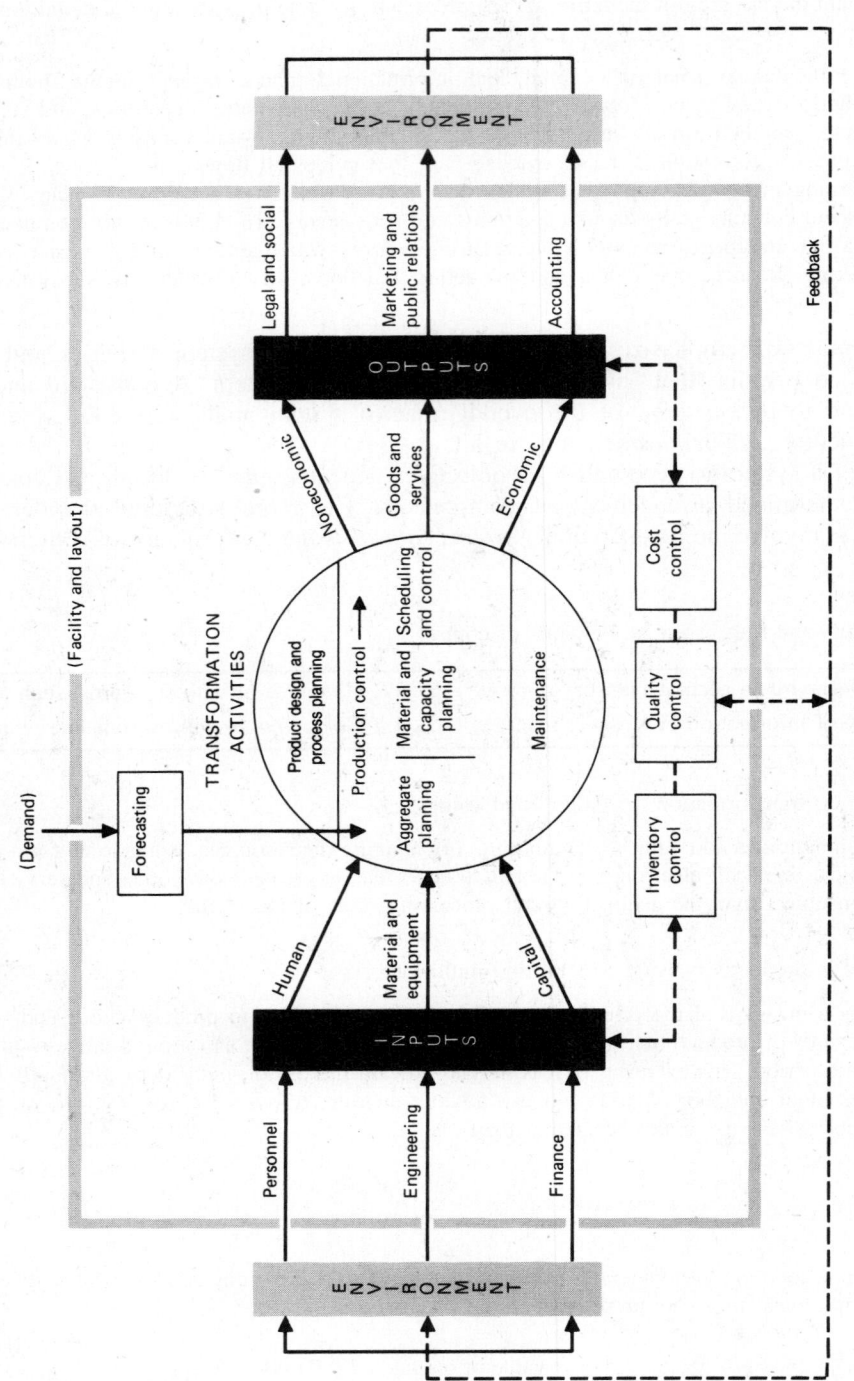

Fig. 1-4 Schematic model of a production system

FACTORS AFFECTING DOMESTIC AND INTERNATIONAL PRODUCTIVITY

As suggested in Fig. 1-1, the United States has experienced a gradual increase in productivity. Prior to the 1950s the average annual growth in output per labor hour of the composite of U.S. businesses is estimated to have increased at an annual rate of roughly 3 percent for many years. However, the rate dropped to about 2 1/2 percent during the 1960s, below that during the 1970s, and to less than 1 percent in some years during the 1980s. The early 1990s witnessed some recovery in manufacturing productivity, but the lower rate of productivity growth in services continued to adversely affect overall productivity. For some time U.S. productivity growth trailed that of Japan, Korea, Great Britain, Germany, Sweden, France, Canada, and some other industrialized countries. By the mid-1990s, the effects of improved inventory management, corporate "rightsizing," innovations in the computer/communications, and other factors (e.g., interest rates) began to have a more positive effect on U.S. productivity, and the growth rate improved.

Question: Identify some of the principal factors influencing productivity changes.

1. Investment (capital/labor ratio)	5. Regulatory and bargaining effects
2. Resource availability/scarcity	6. Mix of goods versus services produced
3. Educational and skill level of people	7. Propensity to save versus spend
4. Innovation and technology employed	8. Quality and global competitiveness

MANUFACTURING AND SERVICE SYSTEMS

Services employ approximately three-fourths of all U.S. workers and generate over half the gross national product. Many services (entertainers, travel agents, stockbrokers, lawyers) depend largely upon the performance of people, while others (phone companies, utilities) rely more heavily upon the use of equipment or facilities. The people-equipment mix is significant because people are better at devising improved ways of doing jobs, but machines are more predictable and measurable. Of most significance, however, is that manufacturing systems deal primarily with planning, scheduling, and controlling *physical materials*. With service systems, the production control efforts concentrate largely upon the flow of (human) *customers*.

Question: What are some common characteristics of most service systems?

1. Demand pattern is often variable.	4. No inventory is accumulated.
2. Labor content is frequently high.	5. Output is intangible, often customized.
3. There is extensive interaction with customer.	6. Output quality is often variable.

Figure 1-5 illustrates the difference (tangible materials versus customer focus) in goods-producing versus services-producing facilities.

OPERATIONS STRATEGY

A *firm's business plan* can be envisioned as consisting of a group of subplans (e.g., marketing, finance, production), each of which can employ different strategies. In general, the strategies it uses to achieve its business plan will depend upon (1) the priorities coming from the (external) environment and the (2) capabilities resident in the (internal) organization itself.

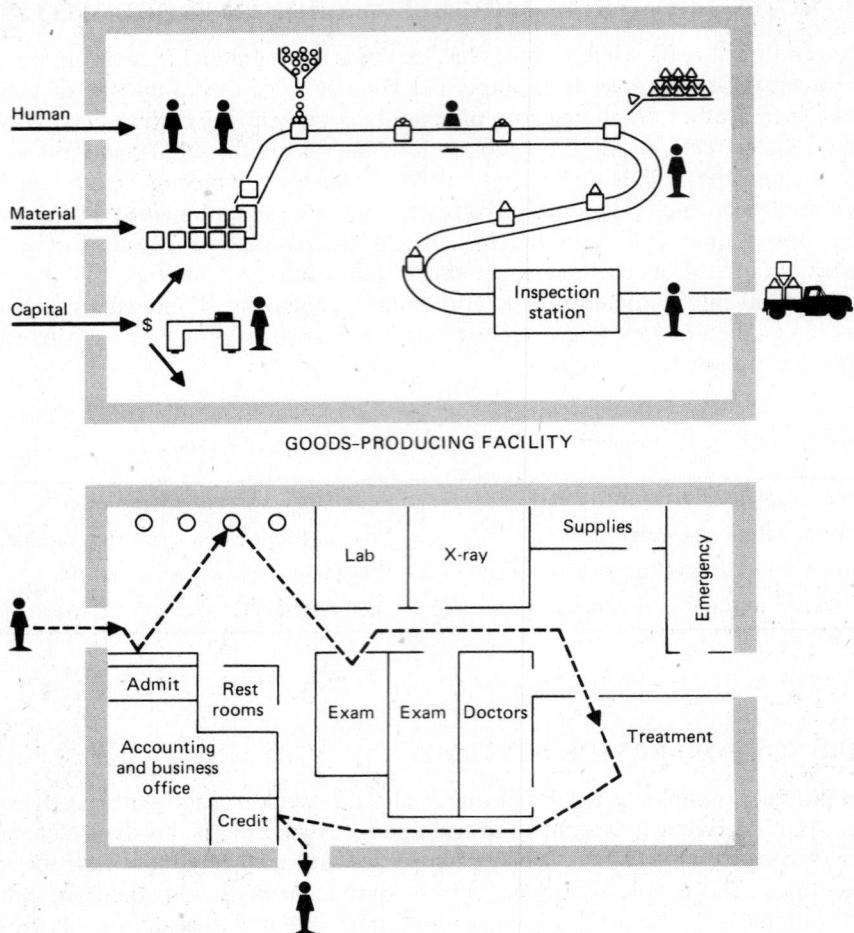

Fig. 1-5 Goods- and services-producing facilities

Question: What types of (*a*) external and (*b*) internal conditions influence an organization's overall competitive strategy?

(*a*) *External conditions* include the social, political, economic (market), legal, and technological environment.

(*b*) *Internal conditions* include human resource skills, facilities available, financial strength, existing product base, technical expertise, and supplier/customer relationships.

Question: Firms use a variety of *operations strategies* to help them achieve their production plans. Give some examples of operations strategies available to competitive organizations.

(1) *Product positioning,* e.g., whether standardized or customized, made-for-stock or made-for-order, focused upon a narrow product line or a broadly competitive product mix

(2) *Facilities/resources,* e.g., whether domestic or international location is better, smaller versus larger capacity, facility designed around process or around product, what is best mix of available personnel, equipment, and capital

(3) *Product development,* e.g., whether to proceed cautiously with prototype and market testing or at an accelerated pace with concurrent engineering, vendor involvement, and reduced time-to-market

(4) *Process development,* e.g., whether to use labor-intensive process or capital-intensive high-tech and computer-integrated systems

Questions and Solved Problems

1.1   In what way did Adam Smith contribute to the development of productive systems?

Adam Smith's book, *The Wealth of Nations* (1776), publicized the advantages of the division of labor, which included skill development, time savings, and the use of specialized machines.

1.2   Identify some significant trends in the management of production systems.

(1)   *Rightsizing* of operating units into more flexible, market-oriented facilities

(2)   Shift in focus from product produced to *continuous improvement* of the process

(3)   *Reduction in the time-to-market* (from initial product design to marketable product)

(4)   Shift of employees from simpleminded tasks to *more use of knowledge, skill, and experience*

(5)   Shift in work activities from individual accomplishment to *self-managed work teams*

(6)   *Reduction in inventories*, *improvement of quality*, and elimination of defective products

1.3   What are some of the more commonly recognized objectives of most firms?

(1)   Producing quality products

(2)   Providing a profitable return to stockholders

(3)   Enhancing the firm's competitive position

(4)   Satisfying the needs of customers

(5)   Nurturing the firm's employees

(6)   Being socially responsible

1.4   In what sense is a hospital a "productive" activity?

Hospitals combine the resources of doctors, nurses, and staff with facilities and equipment to provide health care (service) to patients in a controlled environment. The medical care adds value (and enhanced existence) to the patients' lives.

1.5   Sterling Savings uses customers served per hour by a teller as a measure of its labor productivity. During the past week, the savings institution's five tellers served a total of 2,395 customers. Business hours are 10 a.m. to 4 p.m. Compute a productivity measure of its tellers.

$$\text{Productivity} = \frac{\text{output of number of customers served}}{\text{input of number of hours worked}} = \frac{2{,}395 \text{ customers/wk}}{(5 \text{ tellers})(30 \text{ hr/wk/teller})}$$

$$= 16 \text{ customers/hr}$$

1.6   Explain how the factors cited earlier in this chapter have influenced U.S. productivity.

A low savings rate in the United States relative to other countries has resulted in a decreased level of investment and a low capital/labor ratio. National resources are increasingly scarce (and costly), while the educational competency level of the U.S. work force has declined vis-a-vis some other industrialized countries. New ideas and advanced technology have helped offset some productivity declines, but the costs of regulatory measures (e.g., safety and environmental) and collective bargaining agreements have put U.S. industry at a competitive disadvantage. The shift from manufacturing to service activities is thought to have a detrimental impact upon productivity because productivity in services is generally assumed to be lower than in goods production—though some improvements (e.g., bar coding) have offset this. Finally, the national focus upon improved quality has been a positive influence, as firms realize the benefits from better products and fewer rejects, plus the advantages of cooperative and team-oriented activities.

1.7   How might the terms mission, strategy, and tactic apply to a medical equipment supplier?

(a)   A firm's *mission* might be to become recognized as a leading supplier of medical equipment in its region.

(b)   One *strategy* to achieve the goal might be to increase its market share of heart monitoring equipment to 40 percent.

(c)   A *tactic* to increase its market share might be to form a production team that will redesign an existing product to make it competitively priced and market it by March 1.

1.8   Explain what is meant by the "horizontal corporation."

Although many corporations retain a traditional hierarchical structure, that convention is changing in many corporations, such as Hewlett-Packard, DuPont, and Motorola. Some of the key elements of the horizontal corporation approach include (1) reducing the levels of hierarchy, (2) giving top priority to customer satisfaction, (3) organizing all activities according to processes, (4) giving better training to employees and empowering them to make decisions, and (5) more use of teams, with rewards for teamwork.

1.9   Explain what is meant by the two strategies (a) time-based competition and (b) concurrent engineering.

(a)   *Time-based competition* is a strategic attempt to improve a firm's competitive position by reducing time, e.g., time required to develop a product and deliver it to the customer.

(b)   *Concurrent engineering* is an effort to overlap some engineering and/or other activities so as to bring a competitive and quality product to the market at a lower price in a significantly shorter lead time. (It could be considered a tactic for implementing the time-based competition strategy.)

1.10   Contrast "output" from manufacturing versus service delivery systems.

Manufacturing produces tangible goods that can be measured, stored, and consumed at a later date. Manufacturing systems produce both make-to-stock goods (such as appliances) and make-to-order goods (such as large power transformers).

Most services provide intangible products that convey value directly to the consumer as they are produced (e.g., counseling, medical care). Some service systems also deal with tangible goods (warehousing, distribution, auto repair), but the *focus of most service activities is on the delivery* of the service (e.g., availability, speed of delivery, adequacy of repair). In service activities, when the customer is a participant, production and consumption occur simultaneously and no inventory accrues (e.g., airline travel).

1.11   What are "distinctive competencies"?

Distinctive competencies are the specific abilities, or "weapons," an organization has that give it a competitive advantage. They may include such advantages as (a) convenient location, (b) low production cost, (c) high-quality products, (d) fast time-to-market, (e) on-time shipments, (f) flexible production system, or (g) superior service.

Supplementary Questions and Problems

1.12   Identify the four major eras of economic/industrial development in the United States.
Ans.   (a) Handicraft era, (b) industrial revolution, (c) scientific management era, (d) automation era and computer-integrated systems.

1.13   What factors account for the early growth of the factory system in the United States?
Ans.   Factors include the political structure (constitution), the availability of mechanical power, the availability of labor, railroad expansion, and the absence of other systems.

1.14 What is the first activity essential to any control system?   *Ans.*   Measurement

1.15 Which approach to management holds that managers plan, organize, direct, and control the activities of an organization?   *Ans.*   Functional

1.16 What key concepts are included in the definition of operations management?
Ans.   Include resources, systems, transformation, value-added activities, objectives.

1.17 Productivity expresses the ratio of what two elements?   *Ans.*   Value of outputs to cost of inputs.

1.18 In what sense are the following considered to be "production" activities? (*a*) airline, (*b*) farm, (*c*) restaurant, (*d*) university.
Ans.   Each transforms inputs into higher-valued outputs within a controlled system.

1.19 If a counseling agency were to produce services valued at $4,000 per day and to incur total costs of $3,000 per day, what would be a basic measure of its productivity?   *Ans.*   1.33

1.20 Identify a distinguishing feature of the production activity (goods or services) that employs the larger percentage of U.S. workers.   *Ans.*   Services; they are less tangible than goods.

1.21 Suggest one or two firms that possess each of the "distinctive competencies" referenced in the chapter.   *Ans.*   Refer to firms such as McDonalds, Federal Express, Domino's, Toyota.

1.22 In addition to those factors mentioned in the chapter, numerous other factors are frequently associated with changes in productivity. For example, how are factors such as team building and interest rates incorporated into the factors mentioned in the chapter?
Ans.   Team building is an innovative work methodology (#4) and interest rates affect the level of investment (#1).

1.23 In Frederick Taylor's view, what was the critical determinant of how much work could be accomplished by a person in a day?   *Ans.*   Scientific laws

1.24 Give some examples to show how the transformation process adds value by (*a*) changing, (*b*) combining, (*c*) transporting, and (*d*) preserving resources.
Ans.   (*a*) Chemical changes in the manufacture of plastics and physiological changes in medical care, (*b*) assembling parts of an automobile, (*c*) moving people from one airport to another or facilitating the exchange of a product, (*d*) storing inventory in a warehouse.

1.25 Distinguish between (*a*) single-factor and (*b*) multi-factor productivity.
Ans.   (*a*) Single-factor productivity is a measure of the relationship between a unit of output and the amount of an input needed to produce it. The input most commonly used is labor, which represents about 70% of the cost of value-added in U.S. manufacturing, and is the only measure available in many industries. (*b*) Multi-factor productivity expresses the value of the output in relation to the combined costs of labor, capital, and intermediate purchases. Capital includes land, equipment, and inventories whereas intermediate purchases include materials, energy, and purchased services.

Chapter 2

Quality Management of Competitive Operations

DEFINITION OF QUALITY

Prior to the industrial revolution, the workmanship inherent in a product was largely a function of the craftsperson producing it. As mass production activities ensued, products began flowing from less identifiable groups of workers and machines, and responsibility for the degree of excellence of a product was shifted to supervisors. In 1924, Walter Shewhart introduced statistical methods of monitoring production, followed in the 1930s by (Dodge and Romig's) acceptance sampling procedures—which were used extensively during World War II. Edwards Deming went on to promote the use of statistical methods to correct quality problems among Japanese manufacturers in the 1950s.

Question: Define quality (in a narrow sense).

Quality, in a narrow sense, is a measure of the degree to which the product meets its design standards, which may relate to materials, performance, reliability, time, or any quantifiable characteristic.

The above definition corresponds closely to today's concept of *quality control,* which is the subject of Chap. 18. In the early 1960s Armand Feigenbaum's book, *Total Quality Control*, did much to extend the concept of quality beyond the notion of finding and correcting defects. His work—along with that of Deming, Juran, and others—has broadened the concept of quality.

Question: Define quality (in a broad sense).

Total quality is the ability of a good/service to satisfy customer expectations with respect to:

1. the product's *design*
2. how well the design takes the *user* into account
3. how closely the product *conforms* to the design standards
4. what additional *service* is needed to keep the product operational

---

The (*a*) *design* of a good or service should take account of both the product itself (including aesthetics) and the process used to produce it. (*b*) *User* needs reflect the purpose of the product, as well as safety, liability, and environmental considerations. The (*c*) *conformance* element of the definition incorporates the (essential) control standards that are referred to in the narrow definition of the term. (*d*) *Service* relates to the intended and actual usefulness of the product over its expected life.

Question: To what extent is social responsibility implied in the broad definition of quality?

The extension of product quality perception to safety, liability, and environmental impact implies a degree of social responsibility and is prompting many firms (e.g., banks, food processors, oil companies) to give more attention to the societal impact of their products and processes—in both domestic and foreign operations. This concern has grown as the media has brought more consumer awareness of the social and/or environmentally disruptive operations of some firms (e.g., United Fruit Company's role in Guatemala, Union Carbide's chemical plant explosion in India, Exxon's massive oil spill in Alaska).

Responsibility varies with the firm and the location. Royal Dutch/Shell completely restored a scenic landscape where it laid a pipeline in Wales. But in Nigeria, oil spills and contamination continue to take a health toll on the desperately poor and ailing native population who live in mud huts on oil-saturated land near the company's many wells and refineries.

ELEMENTS OF A TQM SYSTEM

As global competition developed in the 1980s and 1990s, many firms were forced to reassess the quality level of their entire goods/services production systems. This spawned the "Total Quality Revolution"—a massive national focus on all activities that affect the perception of a product's quality, from raw material acquisition to the ultimate customer's level of satisfaction.

Question: Explain what is meant by the term total quality management.

Total Quality Management (TQM) is a managerial philosophy that focuses attention upon identifying and producing a product that meets (or exceeds) customer expectations within an environment that fosters the cooperative effort of knowledgeable employees who are empowered to strive for continuous improvement of the organization's products and processes.

Question: What are the essential elements of a TQM system?

1. A quality vision and goals (e.g., a defect-free or competitively benchmarked product)
2. Strong customer focus (in partnership with suppliers)
3. Thorough knowledge of company processes and customer needs
4. Continuous improvement (via team approach, using empowered employees, etc.)
5. Performance measures to track results

THE PDCA CYCLE, CHARTS, AND DIAGRAMS

Among the many tools available to support continuous improvement are (*a*) the PDCA cycle, (*b*) lists and diagrams, (*c*) flowcharts, and (*d*) cause-and-effect (or fish bone) diagrams. The latter help to solve problems by grouping various causes according to general categories, such as personnel, materials, equipment, and methods.

Question: What is the PDCA cycle?

The *plan-do-check-act (PDCA) cycle* is a convenient mechanism, promoted by Edwards Deming, for fostering continuous improvement, or *kaizen*. It is typically depicted as a circle that begins at the top with (1) the *plan* that includes problem identification, data collection and analysis, and a plan for improvement. The (2) *do* element represents a trial implementation of the plan, and (3) *check* is the evaluation of data to see if the plan is fulfilling expectations. Finally, (4) *act* symbolizes either the full adoption of the plan or adjustments to it.

Question: Identify some of the basic tools used to identify and solve quality problems.

Among the tools advanced by Robert Reid Associates, Inc., a quality management consultant, are (*a*) bug lists, (*b*) why-why diagrams, and (*c*) how-how diagrams. Figure 2-1 illustrates.

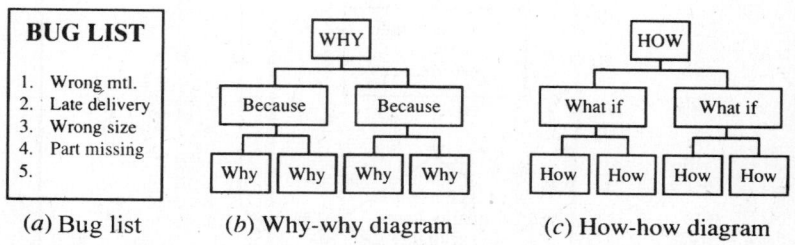

Fig. 2-1   Some tools used to identify and solve quality problems

(*a*)   *Bug lists* simply identify the problems, e.g., wrong material, late delivery, incorrect price. Once the problems are isolated, they can be tackled in a systematic manner.

(b)   *Why-why diagrams* are useful for working back to the cause of a problem. For example, if deliveries are late, it may be because the night shift was overloaded. Then the question is, ''Why was it overloaded?''

(c)   *How-how diagrams* help to suggest solutions to quality problems. For example, if the problem is to improve delivery from an overloaded plant, one might ask, ''What if we subcontract some of the work?''

Question: Use a flowchart to illustrate some steps in the implementation of a TQM activity.

See Fig. 2-2.

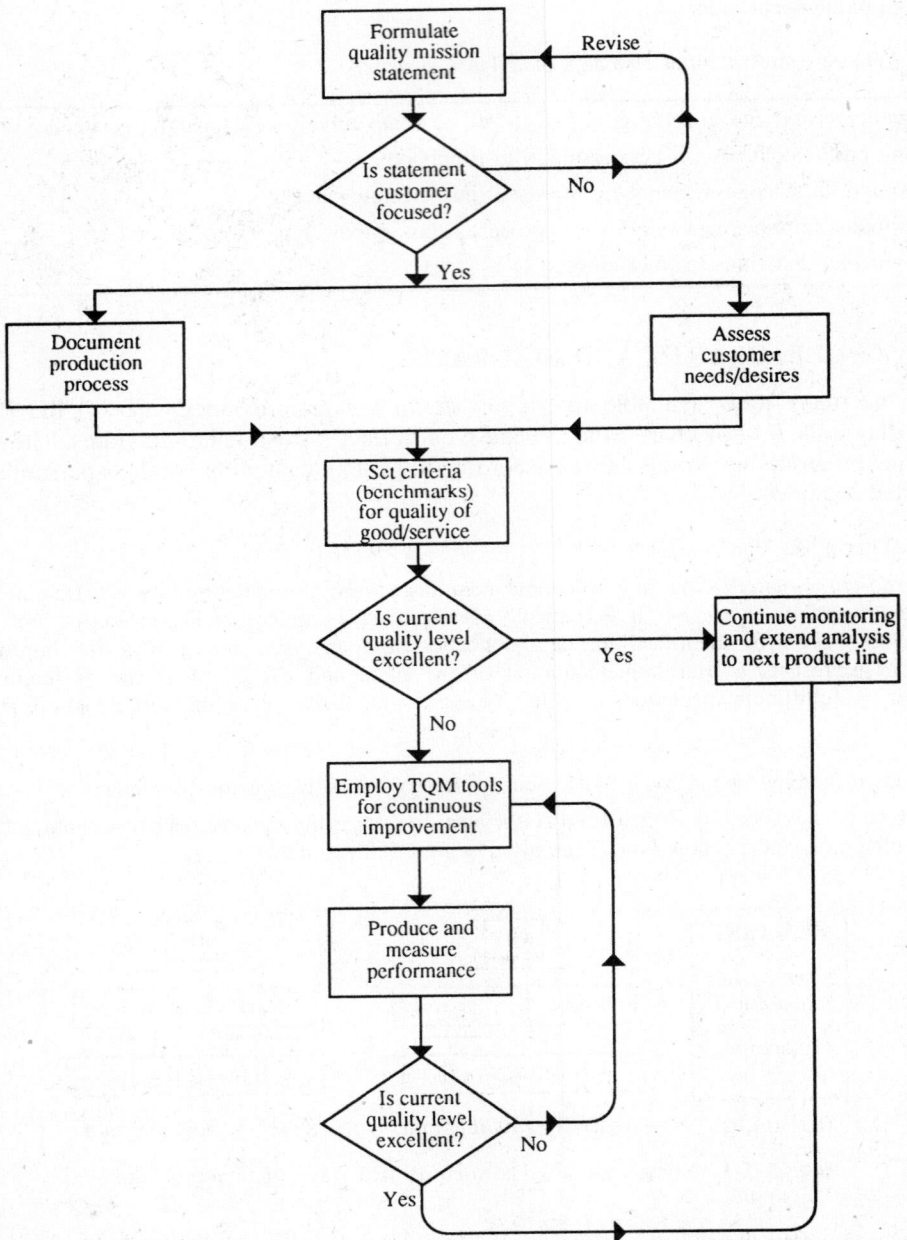

Fig. 2-2   Flowchart for implementing a TQM activity

Question: Prepare a cause-and-effect diagram to analyze the possible causes of dissatisfied customers at a chain offering discounted haircuts.

See Fig. 2-3.

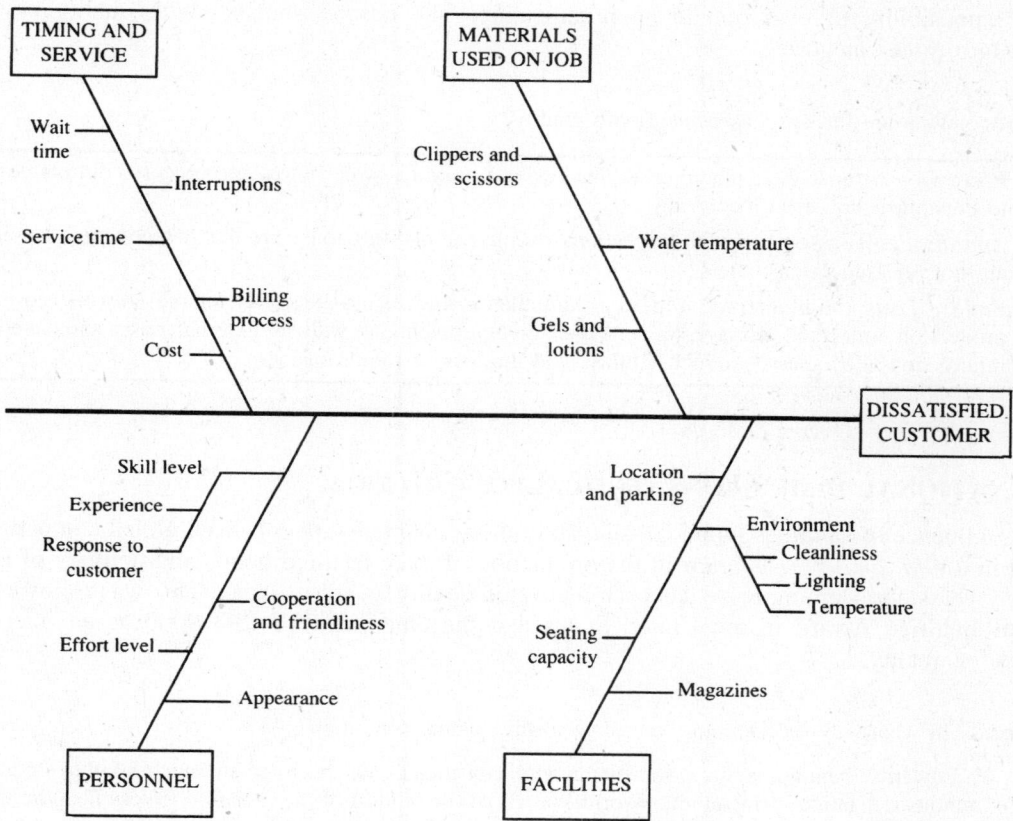

Fig. 2-3   Cause-and-effect diagram for dissatisfied customers at hair salon

BENCHMARKING AND QUALITY CIRCLES

Among the many innovations frequently used as part of TQM systems are benchmarking and quality circles, plus the use of traditional employee suggestion programs and partnering arrangements.

Question: Explain the concept of benchmarking.

A *benchmark* is a comparative standard. Companies using competitive benchmarking typically select an organization or some operational characteristic of that organization that they feel is the best in their field. Then they use that standard as the goal for improving their own activities. Benchmarking can apply to strategies, processes, and procedures, as well as to goods/services.

Question: What is a quality circle?

A *quality circle* is a small group (e.g., 8 to 10) of employees who meet voluntarily on a regular basis to share ideas in an attempt to identify, analyze, and solve quality or job-related problems. In many companies, their recommendations have resulted in a reduced number of defects, reduced absenteeism, and improvements in productivity and job satisfaction. However, some companies have also abandoned such programs.

COSTS OF QUALITY

Quality is a value that must be built into a good/service by a team effort, so it is widely recognized as the responsibility of everyone in an organization. The costs associated with quality have been grouped into three categories.

Question: What are the costs associated with quality?

1. *Prevention costs* such as planning, design reviews, training, procedures, and process controls designed to prevent defects from occurring
2. *Appraisal costs* associated with inspection, testing, and auditing to ensure that defective goods/services are not released to customers
3. *Failure costs* (*both internal and external*) that include such items as internal rework costs, lost production time, and damage to personnel or equipment, as well as external costs associated with returns and allowances, product liability, and the loss of future business

INTERNATIONAL DIMENSIONS OF QUALITY CRITERIA

The influence of Deming, Juran, Crosby, and others, coupled with enhanced global competition, has ushered in a *total quality revolution* in the production of manufactured goods and delivery of services. For many years, the Deming Prize has been a coveted quality award in Japan. This was followed by the Malcolm Baldrige Award in the United States and the emergence of ISO 9000 as an international standard for quality.

Question: In what way did Deming contribute to the quality revolution?

Dr. W. Edwards Deming, a U.S. statistician and consultant, has been given much of the credit for the successful reindustrialization of Japan after World War II. At the request of the Japanese government, he initiated a series of statistically based programs that significantly improved the productivity and quality of Japanese manufacturing activities. In 1951, the Japanese government established the Deming Prize that is presented annually to a company that has most distinguished itself in industrial quality control. Deming's philosophy was not fully accepted in the United States until the 1980s—when General Motors, Ford, and other firms began to give quality a top priority. Deming died in 1992, after his efforts were finally recognized in his own country.

Question: What is the purpose of the Malcolm Baldrige Award?

The *Malcolm Baldrige Award* was established in 1987 to recognize outstanding quality achievements of U.S. firms and, by publicizing them, to encourage quality awareness in other firms. Six awards are presented annually with two awards in each of three categories (large manufacturers, large service organizations, and small business). Competition for an award is extremely keen, and the number of firms applying for an award increases annually.

Question: What is ISO 9000, and why is it important?

ISO 9000 is an internationally recognized series of guidelines that reflect agreed-upon standards by over 90 countries. These guidelines pertain largely to the documentation of an organization's quality processes. After a firm has meticulously documented its quality procedures and successfully passed an on-site assessment, it becomes certified for registration in an ISO directory. This registration is vital to many firms engaged in international operations.

COMPETITIVENESS ON A GLOBAL BASIS

Many organizations have multinational operations and compete on a global basis with goods and services from Canada, Mexico, Japan, Germany, Great Britain, France, Taiwan, Spain, Malaysia, Korea, Brazil, and a host of other countries. When production operations cross international borders, attention must be given to all aspects of operations, from the sourcing of resources to the delivery of products to the market. International trade agreements can have a significant effect on operations.

Question: What is typically required for a firm to achieve world-class performance?

This normally entails (*a*) an initial documentation of the strengths and weaknesses of the firm; (*b*) a clear delineation of what the firm elects to achieve in order to compete (e.g., cycle time, quality level, inventory turns); (*c*) enhancing the systems, procedures, and controls needed to meet specified goals; (*d*) collection and use of data for continuous improvement by means of JIT, MRP-II, CIM, TQM, and other available technology.

Question: What results might be expected from a total quality integrated organization?

1. Customer needs known and consistently satisfied

2. Continuous improvement tools functioning (e.g., waste reduction, statistical controls)

3. Involved and empowered employees who take action without prior approval

4. Database linkage of suppliers, through company processes to customers

5. Meaningful performance improvements realized on an ongoing basis

Question: In what ways do trade agreements influence global operations?

Trade agreements, such as the European Common Market and the North American Free Trade Agreement (NAFTA), reduce the barriers to production and distribution in the associated countries by encouraging the free flow of raw materials and finished goods. The agreements often limit (or abolish) tariffs on designated products and foster competition. This can result in some shifts in employment from regions that have less efficient (or subsidized) operations to other regions that offer the promise of higher levels of productivity.

Firms expanding into a foreign location should have a thorough knowledge of that country's political, economic, and legal systems and must be prepared for the educational and cultural norms of the new environment. Although the work activities may be similar in foreign operations, behaviors that are commonly accepted in one culture are not necessarily appropriate in others. Special attention should be given to ensuring that high ethical and moral standards of conduct are consistently maintained.

Question: Identify some characteristics that in the past have distinguished Japanese firms from U.S. firms.

1. *Corporate objectives.* Employees and customers given priority over shareholders.

2. *Financing.* More use of debt capital.

3. *Long-term perspective.* Viability in the long run more important than short-term profits. Employees evaluated on long-term performance.

4. *Emphasis on training.* Employees thoroughly trained and rotated to learn a diversity of skills; less emphasis on job descriptions.

5. *Employment relations.* Long-term employment of loyal workers. Employees paid on the basis of employee worth and needs. Unions cooperate to benefit firm.

6. *Widespread worker participation.* Direct involvement of employees in productivity improvements via suggestions, quality circles, consultation.

Questions and Solved Problems

2.1 Identify some of the major consequences of poor quality.

    (1) Costs of *increased material and labor* due to scrap loss and rework time

    (2) Costs of production *downtime* as a result of a failure (within own or other organization)

    (3) Costs for *extra inspection, testing, and prevention* mechanisms

    (4) Costs stemming from *damages and legal liabilities*

    (5) *Loss of future business* stemming from dissatisfied customers or loss of goodwill

2.2 Summarize Edwards Deming's 14 points for improving quality.

1. Establish a goal of *continuous innovation* and improvement.
2. Adopt a philosophy that does *not tolerate mistakes*, delays, or defects.
3. Cease dependence upon mass inspection; *require statistical evidence* of quality.
4. *End the practice* of awarding business on the basis of price.
5. Search continuously to *uncover problems* (e.g., using statistical methods).
6. Institute modern methods of *training on the job*.
7. Refocus supervisors' attention *from quantity to quality* that will improve productivity.
8. *Drive out fear*, so that everyone feels secure and encouraged to seek improvement.
9. *Break down departmental barriers* and those with suppliers and customers.
10. *Eliminate posters and slogans* that don't actually help people solve problems.
11. *Eliminate work standards that prescribe numerical quotas*.
12. *Remove barriers* that stand between workers and their right to pride in workmanship.
13. Institute a vigorous *retraining program* to keep up with changes.
14. Create a top management structure that will *push every day for the above 13 points*.

2.3 Use a sketch to illustrate the continuous nature of the PDCA cycle.

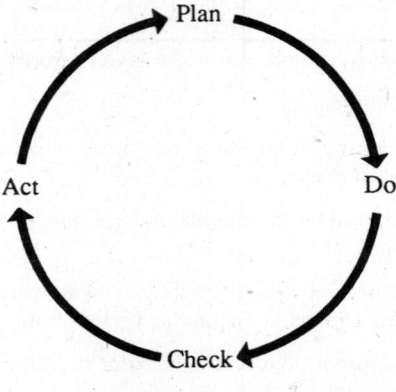

Fig. 2-4

2.4   What is the purpose of "partnering" and "alliances"?

Partnering is a mutual relationship between two parties, such as a firm and an educational institution, or union and management, that allows both to focus on the needs of a common goal, sharing the resultant risks and the benefits. *Alliances* are typically agreements among competitors to share technology that can benefit all involved. For example, in the early 1990s Apple, IBM, and Motorola formed an alliance to develop a power PC computer chip that would compete with a highly successful microprocessor developed by Intel.

2.5   What criteria are used to select winners of the Malcolm Baldrige National Quality Award?

This award recognizes outstanding achievement as determined by points allowed in each of seven categories, with a total possible of 1,000 points.

(1) Leadership (100 points)

(2) Information and analysis (60 points)

(3) Strategic quality planning (90 points)

(4) Human resource management (150 points)

(5) Quality assurance of products and services (150 points)

(6) Quality results (150 points)

(7) Customer satisfaction (300 points)

2.6   Quality is understandably difficult to measure—especially as one attempts to compare activities in the manufacturing, services, and public service sectors. Are there any measures that can be used to position or compare the quality level of goods/services from the standpoint of the level of satisfaction they provide to customers?

Yes. Some countries (e.g., Sweden, Germany, U.S.) have fairly sophisticated quality indexes. The American Customer Satisfaction Index (ACSI), is an attempt to augment the productivity indexes that have been used in the past, with a widely accepted measure of customer satisfaction more in keeping with the emphasis on continuous improvement. It is a response to the shift that has taken place from mass production to customized products in the U.S. economy. The ACSI index, developed at the University of Michigan, provides a quarterly measurement of the satisfaction level of approximately 46,000 customers with the goods and services provided by slightly over 200 companies in the U.S. economy. The index began with an initial value of 74.5 in 1994, and correlates well with employment, stock market, and other indices of economic well-being. Comparative values are provided for (*a*) manufacturing, (*b*) services, and (*c*) public service sectors of the economy, where the indexes tend to be high, medium, and low respectively.

2.7   Explain what is meant by the Taguchi method.

Taguchi's work is credited with speeding and perfecting the product and process design process by categorizing the relevant variables into those that are controllable and those that are uncontrollable (or too expensive to control). By concentrating upon the controllable variables, fewer experiments are needed to arrive at the best product design (than if all variables were considered equally).

Supplementary Questions and Problems

2.8   What managerial concept incorporates elements of customer focus, benchmarking, empowerment, and continuous improvement?   *Ans.*   Total quality management

2.9   To what does the term *kaizen* refer, and how does it relate to the PDCA cycle?
Ans.   It refers to continuous improvement, a vital element in Japan's industrial success. Numerous tools are used to implement continuous improvement, including the PDCA cycle.

2.10 Design a flowchart for randomly checking the accuracy of bathroom scales. (*Note:* Assume the scales are randomly taken from an assembly process before they are packaged for shipment. If the test is unsatisfactory, the scales must be readjusted and returned to the packaging line.)
    Ans. Charts may vary but should be similar in format to Fig. 2-2.

2.11 Some of the methods and tools used to support the continuous improvement process include (*a*) activity diagrams, (*b*) tree diagrams, (*c*) matrix diagrams, (*d*) flowcharts, (*e*) various measurement and tracking systems, (*f*) benchmarking, (*g*) quality circles, (*h*) control charts, (*i*) why-why and how-how diagrams, (*j*) cause-and-effect (fish bone) diagrams, and (*k*) Pareto analysis. Using the journals and/or trade publications in your local library, find an example of one or more of these tools and prepare a brief statement telling how its use can assist in the maintenance of quality.   *Ans.*   Answers will vary.

2.12 A doctor's office, which is supposed to operate from 9 a.m. to 4 p.m., finds that patients are frequently not through the system until 6 p.m. Prepare a cause-and-effect diagram to suggest possible causes for the late closure time.
    Ans. Among the possible causes are the scheduling of too many patients, variability in patient times required, and the need to handle emergencies. Add other causes, and complete the diagram in a manner similar to Fig. 2-3.

2.13 Prepare a cause-and-effect diagram to examine the problem of getting a low exam score.
    Ans. Among the possible causes are (1) People (teacher, students), (2) Materials (book, notes), (3) Methods (lectures, homework, notetaking), and (4) Situation (stress, room, noise). Expand on causes/subcauses and complete the diagram in a manner similar to Fig. 2-3.

2.14 Is the international importance of quality (*a*) increasing or (*b*) decreasing? Explain.
    Ans. Increasing. As standards like ISO 9000 become a ''requirement'' for participation in competitive bidding, firms that are not certified are unable to compete for international business.

2.15 In what sense might quality be considered a less important competitive variable?
    Ans. As more firms reach a high quality standard (e.g., satisfying ISO 9000 standards) the quality advantage one firm has over another may be less apparent. This, in turn, may cause a shift in emphasis to other strategic variables, such as time-to-market. Quality remains important, but not so much as a distinguishing feature.

2.16 In what ways does ''agile manufacturing'' enhance competitiveness?
    Ans. Agile factories attempt to be very highly responsive to customer requests by speeding the delivery of a desired product (e.g., a 3-day car). This requires an extremely dynamic scheduling ''engine'' that can track work-in-process (WIP) and continuously reallocate material/capacity already planned for other orders.

Chapter 3

Operations Decision Making
Break-Even Analysis, Decision Trees, and Statistical Methods

THE DECISION-MAKING PROCESS

An operations manager's major responsibility is to make decisions and take action to guide the operations of an organization toward its mission. Trivial or routine problems may best be handled by simple judgmental choices. However, complex problems that involve interdependent variables, such as major cash flows or plant expansions, usually require more systematic decision-making processes. Scientific decision-making rests upon (*a*) organized principles of knowledge and depends largely upon the (*b*) collection of empirical data and (*c*) analysis of the data in a way that (*d*) repeatable results will be obtained.

Question: List the steps in a systematic decision-making process.

> 1. *Define* the problem and its parameters (relevant variables).
> 2. Establish the decision *criteria* (objectives).
> 3. Relate the parameters to the criteria (i.e., *model* the problem).
> 4. Generate *alternatives* by varying the values of the parameters.
> 5. *Evaluate* the alternatives, and choose the one that best satisfies the criteria.
> 6. *Implement* the decision, and *monitor* the results.

MODEL BUILDING

Models (step 3 of the decision-making process) describe the essence of a problem or relationship by abstracting the relevant variables from the real-world situation and expressing them in simplified form so that the decision maker can study the underlying relationships in isolation.

Question: Identify some major advantages of using models.

> 1. Necessitate a good understanding of the problem.
> 2. Require a recognition of all relevant (controllable and uncontrollable) variables.
> 3. Help identify the relationships, costs, and trade-offs that exist among the variables.
> 4. Permit easy manipulation of variables and testing of alternative courses of action.

Question: What types of models are most useful for operations management decision-making?

(*a*) Verbal models (words and descriptions)

(*b*) Physical models (scale)

(*c*) Schematic models (diagrams and charts)

(*d*) Mathematical models (equations)

Mathematical (and statistical) models are the most abstract—and often the most useful. They can succinctly describe a problem, are readily computerized, and are easily manipulated to test different parameter values.

Example 3.1 Asia Plastics uses a simple linear model to estimate the next period's production requirements.

$$P_{t+1} = D_t - (I + P_{t-1})$$

where P_{t+1} = units of production (P) required next period ($t + 1$)

D_t = estimated current period demand (an unknown and uncontrollable variable)

I = present inventory level

P_{t-1} = units produced in previous period

Last period's production was 20 units, the present inventory level is 5 units, and the current period demand is estimated at 40 units plus or minus 10 percent. Use the model to develop an interval estimate of next period's production requirements.

For the *minimum* estimate, let $D_t = 40 - .10 (40) = 36$ units.

$$P_{t+1} = 36 - (5 + 20) = 11 \text{ units}$$

For the *maximum* estimate, let $D_t = 40 + .10 (40) = 44$ units.

$$P_{t+1} = 44 - (5 + 20) = 19 \text{ units}$$

Using this model, the estimated production required will be from 11 to 19 units.

DECISION METHODOLOGY

The choice of a model depends upon the characteristics of the decision (i.e., significance, time and cost, and complexity). Decisions are more complex when the data describing the variables are incomplete or uncertain. The degree of certainty is classified as:

(1) *Complete certainty.* All relevant information about the decision variables and outcomes is known (or assumed).

(2) *Risk and uncertainty.* Information about the decision variables or the outcomes is probabilistic. Objective data (from large samples) lend more certainty than subjective data.

(3) *Extreme uncertainty.* No information is available to assess the likelihood of alternative outcomes.

Figure 3-1 illustrates some useful quantitative methods that are classified according to the amount of certainty that exists with respect to the decision variables and possible outcomes. These analytical techniques often serve as the basis for formulating models to help reach operational decisions.

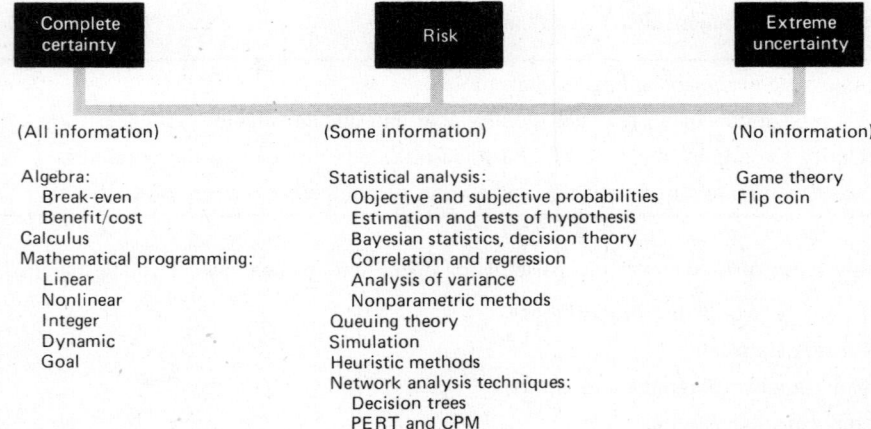

Fig. 3-1   Quantitative methods as a function of degree of certainty

Example 3.2   *Complete Certainty:* Security Storage Inc. leases document storage facilities to government agencies on a multiyear contract basis. The company is considering three potential locations (A, B, and C) for a new facility that will have (contractually guaranteed) annual costs of $27,500, $25,400, and $29,000, respectively. The annual revenue from leasing the facility to a county government is known in advance to be $40,000 for location A, $36,000 for B, and $42,000 for C. Which location will maximize the net return per year?

$$\text{Net return} = \text{revenue/yr} - \text{costs/yr}$$

$$\textit{For A:} \quad = \$40,000 - \$27,500 = \$12,500$$
$$B: \quad = \$36,000 - \$25,400 = \$10,600$$
$$C: \quad = \$42,000 - \$29,000 = \$13,000$$

Therefore, C is optimal.

BREAK-EVEN ANALYSIS (*assumed certainty*)

Question: What is break-even analysis?

Break-even (or cost-volume) analysis is an algebraic and/or graphic model for describing the relationship between costs and revenues for different volumes of production. It is especially useful for identifying the volume at which operations change from being a loss to being profitable.

In break-even analysis, costs are generally assumed to be known and are classified as either fixed (FC) or variable (VC), depending upon whether they vary with the volume of output (Q). Profits occur when total revenues (TR) exceed total costs (TC), where TC = fixed costs (FC) plus total variable costs (TVC).

$$\text{Profits} = \text{TR} - (\text{FC} + \text{TVC}) \tag{3.1}$$

Figure 3-2a illustrates the profit concept, and Fig. 3-2b identifies the quantity at the break-even point, Q_{BEP}. At the break-even point (BEP), the profit is zero and TR = TC. Recognizing that revenues reflect the selling price per unit (P) times the quantity sold (Q), we restate the TR = TC expression as:

$$PQ = \text{FC} + \text{VC} \cdot Q$$

where VC is the variable cost per unit. The quantity at the break-even point is then

$$Q_{\text{BEP}} = \frac{\text{FC}}{P - \text{VC}} \tag{3.2}$$

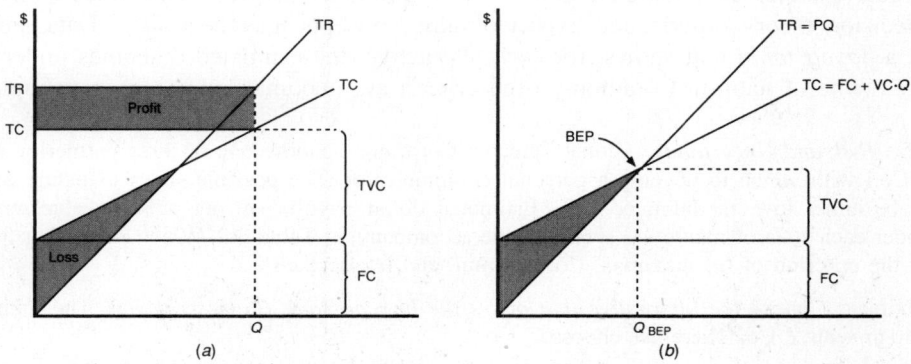

Fig. 3-2 Profit and break-even point analysis

Example 3.3 Annual fixed costs at a small textile shop are \$46,000, and variable costs are estimated at 50 percent of the \$40-per-unit selling price. (*a*) Find the BEP. (*b*) What profit (or loss) would result from a volume of 3,000 units?

(*a*)
$$Q_{\text{BEP}} = \frac{\text{FC}}{P - \text{VC}} = \frac{\$46,000}{\$40 - (.50)(40)} = 2,300 \text{ units}$$

(*b*)
$$\text{Profit} = \text{TR} - (\text{FC} + \text{TVC}) = PQ - (\text{FC} + \text{VC} \cdot Q)$$
$$= (\$40)(3,000) - [\$46,000 + \$20(3,000)] = \$14,000$$

Break-even analysis is a useful model, especially for a single product. But it usually assumes certainty conditions, which limits its applicability. Table 3-1 summarizes some characteristics.

Table 3-1   Assumptions and Advantages of Break-Even Analysis

Assumptions	Advantages
1. All costs and volumes are known.	1. Simple and easy to visualize.
2. Cost-volume relationships are linear.	2. Focuses upon profitability.
3. All output can be sold.	3. Uses algebraic and/or graphic display.

Question: What is contribution?

Contribution is a companion measure of value that tells how much of the revenue from the sale of one unit of a product will contribute to cover fixed costs with the remainder going to profit.

The per-unit contribution C of a product is determined by subtracting the per-unit variable cost (VC) from the price (P).

$$C = P - \text{VC} \tag{3.3}$$

Example 3.4   Find the contribution for the textile shop product of Example 3.3.

$$C = P - \text{VC} = \$40 - (.50)(\$40) = \$20 \text{ per unit}$$

All of the contribution from a product is absorbed in paying for fixed costs up to the break-even point. Beyond that, the contribution is all profit.

DECISION THEORY AND EXPECTED-VALUE CRITERIA

Whereas problems of a complete certainty nature (or assumed certainty) are typically solved in a simple algebraic manner as illustrated above, most problems involve some uncertainty. For these situations, decision theory criteria and expected-value concepts may be useful. Data are sometimes presented in a *payoff table* that shows, for each alternative, the estimated outcomes under each future condition (i.e., state of nature). Commonly used criteria are maximax, maximin, and Laplace.

Example 3.5   *Risk and Uncertainty:* Global Telecom Corp. must choose one of three partnering firms (X Co., Y Co., or Z Co.) with which to develop a personal communicator. The possible states of nature are that future demand may be either low, medium, or high. Estimated dollar payoffs (in net present value terms) for each alternative under each state of nature are shown in the accompanying Table 3-2. Which partnering firm should be chosen under the criterion of (*a*) maximax, (*b*) maximin, and (*c*) Laplace?

   (*a*)   Maximax: *Choose the alternative that offers the best possible (highest) payoff.* The highest payoff is \$140 m with Z Co. Therefore choose Z Co.

   (*b*)   Maximin: *For each alternative, locate the worst possible payoff. Then choose the best of these "worst" values.* The best of the worst is X Co. with a minimum profit of \$10 m.

(c)   Laplace: *Find the average payoff for each alternative, and choose the alternative with the best.* The highest average is Z Co. [(0 + 20 + 140)/3 = \$53.3 m]. Choose Z Co.

Table 3-2   Payoff Table

	Profit (\$m) If Future Demand Is		
Alternatives	Low	Medium	High
X Co. partner	10	50	70
Y Co. partner	−10	44	120
Z Co. partner	0	20	140

When probability values, $P(X)$, are assigned to the states of nature of a payoff table, two additional decision criteria can be considered: maximum probability and expected monetary value. The expected monetary value approach utilizes the highest average, or expected value, $E(X)$.

$$E(X) = \Sigma [X \cdot P(X)] \tag{3.4}$$

Example 3.6   *Risk and Uncertainty:* For the data in Table 3-2, assume probabilities of:

$$P(\text{low demand}) = .3 \qquad P(\text{medium demand}) = .5 \qquad P(\text{high demand}) = .2$$

Which partnering firm should be chosen under the criterion of (a) maximum probability and (b) Expected Monetary Value (EMV)?

(a)   Maximum probability: Find the most likely state of nature, and choose the best alternative within that state. Most likely is [$P(\text{medium demand}) = .5$]. Choose X Co., where payoff is \$50 m.

(b)   Expected monetary value: *For each alternative, multiply the value of each possible outcome (X) by its probability of occurrence, $P(X)$. Then sum across all outcomes, i.e., $\Sigma [X \cdot P(X)]$ to get the expected value, $E(X)$, of each alternative.* The alternative with the highest expected outcome is designated the expected monetary value, EMV*.

$$E(\text{X Co.}) = \quad 10(.3) + 50(.5) + \quad 70(.2) = 42$$
$$E(\text{Y Co.}) = -10(.3) + 44(.5) + 120(.2) = 43 \leftarrow \text{EMV*} = \$43 \text{ m}$$
$$E(\text{Z Co.}) = \quad\quad 0(.3) + 20(.5) + 140(.2) = 38$$

DECISION TREES

Question: What is a decision tree?

A decision tree is a schematic diagram that shows the alternative outcomes and interdependence of choices in a multiphase, or sequential, decision process. The treelike diagram is constructed from left to right, using square boxes for controllable (decision) points and circles for uncontrollable (chance) events. Each branch leads to a payoff that is stated in monetary terms on the right.

Decision trees are analyzed backward (from right to left) by multiplying the payoffs by their respective probabilities (which are assigned to each chance event). The highest expected value then identifies the best course of action and is entered at the preceding decision point. This then becomes the expected value for the next higher-order expectation, as the analyst continues to work back to the trunk of the tree.

Example 3.7   A manufacturer of small power tools is faced with foreign competition, which necessitates that it either modify (automate) its existing product or abandon it and market a new product. Regardless of which course of action it follows, it will have the opportunity to drop or raise prices if it experiences a low initial demand.

Payoff and probability values associated with the alternative courses of action are shown in Fig. 3-3. Analyze the decision tree, and determine which course of action should be chosen to maximize the expected monetary value. (Assume monetary amounts are present-value profits.)

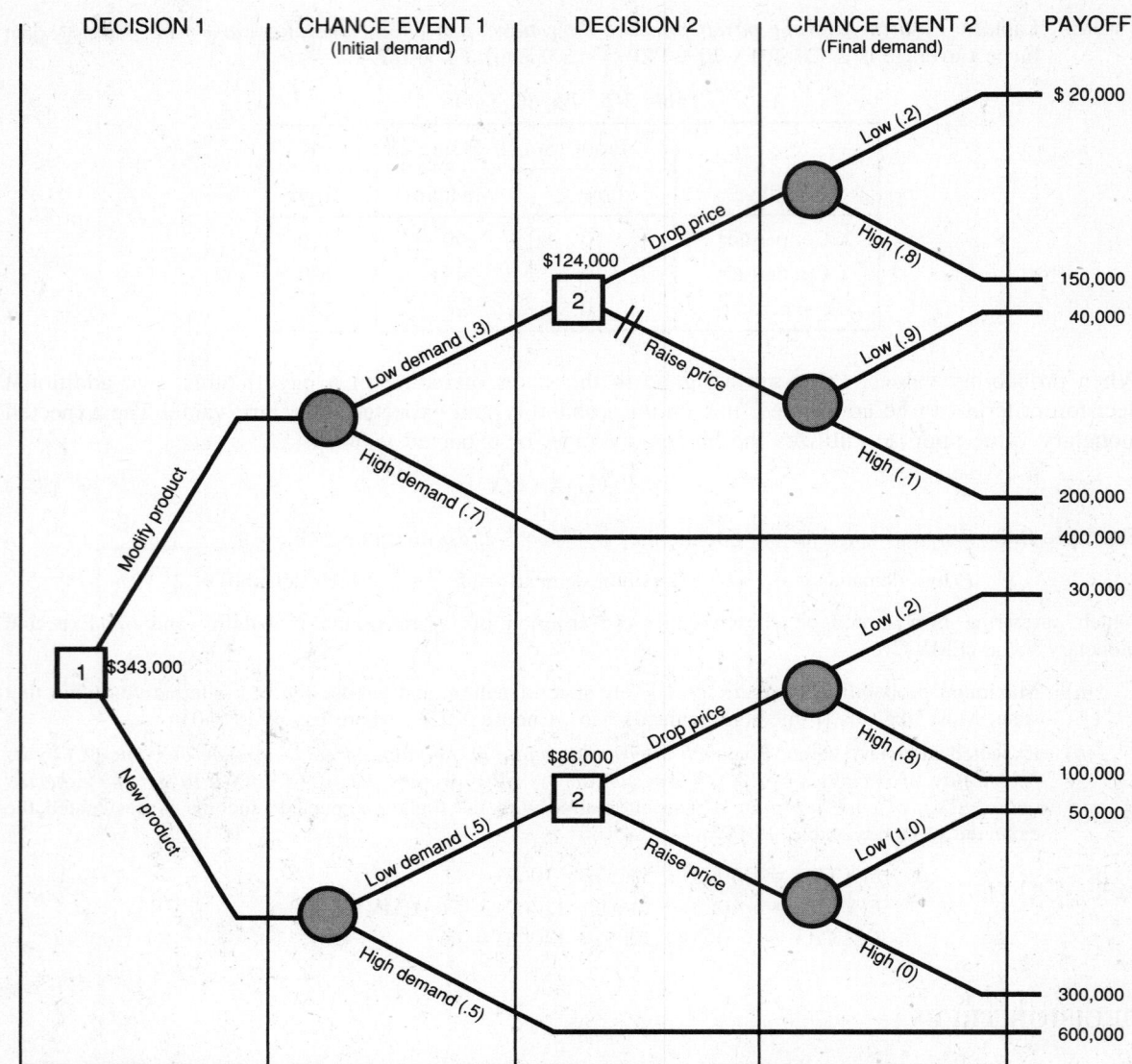

Fig. 3-3 Decision tree diagram

Analyze the tree from right to left by calculating the expected values for all possible courses of action and choosing the branch with the highest expected value. Begin with the top (modify product) branch.

At chance event 2

Drop price branch: $E(X) = \$20,000(.2) + \$150,000(.8) = \$124,000$

Raise price branch: $E(X) = \$40,000(.9) + \$200,000(.1) = \$56,000$

Therefore, choose to drop price and use $124,000 as the value of this branch at Decision 2. *Note:* The $124,000 is an expected monetary value (EMV) and can be entered above the square box under Decision 2. Place slash marks through the other (nonusable) alternative.

At chance event 1

If low demand: $\$124,000(.3) = \$\ \ 37,200$

If high demand: $400,000(.7) = \underline{\ \ 280,000}$

                                        $E(X) = \$317,200$

Therefore, use $317,200 as the value of this branch at Decision 1. Similarly, for the bottom (new product) branch, the values are $86,000 at Decision 2 and [$86,000 (.5) + $600,000 (.5) =] $343,000 at Decision 1. The new product branch thus has a higher expected value and is selected as the best course of action under the expected value criteria.

Decision trees help to structure decisions in an objective way, force an explicit identification of alternatives, and foster a clear distinction between controllable and uncontrollable variables. They also permit us to incorporate uncertainty in a systematic, objective way. But monetary and probability values must still be estimated. Also, the expected-value approach may not be the best for a given situation; other approaches—such as maximax, maximin, or maximum probability—may be preferred.

STATISTICAL MODELS

Business decision makers that must rely upon incomplete information often utilize probabilities, which are the most basic measures of uncertainty. Probabilities attach a quantitative value (between 0 and 1) to the occurrence of an event. Table 3-3 summarizes the rules for applying probabilities.

Table 3-3   Probability Rules

Complement	$P(A) = 1 - P(\bar{A})$	(3.5)
Multiplication	$P(A \text{ and } B) = P(A)P(B\|A)$	
	$= P(A)P(B)$ (if independent)	(3.6)
Addition	$P(A \text{ or } B) = P(A) + P(B) - P(A \text{ and } B)$	
	$= P(A) + P(B)$ (if mutually exclusive)	(3.7)
Bayes' rule	$P(A\|B) = \dfrac{P(A \text{ and } B)}{P(B)}$	
	$= \dfrac{P(A)P(B\|A)}{P(A)P(B\|A) + P(\bar{A})P(B\|\bar{A})}$	(3.8)

Question: Distinguish between (a) classical, (b) empirical, and (c) subjective probabilities.

(a)   Classical probabilities are based upon equally likely outcomes that can be calculated prior to an event on the basis of mathematical logic.

(b)   Empirical probabilities are based upon observed data and express the relative frequency of an event in the long run.

(c)   Subjective probabilities are based upon personal experience or judgment and are sometimes used to analyze one-time occurrences.

Example 3.8   Market research data on a company's product has shown that during the first 3 years of use, 10 percent of the products had a mechanical difficulty and 40 percent had an electrical problem. The probability of an electrical problem, given some mechanical difficulty, is .6. What is the probability that a product will have either a mechanical difficulty or an electrical problem, or both?

$$P(M) = .10 \qquad P(E) = .40 \qquad P(E|M) = .60$$
$$P(M \text{ or } E) = P(M) + P(E) - P(M \text{ and } E)$$
$$\text{where} \quad P(M \text{ and } E) = P(M)P(E|M) = (.10)(.60) = .06$$
$$P(M \text{ or } E) = .10 + .40 - .06 = .44$$

Question: Explain the two concepts underlying statistical inference of (a) a sampling distribution and (b) the Central Limit Theorem.

(a)   *A sampling distribution* is a theoretical distribution of all sample means ($\bar{x}$'s) or all sample proportions (p's) from samples of a given size. The distribution includes all possible values that can occur, along with their probabilities of occurrence.

(b) *The Central Limit Theorem* states that for sufficiently large samples, the distribution of both sample means and sample proportions tends to follow a smooth bell-shaped normal curve.

EQUATIONS FOR DISCRETE AND CONTINUOUS DATA

Frequency data that are grouped into classes and used to express probabilities are either discrete or continuous. *Discrete* data stem from countable populations and are often expressed in terms of proportions (p's). *Continuous* data are obtained from measurable populations and are often classified as variables data (designated x). Discrete probabilities result from summations (Σ) of individual event probabilities, whereas continuous probabilities are obtained from an integration ($\int$) of the area under a continuous probability function. The distinction between discrete and continuous distributions is important because it affects the sample sizes and the risks of error associated with work sampling, quality control, and other productive activities.

Question: Summarize the statistical equations used for computing measures of central value and dispersion for populations, samples, and sampling distributions.

See Table 3-4.

Table 3-4   Statistical Equations for Discrete and Continuous Data

	Discrete (countable: attributes data)	Continuous (measurable: variables data)
Population Central value	Proportion π	Mean μ
Standard deviation	$\sigma = \sqrt{\pi(1-\pi)}$ *(3.9)*	$\sigma = \sqrt{\dfrac{\Sigma(X-\mu)^2}{N}}$ *(3.10)*
Sample Central value	Proportion p	Mean $\bar{X}$
Standard deviation	where $s = \sqrt{pq}$ *(3.11)* $q = 1-p$	$s = \sqrt{\dfrac{\Sigma(X-\bar{X})^2}{n-1}}$ *(3.12)*
Sampling distribution Central value	Average proportion $\bar{p}$	Mean of means $\bar{\bar{X}}$
Standard error	$s_{\bar{p}} = \sqrt{\dfrac{pq}{n}}$ *(3.13)*	$s_{\bar{x}} = \dfrac{s}{\sqrt{n}}$ *(3.14)*

Example 3.9   (Discrete)   In 400 observations of a computer operator, an analyst found him idle 32 times. Find (a) the sample proportion and (b) the standard error of proportion.

(a)
$$\bar{p} = \frac{\text{no. idle}}{\text{total no.}} = \frac{32}{400} = .08$$

(b)
$$s_{\bar{p}} = \sqrt{\frac{pq}{n}} = \sqrt{\frac{(.08)(.92)}{400}} = .014$$

Example 3.10   (Continuous)   In a study to find the time to service customers, a bank teller worked 60 minutes and served 36 customers. A record of the individual service times showed $\Sigma(X-\bar{X})^2 = (.79 \text{ minutes})^2$. Find (a) the sample mean time and (b) the standard error of the mean.

$$(a) \qquad \bar{X} = \frac{\Sigma x}{n} = \frac{60}{36} = 1.67 \text{ min per customer}$$

$$(b) \qquad s_{\bar{x}} = \frac{s}{\sqrt{n}} \qquad \text{where:} \quad s = \sqrt{\frac{\Sigma (X - \bar{X})^2}{n - 1}} = \sqrt{\frac{.79}{36 - 1}} = .15 \text{ min}$$

Therefore, $\qquad s_{\bar{x}} = \frac{.15}{\sqrt{36}} = .025 \text{ min}$

Questions and Solved Problems

THE DECISION-MAKING PROCESS

3.1 Mike Jackson is the operations manager of Supermarket Suppliers Inc., which runs a large old warehouse that services 80 delivery trucks. He must decide how many loading docks to include in a new warehouse. He decided that they should plan on enough capacity to handle average demand plus about 25 percent extra for growth.

To help make his decision, Mike collects some data on the current dock usage and simulates the unloading and loading activities on the company computer. The simulation generates values ranging from 7 to 14 docks. However, 12 loading docks handle average demand, and Mike tells the design engineer to plan for 15. Two weeks later Mike calls the design engineer to make sure everything is working out satisfactorily.

List the sequential steps in the decision process and the corresponding activity from the situation described.

Decision Process Step	Corresponding Activity in Situations
1. Define the problem and its parameters.	1. Problem is to determine number of loading docks required. Parameters are demand and load-unload time.
2. Establish decision criteria.	2. Criterion is capacity to meet average demand plus 25 percent.
3. Construct a model.	3. Model is a simulation model on computer.
4. Generate alternatives.	4. Alternatives range from 7 to 14 docks.
5. Evaluate and choose best alternative.	5. Best alternative is to plan for 15 docks (12 plus 25 percent).
6. Implement and monitor solution.	6. Tell design engineer and follow up 2 weeks later.

BREAK-EVEN ANALYSIS

3.2 For an existing product that sells for $P = \$650$ per unit, FC = $82,000 and VC = $240 per unit. (a) What is the BEP? (b) What volume is needed to generate a profit of $10,250?

$$(a) \qquad Q_{\text{BEP}} = \frac{FC}{P - VC} = \frac{\$82,000}{\$(650 - 240)/\text{unit}} = 200 \text{ units}$$

(b) The volume needed for a specific profit can be computed as:

$$Q_{BEP} = \frac{\text{Profit} + FC}{P - VC} = \frac{\$10,250 + 82,000}{\$(650 - 240)/\text{unit}} = 225 \text{ units}$$

Note: For volumes $>$ BEP, all contribution of ($\$650 - 240 = \$410$/unit) goes into profit. Thus the additional volume needed is $\$10,250 \div \$410$/unit $= 25$ units.

3.3 Process X has fixed costs of $20,000 per year and variable costs of $12 per unit, whereas process Y has fixed costs of $8,000 per year and variable costs of $22 per unit. At what production quantity (Q) are the total costs of X and Y equal?

$$X = Y$$
$$FC_X + VC_X \cdot Q = FC_Y + VC_Y \cdot Q$$
$$\$20,000 + \$12Q = \$8,000 + \$22Q$$
$$\$10Q = \$12,000$$
$$Q = 1,200 \text{ units}$$

3.4 Cover-the-Globe Paint Co. produces 9,000 paint sprayers per year and obtains $675,000 revenue from them. Fixed costs are $210,000 per year, and total costs are $354,000 per year. How much does each sprayer contribute to fixed costs and profit?

$$C = P - VC \quad \text{where} \quad P = \frac{\$675,000 \text{ revenue}}{9,000 \text{ units}} = \$75/\text{unit}$$

$$VC = \frac{TVC}{Q} \quad \text{where } TVC = TC - FC$$

$$= \$354,000 - \$210,000 = \$144,000$$

$$= \frac{\$144,000}{9,000 \text{ units}} = \$16/\text{unit}$$

Therefore, $C = \$75 - \$16 = \$59/\text{unit}$

3.5 A firm has annual fixed costs of $3.2 million and variable costs of $7 per unit. It is considering an additional investment of $800,000, which will increase the fixed costs by $150,000 per year and will increase the contribution by $2 per unit. No change is anticipated in the sales volume or the sales price of $15 per unit. What is the break-even quantity if the new investment is made?

The $2 increase in C will decrease VC to $\$7 - \$2 = \$5$/unit. The addition to FC makes them $3.2 million $+ \$150,000 = \$3,350,000$.

$$Q_{BEP} = \frac{FC}{P - VC} = \frac{\$3,350,000}{\$15/\text{unit} - \$5/\text{unit}} = 335,000 \text{ units}$$

3.6 *Two-Volume, Break-Even Analysis.* A producer of air conditioners sells the industrial model for $175 each. The production costs at volumes of 2,000 and 4,000 units are as shown in Table 3-5. The company does not know the FC at zero volume and realizes that some of its costs are "semivariable." Nevertheless, it wishes to prepare a break-even chart and determine the BEP.

Table 3-5

	2,000 Units	4,000 Units
Labor	$ 40,000	$ 80,000
Materials	90,000	180,000
Overhead	70,000	80,000
Selling and administrative	80,000	90,000
Depreciation and other FC	70,000	70,000
Total	$350,000	$500,000

This is a more realistic situation, because the FC and VC are determined from actual production volumes. Plot the TC at both volumes as seen in Fig. 3-4. The slope of the TC line ($\Delta Y/\Delta X$) is the estimated VC per unit.

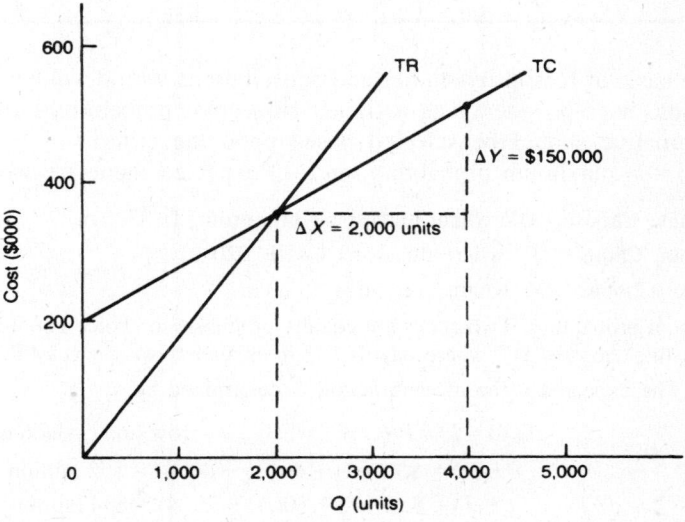

Fig. 3-4

$$\text{VC} = \frac{\text{change in } Y}{\text{change in } X} = \frac{\Delta Y}{\Delta X} = \frac{\$500,000 - \$350,000}{4,000 \text{ units} - 2,000 \text{ units}} = \$75/\text{unit}$$

By subtracting 2,000 units of variable cost from the total cost at 2,000 units, we can evaluate the implied fixed costs as follows:

$$\text{FC} = \text{total cost @ 2,000 volume} - (2,000 \text{ units}) (\text{variable cost/unit})$$

$$= \$350,000 - 2,000 \text{ units } (\$75/\text{unit}) = \$200,000$$

$$Q_{\text{BEP}} = \frac{\text{FC}}{P - \text{VC}} = \frac{\$200,000}{\$175/\text{unit} - \$75/\text{unit}} = 2,000 \text{ units}$$

3.7 A professional sports promoter leases a 40,000-seat stadium for soccer games. Tickets sell for an average of $14 each. If fixed costs per season (four games) are $720,000 and variable costs are $2 per spectator, what is the break-even point in number of seats filled per game?

$$Q_{\text{BEP}} = \frac{\text{FC}}{P - \text{VC}} = \frac{\$720,000/\text{season}}{\$14/\text{seat} - \$2/\text{seat}} = 60,000 \text{ seats/season}$$

$$\text{BEP} = \frac{60,000 \text{ seats}}{4 \text{ games}} = 15,000 \text{ seats/game}$$

DECISION THEORY, EXPECTED VALUE, AND DECISION TREES

3.8   An automobile company is evaluating the prospect of developing fuel cells for cars. As an alternative to financing the research and development (R&D) by itself, the firm is considering joining with an engineering firm. Depending upon the success of the R&D, the automobile company estimates its 10-year, present-value profits (millions $) as shown in Table 3-6.

Table 3-6   Success of R&D

	$\theta_1 = .2$	$\theta_2 = .4$	$\theta_3 = .4$
	Highly Successful	Moderately Successful	Not Successful
D = Develop on own	300	40	−60
J = Joint venture	200	30	−20

On the basis of feasibility studies and consultations with development and marketing groups, the operations vice-president has assigned subjective probabilities of $\theta_1 = .2$, $\theta_2 = .4$, $\theta_3 = .4$. Which alternative should be selected based upon the criteria of (*a*) maximax, (*b*) maximin, (*c*) Laplace, (*d*) maximum probability, and (*e*) expected monetary value?

(*a*)   Maximax: Choose "D" where highest payoff (profit) is $300 m.

(*b*)   Maximin: Choose "J" where the worst loss is $20 m.

(*c*)   Laplace: Choose "D" where average is $93.3 m.

(*d*)   Maximum probability: Two states are equally probable, so choice is not definitive. Under Moderately Successful, choose "D" where payoff is $40 m; under Not Successful, choose "J."

(*e*)   EMV: The expected value of each action is determined by:

$$E(X) = \Sigma\, \theta_{ij} P(\theta_j) \qquad \text{where} \quad i = \text{row and } j = \text{column.}$$
$$E(D) = \$300(.2) + 40(.4) - 60(.4) = \$52 \text{ million}$$
$$E(J) = \$200(.2) + 30(.4) - 20(.4) = \$44 \text{ million}$$

D is the optimal course and is designated with a * or EMV*.

3.9   Depict the previous problem in the form of a decision tree.

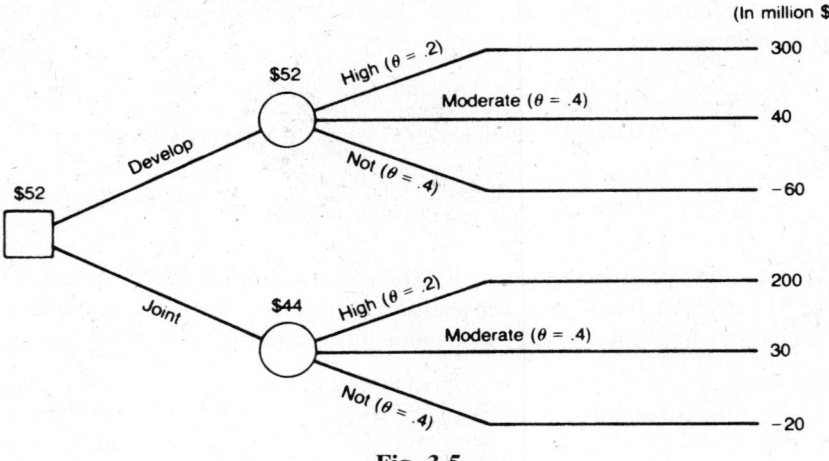

Fig. 3-5

3.10 *Risk:* Ohsaka Games Ltd. is evaluating the cost of producing electronic toys. Analysts are uncertain about the variable costs (VC) and have developed *low*, *most likely*, and *high* estimates to which they have assigned probabilities of .2, .5, and .3, respectively. Develop an expected-value estimate of the cost per unit.

Variable Cost Component	Low	Most Likely	High
Labor cost/unit	$4.10	$4.40	$4.85
Material cost/unit	2.65	2.95	3.10
Variable OH cost/unit	1.80	1.85	2.00
Total VC/unit	$8.55	$9.20	$9.95

$$E(X) = \Sigma\,[X \cdot P(X)]$$
$$E(\text{cost}) = (\text{low \$}) \cdot P(\text{low}) + (\text{most likely \$}) \cdot P(\text{most likely}) + (\text{high \$}) \cdot P(\text{high})$$
$$= \$8.55(.2) + \$9.20(.5) + \$9.95(.3) = \$9.30$$

3.11 A glass factory specializing in crystal is experiencing a substantial backlog, and the firm's management is considering three courses of action: (*A*) arrange for subcontracting, (*B*) begin overtime production, or (*C*) construct new facilities. The correct choice depends largely upon future demand, which may be low, medium, or high. By consensus, management ranks the respective probabilities as .10, .50, and .40. A cost analysis reveals the effect upon profits that is shown in Table 3-7.

Table 3-7

	Profit ($000) If Demand Is		
	Low (P = .10)	Medium (P = .50)	High (P = .40)
A = Arrange subcontracting	10	50	50
B = Begin overtime	−20	60	100
C = Construct facilities	−150	20	200

(*a*) State which course of action would be taken under a criterion of (1) maximax, (2) maximin, (3) maximum probability, (4) maximum expected value.

(*b*) Show this decision situation schematically in the form of a decision tree.

(*a*) *Maximax.* Maximize the maximum profit. Choose *C* in hopes that demand will be high.
Maximin. Maximize the minimum profit. Choose *A*, where the least profit is $10,000.
Maximum probability. Maximize under the most likely state. Choose *B* as the highest payoff under medium demand where *P* = .50.
Maximum expected value. Choose the act with the highest expected value.

$$E(X) = \Sigma\,[X \cdot P(X)]$$
$$E(A) = 10(.10) + 50(.50) + 50(.40) = 46,000$$
$$E(B) = -20(.10) + 60(.50) + 100(.40) = 68,000$$
$$E(C) = -150(.10) + 20(.50) + 200(.40) = 75,000$$

Therefore choose *C*, with expected value of $75,000 profit.

(*b*) For the decision tree, Fig. 3-6, the controllable (choice) decision variables are *A*, *B*, and *C*, and the uncontrollable variable is demand. We begin on the left by showing the decision choices first, followed by the chance alternatives of demand. The payoff value under each alternative is shown at the right. The expected value of each branch is then computed by summing the profit times the probability for each. For example, for *A*

$$E(A) = 10(.10) + 50(.50) + 50(.40) = \$46(000)$$

The best choice here, based upon the expected-value criteria, is the construction of new facilities, C.

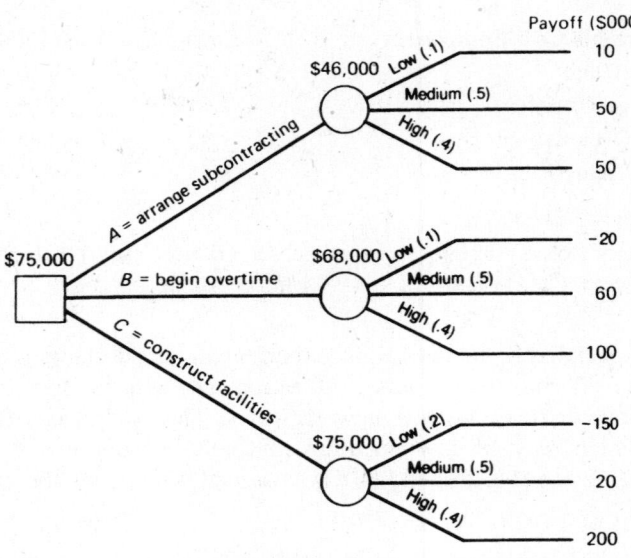

Fig. 3-6   Decision tree

STATISTICAL MODELS

3.12   Three molding machines (X, Y, and Z) are used to produce 600 computer terminal keys that are rushed (without inspection) to a customer. The number of good (G) and defective ($\bar{G}$) keys from each machine are as shown in Table 3-8.

Table 3-8   Quality of Output

	Machine X	Machine Y	Machine Z	Row Total
Good (G)	45	225	270	540
Not good ($\bar{G}$)	5	25	30	60
Total	50	250	300	600

When the customer receives the keys, they are randomly selected for installation. What is the probability that a key selected (*a*) is defective, (*b*) was produced by machine Z and good, (*c*) was either produced by machine Z or is good? (*d*) Is the probability of selecting a good key independent of the machine from which the key was made?

Given the data, we can estimate the empirical probabilities as follows:

$$P(G) = \frac{\text{number of good}}{\text{total number keys}} = \frac{540}{600} = .900$$

$$P(Z) = \frac{\text{number from Z}}{\text{total number keys}} = \frac{300}{600} = .500$$

$P(G|Z)$, which is read "Good, given it is from Z," is found as follows:

$$P(G|Z) = \frac{\text{number from Z that are good}}{\text{total number from Z}} = \frac{270}{300} = .90$$

Now, using the rules of probability we have:

(a) $$P(\bar{G}) = 1 - P(G) = 1 - .90 = .10$$

(b) $$P(Z \text{ and } G) = P(Z)P(G|Z) = (.50)(.90) = .45$$

(c) $$P(Z \text{ or } G) = P(Z) + P(G) - P(Z \text{ and } G) = .50 + .90 - .45 = .95$$

(d)   The $P(G)$ does not depend on whether the key is from machine X, Y, or Z:

$$P(G) = P(G|X) = P(G|Y) = P(G|Z) = \frac{540}{600} = \frac{225}{250} = \frac{45}{50} = .90$$

3.13. *Bayes' Rule.* Let θ represent the probability of defective wiring and A represent an accidental fire. In a large old factory, spot checks have established that $P(\theta) = .20$. Given that a plant has defective wiring, the probability of a fire occurring at some time during the year is .7 (that is, $P(A|\theta) = .7$), and if the wiring is not defective, the chance of a fire is reduced to .1 (that is, $P(A|\bar{\theta}) = .1$). A recent fire burned one employee severely and caused \$90,000 damage. Although evidence is destroyed, the operations manager has been asked by an insurance company to estimate the likelihood that the fire was caused by defective wiring.

$$P(\theta) = .2 \quad \text{thus} \quad P(\bar{\theta}) = 1 - .2 = .8$$
$$P(A|\theta) = .7 \quad \text{thus} \quad P(\bar{A}|\theta) = 1 - .7 = .3$$
$$P(A|\bar{\theta}) = .1 \quad \text{thus} \quad P(\bar{A}|\bar{\theta}) = 1 - .1 = .9$$

We wish to find the probability of defective wiring θ given the occurrence of the recent fire A. Using Bayes' rule we have:

$$P(\theta|A) = \frac{P(\theta)P(A|\theta)}{P(\theta)P(A|\theta) + P(\bar{\theta})P(A|\bar{\theta})} = \frac{(.2)(.7)}{(.2)(.7) + (.8)(.1)} = .64 \text{ or } 64 \text{ percent chance}$$

3.14 *Continuous Data.* Ten observations taken of the time required to assemble a sofa frame in a furniture plant were as shown below (min). Find the (a) mean time, (b) standard deviation, and (c) standard error of the mean.

Table 3-9

Observation	4.7	4.2	5.1	4.8	5.5	5.4	5.8	4.8	5.0	4.7	Total 50.0
$(X - \bar{X})$	$-.3$	$-.8$	$.1$	$-.2$	$.5$	$.4$	$.8$	$-.2$	$.0$	$-.3$	
$(X - \bar{X})^2$	.09	.64	.01	.04	.25	.16	.64	.04	.00	.09	1.96

(a) $$\bar{X} = \frac{\sum X}{n} = \frac{50}{10} = 5.0$$

(b) $$s = \sqrt{\frac{\sum (X - \bar{X})^2}{n - 1}} = \sqrt{\frac{1.96}{9}} = .47$$

Note: An equivalent equation is

$$s = \sqrt{\frac{\sum X^2 - (\sum X)^2/n}{n-1}} = \sqrt{\frac{252 - (50)^2/10}{10-1}} = .47$$

(c)
$$s_{\bar{x}} = \frac{s}{\sqrt{n}} = \frac{.47}{\sqrt{10}} = .15$$

Supplementary Questions and Problems

3.15 Fixed costs are $40,000 per year, variable costs are $50 per unit, and the selling price is $90 each. Find the BEP.   *Ans.* 1,000 units

3.16 Nationwide Survey Co. has fixed costs of $20,000 per year, variable costs of $3 per survey, and charges $5 per survey. What is the break-even point in number of surveys?   *Ans.* 10,000 surveys

3.17 Florida Citrus produced 40,000 boxes of fruit that sold for $3 per box. The total variable costs for the 40,000 boxes were $60,000, and the fixed costs were $75,000. (*a*) What was the break-even quantity? (*b*) How much profit (or loss) resulted?   *Ans.* (*a*) 50,000 boxes (*b*) $15,000 loss

3.18 If the sales price of a product is $8 and the variable cost is $2, what is the contribution?   *Ans.* $6

3.19 A travel agency has an excursion package that sells for $125. Fixed costs are $80,000; and at the present volume of 1,000 customers, variable costs are $25,000 and profits are $20,000. (*a*) What is the break-even point volume? (*b*) Assuming that fixed costs remain constant, how many additional customers will be required for the agency to increase profit by $1,000?   *Ans.* (*a*) 800 units (*b*) 10 customers

3.20 A nonprofit municipal water department has variable costs (direct labor) of $5 million per year. Current revenue, based upon the service of 200,000 accounts, is $20 million. The water-production manager wishes to add equipment that will raise the yearly fixed costs by $1 million and reduce the current and future direct labor costs by 20 percent. What volume of account services would be required to justify the change economically? The price paid per customer is to be held constant.   *Ans.* At least 200,000 accounts

3.21 Last year, Dever Furniture Co. produced 200 maple dressers (pattern 427) that sold for $210 each. The company incurred labor costs of $42 per unit and material costs of $18 per unit, and it allocated $80 per unit of overhead costs to each dresser. Cost records reveal that overhead costs are 60 percent fixed and 40 percent variable. What was the total annual contribution from pattern 427?
Ans. $118/unit, or $23,600 total

3.22 Madison Industries has the following data (Table 3.10) on costs at two volumes of production for a product that sells for $50. (*a*) Construct a two-volume, break-even chart. (*b*) Compute the variable cost, the contribution, and the BEP. (*c*) Using the contribution from (*b*), estimate the profit at a volume of 8,000 units.
Ans. (*a*) TC line is plotted through $X = 6,000$, $Y = $230,000$, and $X = 10,000$, $Y = $300,000$ (*b*) $17.50/unit, $32.50/unit, 3,846 units (*c*) $135,005

Table 3-10

	Labor	Material	Overhead	Other FC	Total
6,000 unit volume	$ 60,000	36,000	54,000	80,000	$230,000
10,000 unit volume	100,000	60,000	60,000	80,000	300,000

3.23 Given the payoff table below showing the profit (present value $m), a firm might expect in a foreign country for three alternative factory investments (X, Y, and Z) under different levels of inflation. Economists have assigned probabilities of .2, .3, .4, and .1 to the possible inflation levels A, B, C, and D, respectively. Find the preferred investment alternative using criteria of (a) maximax, (b) maximin, (c) Laplace, (d) maximum probability, and (e) expected monetary value. Finally, (f) use your "judgment."

Table 3-11

	State of Nature: Amount of Inflation			
	A = 2%	B = 5%	C = 10%	D = 15%
Build factory X	10	30	50	120
Build factory Y	40	50	60	70
Lease plant Z	10	40	80	10

Ans.   (a) X @ $120 m, (b) Y @ $40 m, (c) Y @ $55.0 m, (d) Z @ $80 m, (e) Y, where EMV* = $54 m.
(f) Y is relatively "safe," and yet has a high probability (.8) of $\geq$ $50 m.

3.24 Frozen Pizza Co. is considering whether it should allocate funds for research on an instant freeze-dry process for home use. If the research is successful (and the R&D manager feels there is a 75 percent chance it will be), the firm could market the product at a $4 million profit. However, if the research is unsuccessful, the firm will incur a $6 million loss. What is the expected monetary value (EMV) of proceeding with the research?   *Ans.* $1.5 million

3.25 Company safety records show that 40 percent of all accidents occur when new employees (those with less than 1 year's service) are operating equipment, and 60 percent occur when the more-experienced employees are operating it. The firm averages six accidents over a 300-workday year. What is the chance that on any given day during the year an accident will happen to (a) a new employee, (b) an experienced employee?   *Ans.*   (a) .008 (b) .012

3.26 A stockroom clerk at an auto assembly plant has absentmindedly mixed the stock of arm-support brackets for the left- and right-front doors of automobiles and cannot readily tell them apart, nor can she distinguish the type of mounting (that is, A, B, or C). She receives an urgent request for a right-front bracket, type A, randomly selects one, and rushes it to the assembly area

Suppose the stock records show that she has 500 brackets (total) on hand, in the quantities shown in Table 3-12.

Table 3-12

	Right Front	Left Front
Type A mount	274	146
Type B mount	0	50
Type C mount	26	4

(a) What is the probability that she chose a correct bracket? (b) Suppose she could identify the type of mounting but not whether it was for a right or left door. What would be the probability of a correct choice in this case?   *Ans.*   (a) .55 (b) .65

3.27   In a chemical plant, the probability of any employee being injured from a fall is $P(F) = .005$; from chemical inhalation the probability is $P(C) = .020$. If a worker falls, the probability of injury from chemical inhalation increases to $P(C|F) = .100$. What is the probability that an employee will be injured (a) by *both* a fall and chemical inhalation, and (b) by *either* a fall or chemical inhalation?
Ans.   (a) .0005 (b) .0245

3.28   The operations manager of a large airport is concerned with the problem of having adequate personnel to offer individual assistance to handicapped passengers during rush hours. Data were collected on the number of requests for assistance during 20 randomly selected rush hours and revealed the information shown in Table 3-13.

Table 3-13

Hour no.	1	2	3	4	5	6	7	8	9	10
No. of requests	40	42	42	30	38	48	42	44	37	38
Hour no.	11	12	13	14	15	16	17	18	19	20
No. of requests	49	47	34	57	42	52	56	44	50	48

Determine (a) $\bar{X}$, (b) s, and (c) $s_{\bar{X}}$.   *Ans.*   (a) 44 requests per hour (b) 7 requests per hour (c) 1.57

Chapter 4

Forecasting
Statistical Methods

FORECASTING: DEFINITION, TYPES, AND USES

Forecasts are essential for the smooth operations of business organizations. They provide information that can assist managers in guiding future activities toward organizational goals.

Question: What are forecasts?

Forecasts are estimates of the occurrence, timing, or magnitude of uncertain future events.

Operations managers are primarily concerned with forecasts of demand—which are often made by (or in conjunction with) marketing. However, managers also use forecasts to estimate raw material prices, plan for appropriate levels of personnel, help decide how much inventory to carry, and a host of other activities. This results in better use of capacity, more responsive service to customers, and improved profitability.

COSTS OF FORECASTING

Organizations that pay no attention to forecasting are implicitly assuming that what has happened in the past will continue in the future.

Question: What are the costs associated with forecasting—or not forecasting?

Inadequate forecasts can result in excessive labor, material, or capital costs, as well as expediting costs and lost revenues. On the other hand, as forecasting activity increases, the costs of collecting and analyzing data increase. The optimal level implies a balance. (See Figure 4-1.) Many large organizations find it worthwhile to subscribe to a professional forecasting service.

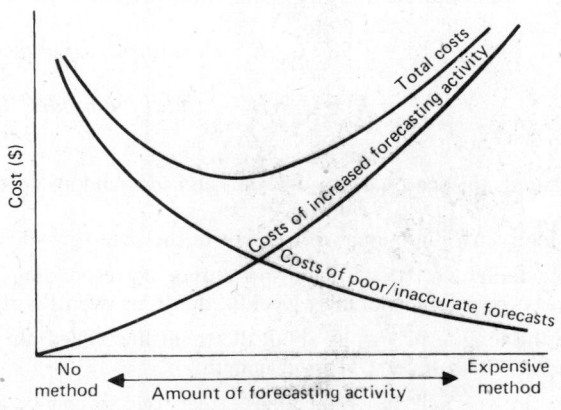

Fig. 4-1 Costs of forecasting

39

FORECASTING DECISION VARIABLES AND METHODOLOGY

Forecasting activities are a function of (1) the type of forecast (e.g., demand, technological), (2) the time horizon (short, medium, or long range), (3) the database available, and (4) the methodology employed (qualitative or quantitative). Forecasts of *demand* are based primarily on nonrandom trends and relationships, with an allowance for random components. Forecasts for groups of products tend to be more accurate than those for single products, and short-term forecasts are more accurate than long-term forecasts (greater than five years). Quantification also enhances the objectivity and precision of a forecast.

Question: Summarize the key features of the more commonly used forecasting methods. (See Table 4-1.)

OPINION AND JUDGMENTAL METHODS

Some opinion and judgment forecasts are largely intuitive, whereas others integrate data and perhaps even mathematical or statistical techniques. *Judgmental forecasts* often consist of (1) forecasts by individual salespeople, (2) forecasts by division or product-line managers, and (3) combined estimates of the two. *Historical analogy* relies on comparisons, *Delphi* relies on expert opinions, and *market surveys* rely on consumer response—which is not always reliable. Another method (focus) relies on the best method from a group of forecasts. All these methods can incorporate experiences and personal insights. However, results may differ from one individual to the next and they are not all amenable to analysis, so there may be little basis for improvement over time.

TIME SERIES METHODS

Question: What is a time series, and what are the components of a time series?

A *time series* is a set of observations of a variable at regular intervals over time. In *decomposition analysis*, the components of a time series are generally classified as trend T, cyclical C, seasonal S, and random or irregular R. (*Note:* Autocorrelation effects are sometimes included as an additional factor.)

Time series are usually tabulated or graphed to show the nature of the time dependence. The forecast value (Y_c) is commonly expressed as a multiplicative or additive function of its components; examples here will be based upon the commonly used multiplicative model.

$$Y_c = T \cdot S \cdot C \cdot R \qquad \textit{multiplicative model} \qquad (4.1)$$

$$Y_c = T + S + C + R \qquad \textit{additive model} \qquad (4.2)$$

Question: Explain the (*a*) trend, (*b*) seasonal, (*c*) cyclical, and (*d*) random components of a series.

(*a*) *Trend* is a gradual long-term directional movement in the data (growth or decline).

(*b*) *Seasonal effects* are similar variations occurring during corresponding periods, e.g., December retail sales. Seasonals can be quarterly, monthly, weekly, daily, or even hourly indexes.

(*c*) *Cyclical factors* are the long-term swings about the trend line. They are often associated with business cycles and may extend out to several years in length.

(*d*) *Random components* are sporadic (unpredictable) effects due to chance and unusual occurrences. They are the residual after the trend, cyclical, and seasonal variations are removed.

Table 4-1　Summary of Forecasting Methods

Method	Description	Time Horizon	Relative Cost
Opinion and judgment (qualitative)			
Sales force composites	Estimates from field salespeople are aggregated	SR–MR	L–M
Executive opinion (and/or panels)	Marketing, finance, and production managers jointly prepare forecast	SR–LR	L–M
Field sales and product-line management	Estimates from regional salespeople are reconciled with national projections from product-line managers	MR	M
Historical analogy	Forecast from comparison with similar product previously introduced	SR–LR	L–M
Delphi	Experts answer a series of questions (anonymously), receive feedback, and revise estimates	LR	M–H
Market surveys	Questionnaires/interviews for data to learn about consumer behavior	MR–LR	H
Time series (quantitative)			
Naive	Forecast equals latest value or latest plus or minus some percentage	SR	L
Moving average	Forecast is average of *n* most recent periods (can also be weighted)	SR	L
Trend projection	Forecast is linear, exponential, or other projection of past trend	MR–LR	L
Decomposition	Time series is divided into trend, seasonal, cyclical, and random components	SR–LR	L
Exponential smoothing	Forecast is an exponentially weighted moving average, where latest values carry most weight	SR	L
Box-Jenkins	A time-series–regression model is proposed, statistically tested, modified, and retested until satisfactory	MR–LR	M–H
Associative (quantitative)			
Regression and correlation (and leading indicators)	Use one or more associate variables to forecast via a least-squares equation (regression) or via a close association (correlation) with an explanatory variable	SR–MR	M–H
Econometric	Use simultaneous solution of multiple-regression equations that relate to broad range of economic activity	SR–LR	H

Key: L = low, M = medium, H = high, SR = short range, MR = medium range, LR = long range.

Question: What steps are involved in using time series data to make a forecast?

Table 4-2   Forecasting Procedure for Using Time Series

> 1. Plot historical data to confirm relationship (e.g., linear, exponential).
> 2. Develop a trend equation (T) to describe the data.
> 3. Develop a seasonal index (*SI*, e.g., monthly index values).*
> 4. Project trend into the future (e.g., monthly trend values).
> 5. Multiply trend values by corresponding seasonal index values.
> 6. Modify projected values by any knowledge of:
>   (*C*)   cyclical business conditions
>   (*R*)   anticipated irregular effects

Trend.   Three methods of describing a trend are (1) moving average, (2) hand fitting, and (3) least squares.

A centered *moving average* (MA) is obtained by summing and averaging the values from a given number of periods repetitively, each time deleting the oldest value and adding a new value. Moving averages can smooth out fluctuations in any data, while preserving the general pattern of the data (longer averages result in more smoothing). However, they do not yield a forecasting equation, nor do they generate values for the ends of the data series.

$$MA = \frac{\Sigma X}{\text{number of periods}} \qquad (4.3)$$

A *weighted moving average* (MA_{wt}) allows some values to be emphasized by varying the weights assigned to each component of the average. Weights can be either percentages or a real number.

$$MA_{wt} = \frac{\Sigma (wt)X}{\Sigma wt} \qquad (4.4)$$

Example 4.1   Shipments (in tons) of welded tube by an aluminum producer are shown in Table 4-3.

Table 4-3   Shipments of Welded Aluminum Tube

Year	1	2	3	4	5	6	7	8	9	10	11
Tons	2	3	6	10	8	7	12	14	14	18	19

(*a*)   Graph the data, and comment on the relationship. (*b*) Compute a 3-year moving average, plot it as a dotted line, and use it to forecast shipments in year 12. (*c*) Using a weight of 3 for the most recent data, 2 for the next, and 1 for the oldest, forecast shipments in year 12.

---

* If seasonal factors are significant, some analysts recommend deseasonalizing the demand values (i.e., by dividing the actual by the *SI*), and recalculating the trend equation using the deseasonalized values.

Table 4-4   3-Year Moving Average

Year	Shipments (tons)	3-Year Moving Total	3-Year Moving Average
1	2	—	—
2	3	----- 11 ----→ ÷ 3 =	3.7
3	6	19 ÷ 3 =	6.3
4	10	24	8.0
5	8	25	8.3
6	7	27	9.0
7	12	33	11.0
8	14	40	13.3
9	14	46	15.3
10	18	------- 51 ----→ ÷ 3 =	17.0
11	19	—	—

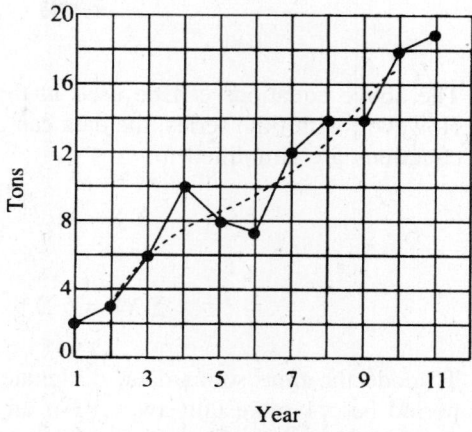

Fig. 4-2   Tube shipments (tons per year)

(a) The data points appear relatively linear. (b) See Table 4-4 for computations and Fig. 4-2 for plot of the MA. The MA forecast for year 12 would be that of the latest average, 17.0 tons.

(c)
$$MA_{wt} = \frac{\sum (wt)X}{\sum wt} = \frac{(1)(14) + (2)(18) + (3)(19)}{1 + 2 + 3} = 17.8 \text{ tons}$$

A *hand fit or freehand curve* is simply a plot of a representative line that (subjectively) seems to best fit the data points. For linear data, the forecasting equation will be of the form:

$$Y_c = a + b(X) \quad \text{(signature)} \tag{4.5}$$

where Y_c is the trend value, a is the intercept (where line crosses the vertical axis), b is the slope (the rise, Δy, divided by the run, Δx), and X is the time value (years, quarters, etc.). The "signature" identifies the point in time when $X = 0$, as well as the X and Y units.

Example 4.2 (a) Use a hand fit line to develop a forecasting equation for the data in Fig. 4-2. State the equation, complete with signature. (b) Use your equation to forecast tube shipments for year 12.

(a) Select points some distance apart. A straight line connecting the values for years 3 and 8 might be a good freehand representation of the data. From this we can determine the slope and intercept:

$$\text{Slope: } b = \frac{\Delta Y}{\Delta X} = \frac{Y_2 - Y_1}{X_2 - X_1} = \frac{14 - 6}{8 - 3} = 1.6 \text{ tons}$$

Intercept: $a = .5$ tons (*Note:* This is the estimated Y value at $X = 0$ from graph.)

signature

Equation: $Y_c = .5 + 1.6X$ (yr 0 = 0, X = yrs, Y = tons)

(b) For year 12: $Y_c = .5 + 1.6(12) = 19.7$ tons

Least squares is a mathematical technique of fitting a trend to data points. The resulting *line of best fit* has the following properties: (1) the summation of all vertical deviations about it is zero, (2) the summation of all vertical deviations squared is a minimum, and (3) the line goes through the means $\overline{X}$ and $\overline{Y}$. For linear equations, the line of best fit is found by the simultaneous solution for a and b of the following two *normal equations:*

$$\Sigma Y = na + b \Sigma X$$

$$\Sigma XY = a \Sigma X + b \Sigma X^2 \qquad (4.6)$$

The above equations can be used in the form shown above and are used in that form for regression. However, with time series, the data can also be coded so that $\Sigma X = 0$. Two terms then drop out, and the equations are simplified to:

$$\Sigma Y = na \qquad\qquad a = \frac{\Sigma Y}{n} \quad \text{(coded data only)}$$

$$\Sigma XY = b \Sigma X^2 \qquad b = \frac{\Sigma XY}{\Sigma X^2} \quad \text{(coded data only)} \qquad (4.7)$$

To code the time series data, designate the center of the time span as $X = 0$ and let each successive period be ± 1 more unit away. (For an even number of periods, use values of $\pm .5$, 1.5, 2.5, etc.)

Example 4.3   Use the least-squares method to develop a linear trend equation for the data of Example 4.1. State the equation complete with signature, and forecast a trend value for year 16.

See Table 4-5 for the data.

Table 4-5

Year	X Year Coded	Y Shipments (tons)	XY	X^2
1	−5	2	−10	25
2	−4	3	−12	16
3	−3	6	−18	9
4	−2	10	−20	4
5	−1	8	−8	1
6	0	7	0	0
7	1	12	12	1
8	2	14	28	4
9	3	14	42	9
10	4	18	72	16
11	5	19	95	25
	0	113	181	110

From Eq. 4.7, we have:

$$a = \frac{\Sigma Y}{n} = \frac{113}{11} = 10.3 \qquad b = \frac{\Sigma XY}{\Sigma X^2} = \frac{181}{110} = 1.6$$

The forecasting equation is of the form $Y = a + bX$.

$$Y = 10.3 + 1.6X \quad \text{(yr 6} = 0, X = \text{years}, Y = \text{tons)}$$

Since year 16 is 10 years distant from the origin (year 6), we have

$$Y = 10.3 + 1.6(10) = 26.3 \text{ tons}$$

Seasonal Indexes.   A *seasonal index* (*SI*) is a ratio that relates a recurring seasonal variation to the corresponding trend value at the given time. In the *ratio-to-moving-average* method of calculation, monthly (or quarterly) data are typically used to compute a 12-month (or 4-quarter) moving average.

(This dampens out all seasonal fluctuations.) Actual monthly (or quarterly) values are then *divided by the moving average value* centered on the actual month. In the *ratio-to-trend* method, the actual values are *divided by the trend value* centered on the actual month. The ratios obtained for several of the same months (or quarters) are then averaged to obtain the seasonal index values. The indexes can be used to obtain seasonalized forecast values, Y_{sz} (or to deseasonalize actual data).

$$Y_{sz} = (SI)Y_c \qquad\qquad (4.8)$$

Example 4.4   Snowsport Int'l has experienced low snowboard sales in July, as shown in Table 4-6. Using the ratio-to-trend values, calculate a seasonal index value for July and explain its meaning.

Table 4-6   Snowboard Sales

	Yr 5	Yr 6	Yr 7	Yr 8	Yr 9	Yr 10	Yr 11	Yr 12	
July actual sales	22	30	18	26	45	36	40		
July trend value, Y_c	170	190	210	230	250	270	290		
Ratio (actual ÷ trend)	.13	.16	.09	.11	.18	.13	.14		Total = .94

(a)   A third row has been added to Table 4-6 to show the ratio of actual to trend values for July. Using a simple average, the July index is $SI_{July} = .94 \div 7 = .13$. This means that July is typically only 13 percent of the trend value for July in any given year. Winter months are likely quite high.

Example 4.5   The forecasting equation for the previous example, centered in July of year 4 with X units in months, was $Y_c = 1800 + 20X$ (July 15, Yr 4 = 0, X = mo, Y = units/yr).* Use this equation and the July seasonal index of .13 to compute (a) the trend (deseasonalized) value for July of year 12 and (b) the forecast of actual (seasonalized) snowboard sales in July of year 12.

(a)   July of year 12 is $(8)(12) = 96$ months away from July of year 4, so the trend value is:

$$Y_c = 1800 + 20(96) = 3{,}240 \text{ units/yr or } 3{,}240 \text{ units/yr} \div 12 \text{ mo/yr} = 310 \text{ units/mo}$$

(b)   The actual (seasonalized) forecast is $Y_{sz} = (SI)\ Y_c = (.13)(310) = 40$ units.

EXPONENTIAL SMOOTHING

Question: What is exponential smoothing?

Exponential smoothing is a moving-average forecasting technique that weights past data in an exponential manner so that the most recent data carry more weight in the moving average.

With simple exponential smoothing, the forecast F_t is made up of the last period forecast F_{t-1} plus a portion, α, of the difference between the last period actual demand A_{t-1} and the last period forecast F_{t-1}.

$$F_t = F_{t-1} + \alpha(A_{t-1} - F_{t-1}) \qquad\qquad (4.9)$$

Example 4.6   A firm uses simple exponential smoothing with $\alpha = .1$ to forecast demand. The forecast for the week of February 1 was 500 units, whereas actual demand turned out to be 450 units.

---

* Thus the trend value for July of year 5 was $Y_c = 1800 + 20(12) = 2040 \div 12$ mo = 170 units. See solved problems for converting X units from years to months and applying correction factors.

(a) Forecast the demand for the week of February 8.

(b) Assume that the actual demand during the week of February 8 turned out to be 505 units. Forecast the demand for the week of February 15. Continue forecasting through March 15, assuming that subsequent demands were actually 516, 488, 467, 554, and 510 units.

(a)
$$F_t = F_{t-1} + \alpha(A_{t-1} - F_{t-1})$$

$$= 500 + .1(450 - 500) = 495 \text{ units}$$

(b) Arranging the procedure in tabular form, we have Table 4-7.

Table 4-7

Week	Actual Demand A_{t-1}	Old Fore-cast F_{t-1}	Forecast Error $A_{t-1} - F_{t-1}$	Correction $\alpha(A_{t-1} - F_{t-1})$	New Forecast (F_t) $F_{t-1} + \alpha(A_{t-1} - F_{t-1})$
Feb. 1	450	500	−50	−5	495
8	505	495	10	1	496
15	516	496	20	2	498
22	488	498	−10	−1	497
Mar. 1	467	497	−30	−3	494
8	554	494	60	6	500
15	510	500	10	1	501

The smoothing constant, α, is a number between 0 and 1 that enters multiplicatively into each forecast but whose influence declines exponentially as the data become older. Typical values range from .01 to .40. A low α gives more weight to the past average and will effectively dampen high random variation. High α values are more responsive to changes in demand (e.g., from new-product introductions, promotional campaigns). An α of 1 would reflect total adjustment to recent demand, and the forecast would be last period's actual demand. A satisfactory α can generally be determined by trial-and-error modeling (on computer) to see which value minimizes forecast error.

Simple exponential smoothing yields only an average. It does not extrapolate for trend effects. No α value will fully compensate for a trend in the data. An α value that yields an approximately equivalent degree of smoothing as a moving average of n periods is:

$$\alpha = \frac{2}{n+1} \tag{4.10}$$

ADJUSTED EXPONENTIAL SMOOTHING

Adjusted exponential smoothing models have all the features of simple exponential smoothing models, plus they project into the future (for example, to time period $t + 1$) by adding a trend correction increment, T_t, to the current period smoothed average, $\hat{F}_t$.

$$\hat{F}_{t+1} = \hat{F}_t + T_t \tag{4.11}$$

Figure 4-3 depicts the components of a trend-adjusted forecast that utilizes a second smoothing coefficient β. The β value determines the extent to which the trend adjustment relies on the latest difference in forecast amounts $(\hat{F}_t - \hat{F}_{t-1})$ versus the previous trend T_{t-1}. Thus:

$$\hat{F}_t = \alpha A_{t-1} + (1 - \alpha)(\hat{F}_{t-1} + T_{t-1}) \qquad (4.12)$$

$$T_t = \beta(\hat{F}_t - \hat{F}_{t-1}) + (1 - \beta)T_{t-1} \qquad (4.13)$$

A low β gives more smoothing of the trend and may be useful if the trend is not well established. A high β will emphasize the latest trend and be more responsive to recent changes in trend. The initial trend adjustment T_{t-1} is sometimes assumed to be zero.

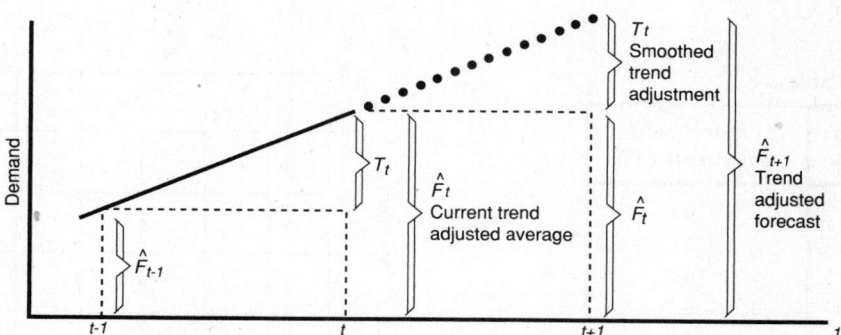

Fig. 4-3   Components of a trend-adjusted forecast

Self-Adaptive Models.   Self-adjusting computer models that change the values of the smoothing coefficients (α's and β's) in an adaptive fashion have been developed; these models help to minimize the amount of forecast error.

REGRESSION AND CORRELATION METHODS

Regression and correlation techniques quantify the statistical association between two or more variables.

Question: Distinguish between (a) simple regression and (b) simple correlation.

(a)   *Simple regression* expresses the relationship between a dependent variable Y and one independent variable X in terms of the slope and intercept of the line of best fit relating the two variables.

(b)   *Simple correlation* expresses the degree or closeness of the relationship between two variables in terms of a correlation coefficient that provides an indirect measure of the variability of points from the line of best fit. Neither regression nor correlation gives proof of a cause-effect relationship.

Multiple regression and correlation analysis (involving more than two variables) and nonlinear models are also useful, but are beyond our scope here.

Regression.   The simple linear regression model takes the form $Y_c = a + bX$, where Y_c is the dependent variable and X the independent variable. Values for the slope b and intercept a are obtained by using the *normal equations* written in the convenient form:

$$b = \frac{\sum XY - n\bar{X}\bar{Y}}{\sum X^2 - n\bar{X}^2} \qquad (4.14)$$

$$a = \bar{Y} - b\bar{X} \qquad (4.15)$$

In Eqs. 4.14 and 4.15, $\bar{X} = (\sum X)/n$ and $\bar{Y} = (\sum Y)/n$ are the means of the independent and dependent variables respectively, and n is the number of pairs of observations made.

Example 4.7 The general manager of a building materials production plant feels that the demand for plasterboard shipments may be related to the number of construction permits issued in the county during the previous quarter. The manager has collected the data shown in Table 4-8.

(a) Review the scatter diagram (Fig. 4-4) to see whether the data can be satisfactorily described by a linear equation.

(b) Compute values for the slope b and intercept a.

(c) Determine a point estimate for plasterboard shipments when the number of construction permits is 30.

Table 4-8

Construction Permits (X)	Plasterboard Shipments (Y)
15	6
9	4
40	16
20	6
25	13
25	9
15	10
35	16

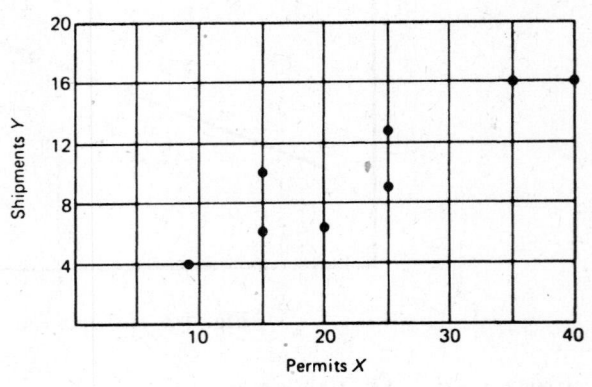

Fig. 4-4

(a) The scatter diagram (Fig. 4-4) shows that the data are not perfectly linear but do approach linearity over this short range.

(b) See Table 4-9 and the accompanying calculations.

Table 4-9

X	Y	XY	X^2	Y^2
15	6	90	225	36
9	4	36	81	16
40	16	640	1,600	256
20	6	120	400	36
25	13	325	625	169
25	9	225	625	81
15	10	150	225	100
35	16	560	1,225	256
184	80	2,146	5,006	950

$n = 8$ pairs of observations

$$\overline{X} = \frac{184}{8} = 23$$

$$\overline{Y} = \frac{80}{8} = 10$$

$$b = \frac{\sum XY - n\overline{X}\,\overline{Y}}{\sum X^2 - n\overline{X}^2} = \frac{2{,}146 - 8(23)(10)}{5{,}006 - 8(23)(23)} = .395$$

$$a = \overline{Y} - b\overline{X} = 10 - .395(23) = .91$$

(c) The regression equation is

$$Y_c = .91 + .395X \qquad (X = \text{permits}, \ Y = \text{shipments})$$

Then, letting $X = 30$,

$$Y_c = .91 + .395(30) = 12.76 \cong 13 \text{ shipments}$$

Standard Deviation of Regression. A regression line describes the relationship between a given value of the independent variable X and the mean $\mu_{Y \cdot X}$ of the corresponding probability distribution of the dependent variable Y. We assume the distribution of Y values is normal for any given X value. The point estimate, or forecast, is the mean of that distribution for any given value of X, as shown in Fig. 4-5.

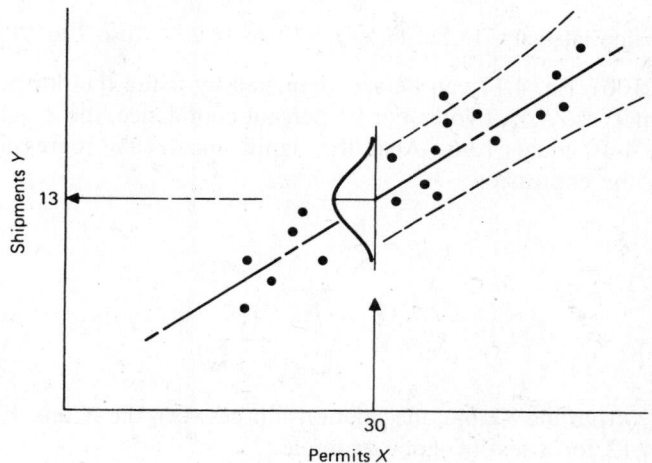

Fig. 4-5   Regression line

The *standard deviation* of *regression* $S_{Y \cdot X}$ is a measure of the dispersion of data points around the regression line (Fig. 4-5). For simple regression, the computation of $S_{Y \cdot X}$ has $n - 2$ degrees of freedom.

$$S_{Y \cdot X} = \sqrt{\frac{\sum Y^2 - a \sum Y - b \sum XY}{n - 2}} \qquad (4.16)$$

Example 4.8   Given the data on permits and shipments in the previous example, compute the standard deviation of regression $(S_{Y \cdot X})$.

$$S_{Y \cdot X} = \sqrt{\frac{\sum Y^2 - a \sum Y - b \sum XY}{n - 2}} = \sqrt{\frac{950 - (.91)(80) - (.396)(2,146)}{8 - 2}} = 2.2 \text{ shipments}$$

Interval Estimate.   A prediction interval can be established for an *individual* forecast value of Y_c by using the expression:

$$\text{Prediction interval} = Y_c \pm t S_{\text{ind}} \qquad (4.17)$$

where $t = $ value from t-distribution table for specified confidence level (see Table 4-10), and

$$S_{\text{ind}} = S_{Y \cdot X} \sqrt{1 + \frac{1}{n} + \frac{(X - \overline{X})^2}{\sum (X - \overline{X})^2}} \qquad (4.18)$$

Example 4.9   Using the data from Examples 4.7 and 4.8, develop a 95 percent prediction interval estimate for the specific number of shipments to be made when 30 construction permits were issued during the previous quarter. *Note:* $\overline{X} = 23$ for the $n = 8$ observations, and $\sum (X - \overline{X})^2 = 774$. Also, from Example 4.7, $Y_c = 13$ shipments, where $X = 30$; and from Example 4.8, $S_{Y \cdot X} = 2.2$ shipments.

$$\text{Prediction interval} = Y_c \pm tS_{\text{ind}}$$

where the t-value (from Table 4-10) for $n - 2 = 8 - 2 = 6$ degrees of freedom $= 2.45$ and where

$$S_{\text{ind}} = S_{Y \cdot X} \sqrt{1 + \frac{1}{n} + \frac{(X - \overline{X})^2}{\Sigma(X - \overline{X})^2}}$$

$$= 2.2 \sqrt{1 + \frac{1}{8} + \frac{(30 - 23)^2}{774}}$$

$$= 2.4 \text{ shipments}$$

$$\therefore \quad \text{Prediction interval} = 13 \pm 2.45(2.4) = 7.1 \text{ to } 18.9 \qquad \text{(use 7 to 19 shipments)}$$

For large samples ($n \geq 100$), Eq. 4.17 can be approximated by using the normal (Z) distribution rather than the t, in the form of $Y_c \pm ZS_{Y \cdot X}$. (*Note:* For 95 percent confidence, the Z value is the same as t with ∞ df, which from Table 4-10 equals 1.96.) Also, the significance of the regression line slope coefficient (b) can be tested using the expression:

$$t_{\text{calc}} = \frac{b}{S_b} \qquad (4.19)$$

where

$$S_b = S_{Y \cdot X} \sqrt{\frac{1}{\Sigma(X - \overline{X})^2}} \qquad (4.20)$$

If the value of $t_{\text{calc}} > t_{\text{df}}$ from the t-table, the relationship between the X and Y variables is statistically significant. See Prob. 4.12 for a test of slope example.

Table 4-10 t-Distribution Values (for 90 Percent and 95 Percent Confidence)

df	5	6	7	8	9	10	12	15	20	30	∞
$t_{.05}$ (90%)	2.02	1.94	1.90	1.86	1.83	1.81	1.78	1.75	1.73	1.70	1.65
$t_{.10}$ (95%)	2.57	2.45	2.37	2.31	2.26	2.23	2.17	2.13	2.08	2.04	1.96

Correlation.   The simple linear *correlation coefficient* r is a number between -1 and $+1$ that tells how well a linear equation describes the relationship between two variables. As illustrated in Fig. 4-6, r is designated as positive if Y increases as X increases, and negative if Y decreases as X increases. An r of zero indicates an absence of any relationship between the two variables.

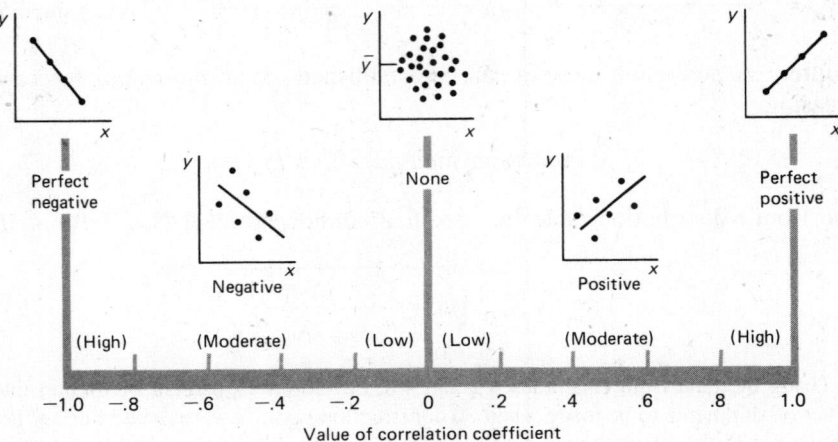

Fig. 4-6   Interpretation of correlation coefficient

The deviation of all points (Y) from the regression line (Y_c) consists of deviation accounted for by the regression line (explained) and random deviation (unexplained). Fig. 4-7 illustrates this for one point, Y. Squaring the deviation we have variation.

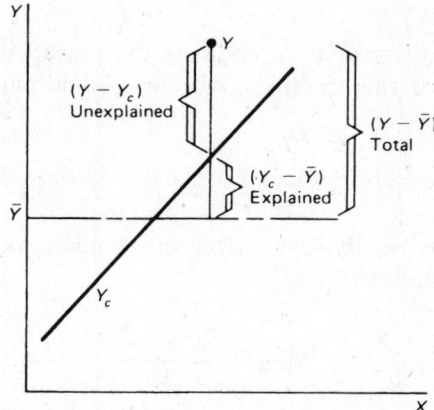

Fig. 4-7　Deviation of dependent variable

Total variation = explained + unexplained

$$\Sigma (Y - \overline{Y})^2 = \Sigma (Y_c - \overline{Y})^2 + \Sigma (Y - Y_c)^2$$

The *coefficient of determination* r^2 is the ratio of explained variation to total variation:

$$r^2 = \frac{\Sigma (Y_c - \overline{Y})^2}{\Sigma (Y - \overline{Y})^2} \tag{4.21}$$

The *coefficient of correlation* r is the square root of the coefficient of determination:

$$r = \sqrt{\frac{\Sigma (Y_c - \overline{Y})^2}{\Sigma (Y - \overline{Y})^2}} \tag{4.22}$$

When the sample size is sufficiently large (e.g., greater than 50), the value of r can be computed more directly from:

$$r = \frac{n \Sigma XY - \Sigma X \Sigma Y}{\sqrt{[n \Sigma X^2 - (\Sigma X)^2][n \Sigma Y^2 - (\Sigma Y)^2]}} \tag{4.23}$$

Example 4.10　A study to determine the correlation between plasterboard shipments X and construction permits Y revealed the following:

$$\Sigma X = 184 \qquad \Sigma Y = 80 \qquad n = 8$$

$$\Sigma X^2 = 5,006 \qquad \Sigma Y^2 = 950 \qquad \Sigma XY = 2,146$$

Compute the correlation coefficient.

$$r = \frac{n \Sigma XY - \Sigma X \Sigma Y}{\sqrt{[n \Sigma X^2 - (\Sigma X)^2][n \Sigma Y^2 - (\Sigma Y)^2]}} = \frac{8(2,146) - (184)(80)}{\sqrt{[8(5,006) - (184)^2][8(950) - (80)^2]}} = \frac{2,448}{\sqrt{7,430,400}} = .90$$

The significance of any value of r can be statistically tested under a hypothesis of no correlation. To test, the computed value of r is compared with a tabled value of r for a given sample size and significance level. If the computed value exceeds the tabled value, the correlation is significant.

FORECAST CONTROLS

A simple measure of forecast error is to compute the deviation of the actual from the forecast values. Deviations will vary from plus to minus, but they should tend to average out near zero if the forecast is on target.

$$\text{Forecast error} = \text{actual demand} - \text{forecast demand} \qquad (4.24)$$

The individual forecast errors are usually summarized in a statistic such as average error, mean squared error, or mean absolute deviation (MAD).

$$\text{MAD} = \frac{\Sigma\,|\text{Error}|}{n} \qquad (4.25)$$

The estimate of the MAD can be continually updated by using an exponential smoothing technique. Thus the current MAD_t is:

$$\text{MAD}_t = \alpha(\text{actual} - \text{forecast}) + (1 - \alpha)\text{MAD}_{t-1} \qquad (4.26)$$

where α is a smoothing constant. Higher values of α will make the current MAD_t more responsive to current forecast errors.

When the average deviation (MAD) is divided into the cumulative deviation [Σ (actual $-$ forecast)], the result is a *tracking signal:*

$$\text{Tracking signal} = \frac{\Sigma\,(\text{actual} - \text{forecast})}{\text{MAD}} \qquad (4.27)$$

Tracking signals are one way of monitoring how well a forecast is predicting actual values. They express the cumulative deviation (also called the running sum of forecast error, RSFE) in terms of the number of average deviations (MADs). Action limits for tracking signals commonly range from three to eight. When the signal goes beyond this range, corrective action may be required.

Example 4.11   A high-valued item has a tracking-signal-action limit of 4 and has been forecast as shown in Table 4-11. Compute the tracking signal, and indicate whether some corrective action is appropriate.

Table 4-11

Period	Actual	Forecast	Error (A − F)	|Error|	(Error)² (A − F)²
1	80	78	2	2	4
2	92	79	13	13	169
3	71	83	−12	12	144
4	83	79	4	4	16
5	90	80	10	10	100
6	102	83	19	19	361
		Totals	36	60	794

$$MAD = \frac{\Sigma |Error|}{n} = \frac{60}{6} = 10$$

$$Tracking\ signal = \frac{\Sigma (actual - forecast)}{MAD} = \frac{36}{10} = 3.6$$

Action limit of 4 is not exceeded. Therefore, no action is necessary.

Control charts are a second way of monitoring forecast error. Variations of actual from forecast (or average) values are quantified in terms of the estimated standard deviation of forecast S_F.

$$S_F = \sqrt{\frac{\Sigma (actual - forecast)^2}{n - 1}} \qquad (4.28)$$

Control limits are then set, perhaps at two or three standard deviations away from the forecast average $\overline{X}$, or the $2S_F$ or $3S_F$ limits are used as maximum acceptable limits for forecast error. Note that the limits are based on individual forecast values, so you assume that the errors are normally distributed around the forecast average. See Fig. 4-8.

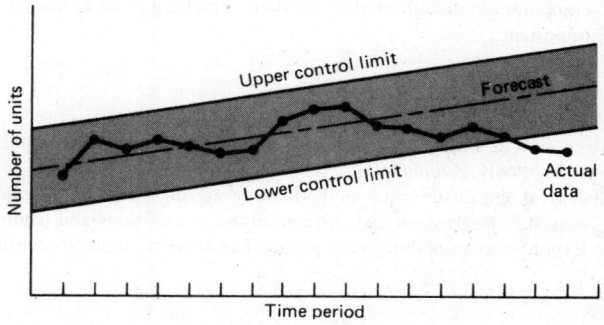

Fig. 4-8   Control limits for forecasts

Example 4.12   (*a*) Compute the $2S_F$ control limits for the data given in Example 4.11. (*b*) Are all forecast errors within these limits?

(*a*)   Control limits about the mean $CL = \overline{X} \pm 2S_F$

$$where \quad \overline{X} = \frac{78 + 79 + 83 + 79 + 80 + 83}{6} = 80$$

$$S_F = \sqrt{\frac{\Sigma (actual - forecast)^2}{n - 1}} = \sqrt{\frac{794}{6 - 2}} = 14$$

Therefore $CL = 80 \pm 2(14) = 52$ to 108 (rounded to integer values)

(*b*)   All forecast errors (as calculated in Example 4.11) are within the ± 28 error limit. *Note:* Since n is less than 30, this distribution of forecast errors does not wholly satisfy the normality assumption. See Prob. 4.13 for use of the T distribution in control limits.

FORECAST APPLICATION

Forecasts should be sufficiently accurate to plan for future activities. Low-accuracy methods may suffice; higher accuracy usually costs more to design and implement. *Long-term forecasts*—used for location, capacity, and new-product decisions—require techniques with long-term horizons. *Short-term forecasts*—such as those for production-and-inventory control, labor levels, and cost controls—can rely more on recent history. Figure 4-9 relates the method to the product life cycle.

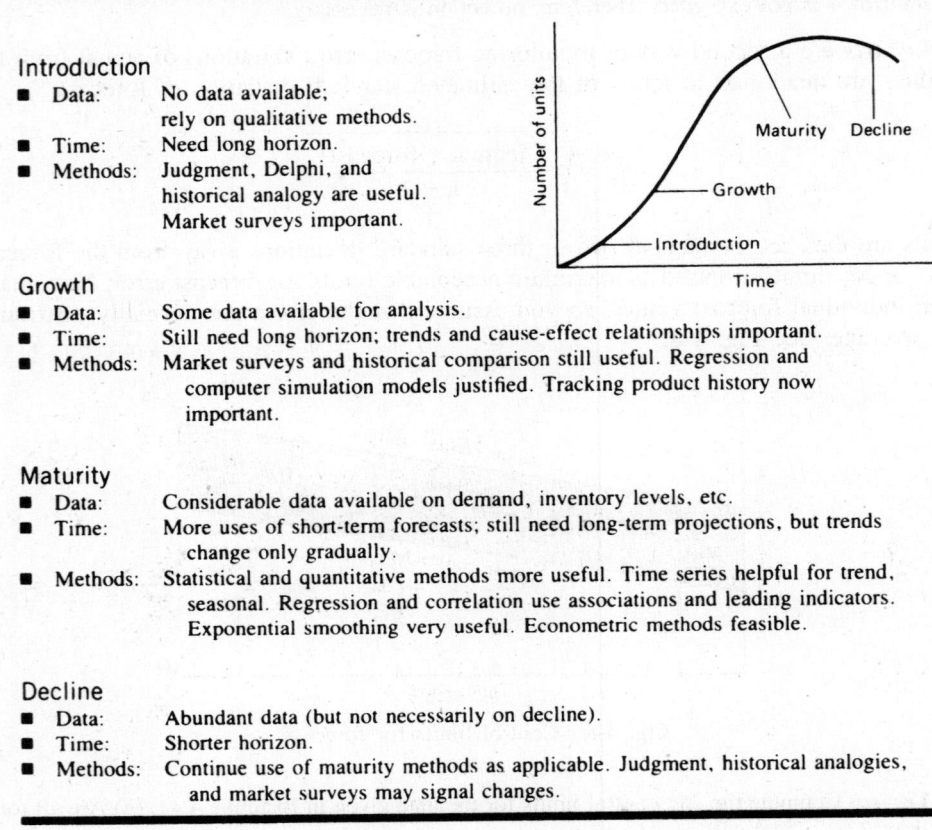

Introduction
- Data: No data available; rely on qualitative methods.
- Time: Need long horizon.
- Methods: Judgment, Delphi, and historical analogy are useful. Market surveys important.

Growth
- Data: Some data available for analysis.
- Time: Still need long horizon; trends and cause-effect relationships important.
- Methods: Market surveys and historical comparison still useful. Regression and computer simulation models justified. Tracking product history now important.

Maturity
- Data: Considerable data available on demand, inventory levels, etc.
- Time: More uses of short-term forecasts; still need long-term projections, but trends change only gradually.
- Methods: Statistical and quantitative methods more useful. Time series helpful for trend, seasonal. Regression and correlation use associations and leading indicators. Exponential smoothing very useful. Econometric methods feasible.

Decline
- Data: Abundant data (but not necessarily on decline).
- Time: Shorter horizon.
- Methods: Continue use of maturity methods as applicable. Judgment, historical analogies, and market surveys may signal changes.

Fig. 4-9   Life cycle effects upon forecasting methodology

Questions and Solved Problems

TIME SERIES METHODS

4.1   (*Moving average*)   A food processing company uses a moving average to forecast next month's demand. Past actual demand (in units) is as shown in Table 4-12.

(a)   Compute a simple 5-month moving average to forecast demand for month 52.

(b)   Compute a weighted 3-month moving average, where the weights are highest for the latest months and descend in order of 3, 2, 1.

Table 4-12

Month	Actual Demand
43	105
44	106
45	110
46	110
47	114
48	121
49	130
50	128
51	137
52	

(a) $MA = \dfrac{\Sigma X}{\text{number of periods}} = \dfrac{114 + 121 + 130 + 128 + 137}{5} = 126$ units

(b) $MA_{wt} = \dfrac{\Sigma (wt)(X)}{\Sigma wt}$

where

$$wt \times value = total$$
$$3 \times 137 = 411$$
$$2 \times 128 = 256$$

and the totals are

$$\dfrac{1}{6} \times 130 = \dfrac{130}{797}$$

Thus,

$$MA_{wt} = \dfrac{797}{6} = 133 \text{ units}$$

4.2   *Equation*   For $n = 7$ years of (coded) time series data, $\Sigma Y = 56$, $\Sigma XY = 70$, and $\Sigma X^2 = 28$.

(a)   Find the intercept and slope of the linear trend line.

(b)   Forecast the Y-value for 6 years distant from the origin.

(a)   Intercept $= \Sigma Y/n = 56/7 = 8.0$ Slope $= \Sigma XY/\Sigma X^2 = 70/28 = 2.5$

(b) $Y_c = a + b(X) = 8.0 + 2.5(6) = 23.0$

4.3   (*Shifting equation*)   The following forecasting equation has been derived by a least-squares method to describe the shipments of welded aluminum tube.

$$Y_c = 10.27 + 1.65X \qquad (1996 = 0,\ X = \text{years},\ Y = \text{tons/yr})$$

Rewrite the equation by (a) shifting the origin to 2001; (b) expressing X units in months, retaining Y in tons per year; (c) expressing X units in months, and Y in tons per month.

(a) $Y_c = 10.27 + 1.65(X + 5)$

$\qquad = 18.52 + 1.65X \qquad$ (yr 2001 = 0, X = years, Y = tons per year)

(b) $Y_c = 10.27 + \dfrac{1.65X}{12}$

$\qquad = 10.27 + .14X \qquad$ (July 1, 1996 = 0, X = months, Y = tons per year)

(c) $Y_c = \dfrac{10.27 + .14X}{12}$

$\qquad = .86 + .01X \qquad$ (July 1, 1996 = 0, X = months, Y = tons per month)

SEASONAL INDEX

4.4   Quarterly trend values for units demanded have been computed as $Q1 = 620$, $Q2 = 655$, $Q3 = 690$, and $Q4 = 725$. The corresponding seasonal indexes for the quarters are .72, 1.33, 1.05, and .90, respectively. Forecast the actual (seasonalized) sales for $Q3$ and $Q4$.

For $Q3$: $Y_{sz} = (SI)Y_c = (1.05)(690) = 725$ units

For $Q4$: $Y_{sz} = (SI)Y_c = (.90)(725) = 653$ units

4.5   (*Correcting index*)   A sportswear manufacturer wishes to use data from a 5-year period to develop seasonal indexes. Trend values and ratios of actual *A* to trend *T* for most months have already been computed as shown in Table 4-13. April and May actual and trend values are shown in Tables 4-14 and 4-15.

Table 4-13

Month	Jan.	Feb.	Mar.	April	May	June	July	Aug.	Sept.	Oct.	Nov.	Dec.
Ratio *A/T*	.72	.58	.85			1.43	1.21	1.05	.98	.92	.88	1.12

Compute the seasonal relatives for April and May, correct the total to equal 12.00, and determine the resulting seasonal indexes. See Tables 4-14 and 4-15.

Table 4-14

Year	1	2	3	4	5
April actual	382	401	458	480	533
April trend	400	436	472	508	544

April A/T   .96   .92   .97   .94   .98
April total = 4.77
April average = 4.77 ÷ 5 = (.95)

Table 4-15

Year	1	2	3	4	5
May actual	485	530	560	592	656
May trend	403	439	475	511	547

May A/T   1.20   1.21   1.18   1.16   1.20
May total = 5.95
May average = 5.95 ÷ 5 = 1.19

Table 4-16

Month	Jan.	Feb.	Mar.	April	May	June	July	Aug.	Sept.	Oct.	Nov.	Dec.	12 mo.
Ratio *A/T*	.72	.58	.85	.95	1.19	1.43	1.21	1.05	.98	.92	.88	1.12	11.88

$$\text{Correction factor} = \frac{12}{11.88} = 1.01$$

Multiplying each month's ratio by the 1.01 correction factor we have Table 4-17.

Table 4-17

Month	Jan.	Feb.	Mar.	April	May	June	July	Aug.	Sept.	Oct.	Nov.	Dec.	12 mo.
SI	.73	.59	.86	.96	1.20	1.44	1.22	1.06	.99	.93	.89	1.13	12.00

4.6   The production manager of the sportswear firm in the previous problem has projected trend values for next summer (June, July, August) of 586, 589, and 592. Using the seasonal indexes given (1.44, 1.22, 1.60), what actual seasonalized production should the manager plan for?

June: $Y_{sz} = \text{SI } Y_c = (1.44)(586) = 844$
July: $Y_{sz} =$ $(1.22)(589) = 719$
August: Y_{sz} $(1.06)(592) = 628$

EXPONENTIAL SMOOTHING

4.7   (*Conversion from moving average*)   Lakeside Hospital has used a 9-month, moving-average forecasting method to predict drug and surgical dressing inventory requirements. The actual demand for one item is as shown in Table 4-18. Using the previous moving-average data, convert to an exponential smoothing forecast for month 33.

Table 4-18

Month	24	25	26	27	28	29	30	31	32
Demand	78	65	90	71	80	101	84	60	73

$$\text{MA} = \frac{\Sigma X}{\text{no. periods}} = \frac{78 + 65 + \cdots + 73}{9} = 78$$

Thus, assume the previous forecast was $F_{t-1} = 78$.

Then estimate α as

$$\alpha = \frac{2}{n+1} = \frac{2}{9+1} = .2$$

so

$$F_t = F_{t-1} + \alpha(A_{t-1} - F_{t-1}) = 78 + .2(73 - 78) = 77 \text{ units}$$

4.8 (*Seasonal adjustment*) A shoe manufacturer, using exponential smoothing with $\alpha = .1$, has developed a January trend forecast of 400 units for a ladies' shoe. This brand has seasonal indexes of .80, .90, and 1.20, respectively, for the first 3 months of the year. Assuming that actual sales were 344 units in January and 414 units in February, what would be the seasonalized (adjusted) March forecast?

(*a*) Deseasonalize actual January demand.

$$\text{Demand} = \frac{344}{.80} = 430 \text{ units}$$

(*b*) Compute the deseasonalized forecast.

$$F_t = F_{t-1} + \alpha(A_{t-1} - F_{t-1})$$

$$= 400 + .1(430 - 400) = 403$$

(*c*) Seasonalized (adjusted) February forecast would be

$$F_{t(sz)} = 403(.90) = 363$$

Repeating for February, we have:

(*a*)

$$\text{Demand} = \frac{414}{.90} = 460 \text{ units}$$

(*b*)

$$F_t = 403 + .1(460 - 403) = 409$$

(*c*)

$$F_{t(sz)} = 409(1.20) = 491$$

4.9 (*Trend adjustment*) Develop an adjusted exponential forecast for the week of 5/14 for a firm with the demand shown in Table 4-19. Let $\alpha = .1$ and $\beta = .2$. Begin with a previous average of $\hat{F}_{t-1} = 650$, and let the initial trend adjustment, $T_{t-1}, = 0$.

Table 4-19

Week	3/19	3/26	4/2	4/9	4/16	4/23	4/30	5/7
Demand	700	685	648	717	713	728	754	762

Using Eqs. 4.11, 4.12, and 4.13, we have:

Week 3/19:
$$\hat{F}_t = \alpha A_{t-1} + (1 - \alpha)(\hat{F}_{t-1} + T_{t-1})$$
$$= .1(700) + .9(650 + 0) = 655.00$$
$$T_t = \beta(\hat{F}_t - \hat{F}_{t-1}) + (1 - \beta)T_{t-1}$$
$$= .2(655 - 650) + .8(0) = 1.0 + 0 = 1.00$$
$$\therefore \quad \hat{F}_{t+1} = \hat{F}_t + T_t = 655 + 1 = 656.00$$

The 656.00 is the adjusted forecast for week 3/26.

Week 3/26:
$$\hat{F}_t = .1(685) + .9(655 + 1.0) = 658.90$$
$$T_t = .2(658.9 - 655) + .8(1.0) = 1.58$$
$$\therefore \quad \hat{F}_{t+1} = 658.9 + 1.58 = 660.48$$

The remainder of the calculations are in Table 4-20. The trend-adjusted forecast for the week of 5/14 is $711.89 \cong 712$ units.

Table 4-20

(1) Week	(2) Previous Average $\hat{F}_{t-1}$	(3) Actual Demand A_{t-1}	(4) Smoothed Average $\hat{F}_t$	(5) Smoothed Trend T_t	(6) Next-Period Projection $\hat{F}_{t+1}$
Mar. 19	650.00	700	655.00	1.00	656.00
26	655.00	685	658.90	1.58	660.48
Apr. 2	658.90	648	659.23	1.33	660.56
9	659.23	717	666.20	2.46	669.06
16	660.20	713	673.09	3.35	676.44
23	673.09	728	681.60	4.39	685.99
30	681.60	754	691.79	5.74	698.53
May 7	692.79	762	704.88	7.01	711.89
14		770			

4.10 Using the demand data from Prob. 4.9 and a value of 770 for the week of May 14,

(a) Compute the mean absolute deviation (MAD) forecast error for the following: (1) simple exponential smoothing ($\alpha = .1$), (2) adjusted exponential smoothing ($\alpha = .1$ and $\beta = .2$), (3) adjusted exponential smoothing ($\alpha = .1$ and $\beta = .8$).

(b) Compare the actual demand and the three forecasts on a graph.

Computations of the forecast values for adjusted exponential smoothing ($\alpha = .1$; $\beta = .2$) are given in the previous example; computations for simple exponential smoothing and for $\alpha = .1$ and $\beta = .8$ are not shown, but the results are given in Table 4-21. The forecast error = actual demand − forecast, and is computed as shown. The best fit of these models is $\alpha = .1$ and $\beta = .8$ for a MAD of 31.6. The graph (Fig. 4-10) confirms that for these data, which have a strong trend, the higher value of β yields better results.

Table 4-21

Week	Actual Demand	Simple: $\alpha = .1$		Adjusted: $\alpha = .1$; $\beta = .2$		Adjusted: $\alpha = .1$; $\beta = .8$	
		Forecast	Error	Forecast	Error	Forecast	Error
3/19	700	650	50	650	50	650	50
3/26	685	655	30	656	29	659	26
4/2	648	658	−10	660	−12	668	−20
4/9	717	657	60	661	56	671	46
4/16	713	663	50	669	44	684	29
4/23	728	668	60	676	52	697	31
4/30	754	674	80	686	68	714	40
5/7	762	682	80	699	63	734	28
5/14	770	690	80	712	58	756	14
	Σ |error|		500		432		284
MAD $= \dfrac{\Sigma \|error\|}{9}$			55.6		48.0		31.6

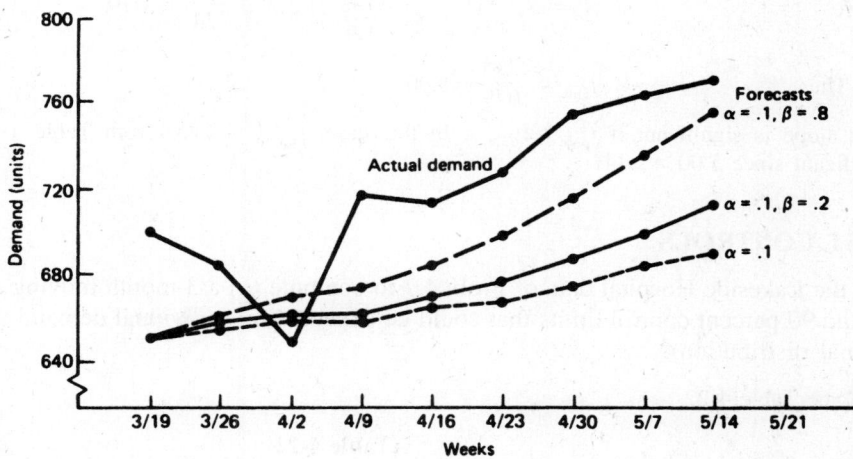

Fig. 4-10 Forecast results using simple and trend-adjusted exponential smoothing

REGRESSION AND CORRELATION

4.11 Given the following:

$$\Sigma X = 80 \qquad \Sigma Y = 1,200 \qquad n = 20 \qquad \Sigma (Y - Y_c)^2 = 800$$

$$\Sigma X^2 = 340 \qquad \Sigma Y^2 = 74,800 \qquad \Sigma XY = 5,000 \qquad \Sigma (Y - \overline{Y})^2 = 2,800$$

Find (a) linear regression equation, (b) $S_{Y \cdot X}$, (c) r.

(a)
$$b = \frac{\sum XY - n\overline{X}\,\overline{Y}}{\sum X^2 - n\overline{X}^2} = \frac{5,000 - (20)(4)(60)}{340 - (20)(4)(4)} = 10$$

$$a = \overline{Y} - b\overline{X} = 60 - 10(4) = 20$$

Thus,
$$Y_c = 20 + 10X$$

(b)
$$S_{Y \cdot X} = \sqrt{\frac{\sum (Y - Y_c)^2}{n - 2}} = \sqrt{\frac{800}{20 - 2}} = 6.67$$

(c)   Because the explained variation is equal to the total minus the unexplained variation, the correlation coefficient is sometimes expressed in the form

$$r = \sqrt{1 - \frac{\text{unexplained variation}}{\text{total variation}}} = \sqrt{1 - \frac{\sum (Y - Y_c)^2}{\sum (Y - \overline{Y})^2}} \qquad (4.29)$$

Thus,
$$r = \sqrt{1 - \frac{800}{2,800}} = .85$$

4.12   Test whether the regression line slope of $b = .395$ developed in Example 4.7 is significant at the 5 percent level. [*Given:* $S_{Y \cdot X} = 2.2$; $\sum (X - \overline{X})^2 = 774$.]

   Note: This test requires use of the t statistical distribution.

$$t_{\text{calc}} = \frac{b}{S_b}$$

where
$$S_b = S_{Y \cdot X} \sqrt{\frac{1}{\sum (X - \overline{X})^2}} = 2.2 \sqrt{\frac{1}{774}} = .079$$

Thus,
$$t_{\text{calc}} = \frac{.395}{.079} = 5.00$$

Test: slope is significant if $t_{\text{calc}} > t_{.05, 6\text{df}}$. In this case, $t_{.05, 6\text{df}} = 2.45$ (from Table 4-10). Thus, slope is significant since $5.00 > 2.447$.

FORECAST CONTROLS

4.13   Use the Lakeside Hospital data of Prob. 4.7 to compute (a) a 3-month moving average (MA) and (b) the 90 percent control limits that could be expected for individual demand values (assuming a normal distribution).

(a)   See Table 4-22.

Table 4-22

Month	Actual Demand	3-Month MA	Forecast Demand	Deviation $(A - F)$	(Deviation)^2
24	78				
25	65	77.7			
26	90	75.3			
27	71	80.3	78	−7	49
28	80	84.0	75	5	25
29	101	88.3	80	21	441
30	84	81.7	84	0	0
31	60	72.3	88	−28	784
32	73		82	−9	81
					1,380

(b)
$$S_F = \sqrt{\frac{\Sigma \, (\text{actual} - \text{forecast})^2}{n - 1}} = \sqrt{\frac{1,380}{6 - 1}} = 16.6$$

Since n is less than 30, we should use the t distribution rather than the Z for the control limits. Referring to any standard statistics text (or Table 4-10), we find that for $n - 1 = 5$ degrees of freedom at the 90 percent level, $t = 2.02$. The mean forecast value is:

$$\bar{X} = \frac{78 + 75 + 80 + 84 + 88 + 82}{6} = 81.2$$

$\therefore$ Control limits $= \bar{X} \pm tS_F = 81.2 \pm 2.02(16.6) = 47.7$ to 114.7

Note: The control limits explicitly recognize the variability in this data and, in turn, the uncertainty associated with trying to forecast it. A larger sample would yield tighter limits.

4.14   The moving-average forecast and the actual demand for a hospital drug are as shown in Table 4-23. Compute the tracking signal, and comment on the forecast accuracy.

Table 4-23

Month	Actual Demand	Forecast Demand	Error $(A - F)$	|Error|
27	71	78	-7	7
28	80	75	5	5
29	101	83	18	18
30	84	84	0	0
31	60	88	-28	28
32	73	85	-12	12
			-24	70

The deviation and cumulative deviation have already been computed above:

$$\text{MAD} = \frac{\Sigma \, |\text{actual} - \text{forecast}|}{n} = \frac{70}{6} = 11.7$$

$$\text{Tracking signal} = \frac{\Sigma \, (\text{actual} - \text{forecast})}{\text{MAD}} = \frac{-24}{11.7} = -2.05 = |2.05|$$

The demand exhibits substantial variation, but a tracking signal as low as 2.05 (that is, ≤ 4) would not suggest any action at this time.

Supplementary Questions and Problems

4.15   The demand over the past 9 months for a new breakfast cereal is shown in Table 4-24. Develop a forecast for November using a 5-period moving average where the weights are (from earliest to latest) 1, 1, 2, 2, and 4.

Table 4-24

Month	Feb.	Mar.	Apr.	May	June	July	Aug.	Sept.	Oct.
Units	70	76	75	80	92	87	93	114	105

Ans.   101 units

4.16   A sugar beet processing cooperative is committed to accepting beets from local producers and has experienced the following supply pattern (in thousands of tons per year and rounded) shown in Table 4-25.

Table 4-25

Year	4	5	6	7	8	9	10	11	12	13
Tons	100	100	200	600	500	400	400	600	800	800

The operations manager would like to project a trend to determine what facility additions will be required by year 18. (*a*) Graph the data, and connect the points by straight line segments. (*b*) Sketch in a freehand curve, and extend it to year 18. What would be your year-18 forecast on the basis of the curve? (*c*) Compute a 3-year moving average, and plot it as a dotted line on your graph.
Ans.   (*a*) Graph should show time on X axis, tons on Y. (*b*) Curves will differ, but forecasts will $\cong$ 1,200 (thousand) tons. (*c*) Averages are 133, 300, 433, 500, 433, 466, 600, 733.

4.17   Use the data of Prob. 4.16 to develop a least-squares line of best fit. Omit year 4. (*a*) State the equation, complete with signature, when the origin is year 9. (*b*) Use your equation to estimate the trend value for year 18.   *Ans.*   (*a*) $Y = 489 + 75X$ (Year 9 = 0, X = years, Y = tons) (*b*) 1,164,000 tons

4.18   The orders booked by an airframe manufacturer are as shown. (*a*) Derive a least-squares linear forecasting equation. (*b*) Use your equation to forecast the trend value for the year 2007.

Table 4-26

Year	1992	1993	1994	1995	1996	1997	1998
Units	5	5	10	12	28	30	15

Ans.   (*a*) $Y_c = 15.0 + 3.5X$ (1995 = 0, X = yrs, Y = units). (*b*) 57 units

4.19   (*Shifting base period*)   A trend equation describing passenger tickets sold was found to be:

$$Y_c = 25,480 + 8,370X \quad (1995 = 0, X = \text{years}, Y = \text{tickets})$$

Shift the equation to a 2002 base.   *Ans.* $Y_c = 84,070 + 8,370X$ (yr 2002 = 0, X = years, Y = tickets)

4.20   (*Converting time units*)   A forecasting equation is of the form:

$$Y_c = 720 + 144X \quad (\text{yr } 2000 = 0,\ X \text{ unit} = 1 \text{ year},\ Y = \text{annual sales})$$

(*a*) Forecast the annual sales rate for yr 2000 and also for 1 year later. (*b*) Change the time (X) scale to months, and forecast the annual sales rate at July 1, 2000, and also at 1 year later. (*c*) Change the sales (Y) scale to monthly, and forecast the monthly sales rate at July 1, 2000, and also at 1 year later.

Ans.   (*a*) 720 units, 864 units (*b*) $Y_c = 720 + 12X$ (July 1, 2000 = 0, X units = 1 month, Y = annual sales rate in units); 720 units per year; 864 units per year (*c*) $Y_c = 60 + 1X$ (July 1, 2000 = 0, X units = 1 month, Y = monthly sales rate in units); 60 units per month; 72 units per month

4.21   Data collected on the monthly demand for a housewares items were as shown in Table 4-27. (*a*) Plot the data. (*b*) Plot a 5-month moving average as a dotted line. (*c*) What conclusion can you draw with respect to length of moving average versus smoothing effect? (*d*) Assume that the 12-month moving average centered on July was 231. What is the value of the ratio to moving average that would be used in computing a seasonal index?

Table 4-27

Month	Jan	Feb	Mar	April	May	June	July	Aug	Sept
Units	100	90	80	150	240	320	300	280	220

Ans.   (*a*) Graph should show time on X axis, units on Y axis. (*b*) First moving average value will be 132 centered on March. (*c*) Longer average yields more smoothing. (*d*) 1.3

4.22   The data shown in Table 4-28 and Fig. 4-11 include the number of lost-time accidents for the Cascade Lumber Co. over the past 7 years. (*Note:* The number of employees is shown for reference only. You will not need it to solve this problem.)

Table 4-28

Year	Number of Employees (000)	Number Accidents
1	15	5
2	12	20
3	20	15
4	26	18
5	35	17
6	30	30
7	37	35
		140

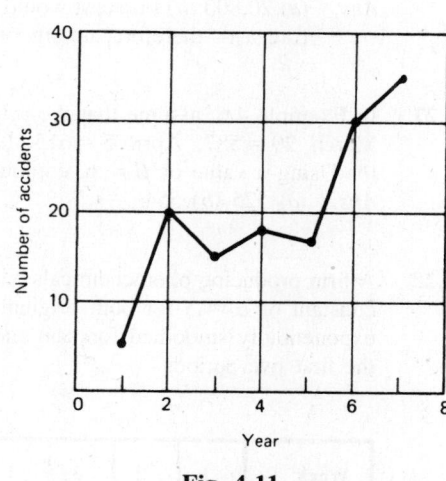

Fig. 4-11

(*a*) Use the normal equations to develop a linear time-series equation for forecasting the number of accidents. State the equation complete with signature. (*b*) Use your equation to forecast the number of accidents in year 10.

Ans.   (*a*) $Y_c = 20 + 4X$ (year 4 = 0, X = years, Y = number of accidents) (*b*) 44

4.23   Forecast demand for March was 950 units, but actual demand turned out to be only 820. If the firm is using a simple exponential smoothing technique with $\alpha = .2$, what is the forecast for April?   *Ans.*   924 units

4.24   Using the results from Prob 4.23, assume the April demand was actually 980 units. Now what is the forecast for May?   *Ans.*   935 units

4.25   A forecaster is using an exponential smoothing model with $\alpha = .4$ and wishes to convert to a moving average. What length of moving average is approximately equivalent?   *Ans.*   4 period

4.26   A university registrar has adopted a simple exponential smoothing model ($\alpha = .4$) to forecast enrollments during the three regular terms (excluding summer). The results are shown in Table 4-29. (*a*) Use the data to develop an enrollment forecast for the third quarter of year 2. (*b*) What would be the effect of increasing the smoothing constant to 1.0?

Table 4-29

Year	Quarter	Actual Enrollment (000)	Old Forecast (000)	Forecast Error (000)	Correction (000)	New Forecast (000)
1	1st	20.50	20.00	.5	.20	20.20
	2d	21.00				
	3d	19.12				
2	1st	20.06				
	2d	22.00				
	3d					

Ans.   (*a*) 20,800 (*b*) Forecast would reflect the total amount of variation of previous demand from previous forecast—therefore, no smoothing.

4.27   In Example 4.6, assume that the actual demand for the next 3 weeks in the sequence is March 22 = 561, March 29 = 587, April 5 = 615. (*a*) Extend the simple exponential forecast to cover these periods. (*b*) Using a value of $\beta = .6$, compute the adjusted exponential forecast for the week of April 12.
Ans.   (*a*) 525 (*b*) 554

4.28   A firm producing photochemicals has a weekly demand pattern as shown in Table 4-30. Using a smoothing constant of $\alpha = .5$ for both original data and trend, and beginning with week 1, (*a*) compute the simple exponentially smoothed forecast and (*b*) compute the trend-adjusted exponentially smoothed forecast for the first five periods.

Table 4-30

Week	1	2	3	4	5	6	7	8	9	10	11	12	13	14	15
Demand	30	34	22	16	10	10	14	20	30	36	30	10	12	20	30

Ans.   (*a*) 24 (*b*) 10

4.29   Find the regression equation resulting from the values $\Sigma X = 70$, $\Sigma Y = 90$, $\Sigma XY = 660$, $\Sigma X^2 = 514$, $n = 10$.   *Ans.* $Y_c = .25 + 1.25X$

4.30   Use the data from Prob. 4.22 to develop a linear-regression equation for forecasting the number of accidents on the basis of the number of employees. Use the equation to forecast the number of accidents when the number of employees is 22(000).   *Ans.*   18 accidents, from equation $Y_c = 3.5 + .66X$

4.31   A producer of roofing materials has collected data relating interest rates to sales of asphalt shingles and found that the unexplained variation = 680, and explained variation = 2840. (*a*) Find the correlation coefficient. (*b*) Explain its meaning.
Ans.   (*a*) $r = .90$ (*b*) 81 percent of the variation in shingle sales is associated with interest-rate levels.

4.32   The Carpet Cleaner Co. is attempting to do a better job of inventory management by predicting the number of vacuums the company will sell per week on the basis of the number of customers who respond to magazine advertisements in an earlier week. On the basis of a sample of $n = 102$ weeks, the following data were obtained.

$$a = 25 \qquad \Sigma (Y - Y_c)^2 = 22,500$$

$$b = .10 \qquad \Sigma (Y - \overline{Y})^2 = 45,000$$

(*a*) Provide a point estimate of the number of vacuums sold per week when 80 inquiries were received in the earlier week. (*b*) Estimate (at the 95.5 percent level) the number of vacuums sold per week when 80 inquiries were received the week earlier. (*c*) State the value of the coefficient of determination. (*d*) Explain the meaning of your r^2 value.
Ans.   (*a*) 33 (*b*) Using the large-sample approximation, the interval is 3 to 63 because $S_{Y \cdot X} = 15$ (*c*) .5
(*d*) 50 percent of the variation in number of vacuums sold is explained by the magazine advertisements.

4.33   A recreation operations planner has had data collected on automobile traffic at a selected location on an interstate highway in hopes that the information can be used to predict weekday demand for state-operated campsites 200 miles away. Random samples of 32 weekdays during the camping season resulted in data from which the following expression was developed:

$$Y_c = 18 + .02X$$

where X is the number of automobiles passing the location and Y is the number of campsites demanded that day. In addition, the unexplained variation is $\Sigma (Y - Y_c)^2 = 1,470$, and the total variation is $\Sigma (Y - \overline{Y})^2 = 4,080$. (*a*) What is the value of the coefficient of determination? (*b*) Explain, in words, the meaning of the coefficient of determination. (*c*) What is the value of the coefficient of correlation?
Ans.   (*a*) .64 (*b*) It tells the percentage of variation in campsites demanded that is associated with automobile traffic at the selected site. (*c*) .80

4.34   Allan's Underground Systems installs septic systems for new houses constructed outside the city limits. To help forecast his demand, Mr. Allan has collected the data shown in Table 4-31 on the number of county building permits issued per month, along with the corresponding number of bid requests he has received over a 15-month period.

Table 4-31

Month	1	2	3	4	5	6	7	8	9	10	11	12	13	14	15
No. building permits	8	20	48	60	55	58	50	45	34	38	10	5	12	29	50
No. bid requests	20	7	8	4	18	40	48	54	47	42	30	22	20	4	3

(*a*) Compute the simple correlation coefficient r between the number of building permits issued and the number of bid requests received in that month. Use all 15 periods of data. (b) Use the first 12 months of data for building permits, and compute r between the number of building permits issued in a month and the number of bid requests received 2 months later (i.e., a 2-month lag). (c) Repeat (b), but use a 3-month lag. (d) Which type of regression model would be best to forecast bid requests: a same-month model, a 2-month lag model, or a 3-month lag model?

Ans.   (a) .08 (b) .84 (c) .96 (d) A 3-month lag model is best. It permits Allan to explain 93 percent of the variation in number of bid requests.

4.35   Two experienced managers have resisted the introduction of a computerized exponential smoothing system, claiming that their judgmental forecasts are "much better than any impersonal computer could do." Their past record of prediction is as shown in Table 4-32.

Table 4-32

Week	Actual Demand	Forecast Demand
1	4,000	4,500
2	4,200	5,000
3	4,200	4,000
4	3,000	3,800
5	3,800	3,600
6	5,000	4,000
7	5,600	5,000
8	4,400	4,800
9	5,000	4,000
10	4,800	5,000

(a) Compute the MAD. (b) Compute the tracking signal. (c) On the basis of your calculations, is the judgmental system performing satisfactorily?   *Ans.*   (a) 570 (b) .53 (c) yes

Chapter 5

Financial Analysis for Operations
Present-Value Criteria

CAPITAL INVESTMENT AND CASH FLOWS

Question: What is capital, and what is the source of capital to an organization?

Capital is a resource of funds owned or used by an organization. It typically comes from equity (stock), from debt (bonds), and from earnings retained from past operations (profits). Contributions and taxes are also an important source of funds for many nonprofit organizations.

Evaluation of investment alternatives requires consideration of the initial investment, plus cash inflows and outflows, depreciation, and taxes over the economic life of the proposed asset. The *net cash flow* is the difference between incoming (revenue) and outgoing (expense) flows and is often expressed in present-value terms.

The *time value of money concept* recognizes that cash available today is more valuable than the same quantity of cash available at a later date. This is important to operations managers because funds have an earning power and can be used to buy factors of production that create even more value.

The difference between present values P and future sums F of money is due to accumulated interest i over the number of periods n. For example, the interest due on \$1,000 borrowed at 10 percent for 1 year would be equal to the (principal)(rate)(time), or (\$1,000)(.10)(1) = \$100. The total amount due at the end of one time period, F_1, consists of the principal amount borrowed P plus the interest on that principal i.

$$F_1 = P + P(i) = P(1 + i)$$

The total amount due at the end of two time periods consists of the amount due at the end of the first period, $P(1 + i)$, plus interest on that amount for the second period.

$$F_2 = P(1 + i) + P(1 + i)i = P(1 + i)(1 + i) = P(1 + i)^2$$

Similarly, at the end of three periods, $F_3 = P(1 + i)^3$, and for n years, the future sum F is:

$$F = P(1 + i)^n \tag{5.1}$$

Example 5.1 Find the principal and the interest due at the end of 2 years if \$1,000 is borrowed at a rate of 10 percent.

$$F_2 = P(1 + i)^n = \$1,000(1 + .10)^2 = \$1,210$$

The accumulation of interest with capital to future values is referred to as *compounding*. Conversely, the present value P of the future sum F *discounted* (reduced in value back to the present) at interest rate i for n periods is

$$P = F \frac{1}{(1 + i)^n} \tag{5.2}$$

Example 5.2 What is the present value of \$1,210 received 2 years from now if the sum is discounted at 10 percent?

$$P = F \frac{1}{(1+i)^n} = \$1,210 \frac{1}{(1+.10)^2} = \$1,000$$

The above expression can be restated in a general formula:

$$P = F \frac{1}{(1+i)^n} = F(\text{PV factor}) = F(\text{PV}_{sp})_{i\%}^{n\ \text{yr}} \tag{5.3}$$

where PV_{sp} is a tabled factor for the present value of a *single payment* made in n years if the interest rate is i percent. This PV_{sp} factor, when multiplied by the amount of the future payment F, will yield a discounted present-value amount P. Appendix E contains PV_{sp} factors for payments of \$1 over a commonly used range of interest rate i and period n values. Many business calculators and computer spreadsheets have been preprogrammed to provide these and other useful features.

Example 5.3   What is the present value of the salvage on a robot if the salvage price 10 years from now is \$9,000 and if the discount rate is 12 percent?

$$P = F(\text{PV}_{sp})_{12\%}^{10\ \text{yr}} = \$9,000(.322) = \$2,898$$

On other occasions, one must determine the present value of a series of *equal payments* made over n years when they are discounted or compounded at an interest rate i. These equal sums paid or received regularly are *annuities* and are usually designated by an R or an A. Appendix F contains present-value factors for annuities (PV_a) of \$1. The present value of an annuity is determined in a manner like that used for single payments, except the factor differs.

$$P = A(\text{PV}_a)_{i\%}^{n\ \text{yr}} \tag{5.4}$$

Example 5.4   Find the present value of a series of \$200 payments, paid at the end of each of 4 years, when the interest rate is 14 percent:

$$P = A(\text{PV}_a)_{14\%}^{4\ \text{yr}} = \$200(2.914) = \$583$$

Some calculators that have present value (P) and future sum (F) keys also have a key for annuities (A or PMT). Appendix G shows the equations for P, F, and A, and includes table values for $i = 10$ percent.

DEPRECIATION

Question: What is depreciation?

Depreciation is an accounting procedure for reducing the book value of an asset by charging it off as an expense over time.

When new equipment is purchased, cash is paid out at the time of purchase. Although no cash actually flows out in later years, the depreciation expense reduces the firm's reported profits and its tax liability. Insofar as no cash actually leaves the firm to cover a depreciation expense, depreciation is deleted from other expenses when computing cash flow.

Example 5.5   Let revenues equal \$10,000; and let labor (\$1,500), materials (\$2,500), and depreciation (\$3,000) be the only expenses. Compute (*a*) the profit and (*b*) the net cash flow.

(*a*)   Profit = revenues − (expenses, including depreciation)

$$= \text{revenues} - (\text{labor} + \text{material} + \text{depreciation})$$

$$= \$10,000 - (\$1,500 + \$2,500 + \$3,000)$$

$$= \$10,000 - \$7,000 = \$3,000$$

(*b*)   Net cash flow = revenues − (expenses, not including depreciation)

$$= \text{revenues} - (\text{labor} + \text{materials})$$

$$= \$10,000 - (\$1,500 + \$2,500)$$

$$= \$10,000 - \$4,000 = \$6,000$$

For internal purposes of measuring the decline in value of production equipment, both *use* and *time* methods of depreciation may be useful to a company. Use methods are often based upon the number of service hours expected from an asset.

Example 5.6   *Use Method:* A \$140,000 molding machine is expected to be capable of producing 400,000 units. What would be the depreciation rate under the use concept?

$$\text{Depreciation} = \frac{\$140,000 \text{ cost}}{400,000 \text{ units}} = \$.35 \text{ per unit}$$

For determining taxable income, firms must compute depreciation on their plant and equipment in accordance with their nation's tax code. Some time-based methods of depreciation are (1) straight-line, (2) double-declining balance, and (3) sum-of-years digits. The following illustrates straight-line depreciation, where the investment amount, less salvage, is divided by the useful life of the item. Note, however, that the depreciation method (and salvage allowed) is a function of tax regulations.

Example 5.7   *Time Method:* A communications company has installed some new plant equipment at a cost of \$3.2 million. The equipment is estimated to have a \$500,000 salvage value after an estimated useful life of 15 years. If depreciation on a straight-line basis with deduction for salvage value is allowed, how much depreciation per year should be deducted as an expense?

$$\text{Depreciation} = \frac{\text{Investment} - \text{Salvage}}{n} = \frac{\$3,200,000 - \$500,000}{15 \text{ yr}} = \$180,000/\text{yr}$$

TAX CONSIDERATIONS

Tax rates vary widely among industrialized nations. Taxes are levied against profits (or in some cases value added) on the basis of rates specified by national and local governments. Remember, *depreciation is an expense* that reduces taxes, even though it does not represent a cash outflow.

$$\text{Tax} = \text{tax rate \% (income} - \text{expense)} \tag{5.5}$$

Following is a simplified illustration of after-tax cash flow where profits are taxed at a rate of 42 percent.

Example 5.8   Telecom Services of Europe has purchased a computer for \$30,000. It is expected to generate an income of \$12,000 per year before operating costs of \$2,000 (year 1 and 2) and \$3,000 (year 3) are deducted, and will be sold after 3 years for \$6,000. Assuming straight-line depreciation and a tax rate of 42 percent, find the after-tax cash flow from this investment.

Table 5-1   After-Tax Cash Flow

		Year 1	Year 2	Year 3
(a)	Revenue	$12,000	$12,000	$12,000
(b)	Less operating costs	2,000	2,000	3,000
(c)	Before-tax cash inflow $[(a) - (b)]$	10,000	10,000	9,000
(d)	Depreciation ($30,000 - $6,000)/3	8,000	8,000	8,000
(e)	Taxable cash inflow $[(c) - (d)]$	2,000	2,000	1,000
(f)	Taxes $[(.42) \times (e)]$	840	840	420
(g)	After-tax cash inflow $[(c) - (f)]$	9,160	9,160	8,580

PAYBACK

Payback (payoff) tells the number of years that an investment takes to pay for itself. Monetary amounts in the payback formula are not normally discounted.

$$\text{Payback} = \frac{\text{Investment} - \text{Salvage}}{\text{Operating Advantage/yr}} = \frac{I - S}{\text{OA/yr}} \qquad (5.6)$$

The *operating advantage* reflects the improvement in cash flows from increased income (e.g., higher volume or prices) or from decreased expenses (e.g., lower labor, material, overhead cost) or both. Thus the denominator is a net cash-flow figure resulting from improved earnings, but does not have depreciation expense deducted from it.

In its simplest form, payback does not consider salvage values or taxes. However, they can be included in the analysis by reducing the investment I by the value of any salvage S and subtracting any tax amounts from the operating advantage.

Example 5.9   Security Equipment Co. is purchasing a drill press that will cost $27,000, last 6 years, and have a guaranteed $3,000 salvage value. It will generate savings of $11,000 per year (before depreciation), but $3,000 of the savings must be paid in taxes. Find the after-tax payback.

$$\text{Payback} = \frac{I - S}{\text{OA/yr} - \text{Tax}} = \frac{\$27,000 - \$3,000}{\$11,000 - \$3,000} = 3.0 \text{ yr}$$

Payback is simple and quick to calculate, easy to understand, and a useful measure of the time required to return an original investment. However, it does not consider the economic life of the investment, the total return on investment, or time value of money.

PRESENT VALUE

Present value tells the worth of future income or expense flows in terms of present dollars. These cash flows are typically the investment value, the maintenance and operating costs, and the income flows. The initial investment is usually already in present-value terms, so there is no need to consider its depreciation and interest charges. Operating expenses, however, do involve cash outlays in the future and must be discounted (reduced) to present values.

$$\text{Present-value cost} = \text{PV investment} + \text{PV other costs} - \text{PV salvage}$$

$$\text{PV}_{\text{cost}} = I(\text{PV}_{\text{sp}}) + \Sigma\, \text{OC}(\text{PV}_{\text{sp}}) - S(\text{PV}_{\text{sp}}) \qquad (5.7)$$

Example 5.10   Sunshine Smelter is considering an investment of $40,000 in a stack filter that is expected to have a salvage value of $10,000 after an economic life of 5 years. Maintenance and operating costs are estimated to be $5,000 the first year and to increase by $1,000 per year thereafter. The firm's cost of capital is 14 percent. Find the present-value cost of this investment.

$$PV_{cost} = I(PV_{sp}) + \Sigma \, OC(PV_{sp}) - S(PV_{sp})$$

where

$$I(PV_{sp})_{14\%}^{0 \text{ yr}} = \$40,000(1.00) = \qquad\qquad \$ \ \ 40,000$$

$$OC(PV_{sp}): \text{year } 1 = (5,000)(.877) = \ \$4,385$$
$$\text{year } 2 = (6,000)(.769) = \ \ \ 4,614$$
$$\text{year } 3 = (7,000)(.675) = \ \ \ 4,725$$
$$\text{year } 4 = (8,000)(.592) = \ \ \ 4,736$$
$$\text{year } 5 = (9,000)(.519) = \ \underline{\ \ \ 4,671}$$
$$\overline{\$23,131} \qquad \underline{\$+23,131}$$
$$\$ \ \ 63,131$$

$$\text{Less } S(PV_{sp})_{14}^{5} = 10,000(.519) = \qquad\qquad \underline{-5,190}$$
$$PV_{cost} = \qquad\qquad\qquad\qquad\qquad\qquad\qquad \$ \ \ 57,941$$

Present value considers the total return, includes time-value considerations, easily handles fluctuations in costs or revenues, and can readily include the effect of taxes, if applicable. However, it does not consider the rate of return or time for an investment to be paid off, and it assumes that cash inflows can be reinvested at the cost of capital. It is widely used.

EQUIVALENT ANNUAL COST

Equivalent annual cost is a time-adjusted method of calculating an equal annual cost over the life of an investment. It permits nonuniform costs to be apportioned equally over the life of the investment. Thus it is especially useful for comparing projects with different economic lives because it offers comparable per-year figures.

The equivalent annual-cost method discounts and compounds cost amounts at a specified interest rate in such a way as to convert them all into annuity amounts. It includes three components: (1) capital recovery and return on the investment, less any salvage, (2) interest on the salvage, and (3) other annual maintenance and operating costs.

$$\begin{pmatrix} \text{Equivalent} \\ \text{annual cost} \end{pmatrix} = \begin{pmatrix} \text{capital recovery} \\ \text{and return} \end{pmatrix} + \begin{pmatrix} \text{interest} \\ \text{on salvage} \end{pmatrix} + \begin{pmatrix} \text{other} \\ \text{costs} \end{pmatrix}$$

$$EAC = (CR \text{ and } R) + i(S) + OC \qquad\qquad (5.8)$$

Apportioning the present-value investment amount into an annuity is the reverse of converting an annuity into a present value, so we use $(1/PV_a)$, which is known as the capital recovery factor (CRF). Thus the CR and R is (investment $-$ salvage) $(1/PV_a)$.

Example 5.11 Sunshine Smelter (of Example 5.10) has received a bid for a stack filter from a second vendor. This proposal would cost \$45,000 and have an estimated salvage value of \$8,000 after an estimated useful life of 6 years. Operating costs are expected to be \$6,000 per year. Use 14 percent as the cost of capital, and find the equivalent annual cost of this investment.

The equivalent annual cost consists of:

$$CR \text{ and } R = (\text{investment} - \text{salvage})\left(\frac{1}{PV_a}\right)$$

$$= (\$45,000 - \$8,000)\left(\frac{1}{3.889}\right) = \$ \ \ 9,514/\text{yr}$$

$$i(S) = (.14)(\$8,000) = \qquad\qquad 1,120/\text{yr}$$

$$OC = \text{maintenance and operation} = \qquad \underline{\ \ 6,000/\text{yr}}$$

$$\$16,634/\text{yr}$$

Example 5.12   Compare the two stack-filter costs from Examples 5.10 and 5.11. Which appears to be the most economical?

Proposal II (Example 5.11) costs $5,000 more initially and has a salvage value of $2,000 less, but the equipment lasts 1 year longer and has a lower average operating cost. It is difficult to decide which proposal is less costly on the basis of these individual variables, especially with the different lifetimes. However, we can compare the two on an equal basis by converting the first proposal into an equivalent annual cost.

$$\text{Proposal I, equivalent annual cost} = PV\left(\frac{1}{PV_a}\right)^{5\,yr}_{14\%} = \$57,941\left(\frac{1}{3.433}\right) = \$16,878/\text{yr}$$

$$\text{Proposal II, equivalent annual cost} = \underline{\hspace{4cm} 16,634/\text{yr}}$$

$$\text{Proposal II advantage} \qquad\qquad 244/\text{yr}$$

The equivalent annual cost has many of the same advantages as present value, for it is readily convertible into a present-value amount (and vice versa). It is especially useful for comparing projects of different lifetimes, but it does not consider the total-cost or income aspects of a project.

INTERNAL RATE OF RETURN (IRR)

The internal rate of return (IRR) is the discount rate that equates an investment cost with its projected earnings. When discounted at the IRR, the present value cash outflow will just equal the present value of cash inflow.

$$\text{IRR} = i \text{ rate, where PV(of cash outflow)} = \text{PV(of cash inflow)} \qquad (5.9)$$

Figure 5-1 depicts the concept of IRR. The problem of finding the IRR is essentially one of finding what interest rate i is realized from an investment I over n years.

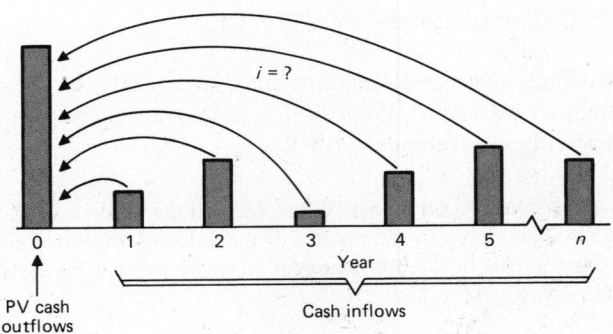

Fig. 5-1   Internal rate of return

If the annual cash inflows are equal annuities A and there is no salvage value, the IRR can easily be determined by using the PV_a table (Appendix F) because:

$$PV_a = \frac{\text{initial investment}}{\text{annuity cash flow}} = \frac{I}{A} \qquad (5.10)$$

Example 5.13   A baggage-handling device will cost $18,000, but will generate savings (positive cash flow) of $4,000 per year for 8 years (no salvage). Find the IRR.

The PV_a ratio associated with this return is:

$$PV_a = \frac{I}{A} = \frac{\$18,000}{\$4,000} = 4.50$$

Referring to the PV table for the row $n = 8$ years, we find the factor $i = 4.639$ for $i = 14$ percent and 4.487 for $i = 15$ percent. The IRR is thus very close to 15 percent.

For nonuniform cash flows, as depicted in Fig. 5-1, the IRR is more difficult to calculate because the PV_{sp} factor must be used, and it differs for each year. The calculation technique involves trial and error, starting first at some arbitrarily selected i rate. If on first try the present value of future earnings is less than the present value of the investment, the true IRR must be lower than that initially tried. With a lower i rate, the present value of a series of earnings will be more—and vice versa. Most business calculators have programs that simplify these IRR calculations.

Example 5.14   A proposed machine costing $20,000 is expected to generate cash inflows of $4,000 in year 1, $6,000 in year 2, $10,000 in year 3, and a salvage value of $10,000 in year 4. Find the IRR.

Try several i rates (for example, 10 percent, 15 percent, and others) until the PV of cash inflow is approximately equal to the PV of cash outflow ($20,000), as shown in Table 5-2.

Table 5-2

Year	Cash Flow	10 Percent		15 Percent		16 Percent	
		PV_{sp} Factor	PV Amount	PV_{sp} Factor	PV Amount	PV_{sp} Factor	PV Amount
1	4,000	.909	$ 3,636	.870	$ 3,480	.862	$ 3,448
2	6,000	.826	4,956	.756	4,536	.743	4,458
3	10,000	.751	7,510	.658	6,580	.641	6,410
4	10,000	.683	6,830	.572	5,720	.552	5,520
			$22,932		$20,316		$19,836

The IRR appears to be a little closer to 16 percent than to 15 percent, but in view of the many uncertainties that accompany such calculations, either value (or perhaps 15 1/2 percent) would be a reasonable estimate.

In addition to providing a time-adjusted rate of profit that can be used for comparison with similar rates from other projects, it is useful to compare the IRR against the cost of capital. If an IRR does not equal or exceed the cost of capital, the firm will lose money on the project. The IRR does not, however, consider the total magnitude of cash flows, and it always requires estimates of returns as well as costs. It also assumes that returns can be reinvested at the same IRR.

INFLATION EFFECTS

Inflation reduces the present value of future cash flows. Thus, when cash is paid out today, but is returned in the future (as inflated revenues), the revenues are really worth less in present-value terms. For example, if a return on investment is calculated to be 16 percent, but 6 percent is due to inflation, the effective rate of return is only 10 percent.

Questions and Solved Problems

CASH FLOWS

5.1 Let $n = 12$ years and $i = 15$ percent. Find (a) the discounted present value of a future sum of $20,000 and (b) the discounted present value of an annuity of $5,000 per year.

(a) $$P = F(\mathrm{PV_{sp}})_{15\%}^{12\ yr} = \$20{,}000(.187) = \$3{,}740$$

(b) $$P = A(\mathrm{PV}_a)_{15\%}^{12\ yr} = \$5{,}000(5.421) = \$27{,}105$$

5.2 Find (a) the future value of $10,000 compounded at 15 percent over a 12-year period and (b) the annuity amount that could be obtained for 12 years from a present value of $30,000 invested at 15 percent.

(a) $$F = P(1 + i)^n = \$10{,}000(1 + .15)^{12} = \$10{,}000(5.35) = \$53{,}500$$

also $$F = P\left(\frac{1}{\mathrm{PV_{sp}}}\right) \qquad\qquad (5.11)$$

$$= \$10{,}000\left(\frac{1}{.187}\right) = \$53{,}476 \text{ (accurate to three decimals)}$$

(b) $$A = \mathrm{CR} \text{ and } R = P\left(\frac{1}{\mathrm{PV}_a}\right) = \$30{,}000\left(\frac{1}{5.421}\right) = \$5{,}535$$

5.3 Operating costs for a machine are estimated at $500 per year for 10 years, plus an additional $1,000 for overhaul at the end of the fifth year. Assuming a 10 percent cost of capital, convert the maintenance and operating cost of the machine to a total present-value amount.

$$\mathrm{PV} \text{ of M and O cost} = \mathrm{PV} \text{ annual operating costs} + \mathrm{PV} \text{ maintenance cost}$$

$$= \$500(\mathrm{PV}_a)_{10\%}^{10\ yr} + \$1{,}000(\mathrm{PV_{sp}})_{10\%}^{5\ yr}$$

$$= \$500(6.145) + \$1{,}000(0.621) = \$3{,}072 + \$621 = \$3{,}693$$

5.4 Let $P = \$30{,}000$, $n = 5$ years, and $i = 10$ percent. Use Appendixes E, F, and G to find (a) F given P, (b) P given F, (c) A given F, (d) A given P, (e) F given A, and (f) P given A.

See Table 5-3 and Appendix G.

Table 5-3

Compound Interest Factors		Calculations for $n = 5$ yr, $i = 10\%$	
To find			
F given P	$F = P(1 + i)^n \quad = P(F\|P)_i^n$	If $P = \$30{,}000$	$F = \$30{,}000(1.611) = \$48{,}315$
P given F	$P = F\dfrac{1}{(1 + i)^n} \quad = F(P\|F)_i^n$	If $F = \$48{,}315$	$P = \$48{,}315(.6209) = \$30{,}000$
A given F	$A = F\dfrac{i}{(1 + i)^n - 1} = F(A\|F)_i^n$	If $F = \$48{,}315$	$A = \$48{,}315(.16380) = \$7{,}914$
A given P	$A = P\dfrac{i(1 + i)^n}{(1 + i)^n - 1} = P(A\|P)_i^n$	If $P = \$30{,}000$	$A = \$30{,}000(.26380) = \$7{,}914$
F given A	$F = A\dfrac{(1 + i)^n - 1}{i} = A(F\|A)_i^n$	If $A = \$7{,}914$	$F = \$7{,}914(6.105) = \$48{,}315$
P given A	$P = A\dfrac{(1 + i)^n - 1}{i(1 + i)^n} = A(P\|A)_i^n$	If $A = \$7{,}914$	$P = \$7{,}914(3.791) = \$30{,}000$

5.5   Depict the P, F, and A values from Prob. 5-4 in schematic form.

See Fig. 5-2.

The annuity payments A at the end of each year necessary to accumulate a future value F (of $48,135) are

$$A = F(A|F)_{10}^{5} = \$48,315 \ (.1638) = \$7,914$$

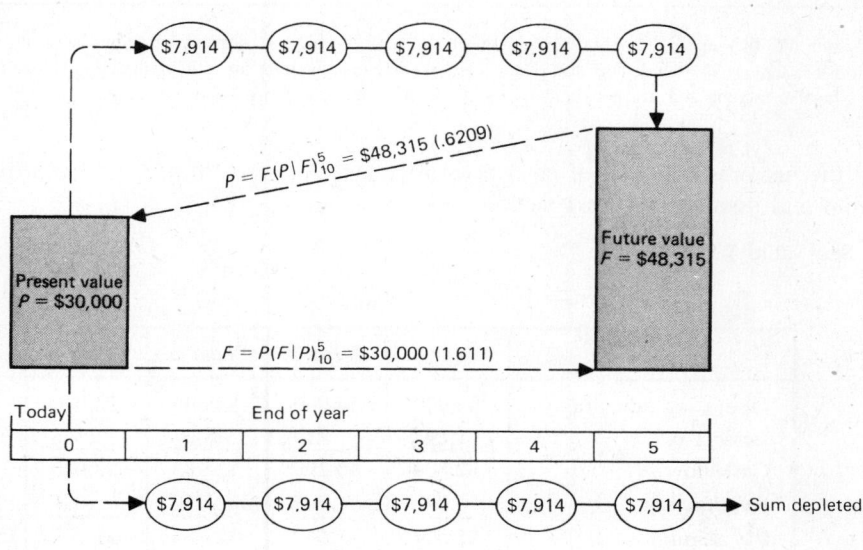

Fig. 5-2

The annuity payments A that could be made at the end of each year from a present value P (of $30,000) are

$$A = P(A|P)_{10}^{5} = \$30,000 \ (.2638) = \$7,914$$

DEPRECIATION AND TAX CONSIDERATIONS

5.6   *Straight Line*. An investment of $12,000 in new equipment was expected to have a salvage value of $2,000 after a 5-year life. Find the straight-line depreciation expense per year.

$$\text{Depreciation} = \frac{I - S}{n} = \frac{\$12,000 - \$2,000}{5 \text{ yr}}$$

$$= \$2,000/\text{yr} \quad (\text{same each year})$$

5.7   Hawthorne Restaurants Inc. has purchased a $7,000 oven that is expected to yield a before-tax operating advantage of $4,000 per year, for 5 years. It will be salvaged during the fifth year for $750 and will be depreciated on an IRS approved schedule as shown in line (*c*) of Table 5-4. Compute the yearly tax amounts, assuming that a 46 percent tax rate applies.

See Table 5-4.

Table 5-4

		Year 1	Year 2	Year 3	Year 4	Year 5
(a)	Operating advantage	$4,000	$4,000	$4,000	$4,000	$4,190*
(b)	Cost basis	7,000	7,000	7,000	7,000	7,000
(c)	Depreciation	1,400	2,240	1,680	1,120	0
(d)	Income − expense [(a) − (c)]	$2,600	$1,760	$2,320	$2,880	$4,190
(e)	Tax @ 46% × (d)	1,196	810	1,067	1,325	1,927

*Using the IRS approved rate, 92 percent of the asset's value is depreciated in years 1 through 4. This leaves (.08)($7,000) = $560 undepreciated. The excess of salvage value over the remaining depreciation ($750 − $560 = $190) is taxable income and, for convenience, is included in the year 5 operating advantage.

5.8   Find the net present value of cash flow after tax (AT) for Prob. 5.7. Use a discount rate of 12 percent and assume that next year is year 1.

See Table 5-5.

Table 5-5

	Year 1	Year 2	Year 3	Year 4	Year 5
Operating advantage	$4,000	$4,000	$4,000	$4,000	$4,000
Less: Tax	1,196	810	1,067	1,325	1,927
Cash flow AT	$2,804	$3,190	$2,933	$2,675	$2,263
PV$_{sp}$ factor	.893	.797	.712	.636	.567
PV amount	$2,504	$2,542	$2,088	$1,701	$1,283

Net PV of cash flow AT = $2,504 + $2,542 + $2,088 + $1,701 + $1,283 = $10,118.

PAYBACK

5.9   A $40,000 banking machine is expected to be obsolete after 10 years, with no salvage value. During its lifetime, it should generate an $8,000-per-year operating advantage, $3,000 of which must be paid in taxes. What is the payoff period?

$$\text{Payoff} = \frac{I - S}{\text{OA/yr}} = \frac{\$40,000 - 0}{\$8,000 - \$3,000} = 8 \text{ years}$$

5.10   A $30,000 investment can be made in either X, Y, or Z, and will yield the estimated cash flows shown in Table 5-6. Find the payback period for each.

Table 5-6

Year	X	Y	Z
1	$20,000	$10,000	$ 0
2	10,000	10,000	5,000
3	5,000	10,000	10,000
4	2,000	10,000	12,000
5	1,000	10,000	10,000

Cumulate the cash flow for each alternative, until the cash flow equals the investment amount of $30,000 as shown in Table 5-7.

Table 5-7

Year	X	Y	Z
1	$20,000	$10,000	$ 0
2	30,000	20,000	5,000
3		30,000	15,000
4			27,000
5			37,000

X's payback is 2 years, Y's is 3 years, and Z's is early in the fifth year. Therefore X appears best, based on payback.

5.11 A proposed new $16,400 automatic machine will have operating costs of $.30 per unit produced, whereas the existing machine costs are $.70 per unit. The existing machine has a market value of $8,700 now and has another 5 years of life. It would cost $500 to remove the existing machine and install the new one. If the firm requires a 3-year payout period, how many units must be produced annually to justify the new machine? Disregard taxes.

$$\text{Payout} = \frac{\text{investment}}{\text{OA/yr}}$$

We have
$$\text{Payout} = 3 \text{ yr}$$
$$\Delta \text{ investment} = \$16,400 - \$8,700 = \$7,700$$
$$\text{Add installation cost:} \quad \underline{\quad 500}$$
$$\text{Total } \$8,200$$
$$\text{OA/unit} = \$.70 - \$.30 = \$.40/\text{unit}$$
$$\text{OA(total)} = \$.40 \ (N \text{ units/yr})$$

Therefore
$$3 \text{ yr} = \frac{\$8,200}{\$.40(N)}$$

and
$$N = \frac{\$8,200}{\$1.20/\text{unit}} = 6,833 \text{ units/yr}$$

PRESENT VALUE

5.12 Internet Services, Inc., offers software services at $1,000 per year for 5 years plus an additional $2,000 at the end of the third year for system facilities update. If a firm contracts for 5 years of services, what is the net present-value cost to the firm? The firm estimates its capital cost at 14 percent and has sales of $3.5 million per year.

The sales data are not relevant to computing the present-value cost.

$$\text{Present-value cost} = \text{PV other costs} = \text{PV (software cost)} + \text{PV (system cost)}$$
$$= \$1,000(PV_a)_{14\%}^{5 \text{ yr}} + \$2,000(PV_{sp})_{14\%}^{3 \text{ yr}}$$
$$= \$1,000(3.433) + \$2,000(.675) = \$3,433 + \$1,350 = \$4,783$$

5.13 An instrument transformer manufacturer in Long Island is considering purchase of an ultrasonic welding machine to replace an existing manually operated machine. Two years ago the existing machine cost $12,000 and has been depreciated down to a $10,000 book value, using a 12-year life and no salvage. However, the market value of the machine is only about $4,000 now. The ultrasonic welder would improve product quality enough to boost revenue from an existing $80,000 per year to $100,000 per year. It would cost $44,000 and have a 10-year life. Any salvage value on it would be consumed in the removal expense. An advantage of the ultrasonic machine is that by reducing annual labor costs, it would cut operating expenses from $8,000 to $3,000 annually. Use a 50 percent tax rate, and estimate the firm's cost of capital at 12 percent. Should the manufacturer purchase the ultrasonic welder?

Determine the after-tax profit under each alternative, and select the most favorable one. It will be most convenient to do calculations on an annual basis and then convert to present value.

Existing machine:

Revenue	$80,000
Less:	
Operating costs	8,000
Depreciation	1,000
Income subject to tax	$71,000
Income tax (@ 50%)	$35,000

Cash inflow = revenue − operating costs − taxes

= $80,000 − $8,000 − $35,500 = $36,500/yr

Present value of cash inflow (AT) = $R(PV_a)_{12\%}^{10 \text{ yr}} = \$36,500(5.65) = \$206,225$

Net PV gain after taxes = PV (cash inflow) − PV(I)

= $206,225 − $4,000 = $202,225

New ultrasonic machine:

Revenue	$100,000
Less:	
Operating costs	3,000
Depreciation	4,400
Income subject to tax	$92,600
Income tax (@ 50%)	$46,300

Cash inflow = $100,000 − $3,000 − $46,300 = $50,700

Present value of cash inflow (AT) = $50,700(5.65) = $286,455

Net PV gain after taxes = $286,455 − $44,000 = $242,455

Note the net PV gain AT from the ultrasonic machine installation exceeds the existing arrangement by $40,230, and thus the new machine should be installed. Note also that the relevant investment cost of the existing machine is the market value, *not the book value*. There is no relevant advantage to be gained from writing off some of the existing machine as a loss, since this write-off should take place whether the new machine is purchased or not. The write-off advantage is not relevant to the decision problem.

EQUIVALENT ANNUAL COST

5.14 Alaska Construction Co. is purchasing a portable generator for $5,026 and plans to finance the purchase from a local bank at 8 percent interest. The contract stipulates that the seller will pay the construction company $1,000 for the used machine after 10 years. What is the equivalent annual purchase cost to Alaska Construction Company?

$$CR \text{ and } R = (I - S)\left[\frac{1}{(PV_a)_{8\%}^{10\ yr}}\right]$$

$$= (\$5{,}026 - \$1{,}000)\left[\frac{1}{6.71}\right] = \$600/yr$$

$$(i)S = (0.08)(\$1{,}000) = \qquad\qquad 80/yr$$

$$\text{Other costs (none considered):} \qquad \underline{00/yr}$$

$$\text{Total } \$680/yr$$

5.15 Olympic Quality Foods plans to sign a 3-year lease for automobiles for its production supervisors at a seafood plant in Norway. The company can obtain car A for $4,000 plus $.25 per mile or car B for $2,400 plus $.45 per mile. If funds cost 18 percent, how many miles must be driven before the use of car A is justified? Use the equivalent annual-cost method.

$$CR \text{ and } R = (I - S)\left[\frac{1}{(PV_a)_{18\%}^{3\ yr}}\right]$$

	Car A	**Car B**
CR and R:	($4,000)(1/2.174) = $1,840/yr	($2,400)(1/2.174) = $1,104/yr
Interest on salvage:	no salvage	no salvage
Mileage charges:	$.25N	$.45N
Total	$1,840 + $.25N	$1,104 + $.45N

Setting the total costs for car A equal to the total costs for car B:

$$TC_A = TC_B$$
$$\$1{,}840 + \$.25N = \$1{,}104 + .45N$$
$$\$.20N = \$736$$
$$N = 3{,}680 \text{ miles}$$

INTERNAL RATE OF RETURN

5.16 An investment of $5,650 is expected to yield an operating advantage (before depreciation and taxes) of $4,000 at the end of the first year, $2,000 at the end of the second year, and $1,000 at the end of the third. What is the time-adjusted rate of return?

Set $PV(\text{income}) = PV(I) = \$5{,}560$.

$$\text{Try 14 percent:} \quad \text{1st-yr earnings} = F(PV_{sp})_{14\%}^{1\ yr} = \$4{,}000(.877) = \$3{,}508$$
$$\text{2d-yr earnings} = F(PV_{sp})_{14\%}^{2\ yr} = \ 2{,}000(.769) = \ 1{,}538$$
$$\text{3d-yr earnings} = F(PV_{sp})_{14\%}^{3\ yr} = \ 1{,}000(.675) = \ \underline{\ 675}$$
$$\$5{,}721$$

$PV(\text{income}) > PV(I)$.

Trying a higher rate of 16 percent: $PV = \$3{,}480 + \$1{,}512 + \$658 = \$5{,}575$.

Since PV (income) is less than $PV(I)$, try a lower rate. Note that the 16 percent rate yields a PV (income) figure $75 below the investment amount, whereas the 14 percent rate yields a figure $71 above. Thus the correct value should be about midway between, or 15 percent.

$$
\begin{aligned}
\text{Try 15 percent:} \quad \text{1st yr} &= \$4,000(.870) = \$3,480 \\
\text{2d yr} &= 2,000(.756) = 1,512 \\
\text{3d yr} &= 1,000(.658) = \underline{658} \\
& \$5,650
\end{aligned}
$$

The (BT)IRR is 15 percent.

Supplementary Questions and Problems

5.17 Bradley Enterprises borrows $60,000 from Suffolk Bank at 14 percent interest per year. How much must the firm repay 2 years later? *Ans.* $78,000 (rounded)

5.18 The maintenance expense on a linotype is expected to be $300 per year over the next 6 years. If the owner must pay 12 percent to borrow money, what is the present-value cost of the maintenance expense? *Ans.* $1,233

5.19 A $20,000 machine will last 10 years, will have no salvage, and will generate a $4,000-per-year operating advantage, $1,000 of which must be paid in taxes. Find the payoff period. *Ans.* 6.7 years

5.20 A new lead-casting furnace costs $30,000 installed and is expected to have a $4,000 salvage value at the end of an 8-year economic life. Operation and maintenance costs will run about $7,000 per year. Using 12 percent interest, what is the present-value cost of the furnace? *Ans.* $63,160

5.21 Rochester Shoe Co. wishes to decide between two automatic machines. Machine X has a net-present-value cost of $25,000 (all costs considered). Machine Y has an initial cost of $14,000 and will have a salvage value of $1,000. The annual labor cost is $3,300 and annual taxes, insurance, and other costs are estimated at 5 percent of the initial cost. Both machines would have the same 4-year life under the heavy use expected. If Rochester Shoe Co. uses an 8 percent interest rate and uses the declining-balance method of depreciation, how would the net present-value cost of machine Y compare with that of machine X? *Ans.* Machine Y is $1,513 more. *Note:* Depreciation method has no impact here.

5.22 Nikko Media Center in Seattle plans to modify its shop layout and is considering the alternatives shown in Table 5-8.

Table 5-8

	Plan 1	Plan 2
New machinery cost	$22,000	$20,000
Installation labor cost	3,000	2,000
Annual savings expected (in operating costs)	8,000	7,000

The new layout is expected to be suitable for 5 years of operation, with no salvage. (*a*) If the company has a 12 percent cost of capital, what is the equivalent annual cost (or savings) for plan 1 and plan 2? (*b*) Which plan should be adopted? *Ans.* (*a*) Plan 1 = $1,065 savings and Plan 2 = $897 savings (*b*) Plan 1

5.23   Andry Corp. has an opportunity to invest $7,680 in a plant modification that is expected to yield an operating advantage of $5,000 at the end of the first year and $1,000 at the end of each of the next 5 years. What is the before-tax internal rate of return?   *Ans.*   12 percent

5.24   Forest Paper Co. is considering the purchase of a $10,000 paperbox press that would be used for 3 years and sold for $1,000 in salvage value. Operating costs are $400 per year. Maintenance costs are $500 the first year and increase by $500 each year thereafter. Production volume is 1,000 units per year, and the firm operates on three-shift-per-day basis. It uses straight-line depreciation, is in a 40 percent tax bracket, and estimates the cost of capital at 10 percent. (*a*) Determine the present-value cost of owning and using the machine before tax is considered. (*b*) Determine what effect taxes have on the present-value cost. (*c*) Assume that the maintenance costs remain constant at $1,000 per year and that all sales and administrative costs are included in the operating costs of $400 per year. If the firm achieves a paperbox sales revenue of $12,000 per year from the press, what is the after-tax payoff period?
   Ans.   (*a*) $12,651, (*b*) Taxes reduce the net profit, but depreciation expenses can be deducted before taxes are computed. If depreciation = $3,000 per year and taxes = 40 percent, then tax = .40 (OA before depreciation − $3,000) = $1,200 less that is paid each year because of the depreciation expense. This $1,200 per year has a present value of ($1,200)(2.487) = $2,984, (*c*) 1.2 years

5.25   The Albi Electronics production manager is evaluating two machines to determine which would be most economical. Both machines are capable of generating the same revenue. The firm's cost of capital is 14 percent. Data on the two machines are shown in Table 5-9.

Table 5-9

	Machine A	Machine B
Initial cost	$20,000	$30,000
Maintenance and operating cost/yr	2,000	3,000
Economic life	4 yrs	4 yrs
Salvage value (estimated)	$ 2,000	$15,000

For machine A find (*a*) the net present value cost and (*b*) the equivalent annual cost.
   Ans.   (*a*) $24,650 (*b*) $8,457 per year

Chapter 6

Facility Location

Transportation Linear Programming

FACTORS AFFECTING LOCATION DECISIONS

Plant expansion and new facility construction are among the most far-reaching decisions an organization must face. Location choices direct the flow of billions of dollars of investment worldwide and commit firms to long-lasting cost, employment, and marketing patterns.

Question: How does an organization go about selecting an optimal facility location?

No location decision process can ensure that an optimal location is chosen; avoiding a troublesome (or disastrous) location may be more important than trying to find an ideal site. Location decisions involve so many factors that a systematic approach (or checklist) is essential to be sure nothing is overlooked.

Question: Some planners recommend a macro-to-micro approach to location that focuses upon inbound and outbound costs. Use a schematic diagram to identify some of the factors involved as an analysis is narrowed down from international to local site considerations.

See Fig. 6-1. Among the major factors influencing location are the incoming raw materials, human resources, and capital. Major outgoing factors include the market and the environment.

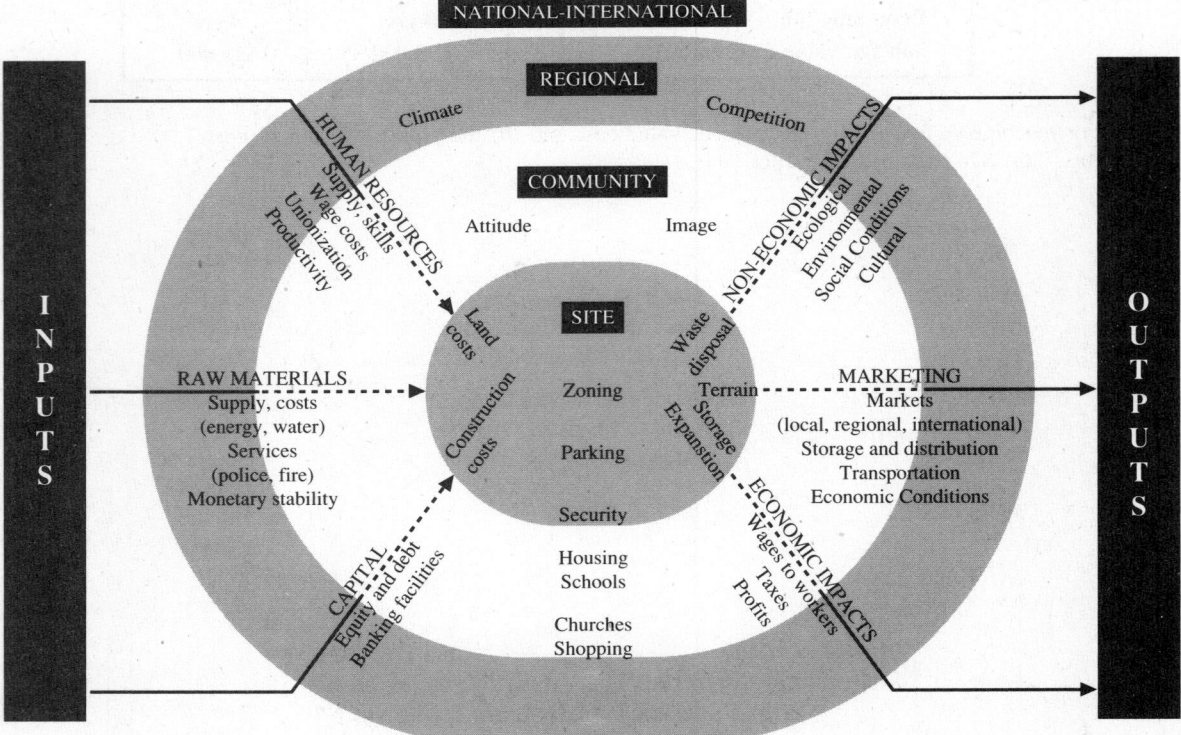

Fig. 6-1 Factors affecting location decisions

THE LOCATION DECISION PROCESS: DOMESTIC AND INTERNATIONAL SITES

Location decisions should not be viewed as static—especially in today's global business environment. Location (and relocation) opportunities should be reviewed as political, social, and economic conditions change. Foreign locations may offer more profit potential, but often at a higher risk due to political instability, currency fluctuations, employment benefit changes, etc.

Question: What location alternatives exist for firms to respond to changing demands?

In addition to (a) maintaining the status quo, firms may have the option of (b) closing existing facilities, (c) expanding existing facilities, or (d) developing new facilities.

Question: What are some of the locational trends in the United States?

(1)   *More diversified geographic location* (influenced by the shift to a service economy, increasing use of computers, electronic mail, faxes, internet, wireless communications, etc.)

(2)   *Movement from large urban to smaller communities* (influenced by taxes and operating costs, quality of life, crime and security considerations, local educational and skill levels)

(3)   *Industrial concentrations* (e.g., sun belt migration to southwest; electronic concentration in Silicon Valley, Boston, and Seattle; manufacturing concentration along Mexican border)

Question: What factors are influencing the trend toward globalization that is taking place in areas such as Mexico/Latin America, Southeast Asia, and Eastern Europe?

(1)   Expanding markets (from rising standards of living in more politically stable economies)

(2)   Reduced trade barriers (e.g., NAFTA, EEC, Southeast Asia agreements)

(3)   Faster flow of capital (within more flexible international monetary system)

(4)   Acceptance of legal and other responsibilities of international citizenship

(5)   Improved flows of goods/information (e.g., via air, satellite, wireless, fiber optics)

Several different methods of analysis are used to narrow down location alternatives from international to specific national, regional, community, and site locations. Some analysts recommend a quantitative analysis first to establish the economic feasibility of each alternative location, followed by an extensive review of qualitative (less tangible) factors. Others focus upon dominant factors first (e.g., labor, markets, zoning), followed by secondary (behavioral) factors; or they use cost-volume analysis or factor rating systems for making a comparative analysis of several sites. *Dun's Review* publishes a *Business Site/Construction Planner* that is helpful.

Question: What steps should be included in making a facility location decision?

Table 6-1   Steps in a Facility Location Decision

1.   Define location *objectives*, *decision criteria*, and *basic requirements*.

2.   Use political, social, and economic (market) data to narrow the potential locations to a few alternatives that satisfy objectives.

3.   Evaluate alternatives against basic (or dominant) requirements, and eliminate unacceptable locations.

4.   Compare alternatives on *quantitative* (i.e., and economic) basis, using cost-volume, linear-programming, center-of-gravity, load-distance, and/or other appropriate models.

5.   Compare alternatives on *qualitative* basis, using factor rating or other subjective means to take less tangible factors into account.

6.   Select the location that best satisfies the quantitative and qualitative criteria, by using a weighted score or group decision process.

TYPE OF FACILITY: GOODS VERSUS SERVICES

Along with the (1) availability of inputs and (2) market for outputs, the (3) type of processing is a key determinant of location. The locational requirements associated with the processing of goods are often quite different from those of services.

Question: What characteristics of goods influence facility location?

Goods can be produced at locations away from the market. Some goods production facilities (e.g., mineral and forest products) are strongly dictated by resource availability (including labor supply), whereas others (light manufacturing) are more flexible. But in all cases, value is stored in the product, which can be transported to customers for use at a later time.

Question: What characteristics of services influence facility location?

Services are typically produced and consumed simultaneously, often in the presence of the customer. Because services cannot normally be stored, many service facility locations (e.g., medical, police, hair salons) are dictated largely by the location of customers. However, the ease/speed of electronic telecommunications has given significant flexibility to the location of information services (e.g., electronic mail, library, brokerage, TV, entertainment). This also enables firms in one country to process work in another (e.g., U.S. credit card billing done in China and secretarial/engineering work faxed to India).

COST-VOLUME (BREAK-EVEN) ANALYSIS OF LOCATIONS

Alternative locations can be compared on an economic basis by (1) estimating the fixed and variable costs associated with each location and (2) graphing (or computing) them for a representative volume. If revenues vary from one location to another, the comparisons should be made on the basis of total revenue (TR) minus total cost (TC) at each location.

Example 6.1 Potential locations in Argentina, Brazil, and Chile have the cost structures shown in Table 6-2 for a product expected to sell for $130. (*a*) Find the most economical location for an expected volume of 6,000 units per year. (*b*) What is the expected profit if the site selected in (*a*) is used? (*c*) For what output range is each location best?

Table 6-2

Potential Location	Fixed Cost/Yr	Variable Cost/Unit
Argentina (A)	$150,000	$75.00
Brazil (B)	200,000	50.00
Chile (C)	400,000	25.00

For each set, plot the fixed costs (costs at zero volume) and total costs (FC + TVC) at the expected volume of output. See Fig. 6-2.

(*a*) $TC = FC + VC(V)$

A: $TC = \$150,000 + \$75\ (6,000) = \$600,000$

B: $TC = \$200,000 + \$50\ (6,000) = \$500,000$

C: $TC = \$400,000 + \$25\ (6,000) = \$550,000$

Therefore the most economical location is B.

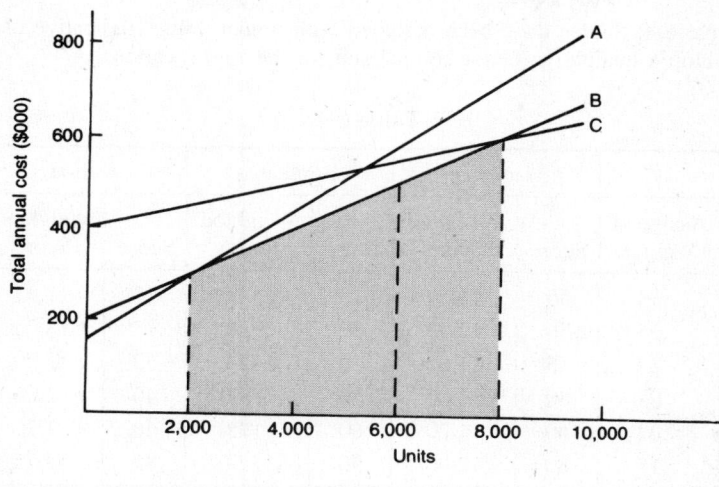

Fig. 6-2

(*b*)   Expected profit (using B)

$$\text{Profit} = \text{TR} - \text{TC} = (\$130/\text{unit})\ (6{,}000\ \text{units}) - \$500{,}000 = \$280{,}000/\text{yr}$$

(*c*)   From the graph (Fig. 6-2), use A for volumes up to 2,000; B for 2,000 to 8,000; and C for volumes greater than 8,000 units.

Locational break-even analysis applies to one-product (or product-line) situations, so if multi-products are involved, their respective cost and volume effects must be appropriately weighted. It also assumes that fixed costs hold constant and that variable costs remain linear. If the expected volume is very close to the crossover point between two locations, other factors may be more influential than costs.

FACTOR RATING SYSTEMS

Factor rating is a means of assigning quantitative values to all the factors related to each decision alternative and deriving a composite score that can be used for comparison. It allows the decision maker to inject his or her own preferences (values) into a location decision, and it can accommodate both quantitative and qualitative factors.

Question: What steps are followed in doing a qualitative factor rating?

Table 6-3   Procedure for Qualitative Factor Rating

1. *Develop a list of relevant factors* (e.g., use a checklist or a model such as Fig. 6-1).
2. *Assign a weight to each factor* to indicate its relative importance (weights may total 1.00).
3. *Assign a common scale to each factor* (e.g., 0 to 100 points), and *designate any minimums*.
4. *Score each potential location* according to the designated scale, and *multiply the scores* by the weights.
5. *Total the points* for each location, and *choose the location* with the maximum points.

Example 6.2   International Glass Co. is evaluating four locations for a new plant and has weighted the relevant factors as shown in Table 6-4. Scores have been assigned with higher values indicative of preferred conditions. Using these scores, develop a qualitative factor comparison for the four locations.

Table 6-4

Relevant Factor	Assigned Weight	Hamburg		Bordeaux		Genoa		Lisbon	
		Score	Weighted Score	Score	Weighted Score	Score	Weighted Score	Score	Weighted Score
Production cost	.33	50	16.50	40	13.20	35	11.55	30	9.90
Raw material supply	.25	70	17.50	80	20.00	75	18.75	80	20.00
Labor availability	.20	55	11.00	70	14.00	60	12.00	45	9.00
Cost of living	.05	80	4.00	70	3.50	40	2.00	50	2.50
Environment	.02	60	1.20	60	1.20	60	1.20	90	1.80
Markets	.15	80	12.00	90	13.50	85	12.75	50	7.50
Totals	1.00		62.20		65.40		58.25		50.70

Weighted scores are computed by multiplying the score times the assigned weight (for example, $50 \times .33 = 16.50$) and summing those products. On the basis of this data, Bordeaux is the preferred location.

CENTER-OF-GRAVITY (GRID) METHOD

Load-distance and grid methods of location have been advanced to help identify the location of a production center or warehouse that will minimize the costs of distributing specified volumes of a product to surrounding market areas. These methods make use of a map, or grid, with some type of coordinate system that shows the relative distances between locations. See Fig. 6-3 where the (X, Y) coordinates are in miles. Any uniform scale can, however, be used for distances.

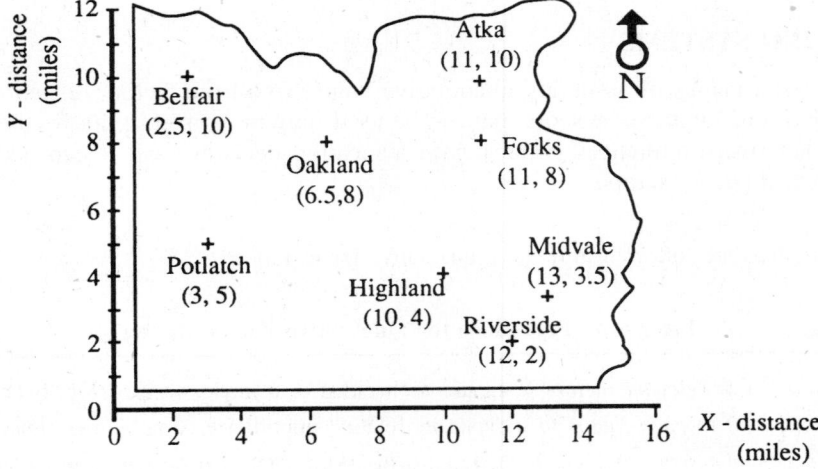

Fig. 6-3   Map showing X- and Y-coordinates

A simplified form of the *center-of-gravity method* assumes that the distribution cost is a function of the volumes shipped and the rectilinear (i.e., X- and Y-coordinate) distances—although straight-line (Euclidean) distances can be used. The distances in each of the X and Y coordinates are averaged, using the volumes as weights. The resultant coordinates then constitute the center of gravity for that grid.

To determine the average X-coordinate, (X_c), the volumes of goods transported to or from each of i destinations (V_i) are multiplied by their respective coordinate distances traveled (x_i). The sum of these is then divided by the total volume shipped, (ΣV_i), to yield the volume-weighted X-coordinate of the center of gravity for the system, X_c. The Y_c coordinate is calculated in similar fashion.

$$X_c = \frac{\Sigma V_i x_i}{\Sigma V_i} \qquad Y_c = \frac{\Sigma V_i y_i}{\Sigma V_i} \qquad\qquad (6.1)$$

Once determined, the X_c and Y_c coordinates constitute a starting point for a new site. Locations in that vicinity may then be evaluated, changes suggested, and perhaps some recalculations done before the final choice is made.

Example 6.3 The grid of Fig. 6-3 shows eight market locations to which a manufacturer of wood windows expects to ship its products. The shipment volumes and X- and Y-coordinates of the locations are shown in Table 6-5. Using the center-of-gravity method, (a) find the volume-weighted X_c and Y_c coordinates, and (b) suggest a possible warehouse location.

Table 6-5

Market Area	Vol	X_i	Y_i	$V_i x_i$	$V_i y_i$
Belfair	8	2.5	10	20	80
Potlatch	20	3	5	60	100
Oakland	12	6.5	8	78	96
Atka	10	11	10	110	100
Forks	30	11	8	330	240
Highland	20	10	4	200	80
Midvale	40	13	3.5	520	140
Riverside	30	12	2	360	60
	170			1678	896

(a) Two additional columns have been included to the right of the double line in Table 6-5 to provide values for $V_i x_i$ and $V_i y_i$. The center-of-gravity coordinates are:

$$X_c = \frac{\Sigma V_i x_i}{\Sigma V_i} = \frac{1678}{170} = 9.9 \qquad Y_c = \frac{\Sigma V_i y_i}{\Sigma V_i} = \frac{896}{170} = 5.3$$

(b) Referring back to Fig. 6-3, the coordinates of ($X_c = 9.9$, $Y_c = 5.3$) lie very close to the city of Highland, suggesting that it may be worthy of consideration for the distribution center. Other factors such as warehouse availability, transportation rates, etc., should then be evaluated for that site.

TRANSPORTATION LINEAR PROGRAMMING

Transportation adds no value to a good other than place utility. However, the transportation costs for raw materials and finished goods are often significant and merit special analysis. Before deciding on a plant location, management may want to know which plants will be used to produce what quantities and to which distribution warehouses all quantities should be shipped.

If the location problem can be formulated as one of minimizing a transportation cost, subject to satisfying overall supply and demand requirements, the transportation linear-programming (LP) method may be useful. The transportation model is a variation of the standard linear-programming approach and assumes the following:

(1) The objective is to minimize total transportation costs.

(2) Transportation costs are a linear function of the number of units shipped.

(3) All supply and demand are expressed in homogeneous units.

(4) Shipping costs per unit do not vary with the quantity shipped.

(5) Total supply must equal total demand.

    (a) If *demand* is larger than *supply*, create a dummy supply and assign a zero transportation cost to it so that excess demand is satisfied.

    (b) If *supply* is larger than *demand*, create a dummy demand and assign a zero transportation cost to it so that excess supply is absorbed.

To use the transportation (also called *distribution*) linear-programming format, the demand requirements and supply availabilities are formulated in a rectangular matrix. The transportation costs between the supply and demand points are placed in the upper corner of each cell.

Supply is then allocated to meet demand by placing entries, which express the number of units shipped from a supply source to a demand destination, into the cells. The solution procedure is an iterative one that begins with an initial solution that is feasible, but not necessarily optimal. The solution is progressively tested and improved upon until an optimal solution is reached. The optimal solution satisfies demand at the lowest total cost.

Several methods of obtaining initial and optimal solutions have been developed:

Initial Solutions	**Optimal Solutions**
(1) Minimum cost (intuitive)	(1) Stepping-stone
(2) Northwest corner	(2) Modified distribution (MODI)
(3) Vogel's approximation (VAM)	

The minimum cost method works well for simple problems, but VAM is likely to yield a better initial solution, which is often also the optimal solution. VAM works by sequentially zeroing in on the most cost-advantageous row-and-column combinations. The northwest-corner method does not usually yield as good an initial solution as VAM, but it is extremely easy to apply.

VAM is useful for hand calculation of relatively large-scale problems. However, most large problems are solved by computer, and numerous computer programs are available, so VAM is not covered in the examples that follow. The MODI method is well-suited to computer applications. It is a modified stepping-stone algorithm that uses index numbers to systematically reach an optimum solution. Example 6.4 uses the northwest-corner method for the initial solution and the stepping-stone method for the final solution.

Example 6.4 (*Describes distribution linear-programming methods*) The Milltex Co. has production plants in Albany, Bend, and Corvallis, all of which manufacture similar paneling for the housing market. The products are currently distributed through plants in Seattle and Portland. The company is considering adding another distribution plant in San Francisco and has developed the transportation costs in dollars per unit, shown in Table 6-6.

Table 6-6

Production Plants	Cost to Ship to Distribution Plant in:		
	Seattle	Portland	San Francisco
Albany	$10	$14	$ 8
Bend	12	10	12
Corvallis	8	12	10

The production capabilities at the Albany, Bend, and Corvallis plants are 20-, 30-, and 40-unit loads per week, respectively. Management feels that a San Francisco plant could absorb 20 units per week, with Seattle and Portland claiming 40 and 30 units per week, respectively. Determine the optimal distribution arrangement and cost if the San Francisco site is selected.

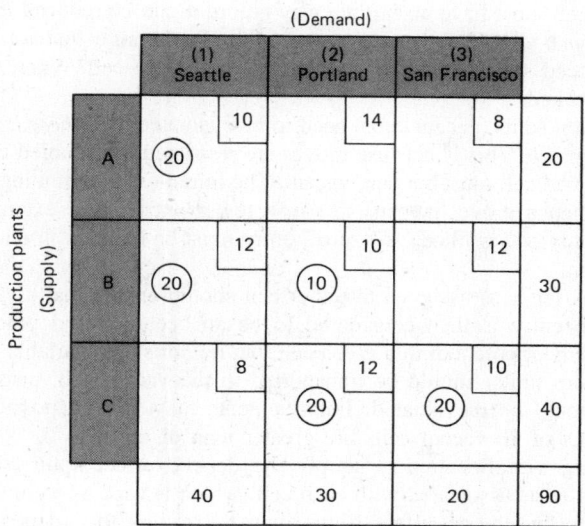

Fig. 6-4 Initial solution to distribution linear-programming matrix

We will use the northwest-corner method for the initial allocation and the stepping-stone method for the final solution. To do this requires that the data be arranged in a matrix. Figure 6-4 shows supply on the horizontal rows, demand on the columns, and unit transportation costs ($) in the small boxes of the matrix.

The initial allocation via the northwest (NW)-corner method is made as follows:

(a) Assign as many units as possible to the NW-corner cell A1 from the total available in row A. Given the 20-unit available supply in row A and the 40-unit demand in column 1, the maximum number of units that can be assigned to cell A1 is 20. This is shown as ⑳, indicating an initial allocation.

(b) Assign additional units of supply from row B (or additional rows) until the demand in column 1 is satisfied. This requires 20 additional units in cell B1 and leaves 10 units of B's unassigned.

(c) Assign remaining units in the subject row to the next column, continuing as above until its demand requirements are satisfied. This means the 10 units left in B are assigned to cell B2. Since this does not satisfy demand in column 2, an additional 20 units are allocated from row C.

(d) Continue down from the NW corner until the whole supply has been allocated to demand. The initial assignment is completed by assigning the 20 units remaining in row C to cell C3.

(e) Check allocations to verify that all supply and demand conditions are satisfied. Since all row and column totals agree, the initial assignment is correct. Also, the number of entries is five. This satisfies the NW-corner requirement of entries equal $R + C - 1$ (rows plus columns minus one) for $3 + 3 - 1 = 5$.

The initial solution is, perhaps obviously, not an optimal (or least-cost) allocation scheme. The transportation cost for this arrangement is:

20 units	A to Seattle @ $10/unit =	$200
20 units	B to Seattle @ $12/unit =	240
10 units	B to Portland @ $10/unit =	100
20 units	C to Portland @ $12/unit =	240
20 units	C to San Francisco @ $10/unit =	200
	Total	$980

An optimal solution can be obtained by following a stepping-stone approach, which requires calculation of the net monetary gain or loss that can be obtained by shifting an allocation from one supply source to another. The important rule to keep in mind is that every increase (or decrease) in supply at one location must be accompanied by a decrease (or increase) in supply at another. The same holds true for demand. Thus there must be two changes in every row or column that is changed—one change increasing the quantity and one change decreasing it. This is easily done by evaluating reallocations in a closed-path sequence with only right-angle turns permitted and only on occupied cells. Of course, a cell must have an initial entry before it can be reduced in favor of another, but *empty (or filled) cells may be skipped over to get to a corner cell.* To be sure that all reallocation possibilities are considered, it is best to proceed systematically, evaluating each empty cell. When any changes are made, cells vacated earlier must be rechecked.

Only unused transportation paths (vacant cells) need to be evaluated, and *there is only one available pattern of moves to evaluate each vacant cell.* This is because moves are restricted to occupied cells. Every time a vacant cell is filled, *one* previously occupied cell must become vacant. The initial (and continuing) number of entries is always maintained at $R + C - 1$. When a move happens to cause fewer entries (for example, when two cells become vacant at the same time but only one is filled), a "zero" entry must be retained in one of the cells to avoid what is termed a *degeneracy* situation.

The zero entry (or Greek letter ε) assigned to either cell should ensure that a closed path exists for all filled cells. The cell with the zero entry is then considered to be an occupied and potentially usable cell. If a cell evaluation reveals an improvement potential in a given cell, but no units are available because of a zero entry in the path to that cell, the zero (zero units) should be transported to the vacant cell, just as any other units would be shipped. Then the matrix should be reevaluated. Improvements may still be possible until the zero entries are relocated to where evaluations of all vacant cells are greater than or equal to 0.

The criterion for making a reallocation is simply the desired effect upon costs. The net loss or gain is determined by listing the unit costs associated with each cell (which is used as a corner in the evaluation path) and then summing over the path to find the net effect. Signs alternate from + to − depending upon whether shipments are being added or reduced at a given point. A negative sign on the net result indicates that cost can be reduced by making the change. The total savings are, of course, limited to the least number of units available for reallocation at any negative cell on the path.

Evaluate cell A2:
Path: A2 to B2 to B1 to A1 (designated as I in Fig. 6-5)
Cost: $+14 - 10 + 12 - 10 = +6$ (cost increase) Therefore, make no change.

Evaluate cell C1:
Path: C1 to B1 to B2 to C2 (designated as II in Fig. 6-5)
Cost: $+8 - 12 + 10 - 12 = -6$ (cost savings)
Therefore, this is a potential change. Evaluate remaining empty cells to see if other changes are more profitable.

Evaluate cell A3:
Path: A3 to C3 to C2 to B2 to B1 to A1 (not shown in Fig. 6-5)
Cost: $+8 - 10 + 12 - 10 + 12 - 10 = +2$ (cost increase). Therefore, make no change.

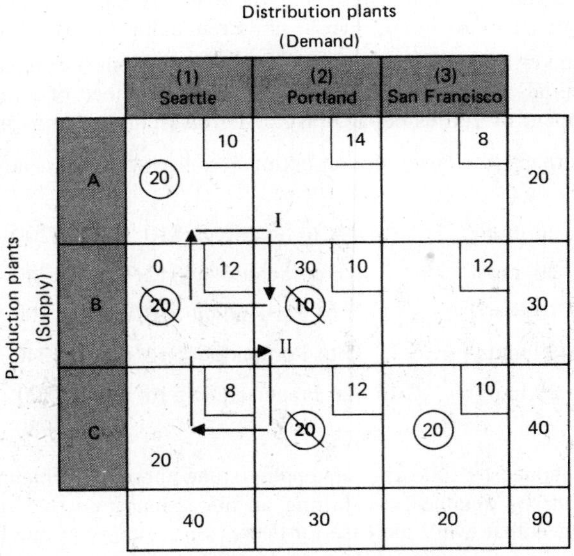

Fig. 6-5 Revision of matrix

Evaluate cell B3:

Path: B3 to C3 to C2 to B2 (not shown in Fig. 6-5)
Cost: $+12 - 10 + 12 - 10 = +4$ (cost increase). Therefore, make no change.

Cell C1 presents the best (only) opportunity for improvement. For each unit from C reallocated to Seattle and from B reallocated to Portland, a $6 savings results. Change the maximum number available in the loop (20) for a net savings of $(\$6)(20) = \$120$. (The maximum number will always be the smallest number in the cells where shipments are being reduced, that is, cells with negative coefficients.) The crossed circles with numbers above on loop II of Fig. 6-5 show that transformations have been made. Note that cells B1 and C2 have both become vacant (a degenerate situation), so a zero has been assigned to one of the vacant cells (B1) to maintain the $R + C - 1$ requirement of 5.

Because a reallocation was made, the empty cells are again evaluated for further improvement:

Cell A2: $A2-B2-B1-A1 = +6$ (no change)
Cell C2: $C2-C1-B1-B2 = +6$ (no change)
Cell A3: $A3-C3-C1-A1 = -4$ (a possibility)
Cell B3: $B3-C3-C1-B1 = -2$ (a possibility)

Cell A3 has the greatest potential for improvement. (Note that the loop evaluating cell B3 has zero units available for transfer from cell B1, so no reallocation could take place without first locating another route to B3. This would be done by relocating the zero. However, in this example cell A3 offers the best improvement, so we capitalize upon the opportunity to load cell A3.) A reallocation of 20 units to cell A3 results in the matrix shown in Fig. 6-6. Note that a zero has again been retained in one of the vacated cells (C3) to satisfy the $R + C - 1$ constraint.

Further evaluation of the cells reveals that no additional savings can be achieved. The optimal solution is shown in Fig. 6-6. The transportation cost for this arrangement is:

40 units	C to Seattle @	$8/unit = $320
30 units	B to Portland @	$10/unit = 300
20 units	A to San Francisco @	$8/unit = 160
		Total $780

Net savings over the initial allocation is $\$980 - \$780 = \$200$/week.

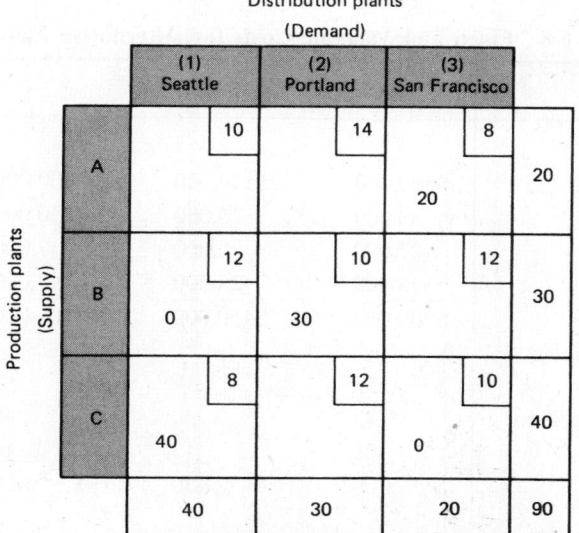

Fig. 6-6   Optimal solution

Questions and Solved Problems

6.1   Distinguish between (a) locational break-even analysis and (b) factor rating.

(a)   *Locational break-even analysis* is an economic comparison of total costs (FC + TVC) at an expected volume. The location having the lowest total cost can be selected.

(b)   *Factor rating* yields a composite score that reflects all factors relevant to a location alternative (qualitative and quantitative). From this, the preferred location can be selected.

COST-VOLUME (BREAK-EVEN) ANALYSIS OF LOCATIONS

6.2   A firm is considering four alternative locations for a new plant and has researched the costs shown in Table 6-7. The firm will finance the new plant from bonds bearing 10 percent interest. Determine the most suitable location (economically) for output volumes in the range of 50,000 to 130,000 units per year.

Table 6-7   Cost Data for Alternative Plant Sites

	A	B	C	D
Labor (per unit)	$ .55	$ 1.10	$ .80	$ .45
Plant construction cost (million $)	5.00	3.90	4.00	4.85
Materials and equipment* (per unit)	.43	.60	.40	.30
Electricity (per yr)	30,000	26,000	30,000	28,000
Water (per yr)	27,000	6,000	7,000	7,000
Transportation (per unit)	.02	.10	.10	.05
Taxes (per yr)	43,000	28,000	63,000	55,000

*This cost includes a projected depreciation expense, but no interest cost.

See Table 6-8 for cost allocation into fixed and variable portions. Note that the fixed costs include the 10 percent charge needed to retire the bonds.

Table 6-8   Fixed and Variable Costs for Alternative Plant Sites

Costs	A	B	C	D
Fixed costs (per yr)				
10% of investment	$500,000	$390,000	$400,000	$485,000
Electricity	30,000	26,000	30,000	28,000
Water	27,000	6,000	7,000	7,000
Taxes	43,000	28,000	63,000	55,000
Total	$600,000	$450,000	$500,000	$575,000
Variable costs (per unit):				
Labor	$ .55	$ 1.10	$ .80	$ .45
Materials and equipment	.43	.60	.40	.30
Transportation	.02	.10	.10	.05
Total	$ 1.00	$ 1.80	$ 1.30	$ .80
Total costs	$600,000 +	$450,000 +	$500,000 +	$575,000 +
	$1.00/unit	$1.80/unit	$1.30/unit	$ .80/unit

The points for a plant location break-even analysis chart are as follows. At zero units of output, use fixed-cost values. At an arbitrarily chosen output of 100,000 units, total costs are:

$$A = \$600,000 + 100,000(\$1.00) = \$700,000$$
$$B = \$450,000 + 100,000(\$1.80) = \$630,000$$
$$C = \$500,000 + 100,000(\$1.30) = \$630,000$$
$$D = \$575,000 + 100,000(\$ .80) = \$655,000$$

For minimum cost, use site B for a volume of 50,000 to 100,000 units; use site C for volumes of 100,000 to 130,000 units, as shown in Fig. 6-7.

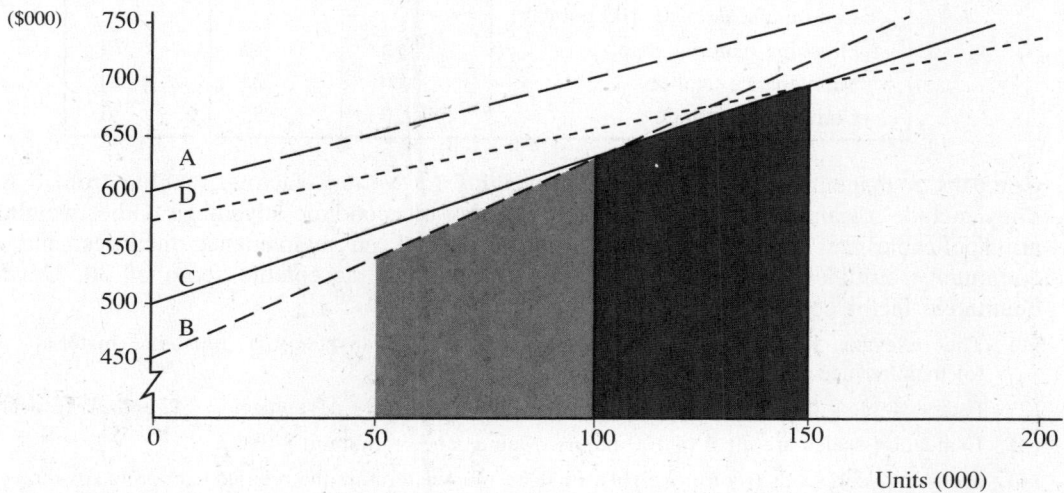

Fig. 6-7   Cost-volume chart for four plant locations

6.3   Using the data from Prob. 6.2, assume that the market research department of the firm has estimated the market volume for the product per year over the next 10 years. For volumes (in thousands) of 50, 75, 100, and 200 units, the probabilities are .2, .4, .1, and .3, respectively. What is the most suitable location on the basis of an expected-value criterion?

Set up Table 6-9 as shown to determine the expected volume, from $E(X) = \Sigma [X \cdot P(X)]$

Table 6-9

Volume X	Probability P(X)	Expected Value XP(X)
50,000	.20	10,000
75,000	.40	30,000
100,000	.10	10,000
200,000	.30	60,000
Expected demand		110,000

Select site C, which (from the chart) has the lowest total cost for a volume of 110,000 units.

FACTOR RATING SYSTEMS

6.4   The Australian Tanning Co., which has distribution plants in Sydney and Melbourne, is considering adding a third plant in either Adelaide, Bourke, or Cooktown. The company has collected the economic and noneconomic data shown in Table 6-10.

Table 6-10

Factor	Adelaide	Bourke	Cooktown
Transportation cost/day	$ 780	$ 640	$ 560
Labor cost/day	$1,200	$1,020	$1,180
Selected criteria scores (based on a scale of 0–100 points):			
Finishing material supply	35	85	70
Maintenance facilities	60	25	30
Community attitude	20	85	70

Company management has preestablished weights for various factors, ranging from 0 to 1.0. They include a standard of .2 for each $10 per day of economic advantage. Other weights that are applicable are .3 on finishing material supply, .1 on maintenance facilities, and .4 on community attitudes. Maintenance also has a minimum acceptable score of 30. Develop a qualitative factor comparison for the three locations.

(1)   The relevant factors are (*a*) relative economic advantage, (*b*) finishing material supply, (*c*) maintenance facilities, and (*d*) community attitude.

(2)   Factor weights for (*a*), (*b*), (*c*), and (*d*) are .2 per $10 daily advantage, .3, .1, and .4, respectively.

(3)   Evaluation scales are all 0 to 100 points. Maintenance minimum = 30.

(4)   Weighted scores = Σ (score) (weight). First we must determine the relative economic advantage score as shown in Table 6-11.

Table 6-11

	Adelaide	Bourke	Cooktown
Cost/day (transportation + labor)	$1,980	$1,660	$1,740
Relative economic advantage (highest cost − cost/day)	0	320	240
Economic advantage score in $10 units	0	32	24

(5)   The Bourke site (Table 6-10), with a score of 25, does not meet the maintenance minimum (or threshold) of 30. Cooktown has the highest total points (see Table 6-12) and so would be recommended on the basis of this limited analysis (even though Bourke has a lower cost structure).

Table 6-12

Factors	Adelaide	Bourke	Cooktown
Economic	0(.2) = 0	32(.2) = 6.4	24(.2) = 4.8
Material supply	35(.3) = 10.5	85(.3) = 25.5	70(.3) = 21.0
Maintenance	60(.1) = 6.0	25(.1) = 2.5	30(.1) = 3.0
Community	50(.4) = 20.0	65(.4) = 26.0	70(.4) = 28.0
Total	36.5	60.4	56.8

TRANSPORTATION LINEAR PROGRAMMING

6.5   A lighting equipment manufacturer in Italy is considering locating two warehouses capable of absorbing 80 units (total) per week from the firm's plants. If unit transportation costs are as shown ($), what is the total transportation cost for an optimal allocation? Apply the northwest-corner and stepping-stone methods to Fig. 6-8.

Warehouse site

Production plant	Venice	Split	
Rome	10	12	40
Milan	12	15	40
	30	50	80

Fig. 6-8   Initial matrix

(a)   Beginning in the NW corner of Fig. 6-8, we assign 30 units to Venice and the remaining 10 to Split. This exhausts our supply from Rome.

(b)   Going to row 2 we assign all 40 units to Split. This completes the initial allocation. All row and column totals agree, and we have $R + C - 1 = 2 + 2 - 1 = 3$ entries.

(c)   Evaluate cell Milan-Venice.

Path: Milan-Venice to Rome-Venice to Rome-Split to Milan-Split
Cost: $+ 12 - 10 + 12 - 15 = -1$ ($1 cost decrease)

Therefore, change 30 units, as shown in Fig. 6-9.

(d)   No other changes can be made to improve the allocation. Cost of the optimal solution is:

40 units from Rome to Split @ $12   = $480
30 units from Milan to Venice @ 12 = 360
10 units from Milan to Split @ 15   = 150
                          Total cost $990

Warehouse site

Production plant	Venice	Split	
Rome	(30) 10	(10) 40 12	40
Milan	30 12	(40) 10 15	40
	30	50	80

Fig. 6-9   Solution

6.6   The transportation LP cost analysis for a plant-location study covering possible sites X, Y, and Z is complete except for final evaluation of cell AY in Fig. 6-10. Complete the evaluation of cell AY, make any changes justified, and compute the optimal cost.

Demand

Supply	X	Y	Z	
A	−[5] 10	+[2]	[9]	10
B	[10]	−[6] 15	+[3] 3	18
C	+[9] 5	[8]	−[6] 7	12
	15	15	10	40

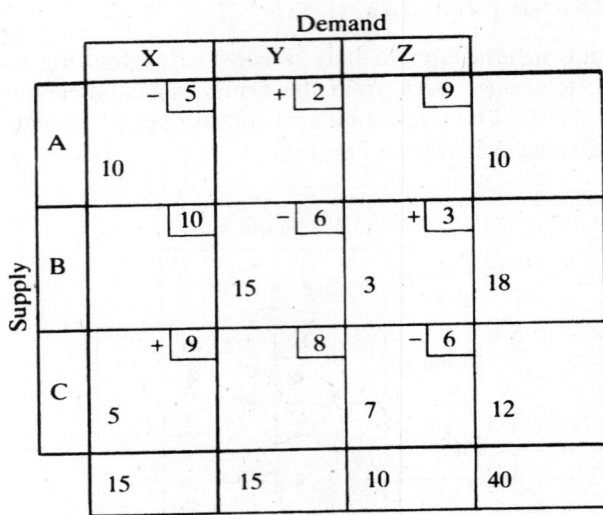

Fig. 6-10 Nonoptimal solution

Evaluate cell AY as shown in Fig. 6-11.

Path: AY to BY to BZ to CZ to CX to AX to AY

Cost: $+2-6+3-6+9-5=-3$ ($3 cost savings)

Therefore, change 7 units (the smallest entry in the negative cells).

Cost: $3(5)+12(9)+7(2)+8(6)+10(3)=\$215$

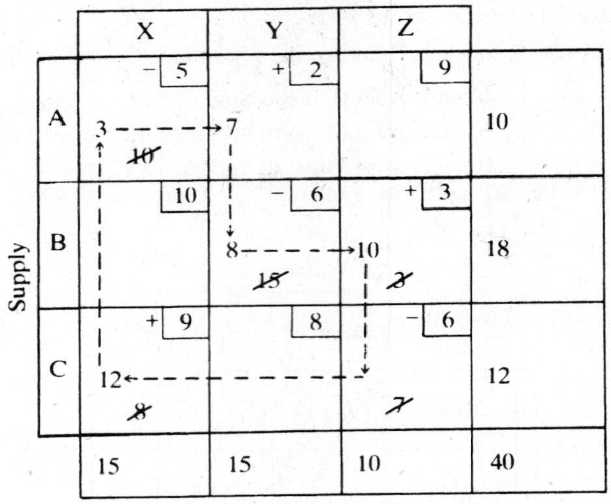

Fig. 6-11 Optimal solution

6.7 A plant-location study yielded the matrix shown in Fig. 6-12 as one iteration in a stepping-stone solution. (*a*) Evaluate the empty cells, make whatever improvements are possible, and show the optimal solution in the matrix on the right. (*b*) Compute the optimal matrix cost.

(*a*) Evaluation of all vacant cells is not shown. However, for cell D1 the costs are:

$$+4-6+6-16+8-10=-14 \text{ (a cost savings)}$$

Therefore, change = 40 units. The solution is shown in Fig. 6-13.

(b)　　　　　　　Cost = 30(6) + 80(4) + 40(4) + 80(6) + 20(16) + 100(8) = $2,260

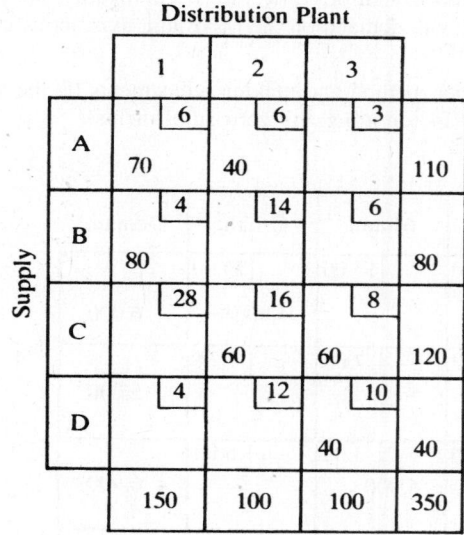

Fig. 6-12　Nonoptimal solution　　　　　　　　**Fig. 6-13**　Optimal solution

6.8　Plastic Cabinet Supply Company (PCS) is a wholly owned subsidiary of an international conglomerate firm that has major interests in the housing industry. PCS has cabinet plants located in Boston, Seattle, and Miami. The plants produce prefabricated housing components that are delivered to other company assembly plants in Chicago, Denver, and Nashville. Demand has grown to the point where PCS can justify construction of another plant. The immediate problem is determining a location that will minimize production and transportation costs to the existing assembly plants. In order to be close to raw material supply and to service other potential markets, the alternative plant locations have been narrowed down to Omaha and Phoenix. Cost, demand, and production data on the various alternatives are shown in Tables 6-13a and b.

Table 6-13a

	PCS Production Data			Assemblies Demand	
Plants	Units per Month	Cost per Unit		Plant	Units per Month
Boston (B)	2,000	$7.00		Chicago (C)	6,000
Seattle (S)	6,000	7.08		Denver (D)	5,000
Miami (M)	5,000	6.90		Nashville (N)	6,000
Omaha (O)	4,000	6.90 (anticipated)			
Phoenix (P)	4,000	6.20 (anticipated)			

Table 6-13b　Transportation Cost, $/unit

	From				
To	Boston	Seattle	Miami	Omaha	Phoenix
Chicago	$5.00	$7.00	$5.00	$4.00	$6.00
Denver	6.00	4.00	7.00	3.00	4.50
Nashville	5.50	7.00	3.00	5.00	5.00

Which of the two plant locations (Omaha or Phoenix) is more desirable from an economic standpoint?

It makes no difference whether supply is on the horizontal or vertical axis. The major concern is that row and column totals agree. Since the data are given with demand on the horizontal axis, let us use it that way.

(a) *Using Omaha.* Allocating via the northwest-corner method and making adjustments by the stepping-stone method, we arrive at the matrix in Fig. 6-14 (omitting any zero adjustments).

	Boston	Seattle	Miami	Omaha	Demand
Chicago	$5.00 2,000	$7.00	$5.00	$4.00 4,000	6,000
Denver	6.00	4.00 5,000	7.00	3.00	5,000
Nashville	5.50	7.00 1,000	3.00 5,000	5.00	6,000
Supply	2,000	6,000	5,000	4,000	17,000

Fig. 6-14

Transportation cost calculation:

$$2,000 \times 5.00 = \$10,000$$
$$4,000 \times 4.00 = 16,000$$
$$5,000 \times 4.00 = 20,000$$
$$1,000 \times 7.00 = 7,000$$
$$5,000 \times 3.00 = \underline{15,000}$$
$$\$68,000 \longrightarrow \$68,000$$

Add production costs (Omaha): $6.90/unit $\times$ 4,000 = $\underline{27,600}$

$$\$95,600$$

(b) *Using Phoenix.* Figure 6-15 gives the information for Phoenix.

	Boston	Seattle	Miami	Phoenix	Demand
Chicago	$5.00 2,000	$7.00 1,000	$5.00	$6.00 3,000	6,000
Denver	6.00	4.00 5,000	7.00	4.50	5,000
Nashville	5.50	7.00	3.00 5,000	5.00 1,000	6,000
Supply	2,000	6,000	5,000	4,000	17,000

Fig. 6-15

Cost calculation:

$$2,000 \times 5.00 = \$10,000$$
$$1,000 \times 7.00 = 7,000$$
$$3,000 \times 6.00 = 18,000$$
$$5,000 \times 4.00 = 20,000$$
$$5,000 \times 3.00 = 15,000$$
$$1,000 \times 5.00 = \underline{5,000}$$
$$\$75,000 \longrightarrow \$75,000$$

Add production costs (Phoenix): $6.20/unit $\times$ 4,000 = $\underline{24,800}$

$$\$99,800$$

Omaha is the best choice since it shows the least cost per month.

DUMMY VARIABLE, DEGENERATE SOLUTION, AND REQUIRED SHIFTING ZEROS

6.9   A transportation LP problem has the rim requirements (supply and demand) and cost coefficients shown in Fig. 6-16. Use the northwest-corner and stepping-stone methods to obtain an optimal solution.

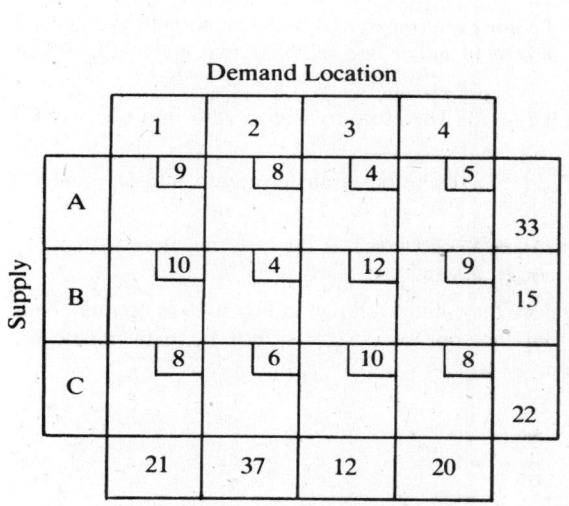

Fig. 6-16

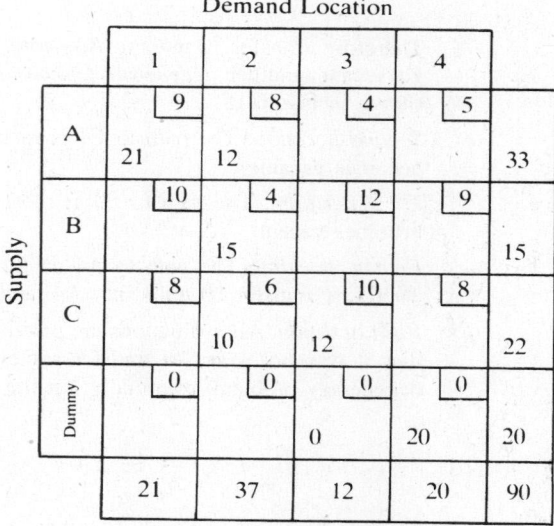

Fig. 6-17   Initial allocation

(a)   First note that the demand exceeds the supply, so add a dummy supply and assign zero transportation costs to it as shown in Fig. 6-17.

(b)   Allocate the units via northwest corner beginning with 21 units in cell A1, 12 in A2, etc.

(c)   Check the $R + C - 1$ constraint; we should have $4 + 4 - 1 = 7$ entries. Because we have only 6 (a *degenerate* situation), assign a zero to one of the cells that could normally have an allocation under the northwest-corner method, such as D3.

(d)   *First iteration*

B1: $+10 - 9 + 8 - 4 = +5$	B3: $+12 - 10 + 6 - 4 = +4$
C1: $+8 - 9 + 8 - 6 = +1$	A4: $+5 - 0 + 0 - 10 + 6 - 8 = -7$
D1: $+0 - 9 + 8 - 6 + 10 - 0 = +3$	B4: $+9 - 0 + 0 - 10 + 6 - 4 = +1$
D2: $+0 - 6 + 10 - 0 = +4$	C4: $+8 - 0 + 0 - 10 = -2$
A3: $+4 - 10 + 6 - 8 = -8$ (most negative)	

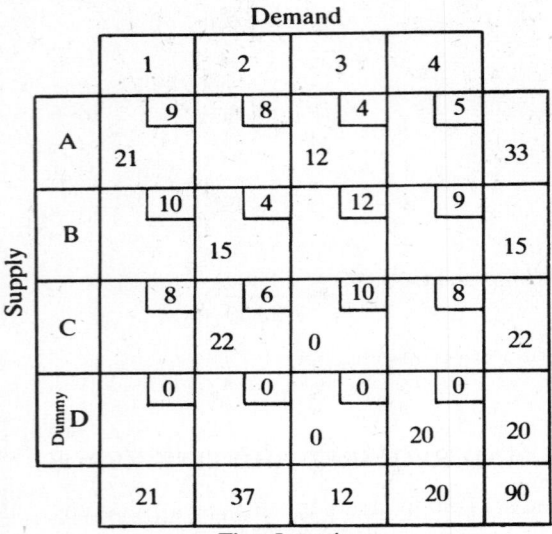

Fig. 6-18 — First Iteration

Supply \ Demand	1	2	3	4	Supply
A	9 — 21	8	4 — 12	5	33
B	10	4 — 15	12	9	15
C	8	6 — 22	10 — 0	8	22
Dummy D	0	0	0 — 0	0 — 20	20
	21	37	12	20	90

Fig. 6-19 — Fifth Iteration

Supply \ Demand	1	2	3	4	Supply
A	9 — 1	8	4 — 12	5 — 20	33
B	10	4 — 15	12	9	15
C	8 — 0	6 — 22	10	8	22
Dummy D	0 — 20	0	0	0	20
	21	37	12	20	90

Therefore, transfer 12 units to A3. *Note:* When 12 units are transferred to A3, both cells A2 and C3 go vacant (another *degenerate* situation). Add a zero to either one of those two cells, e.g., C3 as shown in Fig. 6-18.

(e)   *Second iteration.* The path to C1 is most negative (-7). Therefore transfer a zero into C1, and C3 becomes vacant.

(f)   *Third iteration.* The path to D1 is most negative (-5). Therefore, transfer a zero into D1, and D3 becomes vacant.

(g)   *Fourth iteration.* The path to A4 on the loop $A4 - D4 - D1 - A1$ is the only negative path (-4). Therefore transfer 20 units into A4, and D4 becomes vacant.

(h)   *Fifth iteration.* All evaluations are positive; therefore the solution shown in Fig. 6-19 is optimal. Note that it was necessary to transfer some zeros and that one of the zeros (put in to overcome the degeneracy problem) eventually left the solution.

Supplementary Questions and Problems

6.10   A medical products firm considering locating in one of four cities has estimated costs as shown in Table 6-14.

Table 6-14

	Lille	Bologna	Córdoba	Dublin
Fixed costs $/yr	80,000	70,000	60,000	92,000
Variable costs $/unit	32	35	40	38

On the basis of a locational break-even analysis, at what volume of production does the Bologna location becomes less costly than the Córdoba location?

Ans. By setting the total cost equation for Bologna ($70,000 \times 35X$) equal to the cost equation for Córdoba ($60,000 + 40X$) and solving for X, we obtain a volume of 2,000 units. Córdoba has a lower fixed cost, so it would be least expensive for low volumes.

6.11 Tractorboy Products Company is evaluating three different cities for a new plant designed to produce lawn mowers that will wholesale for $145 each. The economic portion of a plant-location study is shown in Table 6-15.

Table 6-15

Cost Data	Panama	Bogotá	Caracas
Fixed cost/yr	$300,000	$200,000	$75,000
Variable costs/unit	30	45	70

Demand is uncertain, but probabilities of .1, .3, and .6 have been estimated for market volumes of 4,500, 5,500, and 6,500, respectively.

(*a*) On the basis of maximizing an economic expected value, graph the plant-location cost curve using appropriate scales. (*b*) Which city should be selected on the basis of the given volume estimate? (Use your graph.) (*c*) What is the break-even volume for the city selected?

Ans.   (*a*) Volume range 0–10,000; Dollar range $0–$800,000 (*b*) Bogotá (*c*) 2,000 units

6.12 Coombes Container Co. is considering three potential locations for a new aluminum can plant, and management has assigned the scores shown in Table 6-16 to the relevant factors on a 0 to 100 basis (100 is best).

Table 6-16   Score of Relevant Factor for Plant Location

	Hong Kong	Manila	Honolulu
Material supply	50	90	80
Labor cost	90	80	40
Regulations	100	60	30
Distribution	30	80	70

The relevant factors have been assigned the following weights: material supply = .3, labor cost = .3, regulations = .2, distribution = .2. Using a factor-rating analysis, which location would be preferred?

Ans.   Manila (with 79 points)

6.13 A large winery has markets in six metropolitan areas in France and is considering locating a warehouse in the area. The distributor has overlaid a grid on a map of the area and has collected data on the number of truckload containers shipped to each market area per month. The grid volumes shipped to the respective grid coordinates are shown in Table 6-17. Use the center-of-gravity method to identify the metropolitan area that should serve as a starting point for possible location of the warehouse.

Table 6-17   Volume and Grid Coordinates for Market Areas

	Area A	Area B	Area C	Area D	Area E	Area F
X-coordinate	2	4	8	10	12	14
Y-coordinate	4	9	7	4	7	1
Vol (# containers)	120	650	400	90	850	200

Ans. $X_c = 19,940/2310 = 8.6$, $Y_c = 15,640/2310 = 6.8$. Locate warehouse in area C.

6.14 A firm producing appliances at plants #1 and #2 is considering locating distribution centers in Kent and London. If transportation costs are as shown (in £) in Fig. 6-20, what is the total transportation cost for an optimal allocation?   *Ans.* £6,360

	SF	NY	
#1	40	50	60
#2	50	62	60
	30	90	

Fig. 6-20

6.15 A materials manager is considering locating warehouses in Knoxville and Jersey City capable of absorbing 30 and 60 units per day, respectively, from the firm's two plants, each of which can produce 45 units per day. Unit transportation costs ($) are shown in Table 6-18.

Table 6-18

	To Knoxville	To Jersey City
From Plant #1	$ 9	$11
From Plant #2	11	14

(*a*) Show the northwest-corner allocation in an initial matrix. (*b*) Show the optimal allocation in a final matrix. (*c*) Compute the optimal transportation cost.
Ans.   (*a*) #1K = 30, #1JC = 15, #2JC = 45 (*b*) Allocations are #1JC = 45, #2K = 30, #2JC = 15
(*c*) $1035/day

6.16 A building-materials firm with production plants in Recife, Salvador, and Laguna ships freight cartons to three distribution centers with costs per carton as shown in Table 6-19.

Table 6-19   Cost per Carton ($00) to Ship to Three Centers

From:	Center 1 (30-carton monthly demand)	Center 2 (30-carton monthly demand)	Center 3 (35-carton monthly demand)
Recife (25-carton monthly productive capacity)	$3	$3	$2
Salvador (40-carton monthly productive capacity)	4	2	3
Laguna (30-carton monthly productive capacity)	3	2	3

Use the northwest-corner and stepping-stone methods to determine the optimal allocation to minimize costs. Find the optimal cost. *Ans.* $23,000

6.17 Comfort Zone Furniture has plants in Boston, Dallas, and Seattle, which ship to four demand locations, with transportation costs as shown in Fig. 6-21. Use the northwest-corner and stepping-stone methods to determine the optimal transportation cost. Show your initial and final solutions.
Ans. The northwest-corner allocation is shown in Fig. 6-21. The final solution is not shown in Fig. 6-22, but has an optimal cost of $610.

Initial Matrix

Demand Location

Supply Plants	1	2	3	4	
Boston	[3] 20	[5] 30	[7]	[4]	50
Dallas	[6]	[8] 30	[5] 20	[2]	50
Seattle	[1]	[9]	[7] 10	[3] 40	50
	20	60	30	40	150

Fig. 6-21

Final Matrix

Demand Location

Supply Plants	1	2	3	4	
Boston	[3]	[5]	[7]	[4]	50
Dallas	[6]	[8]	[5]	[2]	50
Seattle	[1]	[9]	[7]	[3]	50
	20	60	30	40	150

Fig. 6-22

6.18 A large copper producer has refineries in Magna, Utah; Yuma, Arizona; and Grants, New Mexico—all of which receive ore from mines identified as MX-1, MX-2, MX-3, and MX-4—located in the Four Corners Area. The mine supply and mill capacities (units per day) and shipping cost (dollars per unit-load) data are shown in Table 6-20.

The vice-president of operations has asked you to analyze the transportation costs and determine an optimal distribution. *Ans.* More than one solution may be optimal at a transportation cost = $7,400.

Table 6-20

	Refineries			
	Magna	Yuma	Grants	Supply
MX-1	3	5	5	400
MX-2	5	7	8	500
MX-3	2	9	5	200
MX-4	10	7	3	700
Capacity	500	500	800	1,800 units/day

6.19 Suppose that in Prob. 6.18 the variable production costs (in dollars per unit-load) for mines MX-1, MX-2, MX-3, and MX-4 are $2, $1, $3, and $1, respectively. What is the optimal distribution, taking production as well as distribution costs into account?
Ans. More than one solution may be optimal at a production and transportation cost equal to $10,000.

Chapter 7

Layout of Facilities

Line Balancing

DESIGN AND SYSTEM CAPACITY

Design and capacity decisions link the location and layout considerations. For operations management purposes, capacities are often stated in physical units, customers served, service times, or work-center hours rather than in dollar volume of sales.

The *design capacity* of a facility is the engineered rate of output of standardized products under normal operating conditions. However, economic, competitive, and (uncontrollable) market conditions often cause output to vary from the design. *System capacity* is the maximum output of a specific product or product mix that the system of workers and equipment is capable of producing as an integrated whole. Figure 7-1 illustrates the relationship between design capacity, system capacity, and actual output. The system efficiency (SE) is a measure of the actual output of goods or services as a percentage of system capacity.

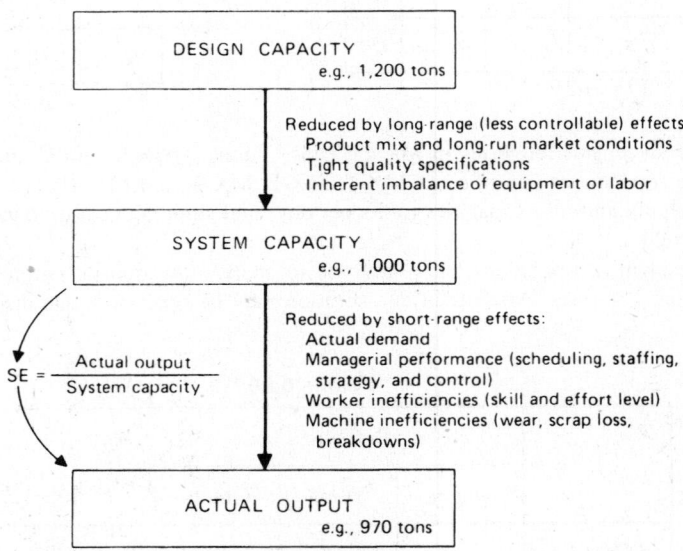

Fig. 7-1 Relationship between capacities and output

Example 7.1 A large title insurance company processes all titles sequentially through four centers (A, B, C, D), which handle the search and recording activities. The individual work-center capacities and actual average output in titles processed per day are as shown. Find (*a*) the system capacity and (*b*) the system efficiency.

$$A \rightarrow B \rightarrow C \rightarrow D \rightarrow \text{Actual Output}$$

$$\boxed{24} \rightarrow \boxed{30} \rightarrow \boxed{22} \rightarrow \boxed{40} \rightarrow \boxed{18 \text{ titles/day}}$$

(*a*)   System capacity = capacity of most limited component in the line = 22 titles/day

(*b*)
$$\text{SE} = \frac{\text{Actual Output}}{\text{System Capacity}} = \frac{18}{22} = .82 = 82\%$$

104

COMPUTATION OF EQUIPMENT REQUIREMENTS

If the actual output is specified (e.g., by design), the amount or size of equipment required to deliver that output can often best be determined by working backward to allow for system losses and inefficiencies.

Example 7.2 An automobile equipment supplier wishes to install a sufficient number of ovens to produce 400,000 good castings per year. The baking operation takes 2.0 minutes per casting, but the oven output is typically about 6 percent defective. How many ovens will be required if each one is available for 1,800 hours (of capacity) per year?

$$\text{Number ovens required} = \frac{\text{Required systems capacity}}{\text{Individual oven capacity}}$$

$$\text{where:} \quad \text{Required system capacity} = \frac{\text{Actual (good) Output}}{\text{SE}}$$

$$= \frac{400,000}{.94} = 425,532 \text{ units/yr}$$

$$= \frac{425,532 \text{ units/yr}}{1,800 \text{ hrs/yr}} = 236 \text{ units/hr}$$

$$\text{Individual oven capacity} = \frac{60 \text{ min/hr}}{2.0 \text{ oven-min/unit}} = 30 \text{ units/oven-hr}$$

$$\text{Number ovens required} = \frac{236 \text{ units/hr}}{30 \text{ units/oven-hr}} = 7.9 \text{ (8) ovens}$$

LAYOUT DETERMINANTS AND OBJECTIVES

Facility layouts are the arrangements of production, support, customer service, and other work centers that support the operational strategies of an organization. Layouts affect materials handling, capital equipment utilization, inventory storage levels, worker productivity, and even group communications and employee morale.

Question: What determines the type of layout used by an organization?

Determinants of Facility Layout

1. *Type of product* (i.e., whether a good or service, product design, size/shape, quality)
2. *Type of process* (i.e., technology employed, materials/services used, sequencing)
3. *Volume of production* (i.e., high volume/continuous flow versus low volume/intermittent flow)

Question: What types of concerns are addressed in a layout design?

Designers deal with issues like (*a*) *relative importance* (what work centers should be included in the facility), (*b*) *size* (how much space is needed in each work area), (*c*) *proximity* (where each work area should be located), (*d*) *effectiveness* (how each work area should be arranged to achieve maximum efficiency), and (*e*) *support* (what electronic, storage, and distribution facilities are needed).

Question: Identify some of the key differences in the objectives (i.e., performance criteria) for (*a*) manufacturing, (*b*) warehouse, (*c*) retail (service), and (*d*) office layouts.

Common goals of all layouts usually include cost effectiveness, high productivity, efficiency, and flexibility. However, the actual layout of a productive facility can differ depending upon the end use of the facility.

(*a*) *Manufacturing:* minimize the cost of transporting, processing, and storing materials.

(b) *Warehouse:* maximize the efficiency of receiving, storing, shipping, and tracking materials.

(c) *Retail (service):* enhance the appeal of the product and comfort/convenience of the customer.

(d) *Office:* promote worker effectiveness by fostering communication and efficient flow of work.

Question: How do today's manufacturing layouts differ from those of the past?

Manufacturing facilities today, compared with those of the past, are typically:

(1) *Smaller*, with open design, fewer partitions, and environmentally attractive facilities

(2) More product focused, but with *considerable flexibility* within the product line

(3) More *highly automated*, with a higher investment in machines and *fewer workers*

(4) More *easily adaptable* to new configurations, less space reserved for inventories

(5) Equipped with *computer-guided handling equipment*, automated storage systems

(6) Designed to *facilitate communication and interaction* for team and group efforts

TYPES OF LAYOUTS

Question: What are the basic types of facility layouts?

> 1. Process (functional) layouts
> 2. Product (line) layouts
> 3. Fixed-position layouts
> 4. Manufacturing cells
>   (plus a number of other hybrid arrangements)

Question: Describe a process layout, and provide some examples

Process layouts are arrangements that group together the workers and equipment that perform similar functions, such as painting, testing, or data processing. They lend themselves to low volumes of customized jobs (i.e., are job shops) using skilled employees and a variety of relatively general-purpose equipment. Work flow is intermittent and guided by individual work orders. Machine shops, hospitals, and tax consultants are examples. Figure 7-2 illustrates a process layout.

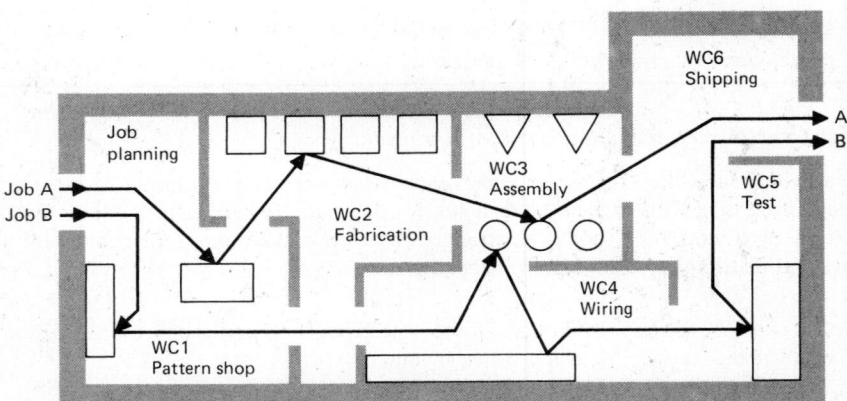

Fig. 7-2 Process layout for time and temperature sign production

Question: What are the major advantages and disadvantages of process layouts?

Advantages and Disadvantages of Process (Functional) Layouts

Advantages	Disadvantages
1. Geared to produce customized work	1. Higher materials-handling costs per job
2. Flexible equipment and personnel	2. Low volumes and equipment utilization
3. Smaller investment in equipment	3. Higher cost in skilled labor
4. Less vulnerability from breakdowns	4. More complex production control
5. Enhanced job satisfaction (via more diversity and challenge)	5. Higher supervision cost per employee

Question: Describe a product layout, and provide some examples.

Product (or line) layouts are arrangements that group the workers and equipment according to the sequence of operations performed on the product or customer. They lend themselves to the use of (assembly-line) conveyors to produce large volumes of similar models where work flow is guided by standardized instructions. However, modern systems can be used to produce customized products at assembly-line volumes by using flexible and computer-assisted manufacturing (CAM) systems. Automobiles and computers are manufactured on line layouts. Figure 7-3 shows a typical product layout.

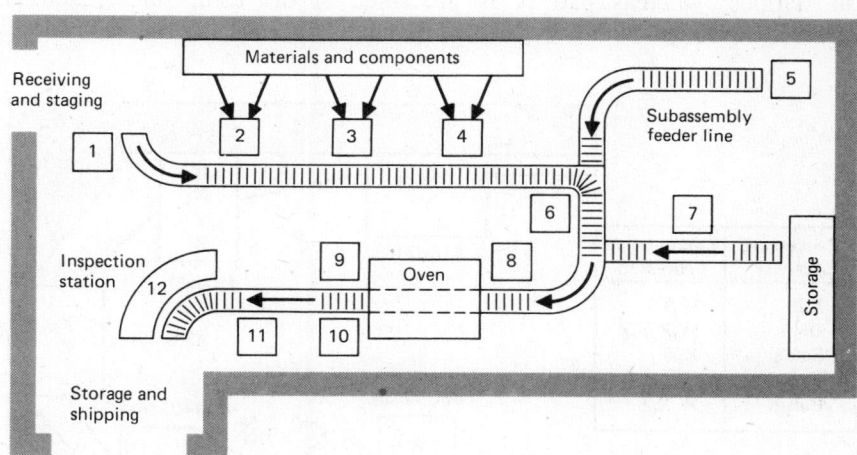

Fig. 7-3   Product layout for electronic toy manufacturing

Traditional product layouts are often highly structured (i.e., hard automation) leaving workers with mundane and repetitive tasks. *Flexible production systems* relegate the more repetitive tasks to numerically controlled machines and robots that can produce in large lot sizes or in lot sizes as small as one. This often results in a need for fewer (but more highly skilled) workers who have more challenging planning and control tasks.

Question: What are the major advantages and disadvantages of product (line) layouts?

Advantages and Disadvantages of Product (Line) Layouts

Advantages	Disadvantages
1. High utilization of people and equipment	1. Inflexible system (unless designed for flexibility)
2. Low materials-handling cost	2. High-cost specialized equipment
3. Low-cost unskilled labor	3. Interdependent operations
4. Less work-in-process inventory	4. Dull, monotonous jobs (unless products are customized)
5. Simplified production control	

Question: Describe a fixed-position layout, and provide some examples.

Fixed-position layouts are arrangements where labor, materials, and equipment are brought to the job site. They apply to construction, farming, mining, and other activities that must be completed in place. Homebuilding and relay station construction are other examples.

Question: Identify some advantages and disadvantages of fixed-position layouts.

Fixed-position layouts minimize the materials-handling cost of the end product and enable managers to take advantage of project management methods. However, the cost of attracting skilled personnel to the job site may be high, and equipment is not always fully utilized.

Question: What is a manufacturing cell?

A manufacturing cell is a close grouping of perhaps 3 to 10 machines (or more) that perform a sequence of operations on a part or on a group of similar parts. The cells are often "islands" of product-line-type automation in the midst of a larger process (job shop) type of layout. They typically constitute only a small percentage of total production activity and employ relatively few workers.

Figure 7-4 illustrates a U-shaped cellular layout where all components (W, X, Y, Z) follow the same path, but processing is selective depending upon the need of the part. Part W is processed at the mill, drill, and polish stations; whereas part X is processed at the drill, solder, and polish stations. Nevertheless, all parts travel the same route.

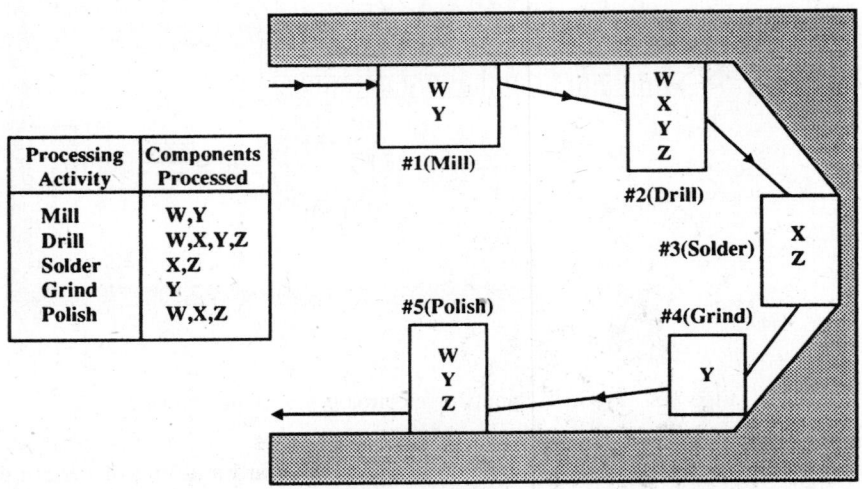

Fig. 7-4   Manufacturing cell layout

Question: Explain the terms (*a*) group technology and (*b*) flexible manufacturing systems (FMS), and tell how these concepts are related to manufacturing cell layouts.

(*a*) *Group technology* is the process used to identify and compare the design and manufacturing characteristics of parts in order to group them into families with similar characteristics. The objective is to find a generally common production sequence. If possible, manufacturing cells can then be designed and assembled to process an entire family of parts.

(*b*) *Flexible Manufacturing Systems* (FMS) are more fully automated and interconnected manufacturing cells. FMSs typically have quite sophisticated computer control that enables them to download individual instructions for specific components to specific cells and to rapidly adjust to the processing needs of different products.

Question: What are some of the advantages and disadvantages of using manufacturing cells?

Manufacturing cells can speed up production, improve quality, reduce materials-handling costs, decrease the need for inventories, and facilitate quick changeovers from one product to another. On the other hand, they usually require a highly trained work force and a large capital outlay, so the demand pattern must be stable enough to give the firm a satisfactory return on its investment.

CIM Systems.   When the raw material supply and the distribution patterns are also integrated (via computer) into the computer-controlled manufacturing process, the system is sometimes referred to as a *Computer Integrated Manufacturing* (*CIM*) system. CIM systems (e.g., Johnson Matthey metals processing facility in Washington) constitute some of the most fully automated systems currently in operation.

PROCESS (FUNCTIONAL) LAYOUT METHODS

The objective of many process layout methods is to locate work centers that have high interaction near to each other, resulting in a minimum flow of material (or personnel) to nonadjacent work centers. Some methods focus on the amount of material moved and the distance; others utilize a *nearness* criteria. Four methods of layout design are (1) the simple graphic approach, (2) operations sequence analysis, (3) load-distance analysis, and (4) systematic layout planning.

A *simple graphic approach* uses trial and error to minimize nonadjacent flows by centrally locating the active departments. A travel chart is first developed to show the number of moves made between departments and to identify the active departments. Then a trial solution is developed using circles to depict work centers and connecting lines to represent the loads transported per time period. Departments next to each other or diagonally across from each other are regarded as adjacent.

Example 7.3   Valley Electronics has a facility with six production areas as shown in Fig. 7-5 below. It proposes to locate six departments (A, B, C, D, E, F), which have the number of moves per day between departments as shown in Table 7-1. Develop a layout that minimizes the nonadjacent flows.

Table 7-1   Travel Chart

			Number of Moves to			
		A	B	C	D	E	F
	A	—	5	10	—	3	2
	B	—	—	—	12	—	—
From	C	10	4	—	8	—	—
	D	—	—	16	—	—	—
	E	—	—	7	—	—	—
	F	—	—	8	—	—	—

1	2	3
4	5	6

Fig. 7-5   Facility outline

First, determine which departments have the most frequent links with other departments. This can be done by totaling the *number of entries* in each row and column. Thus, A has four row entries (B, C, E, and F) and one column entry, for a total of five links.

Department	A	B	C	D	E	F
Number of links	5	3	7	3	2	2

Second, try to locate the most active departments in central positions. Thus, we place departments A and C in locations 2 and 5.

Third, use trial and error to locate the other departments so that nonadjacent flows will be minimized.

Fourth, if all nonadjacent flows are eliminated, the solution is complete (as shown in Fig. 7-6). If nonadjacent flows still exist, try to minimize the *number of units* flowing to nonadjacent areas. In this instance, you would weight the number of flows by the number of distance units, as discussed next.

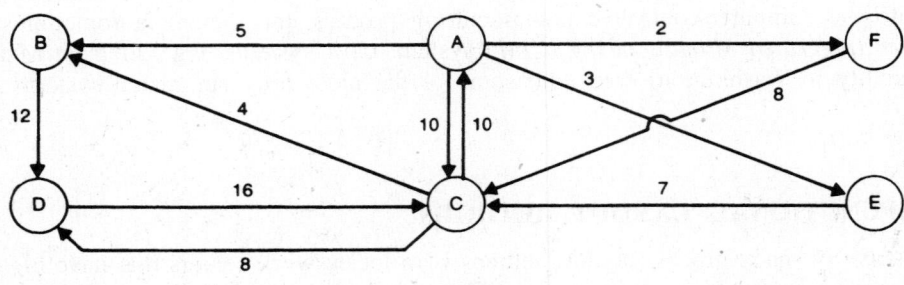

Fig. 7-6

Operations sequence analysis is a refinement over the simple graphic approach by using a weighted load-distance measure. Adjacent departments are assigned a distance factor of 1, and others take on successively higher integer values, depending upon how many rows or columns they are from each other. Departments are then shifted in an effort to minimize the sum of the load times the distance for the entire matrix. The selection of which departments to shift is done by visual inspection.

Load-distance analysis evaluates alternative layouts on the basis of the sum of actual distance (feet) times the load (units) for each alternative. A variation of this is to compute the materials-handling cost directly by multiplying the number of loads by the materials-handling cost per load. The layout with the lowest load times distance total or load times cost total is the best choice. Costs are usually a linear function of distance, unless pickup and unload costs are considered separately.

Example 7.4 A facility that will be used to produce a single product has three departments (A, B, C) that must be housed in the configuration shown in Fig. 7-7 (on left). The interdepartmental workload flows and travel distances between work centers are given in Table 7-2. In addition, two trial-and-error optional layouts are shown. Evaluate the two layouts on a load-distance basis, and identify the preferred layout. Assume that the cost to transport this product is $1 per load-foot.

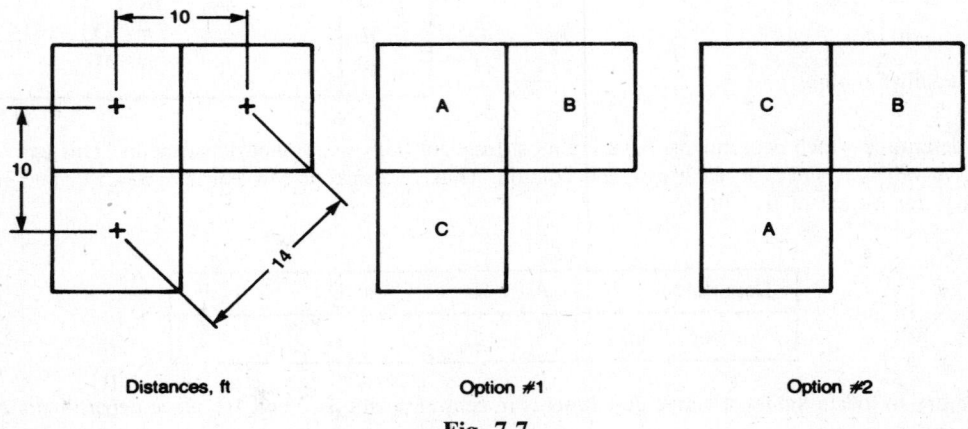

Fig. 7-7

Table 7-2   Interdepartmental Workload per Week

		To: A	B	C
From:	A	—	30	25
	B	20	—	40
	C	15	50	—

No. Loads per Week (both directions)	Option #1 (load) (distance)	Option #2 (load) (distance)
A to B and B to A = 30 + 20 = 50	(50)(10) = 500	(50)(14) = 700
A to C and C to A = 25 + 15 = 40	(40)(10) = 400	(40)(10) = 400
B to C and C to B = 40 + 50 = 90	(90)(14) = 1,260	(90)(10) = 900
	Total 2,160	= 2,000

At \$1 per load-foot, option #2 would be preferred at a total cost of \$2,000. However, 3 factorial (!) or $3 \cdot 2 \cdot 1 = 6$ options are possible, and a different arrangement may be less costly.

Systematic layout planning is a generalized approach to layout, developed by Richard Muther, that utilizes a grid matrix to display ratings of the relative importance of the distance between departments. The importance ratings are indicated by code letters (a, e, i, o, u, and x) in the matrix and range from absolutely necessary (a) to undesirable (x). A reason code (usually a number) can also be assigned. For example, reason 1 might be the use of common personnel, 2 might be noise isolation, and 3 might be safety. Figure 7-8 illustrates the systematic layout planning (SLP) approach as applied to the sign production (process) layout of Fig. 7-2.

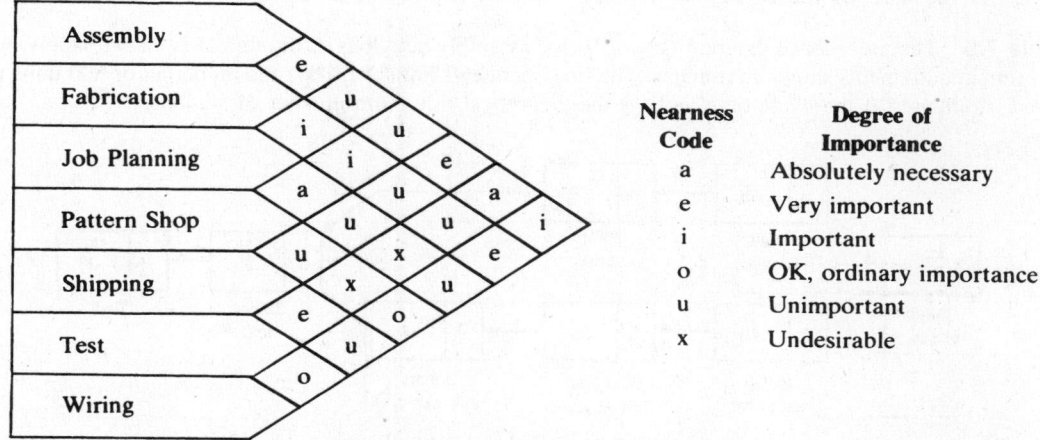

Fig. 7-8   Nearness codes for process layout of Fig. 7-2

Several *computerized approaches* are available for developing and analyzing process layouts. The analytical (software) packages are primarily heuristic, step-by-step (iterative) methods. The ALDEP (automated layout design programs) and CORELAP (computerized relationship layout planning) programs attempt to maximize a nearness rating within the facility dimension constraints. The CRAFT (computerized relative allocation of facilities technique) program attempts to minimize materials-handling costs by calculating costs, exchanging departments, and calculating more costs until a good solution is obtained. None of the methods guarantees optimality.

LINE BALANCING IN PRODUCT LAYOUTS

Line balancing is the apportionment of sequential work activities into workstations to gain a high utilization of labor and equipment and therefore minimize idle time. Compatible work activities are combined into approximately equal time groupings that do not violate precedence relationships. The length of work time that a component is available at each workstation is the cycle time, CT. CT is also the time interval at which completed products leave the line.

$$CT = \frac{\text{Available Time/period}}{\text{Output Desired/period}} = \frac{AT}{\text{Output}} \qquad (7.1)$$

Example 7.5 A production line operating 450 minutes per day is to have an output of 250 units per day. Find the cycle time.

$$CT = \frac{AT}{\text{Output}} = \frac{450 \text{ min/day}}{250 \text{ units/day}} = 1.8 \text{ min/unit}$$

If the time required at any station exceeds that which is available to one worker, or machine, additional resources may have to be added to that station.

The theoretical (ideal) minimum number of stations (or workers at one per station) needed on a line is the product of the total work time needed for each unit produced multiplied by the desired number of units per period, and divided by the available time per period.

$$\begin{array}{l}\text{Theoretical minimum}\\\text{number of stations}\end{array} = \frac{(\text{work time/unit})(\text{output units/period})}{\text{available time/period}} = \frac{\Sigma t}{CT} \qquad (7.2)$$

where Σt is the sum of the actual work time required to complete one unit.

Example 7.6 The precedence diagram (Fig. 7-9) for assembly activities A through H is shown below, with the element time requirements shown in minutes. The line operates 7 hours per day, and an output of 600 units per day is desired. Compute (*a*) the cycle time and (*b*) the theoretical minimum number of stations.

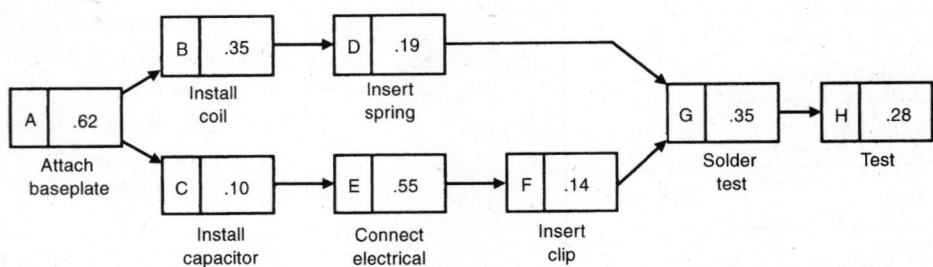

Fig. 7-9 Precedence diagram

$$CT = \frac{\text{Available Time/period}}{\text{Output Desired/period}} = \frac{(7 \text{ hrs/day})(60 \text{ min/hr})}{600 \text{ units/day}} = \frac{420}{600} = .70 \text{ min/unit}$$

$$\text{Theoretical minimum} = \frac{\Sigma t}{CT}$$

where $\Sigma t = .62 + .35 + .10 + .19 + .55 + .14 + .35 + .28 = 2.58$ workstation min/unit

$$\therefore \quad \text{Theoretical minimum} = \frac{2.58 \text{ workstation min/unit}}{70 \text{ min/unit}} = 3.69 \text{ stations}$$

The procedure for analyzing line balancing problems involves (*a*) determining the number of stations and time available at each station, (*b*) grouping the individual tasks into amounts of work at each station, and (*c*) evaluating the efficiency of the grouping or idle time (IT). An efficient balance will minimize the amount of idle time. The balance efficiency (Eff_B) and idle time (IT) can be computed as:

$$\text{Eff}_B = \frac{\text{output of task times}}{\text{input of station times}} = \frac{\Sigma t}{\text{CT}(n)} \qquad (7.3)$$

$$\text{IT} = \text{CT}(n) - \Sigma t \qquad (7.4)$$

where CT is the cycle time per station and *n* is the number of stations. The balance efficiency is also equal to the theoretical minimum number of stations (or workers) divided by the actual number.

The grouping of tasks is done heuristically with the aid of a precedence diagram. Starting with the most upstream activity, designate a work zone on the precedence diagram and move succeeding activities into preceding zones (i.e., to the left) until the time is as fully used as possible. Precedence relationships must be observed, and the available time cannot be exceeded. Two decision rules (of many) for selecting potential activities to move into a work center are:

(1)   Select the succeeding activity that has the *longest work time*. (Then continue to move activities with shorter times into the station to use up the available time.)

(2)   Select the succeeding activity that has the *most successors*.

Example 7.7 (*longest work-time rule*)   Using the data and precedence diagram from Example 7.6, (*a*) group the assembly-line tasks into an appropriate number of workstations using the longest work-time rule, and (*b*) compute the balance efficiency and (*c*) the idle time.   (*Note:* CT = .70 minute, Σt = 2.58 minutes.)

(*a*)   The CT of .70 means that .70 minutes is available at each workstation. Activity A consumes .62 of the .70 minutes available at the first station, but neither the next largest downstream activity (B at .35 minutes) nor the smaller one (C at .10 minutes) can be shifted to station 1 without exceeding the .70 minutes. Using the longest work-time rule, we begin the next station with B at .35 minutes followed by D at .19 minutes and C at .10 for a total of .64 minutes. Station 3 must start with E at .55 minutes and can include F at .14 for a total of .69 minutes. Station 4 then includes activities G and H as shown in Fig. 7-10.

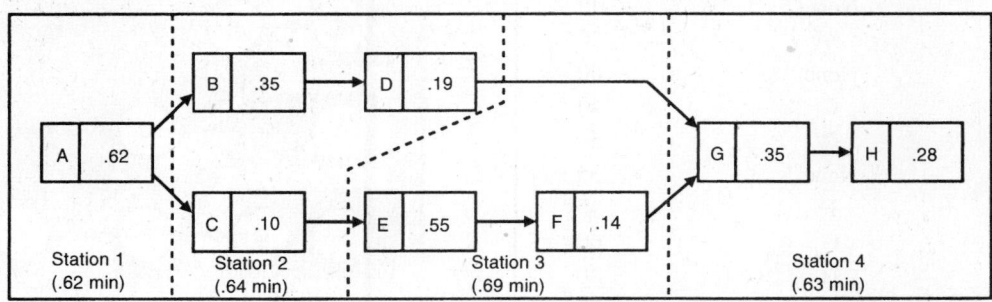

Fig. 7-10

(*b*) $\text{Eff}_B = \dfrac{\Sigma t}{(\text{CT})n} = \dfrac{2.58}{(.70)(4)} = 92\%$

(*c*)   Idle time $= \text{CT}(n) - \Sigma t = (.70)(4) - 2.58 = .22$ min

In the preceding example, the output and activity times specified the production-line output and determined the number of workstations. If, instead of output, the number of workstations *n* is specified, the total processing time requirement can be used to define a target cycle time, CT_t:

$$Ct_t = \frac{\Sigma t}{n} \qquad\qquad (7.5)$$

where Σt is the summation of activity times. The target cycle time represents the minimum average time necessary at a workstation and must be greater than or equal to the longest activity time.

Example 7.8   Suppose the activities shown in Fig. 7-9 are to be grouped into a three-station assembly line. (*a*) What is the target cycle time? (*b*) Which grouping of activities results in the largest output per hour? (*c*) What output will result in a 7-hour day?

(*a*) $CT_t = \dfrac{\Sigma t}{n} = \dfrac{2.58 \text{ min}}{3 \text{ stations}} = .86 \text{ min/station}$

(*b*)   The largest output will result from the smallest CT.
Note: CT is determined by the station requiring the most time.

Trial	Station #1	Station #2	Station #3	CT
1	.62 + .10 = .72	.35 + .19 + .55 = 1.29	.14 + .35 + .28 = .77	1.29
2	.62 + .35 = .97	.10 + .55 + .14 = .79	.19 + .35 + .28 = .82	.97
3	.62 + .10 = .72	.35 + .55 = .90	.19 + .14 + .35 + .28 = .96	.96 (best)

(*c*) $\text{Output} = \dfrac{(7 \text{ hr/day})(60 \text{ min/hr})}{.96 \text{ min/unit}} = 437 \text{ units/day}$

Example 7.9 (*most successors rule*)   Tasks A through I have the predecessor and time requirements shown in Table 7-3. Output is to be 200 units per day, and operating time is 450 minutes per day. Using the *most successors rule* to assign tasks to work centers, and the *longest work-time rule* as a tie breaker, (*a*) group the tasks into work centers, WCs, and compute (*b*) the balance efficiency, (*c*) the idle time, and (*d*) the balance delay.

Table 7-3

Task	Predecessor	Time (sec)
A	None	40
B	A	20
C	None	60
D	C	40
E	D	30
F	None	35
G	F	45
H	G	60
I	H	40
	Total	370

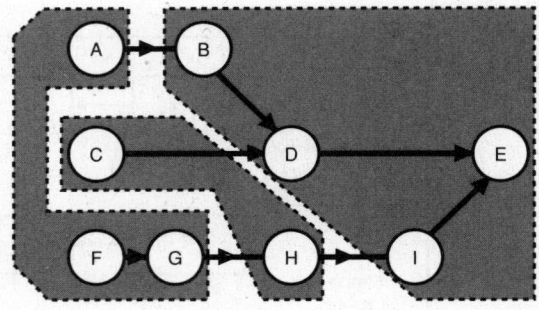

Fig. 7-11   Precedence diagram

(*a*)   First, we must find the cycle time (CT) to know the time available at each workstation.

$$CT = \frac{AT}{\text{output}} = \frac{(450 \text{ min/day})(60 \text{ sec/min})}{200 \text{ units/day}} = 135 \text{ sec/unit}$$

Beginning with the task with the most successors, assign tasks to workstations until the 135 seconds is as fully used as possible. See Table 7-4 where columns have been added to show the time remaining and potential assignments, plus the tasks with the most successors (*primary rule criterion*), and which of those tasks have the longest time requirements (*secondary rule criterion*).

Table 7-4

	Task	Time (sec) Required	Time (sec) Remaining (CT = 135)	Remaining Tasks w/satisfy precedence req't.	Remaining Tasks w/most successors	Remaining Task w/longest oper. time
WC #1	F	35	100	A, C, G	A, G	G
	G	45	55	A, C, H	A	
	A	40	15	None ≤ 15		
WC #2	C	60	75	B, H	B, H	H
	H	60	15	None ≤ 15		
WC #3	B	20	115	D, I	D, I	(tie, use either one)
	D	40	75	I		
	I	40	35	E		
	E	30	5	None left		

Figure 7-11 shows the resultant grouping of the tasks into the three work centers.

(*b*) $\text{Eff}_B = \dfrac{\Sigma t}{\text{CT}(n)} = \dfrac{370}{(135)(3)} = 91.4\%$

(*c*) Idle time $= (\text{CT})n - \Sigma t = (135)(3) - 370 = 35$ sec

(*d*) Balance delay $= \dfrac{\text{IT}}{(\text{CT})n} = \dfrac{35}{(135)(3)} = 8.6\%$ (*Note:* This is also $100\% - \text{Eff}_B$.)

COMPUTERIZED AND MIXED-MODEL LINE BALANCING

Computerized routines are available for testing the multitude of potential workstation configurations that exist for realistic large-scale line balancing problems. Although they utilize heuristic decision rules, they rapidly converge on a reasonably good balance. The General Electric model (ASYBL$), which assigns tasks according to positional weights, gives the designer the option of adding more workstations to gain a better efficiency. In line with the movement of firms toward more flexibility, some fixed-model programs have been designed to facilitate model changeovers, permitting different lot sizes, different setup times, and some task variability.

Questions and Solved Problems

DESIGN AND SYSTEM CAPACITY

7.1 (*office capacity*) A transfer operation at a title company involves four tasks that must be done in sequence (i.e., A = locating property, B = verifying ownership, C = recording data, and D = typing forms). Given the accuracy requirements, the maximum number of titles per hour that can be expected from well-trained workers in each of these positions is 75, 50, 70, and 60, respectively. However, workers A and D operate at 70 percent efficiency, and workers B and C operate at 90 percent efficiency. Assuming the workload is greater than the capacity, find (*a*) the effective capacity of the system and (*b*) the actual output that can be expected from the current workers.

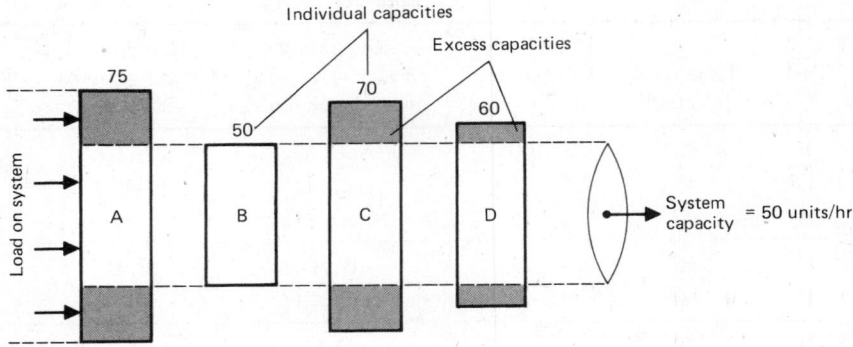

Fig. 7-12

(*a*) Fig. 7-12 is a schematic representation of the system. The effective capacity is limited by the individual capacity of B. Therefore, the system capacity is 50 units per hour.

(*b*) If, in this human system, each worker's efficiency were judged in relation to his or her own individual capability, the potential output from each worker would be:

$$A: \ (75 \text{ units/hr})(.70) = 52.5 \text{ units/hr}$$

$$B: \ (50 \text{ units/hr})(.90) = 45.0 \text{ units/hr}$$

$$C: \ (70 \text{ units/hr})(.90) = 63.0 \text{ units/hr}$$

$$D: \ (60 \text{ units/hr})(.70) = 42.0 \text{ units/hr}$$

Thus D would be the limiting component at 42 units per hour. Notice, however, that if B produced only 45 units per hour for D to work on, and if D's efficiency of 70 percent were applied to that input, the actual output from D would be only (45 units/hr)(.70), or 31.5 units per hour.

EQUIPMENT REQUIREMENTS AND LAYOUT DETERMINANTS

7.2 Silver Valley Smelting is considering the expansion of a production process by adding more 1-ton-capacity curing furnaces. Each batch (1 ton) of ore must undergo 30 minutes of furnace time, including load and unload operations. However, the furnace is used only 80 percent of the time due to power restrictions in other parts of the system. The required output for the new layout is to be 16 tons per shift (8 hours). Plant (system) efficiency is estimated at 50 percent of system capacity. (*a*) Determine the number of furnaces required. (*b*) Estimate the percentage of time the furnaces will be idle.

(*a*) $$\text{Required system capacity} = \frac{\text{actual output}}{\text{SE}} = \frac{16 \text{ tons/shift}}{.50} = 32 \text{ tons/shift}$$

$$= \frac{32 \text{ tons/shift}}{(.8)(8 \text{ hr/shift})} = 5 \text{ tons/hr}$$

$$\text{Individual furnace capacity} = \frac{1 \text{ ton}}{.5 \text{ hr}} = 2 \text{ tons/hr per furnace}$$

$$\therefore \quad \text{Number of furnaces required} = \frac{5 \text{ tons/hr}}{2 \text{ tons/hr per furnace}} = 2.5 \ (3) \text{ furnaces}$$

(b)   Total hours available per shift = 3 furnaces @ 8 hours = 24 furnace hours

Total hours of actual use per shift = 16 tons(.5 hr/ton)   = 8 furnace hours

Idle time = 16 hours

$$\text{Percentage of idle time} = \frac{16 \text{ hours idle}}{24 \text{ hours total}} = 67\% \text{ idle time}$$

7.3   A media center must determine how many digital processors are required to maintain an output of 200 good prints per hour. The set up and exposure can theoretically be done in 2 minutes per print, but operators are on the average only 90 percent efficient, and in addition, 5 percent of the prints must be scrapped and redone. Also, the processors can be utilized only 70 percent of the time. (a) What is the required system capacity in prints per hour? (b) What average output per hour can be expected from each processor, taking its use factor and efficiency into account? (c) How many processors are required?

(a) $$\text{System capacity} = \frac{\text{good output}}{SE} = \frac{200}{.95} = 210.5 \text{ prints/hr}$$

(b)   Output/hr = (unit capacity)(utilization %)(efficiency)

$$= \left(\frac{60 \text{ min/hr}}{2 \text{ min/print}} \right)(.70)(.90) = 18.9 \text{ prints/hr}$$

(c) $$\text{Number processors} = \frac{210.5 \text{ prints/hr required}}{18.9 \text{ prints/hr-cubicle}} = 11.14 \text{ (use either 11 or 12)}$$

7.4   How does the major concern of layout analysis differ for (a) process, (b) product, and (c) manufacturing cell layouts?

(a)   *Process (functional) layout analysis* is concerned largely with minimizing materials-handling costs by arranging departmental sizes and locations according to the volume and flow rate of products between work centers.

(b)   *Product (line) layout analysis* is concerned with maximizing worker or equipment effectiveness by grouping sequential work activities into workstations that yield a high utilization of labor and equipment with a minimum of idle time (i.e., line balancing).

(c)   *Manufacturing cell analysis* is concerned with enhancing the processing speed and flexibility by automating modules of activity that are common to a part or family of parts.

7.5   How are layouts designed in order to achieve optimality?

Facility layouts integrate numerous interdependent variables (materials-handling equipment, work-center locations, storage space, washrooms, offices, etc.). No single technique guarantees optimality. Good layouts minimize the costs of activities that do not add value to a product, such as materials handling, while maximizing the effectiveness of value-adding activities.

PROCESS (FUNCTIONAL) LAYOUT METHODS

7.6   *Load-Distance Analysis with Multiple Products.* Assume that a second product (Table 7-5) is to be produced in the facility described in Example 7.4. Recalculate the load-distance analysis for the two options. (*Note:* Weight demand values by their probabilities to find average demand.)

Table 7-5   Product #2 Workloads/Week

	To: A	B	C
From: A	—	40	5
B	25	—	7
C	10	8	—

No. Loads/Week Product #2 (both directions)	Option #1 (load)(distance)	Option #2 (load)(distance)
A to B and B to A = 40 + 25 = 65	(65)(10) = 650	(65)(15) = 975
A to C and C to A = 5 + 10 = 15	(15)(10) = 150	(15)(10) = 150
B to C and C to B = 7 + 8 = 15	(15)(15) = 225	(15)(10) = 150
	1,025	1,275

Given this second product, also at \$1 per load-foot cost, option #1 is now the lesser cost as seen in Table 7-6.

Table 7-6

Product	Load-Distance Total under	
	Option #1	Option #2
No. 1 (Example 7.4)	2,250	2,050
No. 2 (this example)	1,025	1,275
Total	3,275	3,325

7.7 *Systematic Layout Planning.* Which departments in the Muther grid of Fig. 7-8 must (*a*) be located adjacent to each other and (*b*) be separated?

(*a*)   Critical links are the a links: job planning–pattern shop, testing–assembly

(*b*)   Critical separations are the x links: pattern shop–testing, job planning–testing

Note: The major criteria for the sign production layout is likely to be a smooth flow (material handling) of the job. However, the nearness codes refine that.

7.8 *Systematic Layout Planning.* A chemical fertilizer facility has eight work centers that must be arranged into a 2 row by 4 column facility. The closeness ratings for absolutely necessary (a), very important (e), and undesirable (x) indexes are given in Fig. 7-13. Assign the critical work centers (i.e., a and x) first, and develop a suitable layout.

First: List the critical (a and x) links, and identify the work centers that should be centrally located (have the most links) and those that should be separated.

a links: 1-2, 1-3, 1-6, 3-7, 3-8 (WC3 and WC1 are most common)

x links: 1-8, 2-4, 3-6, 5-8, 6-8 (WC8 is most common)

Second: Form a cluster (or clusters) of the a links beginning with the most common (WC3 and WC1). Also graph the x links (Fig. 7-14).

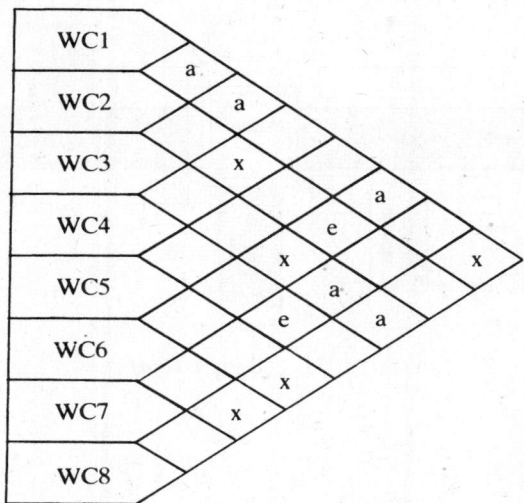

Fig. 7-13

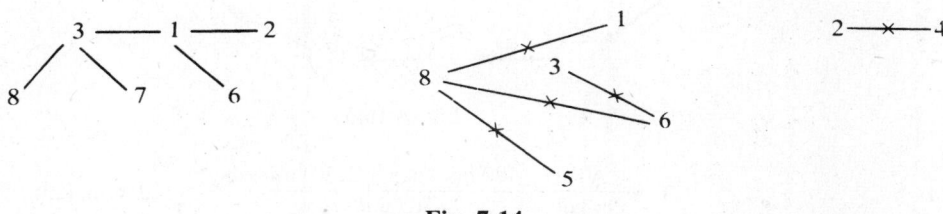

Fig. 7-14

Third: Add on to the cluster to meet other nearness criteria, and rearrange as necessary to fit into the specified 2 × 4 matrix.

    Note: The matrix shown (Fig. 7-15) satisfies are requirements, including the 1 to 6 linkage because of the corner-to-corner contact.

8	3	1	5
4	7	2	6

Fig. 7-15

LINE BALANCING IN PRODUCT LAYOUTS

7.9   An electric appliance assembly area is as shown in Fig. 7-16(*a*) with potential workstations A through F. The tasks that must be done, along with their respective times, are indicated in the precedence diagram, Fig. 7-16(*b*).

    The machine scan is automatic and can come any time after task 2. The manufacturer desires an output of 367 units per 8-hour day and stops the line for a 20-minute break in the middle of the morning and the afternoon. (*a*) Group the assembly-line tasks into appropriate workstations—using the longest time rule, and (*b*) compute the balance efficiency.

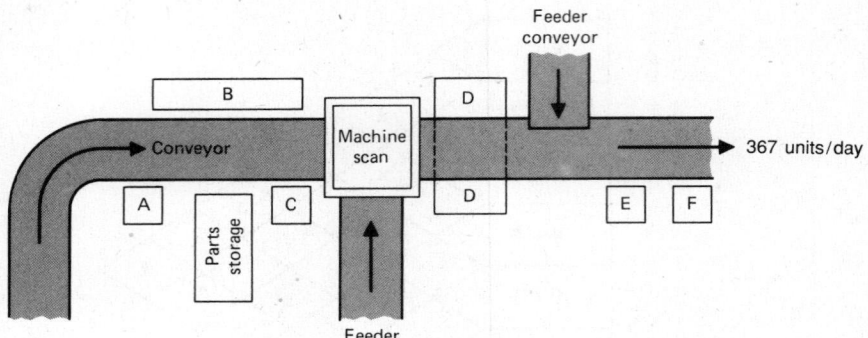

Fig. 7-16(a)

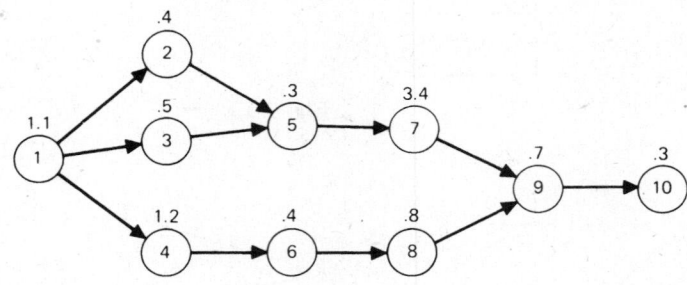

Fig. 7-16(b)

(a)
$$CT = \frac{AT}{Output} = \frac{480 \text{ min/day} - 2(20) \text{ min/day}}{367 \text{ units/day}} = 1.20 \text{ min/unit}$$

Each worker can be scheduled for up to 1.2 minutes of work at a workstation. Grouping the tasks into the maximum amounts of work that can be done at a workstation, we obtain one arrangement as shown in Fig. 7-17. Potential workstations are marked off with dashed lines.

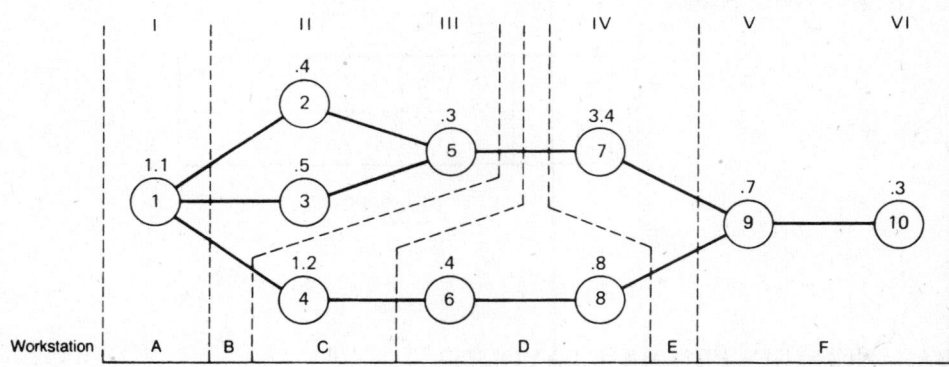

Fig. 7-17

Table 7-7

Workstation	A	B	C	D	E	F	
Tasks	1	2, 3, 5	4	6, 8	7	9, 10	
Actual time (min)	1.1	1.2	1.2	1.2	3.4	1.0	$\Sigma t = 9.1$

In Fig. 7-17, notice that task 7 requires 3.4 minutes of work (Table 7-7). Because the nearest multiple of cycle time is 3.6 (i.e., 3×1.2), this station will require three workers. The addition of two extra workers makes the equivalent number of workstations (or number of workers) eight rather than six, so we will use that for our calculations.

(b)
$$\text{Eff}_B = \frac{\Sigma t}{(\text{CT})n} = \frac{9.1}{(1.2)(8)} = 94.8\%$$

7.10 Overland Motors produces 50 cars per hour (Fig. 7-18) and has a transmission feeder shop with three workstations (A, B, C), which take times of 55, 45, and 60 seconds, respectively. Station C assembly time is normally distributed, with a standard deviation of 5 seconds. (a) If all work arrived at C on time, what proportion of the time would the feeder shop fail to deliver transmissions on time to the main auto assembly line? (b) What is the balance efficiency for the transmission feeder shop?

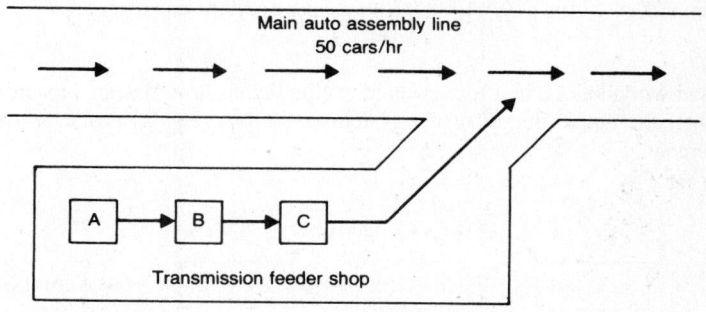

Fig. 7-18

(a)   Main assembly line rate $= \dfrac{60 \text{ min/hr}}{50 \text{ cars/hr}} = 1.2 \text{ min/car} = 72 \text{ sec/car}$

Corresponding Z value: $Z = \dfrac{x - \mu}{\sigma} = \dfrac{72 - 60}{5} = 2.4 \text{ sec}$

From normal distribution: $P(x > 72) = P(Z > 2.4)$
$$= .008 \cong .01 = 1\%$$

See Fig. 7-19.

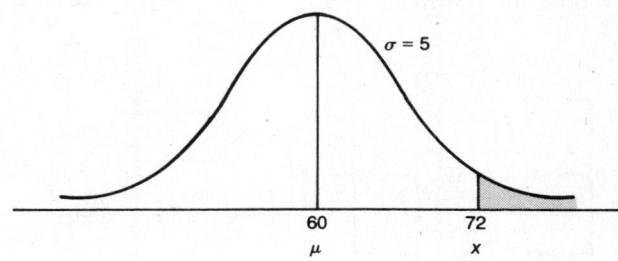

Fig. 7-19

(b)
$$\text{Eff}_B = \frac{\text{output of task times}}{\text{input of station times}} = \frac{\Sigma t}{(\text{CT})n}$$

where cycle time CT is governed by main assembly line output $= 72$ sec/car

Therefore, $\text{Eff}_B = \dfrac{55 + 45 + 60}{(72)(3)} = \dfrac{160}{216} = 74\%$

Supplementary Questions and Problems

7.11 A manufacturer of communicators uses three TR87 electronic chips in each unit produced. Demand estimates for the number of communicators that could be sold next year are shown in Table 7-8. (*Note:* Weight demand values by their probabilities to find average demand.)

Table 7-8

Demand X	20,000	40,000	50,000
$P(X)$	.30	.50	.20

(*a*) Assuming the firm decides to produce on an expected-value basis, how many TR87 chips should it plan to produce for next year's communicator sales? (*b*) What capacity is required to meet 150 percent of expected demand?   *Ans.*   (*a*) 108,000 chips (*b*) 162,000 chips

7.12 The individual workstations in a toy production-line layout have design capacities (units per day) as shown in Fig. 7-20. If the actual output of the system is 80 toys per day, what is the system efficiency? *Ans.*   53 percent

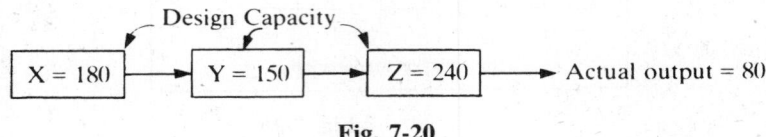

Fig. 7-20

7.13 An existing factory has the equipment arrangement shown in Fig. 7-21. Manufactured fittings must be processed through each of three operations in sequence, but it does not matter which lathe or mill is used. Each lathe is capable of handling 30 fittings per hour, each mill can handle 45 per hour, and the grinder can handle 80 per hour. A different operator runs each group of machines, and because of the workload, the lathe operator can handle an output of 25 fittings per hour from each lathe when they are all operating. The mill and grinder operators can produce 45 per hour (per mill) and 80 per hour (per grinder), respectively. During the past 40-hour week, actual production from this department was 1,000 fittings. Find (*a*) the system capacity and (*b*) the system efficiency.   *Ans.*   (*a*) 75 per hour (*b*) 33.3 percent

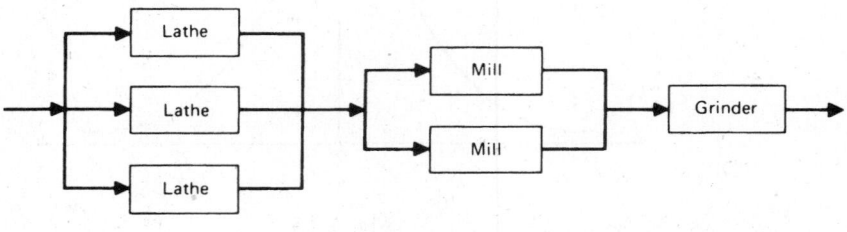

Fig. 7-21

7.14 Your firm must purchase some new plating machines capable of producing 160,000 *good* parts per year. They will become part of a processing line, and you expect that 20 percent of the production will have to be scrapped because of defects. What is the required system capacity in parts per year? *Ans.*   200,000 parts per year.

7.15   Use the data from Prob. 7.14, and assume that plating one part takes 90 seconds. The plant in which the machines will be installed operates 2,000 hours per year. However, the plating machines are used only 50 percent of the time. If the plating machines are 90 percent efficient, what actual plating machine output per hour is achieved?   *Ans.*   18 parts per hour

7.16   Using the data from Probs. 7.14 and 7.15, how many plating machines are required?
Ans.   5.6 (6) machines

7.17   A textile firm wishes to acquire enough stamping machines to produce 30,000 good T-shirts per month. They operate 200 hours per month, but the stamping machines will be used for T-shirts only 70 percent of the time, and the output is 4 percent defective. The stamping operation takes 1 minute per T-shirt. Allowing for adjustments, clean out, and unavoidable downtime, the stamping machines are 90 percent efficient. How many stamping machines are required?   *Ans.*   4.13 (4) machines

7.18   Mohawk Valley Furniture has purchased a plant with six production areas as shown in Fig. 7-22 facility outline. The firm proposes to locate six departments (A, B, C, D, E, F), which have the number of moves per day between departments as shown in Table 7-9's travel chart.

Table 7-9   Travel Chart

		A	B	C	D	E	F
		Number of Moves to					
From	A	—	7	—	—	—	5
	B	—	—	—	4	10	—
	C	—	7	—	—	2	—
	D	—	—	8	—	—	—
	E	4	—	—	—	—	3
	F	—	6	—	—	10	—

1	2	3
4	5	6

Fig. 7-22   Facility outline

Develop a layout of the six departments that minimizes the nonadjacent flows.
Ans.   One arrangement is shown in Fig. 7-23.

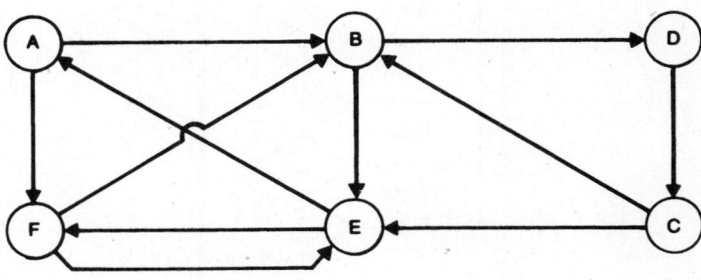

Fig. 7-23

7.19   The distances between centers of departments for one solution of Prob. 7.18 (above) are shown in Fig. 7-24. Diagonal moves are permissible. (*a*) Using daily totals, compute the (load) (distance). (*b*) Compute the materials-handling cost if the cost to move a load is $.025 per foot. (*c*) Use a trial-and-error approach to develop a less costly layout. (*d*) How much does your new layout save?

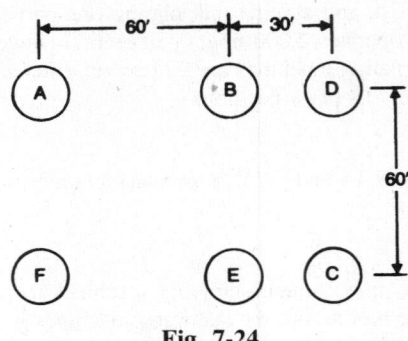

Fig. 7-24

Ans. (*a*) 4,078 foot-loads per day. (*b*) $102 per day. (*c*) More than one answer is possible. (*d*) A savings of over $3,000 per year is possible.

7.20 Arrange six work centers into a 2(row) × 3(column) grid in a layout that satisfies the following: WC1 must adjoin WC4, WC1 must adjoin WC5, WC5 must adjoin WC6, WC2 and WC5 must be separated. *Ans.* One arrangement is

$$\begin{bmatrix} 3 - 4 - 1 \\ 2 - 6 - 5 \end{bmatrix}$$

7.21 Using only the nearness codes a, o, and x, develop a Muther grid for Prob. 7.20. Assume unspecified relationships are all of ordinary importance.
Ans. Show a's in intersections 1 and 4, 5 and 6, 1 and 5; show x in 2 and 5; others are all o.

7.22 Arrange the six work centers shown in Fig. 7-25 into a 3(row) × 2(column) grid that satisfies the nearness criteria shown on the Muther grid.   *Ans.* Departments 3 and 6 must be in the middle row.

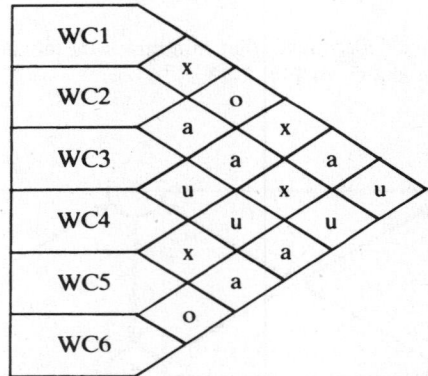

Fig. 7-25

7.23 A line balancing problem involves 10 workstations having a Σ times = 24.0 minutes (where the shortest is 2.1 minutes and the longest is 3.0 minutes). Assuming only one worker is located at each station, and using the longest time as the cycle time, what would be the balance efficiency?   *Ans.* 80 percent

7.24 A line balancing analysis resulted in a precedence grouping as shown in Table 7-10. Find the balance efficiency, assuming the longest actual time is the cycle time.   *Ans.* CT = 1.5, so Eff_B = 84 percent

Table 7-10

Work Center	Activity Numbers	Actual Time (min)
A	1, 2	1.2
B	3, 5, 6	1.4
C	4, 7	.9
D	8, 10, 11	1.3
E	9	1.5

7.25 A furniture-manufacturing activity requires the times shown in Fig. 7-26 to perform five tasks in an assembly line. Operations are to be scheduled for producing six units per hour, and each employee can contribute 48 minutes per hour of productive work. (*a*) What is the cycle time in minutes per unit? (*b*) What is the theoretical minimum number of personnel? (*c*) Combine the tasks into the most efficient grouping of workstations–using the longest-time rule. What is the resulting efficiency of balance? *Ans.* (*a*) 8 minutes per unit (*b*) 3.78 employees (*c*) 94.7 percent

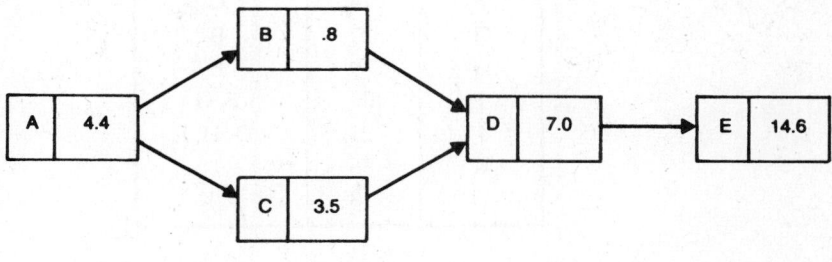

Fig. 7-26

7.26 A toy manufacturer produces doll houses on a product line geared to an output of one per minute. The assembly precedence relationships and activity times (in minutes) are as shown in Fig. 7-27.

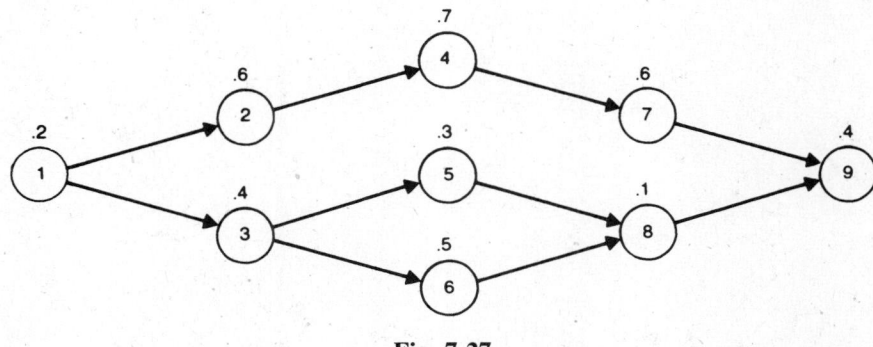

Fig. 7-27

(*a*) Group the activities using the longest-time rule. (*b*) What is the balance efficiency (Eff$_B$)? *Ans.* (*a*) One solution is shown in Table 7-11 (*b*) 76 percent

Table 7-11

Workstation	A	B	C	D	E
Tasks	1, 2	3, 6	4, 5	7, 8	9
Actual times	.8	.9	1.0	.7	.4

7.27 Robotic Controls Corp. uses a robotic-controlled flexible production system to assemble the robots it sells. Five robots are available and must complete the tasks specified in Table 7-12.

Table 7-12

Task	Time (sec)	Preceding Task(s)
A	10	None
B	24	None
C	17	A
D	49	A
E	12	C
F	14	C
G	27	B
H	9	E
I	20	F, G
J	23	D, H, I
K	36	I
L	18	J, K

(*a*) Draw a precedence diagram. (*b*) What is the theoretical minimum (target) cycle time if all five robots are fully utilized in a five-station assembly line? (*c*) Group the tasks into the most efficient five-station assembly line. (*d*) What is the cycle time? (*e*) What is the balance efficiency?

Ans. (*a*) Diagram must meet precedence requirements. (*b*) 51.8 seconds/station (*c*) I = A, C, E, F; II = B, G; III = D; IV = H, I, J; V = K, L (*d*) 54 seconds (from station V) (*e*) 96 percent

Chapter 8

Product Design: Goods and Services

Linear Programming

STAGES IN PRODUCT AND PROCESS DEVELOPMENT

Products are goods and services, and *processes* are the means (skills and equipment) used to produce them. Product decisions are discussed in this chapter, and process decisions in the next.

Figure 8-1 illustrates one of many possible paths from an idea to a finished good or service.

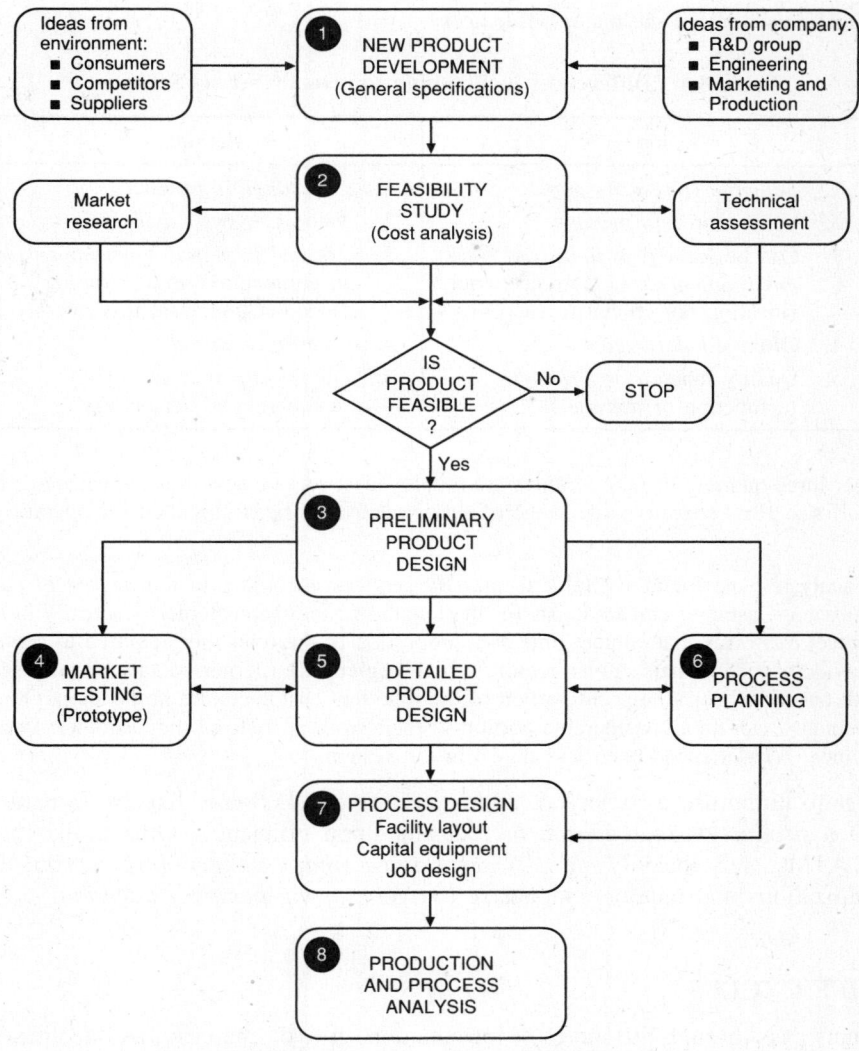

Fig. 8-1 Product and process design and analysis

127

PLANNING STRATEGIES FOR GOODS AND SERVICES

Although there are similarities, planning for goods tends to be more formalized than for services. In addition, the manufacturing environment is typically separated from the consumer, and activities can be tightly controlled. Service activities usually take place in the presence of the consumer and often require more flexibility.

Question: What competitive or performance capabilities are particularly relevant in deciding what goods or services an organization is best suited to produce?

Among the most frequently cited factors identified with a competitive advantage are:

(a) *Cost efficiency*, which permits lower selling prices resulting in improved market share

(b) *Quality*, which yields products that meet specifications and satisfy customer expectations

(c) *Flexibility*, which permits customization and the adaptability to fluctuations in volume

(d) *Dependability*, which helps develop, produce, and deliver products on time

Question: What are some of the other differences in planning for goods versus services?

Table 8-1 highlights some of the major differences.

Table 8-1 Differences in Planning for Goods Versus Services

Goods	Services
1. *Tangible* (physical) product	1. *Less tangible* product
2. Value stored *in product*	2. Value conveyed *as used*
3. Can be located in *industrial environment* away from customer (location not crucial to success)	3. Located in *market environment* in conjunction with customer (location important to success)
4. Often *standardized*	4. Often *customized*
5. Quality inherent in *product* (a function of materials)	5. Quality inherent in *process* (a function of personnel)

Question: Over three-quarters of U.S. workers are employed in service activities, which range from small hair salons to large airlines. How has this wide range of service activities been classified for operations management purposes?

Operations analysts have found it useful to classify services according to the *degree of customer contact* involved in the service. Customer contact exists for the length of time the customer is actually in the system, and the degree of contact expresses that contact time as a proportion of the total time required to serve the customer. Some service providers, such as hair salons, require a high degree of customer contact because service providers must interact with customers for a large proportion of the time that customers are in the shop. Others, such as tax preparation accountants, can do a considerable portion of their work apart from the customer. Users of automatic bank teller machines (ATMs) spend even less time "in the system."

Service organizations use a variety of approaches to excel. Some, like McDonalds Corporation, have taken (1) a *production-line approach*, focusing upon efficiency. Others, like home shopping networks and ATMs, rely heavily on (2) *self-service and customer involvement*. Finally, some hospitality organizations and retailers emphasize (3) very *strong personal attention*.

PRODUCT LIFE CYCLES

Most products pass through the stages of introduction, growth, maturity, and decline, as depicted in Fig. 8-2. Not all products follow the same pattern, but knowledge of the general pattern of goods and services helps planners to forecast demand and to maintain a viable mix of products in the firm's product line.

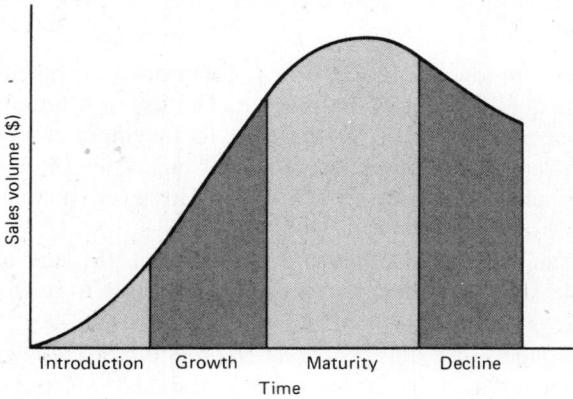

Fig. 8-2   Typical good or service life cycle

Question: As a product moves through its life cycle, what changes can be expected in (a) the variety of designs offered, (b) its competitive factors, and (c) its profitability?

Table 8-2   Some Life Cycle Characteristics in Stage of

		Introduction	Growth	Maturity	Decline
(a)	Design variety	Little variety or customized	Some degree of standardization	A dominant design prevails	Highly standardized
(b)	Competitive emphasis on	Product features	Design/quality availability	Price/delivery reliability	Price
(c)	Profit potential	Loss up to low profits	Good up to high profits	High down to good profits	Declining to low profits

RESEARCH AND DEVELOPMENT (R&D)

Research is a consciously directed investigation to find new knowledge. Research activities have spawned many electronic, medical, communication, transportation, and other products.

Question: Define (a) basic research, (b) applied research, and (c) development.

(a)   *Basic research* is a search for new knowledge without regard to any near-term commercial use, such as that done by universities and foundations.

(b)   *Applied research* is investigative or experimental work directed toward specific problems, products, or processes with the objective of achieving commercial value.

(c)   *Development* is the activity of converting research results into commercial products. Development efforts consume the bulk of research and development (R&D) funds.

PRODUCT DESIGN AND STANDARDIZATION

Product design is the structuring of component parts or activities so that as a unit they can provide a specified value. Product specification is typically an engineering function; detailed drawings or specifications are prepared that give dimensions, weights, colors, and other physical characteristics. In service industries, product specification often consists of an environmental requirement to be maintained or a procedure to be followed, such as operating room procedures in a hospital.

Question: What is meant by the "phases" of a product design?

During their development, products sometimes undergo (a) a *functional design*, wherein a model is developed just to see if it works, then (b) an *industrial design*, to ensure it has the features the user wants, and finally (c) a *design for manufacturability, or delivery*, to ensure the product can be effectively produced and/or delivered. The result is a product specification.

Standardization involves producing items to a commonly accepted standard to ensure the interchangeability and/or the quality level of the product. The use of a limited number of uniform parts reduces the sizes and number of items to be purchased, cuts inventory storage and handling costs, and enables firms to work with larger (and more economical) quantities of fewer items. Standardization makes both mass production and maintenance much easier. However, standardization limits the options available to consumers.

Modular designs also facilitate production and maintenance. Modules are common components grouped into interchangeable subassemblies, and they range in size from microelectronics to prefabricated houses.

Once developed, many products also undergo *value engineering* or *value analysis*. This is an attempt to see if any materials or components can be substituted or redesigned in such a way as to attempt continue to perform the desired function, but at a lower cost. After prototype units are designed and produced, the products are further analyzed and tested to see how well the quality, performance, and costs conform to the design objectives. *Simplification* may take place to reduce unnecessary variety in the *product line* by decreasing the number and variety of products produced.

CAD, CAM, GT, MANUFACTURING CELLS, AND CIM

CAD, CAM, GT, and CIM reflect the trend toward fully automated manufacturing facilities that are designed to integrate product design and manufacturing activities with both the suppliers of materials and components as well as the customers of the firm's products. Although few firms are fully interconnected in this way, portions of these links are operational on a worldwide basis through computers, satellite networks, and global information systems (e.g., Internet).

Question: Describe the process of Computer-Aided Design (CAD).

CAD is a package of computer software that enables designers to input product specifications and create three-dimensional geometric models of a product on computer consoles. The images can be rotated, enlarged, and subjected to cost analysis, reliability, serviceability, and other tests on the computer even before the product is manufactured.

Question: What is Computer-Aided Manufacturing (CAM)?

CAM follows CAD and is the extensive use of computers to actually perform and control production operations. Some major uses of computers here are in (1) numerically controlled (NC) machines, (2) process controllers, (3) systems that are linked by group technology, (4) automatic assembly operations, and (5) computer-aided inspection and testing. *NC machines* convert CAD specifications into precise machine commands and can be integrated into a network that controls production, scheduling, and other information relevant to productivity.

Group technology, manufacturing cells, and computer integrated manufacturing systems extend and enhance the benefits initiated by the CAD and CAM systems. (See Chap. 7.) *Group technology (GT)* helps identify families of parts that can be designed and processed efficiently in *manufacturing cells*, which often constitute "islands of automation." As these cells are linked (via computer) with the demand from customers, and the needed inputs from suppliers, a total *computer integrated manufacturing system (CIM)* begins to take shape. With the use of satellite telecommunications, the technology of globally linking suppliers, manufacturers, and customers is widely available.

PRODUCT RELIABILITY

Question: What is product reliability?

Product reliability is the mathematical probability of a product performing a specific function in a given environment for a specific length of time or number of cycles.

As suggested in Fig. 8-3, early failures (perhaps due to improper assembly or damage in shipment) may tend to follow a negative exponential pattern. During the typical operating lifetime, failures occur on a *rare-event* basis, often described by a Poisson distribution. As components wear out and fail, the products may follow a pattern described by a normal distribution.

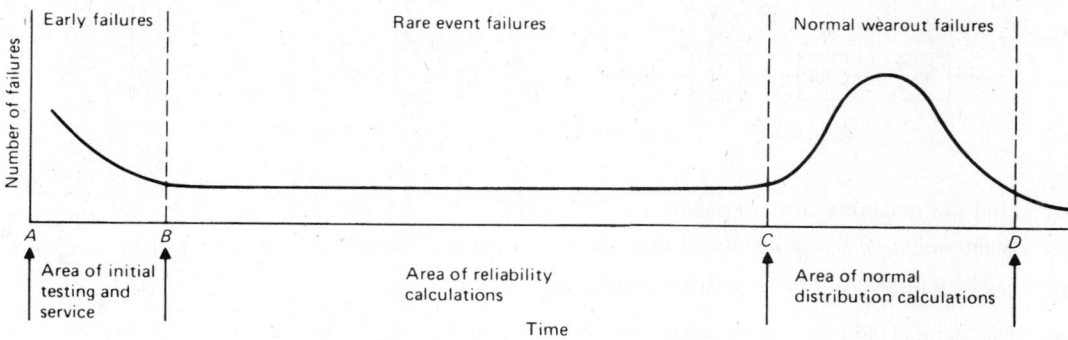

Fig. 8-3   Product failure rates

Example 8.1   The manufacturing area in the plant of a New Jersey drug manufacturer requires 5,000 fluorescent light tubes. The lights have a normally distributed lifetime, with a mean of 4,000 hours and a standard deviation of 120 hours. The plant manager has found that after 10 percent of the lights burn out, the quality of items and the productivity of workers in the plant are affected. He would like to schedule maintenance activities so that all lights are replaced when 10 percent fail. After how many hours of operation should the replacement activities be scheduled?

At the mean lifetime μ, 50 percent of the lights are still operating. We wish to find the earlier time X such that 40 percent more (or 90 percent total) are operating. Since the distribution is normal, we know (from Appendix B) that the number of standard deviations required to include an area of .40 is $Z = 1.28$. See Fig. 8-4.

$$-Z = \frac{X - \mu}{\sigma}$$

$$\therefore \quad X = \mu - Z\sigma = 4,000 - 1.28(120) = 3,846 \text{ hr}$$

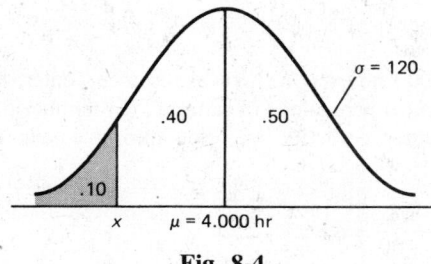

Fig. 8-4

- Improve the design of components.
- Simplify the design of the system.
- Improve production techniques.
- Improve quality control.
- Test components and the system.
- Install parallel systems.
- Perform periodic preventive maintenance.
- Derate components and/or the system.

Fig. 8-5   Ways to improve reliability

Figure 8-5 describes ways to improve product or system reliability. The use of parallel systems is a standard design procedure in many hazardous and capital-intensive applications. The reliability of components in series R_s is:

$$(\text{Series})\ R_s = R_1 \cdot R_2 \cdots R_n \tag{8.1}$$

For parallel circuits the reliability R_p of the system is determined by

$$(\text{Parallel})\ R_p = 1 - (1 - R_{s1})(1 - R_{s2}) \tag{8.2}$$

Example 8.2   An acid control system has three components in series with individual reliabilities (R_1, R_2, and R_3) as shown in Fig. 8-6.

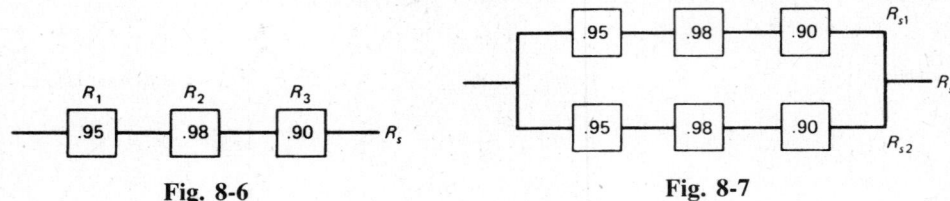

Fig. 8-6   **Fig. 8-7**

(a)   Find the reliability of the system.

(b)   What would be the reliability of the system if a parallel circuit were added?

(a)   Series $R_s = R_1 \cdot R_2 \cdot R_3 = (.95)(.98)(.90) = .84$

(b)   The parallel system design would be as shown in Fig. 8-7, where R_{s1} and R_{s2} are the computed reliabilities of the respective series circuits.

$$\text{Parallel } R_p = 1 - (1 - R_{s1})(1 - R_{s2}) = 1 - (1 - .84)(1 - .84) = .97$$

FAILURE RATES AND MTBF

A *failure* is an event that changes a product from operational to nonoperational. The failure rate (FR) can be expressed as either a percentage of failures among the total number of products tested or in service, or as a number of failures per given operating time.

$$\text{FR}_\% = \frac{\text{number of failures}}{\text{number tested}} \tag{8.3}$$

$$\text{FR}_n = \frac{\text{number of failures}}{\text{operating time}} = \frac{F}{\text{TT} - \text{NOT}} \tag{8.4}$$

where F = number of failures
$\quad\quad$ TT = total time
NOT = nonoperating time

Example 8.3   Fifty artificial heart valves were tested for 10,000 hours at a medical research center, and three valves failed during the test. What was the failure rate in terms of (a) percentage of failures? (b) number of failures per unit-year? (c) On the basis of these data, how many failures could be expected during a year from the installation of these valves in 100 patients?

(a)
$$\text{FR}_\% = \frac{\text{number of failures}}{\text{number tested}} = \frac{3}{50} = 6.0\%$$

(b)
$$\text{FR}_n = \frac{\text{number of failures during period}}{\text{operating time}} = \frac{F}{\text{TT} - \text{NOT}}$$

Note that the operating time is reduced by those units that failed. In the absence of actual data, we assume that failures are averaged throughout the test period. Therefore,

Total time = (10,000 hr)(50 units)   = 500,000 unit-hr

Less: Nonoperating time of 3 failed units for

average of $\dfrac{10,000}{2}$ hr $-15,000$ unit-hr

Operating time   = 485,000 unit-hr

$$FR_n = \frac{3 \text{ failures}}{485,000 \text{ unit-hr}} = .0000062 \text{ failure/unit-hr}$$

or in terms of years,

$$FR_n = (.0000062)\left(\frac{24 \text{ hr}}{\text{day}}\right)\left(\frac{365 \text{ days}}{\text{yr}}\right) = .0542 \text{ failure/unit-yr}$$

(c)   From 100 units,

$$\left(\frac{.0542 \text{ failure}}{\text{unit-yr}}\right)(100 \text{ units}) = 5.42 \text{ failures/yr}$$

The mean time between failure (MTBF) is another useful term in maintenance and reliability analysis. The MTBF is the reciprocal of FR_n:

$$MTBF = \frac{\text{operating time}}{\text{number of failures}} = \frac{TT - NOT}{F} \tag{8.5}$$

Example 8.4   Find the MTBF for the heart valves described in Example 8.3.

$$MTBF = \frac{TT - NOT}{F} = \frac{500,000 - 15,000}{3} = 161,666.67 \text{ unit-hr/failure}$$

$$= \frac{161,666.67}{(24)(365)} = 18.46 \text{ unit-yr/failure}$$

The 18.46 unit-year per failure figure represents the mean service time between failures that might be expected from a group of units during their several years of service. It is not necessarily indicative of the expected life of an individual unit.

The MTBF can be used to express the reliability of a component or a system if the failure rate is constant. If we let R represent the system reliability, and t the time period in question, then

$$R = e^{-(t/MTBF)} \tag{8.6}$$

where $e =$ the base of the natural logarithms, 2.7183.

Example 8.5   Safety valves used in an oil refinery have a constant failure rate with an MTBF of 16 years. What is the probability that a newly installed valve will function without failure for the next 8 years?

$$R = e^{-(t/MTBF)} = e^{-(8/16)} = e^{-1/2} = \frac{1}{e^{1/2}} = \frac{1}{\sqrt{e}} = \frac{1}{\sqrt{2.7183}} = \frac{1}{1.6478} = .6065$$

PRODUCT SELECTION

Product selection decisions are influenced by (1) the firm's resource and technology base, (2) the market environment, and (3) the firm's motivation to use its capabilities to meet the needs of the marketplace. Figure 8-8 depicts the three factors.

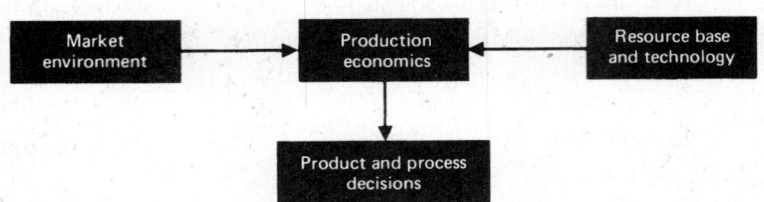

Fig. 8-8　Factors relevant to product selection decisions

PRODUCT-MIX DECISIONS VIA LINEAR PROGRAMMING (GRAPHIC)

Within the product-line groupings, decisions must be made to select which mix of products to produce (or which processes to use) in view of cost, capacity, and other limitations. Linear programming is a useful technique for assisting in the product-mix (and numerous other) decisions. It applies to situations where the firm has a demand for whatever quantities of two or more products it can produce. Another typical application is for the selection of the least costly mix of raw materials or processes to use when several are available.

Linear programming is a mathematical technique for maximizing or minimizing a linear objective function, subject to linear constraints. It assumes that cost and revenue values are known (*certainty*) and that profits from various activities are additive (*additivity*), and it does not allow negative production values (*non-negativity*). We review linear programming here in the context of a product-mix decision. However, it has widespread application to other problems such as capital budgeting, line balancing, planning, and scheduling.

Linear programming problems are expressed in terms of a single linear *objective function* that specifies the benefit or cost associated with each *decision variable*. For example, if the profit (Z) from decision variable X_1 (chairs) is \$20 and from X_2 (tables) is \$70, the linear objective may be to maximize $Z = \$20X_1 + \$70X_2$. *Constraints* express the resource limitations or needs to produce the end products and may be stated as less than or equal to ($\leq$), equal to ($=$), or greater than or equal to ($\geq$) a specified amount. Thus if each chair (X_1) required 10 minutes of assembly time, and each table (X_2) required 20 minutes, the number of chairs and tables that could be assembled would be limited by the total assembly time available, say 420 minutes. The linear equation for the assembly time constraint would then be $10X_1 + 20X_2 \leq 420$. Other constraints (as many as apply) would be formulated in a similar manner. Taken together, the constraints define a *feasible region*, an area within which all possible solution combinations lie. The *optimal solution* (or mix of variables) depends upon the criteria (e.g., profit or cost) expressed in the objective function, but will always be at some intersection of constraints (a *corner*) of the feasible region.

One of the easiest methods of solving two-variable (two-product) problems is the graphic method. Table 8-3 lists the steps of solution.

Table 8-3　Graphic Method of Solving Linear Programming Problems

1. Formulate the problem in terms of a linear objective function and linear constraints.
2. Set up a graph with one decision variable on each axis, and plot the constraints. They define the feasible region.
3. Determine the slope of the objective function, and indicate the slope in the feasible region on the graph.
4. Move the objective function parallel to itself in an optimizing direction until it is constrained.
5. Read off the solution values of the decision variable from the respective axes.

Example 8.6 A chemical firm produces automobile cleaner X and polisher Y and realizes $10 profit on each batch of X and $30 on Y. Both products require processing through the same machines, A and B, but X requires 4 hours in A and 8 in B, whereas Y requires 6 hours in A and 4 in B. During the forthcoming week machines A and B have 12 and 16 hours of available capacity, respectively. Assuming that demand exists for both products, how many batches of each should be produced to realize the optimal profit Z? (*Hint:* Follow the steps outlined in Table 8-3.)

1. The objective function is: Max $Z = \$10X + \$30Y$

    The constraints are: A: $4X + 6Y \le 12$

                                          B: $8X + 4Y \le 16$

                                          Also: $X \ge 0,\ Y \ge 0$

2. The variables are X and Y. The constraints are plotted as equalities. To graph:

    A: If $X = 0$, $Y = 2$ B: If $X = 0$, $Y = 4$

    A: If $Y = 0$, $X = 3$ B: If $Y = 0$, $X = 2$

    Note that the graph (Fig. 8-9) establishes a feasible region bounded by the explicit capacity constraints of A and B and the implicit constraints that production of $X \ge 0$ and production of $Y \ge 0$.

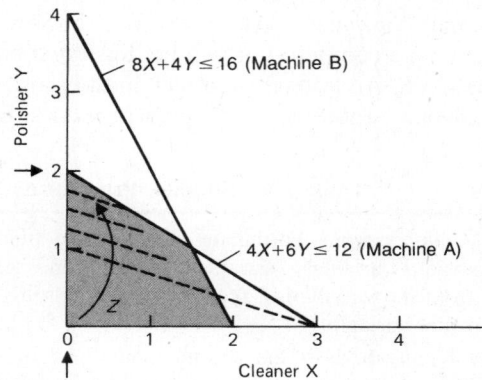

Fig. 8-9 Graphic linear programming solution

3. The slope of the objective function is: $Z = 10X + 30Y$

    The standard slope-intercept form of a linear equation is

    $$Y = mX + b \qquad\qquad (8.7)$$

    where m is the slope of the line (that is, change in Y per unit change in X) and b is the Y intercept. Expressing our objective in this form, we have:

    $$30Y = -10X + Z \qquad \therefore\ Y = -\frac{1}{3}X + \frac{Z}{30}$$

    The slope $= -1/3$; that is, the line decreases one unit in Y for every three positive units of X. This is plotted at any convenient spot within the feasible region (and shown as dashed lines in Fig. 8-9). The dashed line from $Y = 1$ to $X = 3$ illustrates this.

4. The slope of the objective function is moved away from the origin until restrained by the furthermost intersection of constraint A and the implicit constraint $X \ge 0$. The optimal solution will always be at a corner in the feasible region.

5. The arrows point to the solution, which is determined by the X and Y coordinates at the corner. In this

example, $X = 0$ and $Y = 2$, so the firm should produce no cleaner and two batches of polisher for a profit of:

$$Z = \$10(0) + \$30(2) = \$60$$

As can be seen from the graph, the constraint imposed by machine B (that is, that $8X + 4Y \leq 16$) has no effect, for it is the 12 hours of machine A (denoted by $4X + 6Y \leq 12$) that are constraining production of the more profitable polisher. The graph also reveals that profit would continue to increase if more hours could be made available on machine A up to the point of doubling output (to $X = 0$ and $Y = 4$). At this point, the time available from machine B would become constraining.

Example 8.6 assumed that the profit contribution was known and that the constraint amounts, processing time, and available machine time were known with certainty.

LINEAR PROGRAMMING (SIMPLEX METHOD)

Realistic linear programming problems often have several decision variables and many constraint equations. Such problems cannot be solved graphically, so algorithms such as the simplex procedure are used. The *simplex method* is an iterative procedure that progressively approaches and ultimately reaches an optimal solution to linear programming problems. Numerous computer programs are available for both mainframe and personal computers. Although the simplex method is especially useful for large-scale problems (solved with a computer), it will be illustrated below for the same problem that was solved graphically in Example 8.6. Additional examples in the problem section extend its application to three and more decision variables. Table 8-4 outlines a general simplex procedure.

Table 8-4   Simplex Procedure

1. *Set up initial simplex table*. Formulate the objective function and constraints; and enter the decision variables, variables in solution, solution (RHS) values, C (contribution from the variable), Z (cost of introducing the variable), and $C - Z$ (net contribution of the variable).

2. *Select the pivot column*. It is the column with the largest positive number in the bottom ($C - Z$) row. It becomes the new variable in solution.

3. *Select the pivot row*. It is the row with the smallest ratio of RHS value divided by pivot column value. Use only positive numbers. This identifies the variable leaving the solution.

4. *Circle the pivot*. It is at the intersection of the pivot row and pivot column.

5. *Convert the pivot into a 1*. Do this by dividing each value in the pivot row by the pivot value. Enter this new row in a new table.

6. *Generate other rows for the next table with zeros in the pivot column*. This is done by multiplying the new row (from step 5) by the negative of the element in the pivot column that is to be converted and adding the result to the old row. Enter the revised row in the new table, and continue this procedure for each row in the center section of the table.

7. *Test for optimality*. Compute the values of Z and $C - Z$. Z values for each column are Σ(column elements)(C). If all $C - Z$ values are ≤ 0, the solution is optimal. Read the values for the variables in solution from the RHS column and the value of the objective function from row Z in the RHS column. If the solution is not optimal, return to step 2.

Slack Variables.   The simplex method begins with a statement of the objective function and constraint equations. Computerized linear programming (LP) routines will automatically arrange these inputs, but for manual solutions we must construct our own simplex table. This necessitates that the constraints be stated as equalities rather than as inequalities. In maximization problems we accomplish this by adding a *slack variable* (S) to each constraint. The slack represents an unused amount, or the difference between what *is* being used and the limit of what *could be* used. For example, by adding slack variables to the inequality constraints of Example 8.6 we get new equations as shown in Table 8-5. Note that S_1 relates to the machine A constraint and S_2 to machine B.

Table 8-5

Constraint	Inequality	Equation with Slack
Machine A hr	$4X + 6Y \leq 12$	$4X + 6Y + S_1 = 12$
Machine B hr	$8X + 4Y \leq 16$	$8X + 4Y + S_2 = 16$

The machine A constraint now says 4 hours times the number of units of X produced plus 6 hours times the number of units of Y produced plus slack hours equals 12 hours. Thus, if one unit of X and one of Y are produced, we have 2 hours of slack time S on machine A, since $4(1) + 6(1) + 2 = 12$. If no X or Y is produced, we ''produce'' all slack, and $S_1 = 12$.

Initial Solution.   The simplex method always begins with a feasible solution wherein only slack is produced. This corresponds to the origin of the graphic solution, where both X and Y equal zero.

Each simplex table is a solution that graphically corresponds to a corner of the feasible region. We begin with a poor but feasible solution that corresponds to the origin, where only slack is produced, that is, zero profit. Thus the slack variables (for example, S_1 and S_2) are *in the solution*, and the other decision variables (X and Y) are not in the solution (that is, have values of zero).

Example 8.7   Arrange the objective and constraint equations from Example 8.6 into an initial simplex table (Table 8-6).

Objective function: Max $Z = \$10X + \$30Y$
Constraints: • Machine A hr: $4X + 6Y \leq 12$
                             Machine B hr: $8X + 4Y \leq 16$

Table 8-6 Simplex Format

$C \rightarrow$ $\downarrow$	Variables in Solution	10	30	0	0	Solution Values (RHS)
			Decision Variables			
		X	Y	S_1	S_2	
0	S_1	4	6	1	0	12
0	S_2	8	4	0	1	16
	Z	0	0	0	0	0
	$C - Z$	10	30	0	0	0

Elements of the Simplex Table.   The *central portion* of the simplex table consists of the coefficients of the constraint equations from:

$$4X + 6Y + 1S_1 + 0S_2 = 12$$
$$8X + 4Y + 0S_1 + 1S_2 = 16$$

Note that a one (1) has been assigned to the slack variable associated with its own constraint, and a zero (0) has been assigned to the other slack variable.

The variables-in-solution column tells *what variables* are in solution (in this case, only slack), and the solution-values column gives the *amount in solution*. The numbers come from the right-hand side (RHS) of the constraint equations (in this case, 12 hours of slack for machine A and 16 hours of slack for machine B).

The C in the upper left corner is both a row heading and column heading. It specifies the amount of contribution to the objective function from each unit of the variables it refers to. Thus, each unit of X

(cleaner) contributes \$10 to profits, and each unit of Y (polisher) contributes \$30 to profits, but the slack time from machines A and B yields \$0 contribution for both S_1 and S_2.

The Z row in the table shows the *opportunity cost*, or the amount of contribution that must be given up to introduce (or produce) one unit (or one more unit) of the variable in each column. It is computed for each column by multiplying the elements of the column by the contribution in the C column and then adding. For example, the Z value for column X is $(4 \times 0) + (8 \times 0) = 0$. This means that to introduce one unit of X (cleaner) into solution, we must give up 4 hours of slack time on machine A at a cost of \$0 and 8 hours of slack time on machine B, also at a cost of \$0. The Z value for the RHS column represents the *total contribution* from variables currently in solution. Because this (initial) solution is to "produce" 12 hours of slack on machine A (at \$0 contribution) and 16 hours of slack on machine B (at \$0 contribution), our total profit from this initial solution is zero. The Z row in the initial solution always has zeros, but it changes as the solution progresses.

The values in the bottom $(C - Z)$ row represent the *net contribution* from introducing one unit of the column variable into solution. In the initial table, they are simply the coefficients of the objective function followed by zeros for the slack variable columns. Thus, we would increase the value of the objective function by a full \$10 for each unit of X produced and by \$30 for each unit of Y produced, because nothing but worthless slack must be given up to introduce X or Y at this stage. Producing more slack would obviously not improve profits:

Computational Methodology. The solution methodology for maximization problems involves selecting a pivot column and row and revising the table values until all quantities in the bottom row are less than or equal to zero.

Example 8.8   Use the simplex method to solve the linear programming problem of Example 8.7.

We will follow the steps of the simplex procedure listed in Table 8-4.

(1)   The objective and constraints are:

$$\text{Max } Z = \$10X + 30Y$$
$$\text{Subject to:} \quad 4X + 6Y + 1S_1 + 0S_2 = 12$$
$$8X + 4Y + 0S_1 + 1S_2 = 16$$

This yields the simplex table (Table 8-7) developed in Example 8.7.

Table 8-7

$C \rightarrow$ $\downarrow$	Variables in Solution	10 X	30 Decision Variables Y	0 S_1	0 S_2	Solution Values (RHS)	
0	S_1	4	⑥	1	0	12	$\frac{12}{6} = 2$ (minimum)
0	S_2	8	4	0	1	16	$\frac{16}{4} = 4$
	Z	0	0	0	0	0	
	$C - Z$	10	30 ↑	0	0		

(2)   The *pivot column* has the largest positive number (30) in the bottom row.

(3)   The *pivot row* has the smallest ratio:

$$\frac{12}{6} = 2 \qquad \frac{16}{4} = 4$$

Therefore row 1 is the pivot row.

(4)   The pivot is *circled*.

(5) Divide each value in the pivot row by the pivot (6), and enter the values in a new table (Table 8-8).

Table 8-8

	X	Y	S_1	S_2	RHS
Y	$\frac{2}{3}$	1	$\frac{1}{6}$	0	2

(6) Generate other rows for the next table such that all elements in the pivot column equal zero.
We begin with the S_2 row, which has a 4 in the Y column. Multiply the new row (from step 5 above) by the negative of the value we wish to convert (-4), and add it to the old S_2 row. Multiply the new row by -4. The result is shown in Table 8-9.

Table 8-9

	X	Y	S_1	S_2	RHS
	$-4(\frac{2}{3})$	$-4(1)$	$-4(\frac{1}{6})$	$-4(0)$	$-4(2)$
to get result	$-\frac{8}{3}$	-4	$-\frac{2}{3}$	0	-8
Add to old row of	8	4	0	1	16
to get new row	$\frac{16}{3}$	0	$-\frac{2}{3}$	1	8

And enter into new table (Table 8-10):

Table 8-10

$C \rightarrow$		10	30	0	0	
$\downarrow$	Variables in Solution	X	Y	S_1	S_2	RHS
30	Y	$\frac{2}{3}$	1	$\frac{1}{6}$	0	2
0	S_2	$\frac{16}{3}$	0	$-\frac{2}{3}$	1	8
	Z					

If there were more rows to convert, we would repeat this step for the next row. Since there are no more, we go on to compute Z and $C - Z$.

(7) Values in the Z row are Σ (column elements)(C). For example:

$$\text{for } X: \quad Z = (\tfrac{2}{3})(30) + (\tfrac{16}{3})(0) = 20$$
$$\text{for } Y: \quad Z = 1(30) + 0(0) = 30$$
$$\text{for } S_1: \quad Z = \tfrac{1}{6}(30) - \tfrac{2}{3}(0) = 5$$
$$\text{for } S_2: \quad Z = 0(30) + 1(0) = 0$$
$$\text{for RHS}: \quad 2(30) + 8(0) = 60$$

After entering these and the $C - Z$ values in the next matrix (Table 8-11) we have:

Table 8-11

$C \rightarrow$		10	30	0	0	Solution
$\downarrow$	Variables in Solution		Decision Variables			Values
		X	Y	S_1	S_2	(RHS)
30	Y	$\frac{2}{3}$	1	$\frac{1}{6}$	0	2
0	S_2	$\frac{16}{3}$	0	$-\frac{2}{3}$	1	8
	Z	20	30	5	0	60
	$C - Z$	-10	0	-5	0	

Repeat steps 2 through 7 until all values in the bottom row are ≤ 0. Since all values are ≤ 0, the optimal solution is already reached. Variables in solution are identified by columns in the central portion of the table that have one entry of 1 and remaining values of zero. The solution values are given in the right-hand column as seen in Table 8-12.

Table 8-12

	X	Y	S_1	S_2	RHS
	—	1	—	0	2
	—	0	—	1	8
Z	—	—	—	—	60

Therefore,

$$X = \text{not in solution}$$
$$Y = 2 \text{ units}$$
$$Z = \$60$$

Note that the slack variable associated with constraint 2 also has a 1 and zeros, which signifies that we have slack in solution and that the constraint is not binding. Thus we have only one (nonslack) decision variable in the solution (Y) and one binding constraint (number 1). This agrees with the fundamental theorem of linear programming, which states that the number of (nonslack) decision variables in solution always equals the number of constraints that are binding.

This solution is the same as that given for Example 8.6, so you may want to review that example for further interpretation of the output.

Optional Solution. Referring back to the initial solution (step 1 of Example 8.8), note the configuration of 1s and 0s in the two rows directly below the slack variable symbols. They form what is called an *identity matrix*. It is a square array of numbers with 1s on the diagonal and 0s elsewhere. A problem with three constraints would have three rows consisting of 1 0 0 and 0 1 0 and 0 0 1.

In a simplex table, a decision variable column that has a (positive) 1 with zeros elsewhere in that column identifies a variable in solution. In our initial table the 1 and 0 below the slack variables indicated that both S_1 and S_2 were in solution (that is, being produced). The value, or amount, of the variable in solution is given in the RHS column. (The RHS values for variables not in solution are automatically equal to zero.) Thus our initial table (8-7) had 12 hours of machine A slack time and 16 hours of machine B slack time in solution, and no cleaner or polisher was produced (that is, neither X nor Y was in the initial solution, so $X = 0$ and $Y = 0$).

Example 8.9 presents and interprets the optimal solution to the problem posed in Example 8.8. The graphic solution is repeated to aid in understanding the corresponding elements of the simplex solution.

Example 8.9   Interpret the optimal solution to the cleaner-polisher problem in Example 8.8.

The solution shown in Table 8-13 is optimal because all values in the $C - Z$ row are less than or equal to zero.

Table 8-13   Simplex Solution

$C \rightarrow$ $\downarrow$	Variables in Solution	10	30 Decision Variables	0	0	Solution Values (RHS)
		X	Y	S_1	S_2	
30	Y	$\frac{2}{3}$	1	$\frac{1}{6}$	0	2
0	S_2	$\frac{16}{3}$	0	$-\frac{2}{3}$	1	8
	Z	20	30	5	0	60
	$C - Z$	-10	0	-5	0	(Profit)

Variables in solution

Two columns have 1s and 0s.

Y is in the solution with a (RHS) value of 2.

S_2 is in the solution with a (RHS) value of 8.

The value of $Y = 2$ can be read from the RHS column in the table (and agrees with the graph, Fig. 8-10).

$C \rightarrow$ $\downarrow$	Variables in Solution	10 X	30 Y	0 S_1	0 S_2	Solution Values (RHS)
		\multicolumn{4}{c}{Decision Variables}	
30	Y	$\frac{2}{3}$	1	$\frac{1}{6}$	0	2
0	S_2	$\frac{16}{3}$	0	$-\frac{2}{3}$	1	8
	Z	20	30	5	0	60
	$C - Z$	-10	0	-5	0	(Profit)

(*a*) Simplex solution

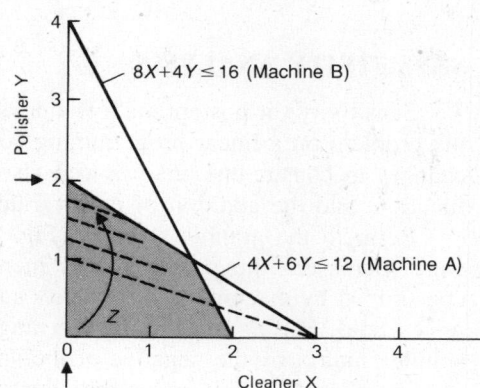

(*b*) Graphic solution

Fig. 8-10

Recall that each unit requires 4 hours of machine B time, so 2 units of Y use 8 hours of the 16 hours available. This leaves 8 hours of machine B slack, as indicated by the RHS value of 8 for S_2, which is also in solution.

Value of the objective function

$Z = \$60$ profit, as shown in the RHS column. This comes from producing two units of Y at \$30 each plus eight units of slack at \$0.

Values in the $C - Z$ row

The figures in the bottom row of Table 8-13 ($-10, 0, -5, 0$) reveal the following:

(1) (-10) To produce one can of X (cleaner) would reduce profits by \$10 because it would take machine A time away from the production of Y.

   Note: The \$10 amount is explained by the X column. Introducing one unit of "out variable" X would:

	X	
Y	$\frac{2}{3}$	←reduce Y by $\frac{2}{3}$ units @ \$30/unit = \$20 reduction
S_2	$\frac{16}{3}$	←reduce S_2 by $\frac{16}{3}$ units @ \$0/unit = \$0 reduction
Z	20	←for a total amount of \$20 − \$0 = \$20 cost
$C - Z$	-10	←which is offset by \$10 profit from each unit of X

The result is a net (loss) contribution of $C - Z = \$10 - \$20 = -\$10$.

(2) (*0*) The first zero indicates that Y is in solution (being produced).

(3) (-5 *and* 0) These two values are referred to as *shadow prices*. *Shadow prices go with constraints* and show the amount of change in the objective function that would result from each unit of change in the constraint. Thus they show the net effect of increasing (or decreasing) the slack or idle time of machines A and B by one unit.

(4) (-5) Since machine A is fully utilized, to take 1 hour out of production and acquire 1 hour of idle time

would reduce profit by \$5. (Profit from Y is \$30 for each 6 hours of work on A, that is, a rate of \$5 per hour.) Conversely, if another hour could be made available, say by shifting a current job from A, the time on A could be profitably utilized at a profit rate of \$5 per hour.

(5) (O) The zero corresponding to the constraint of machine B signifies that machine B already has slack time (see Fig. 8-10). Increasing B's available time (or decreasing it) by one unit would have no effect on profits.

SENSITIVITY ANALYSIS

Sensitivity (or postoptimality) analysis is concerned with the effect of changes in the parameters of the problem on a linear programming solution. While our concern here will be limited to the effect of changes in constraints, analysis can also be made to determine the effect of changes in the objective function, and the addition of new variables and new constraints.

Refer to the graphic solution (Fig. 8-10) of Example 8.6. If the time availability for machine A were increased, the profits would increase (at \$5 per additional hour) until they were ultimately constrained by machine B. Sensitivity analysis enables one to determine the ranges over which shadow prices hold. For $\leq$ constraints, the range can be determined by dividing the RHS value of the final simplex matrix by the negative of the values in the columns with shadow prices. The *smallest positive quotient* then tells how much the constraint can be changed until another constraint becomes binding.

Example 8.10 Determine the effect of changes in the binding constraint shown in the optimal solution of Example 8.9.

Machine A is the only active (explicit) constraint. The sensitivity ratios for this constraint are:

$$\text{For } Y: \qquad \frac{\text{RHS}}{-S_1} = \frac{2}{-1/6} = -12$$

$$\text{For } S_2: \qquad \frac{\text{RHS}}{-S_1} = \frac{8}{2/3} = \frac{24}{2} = 12$$

The smallest positive ratio is the 12 associated with S_2. This suggests that constraint A may be relaxed by 12 hours (to 24 hours) before the machine B constraint begins to limit the solution.

A glance at the graphic solution shows that as constraint A is relaxed (i.e., as more hours are added), the machine B constraint takes effect at $Y = 4$. At that point, the profit would be $Z = \$10X + \$30Y = \$10(0) + \$30(4) = \$120$. Also, at $Y = 4$, both machines would be fully utilized, as can be shown by substituting values for X and Y into the constraint equations.

Old Constraint		**Revised Limit**	**at $X = 0$, $Y = 4$**
Machine A:	$4X + 6Y \leq 12$	$4X + 6Y \leq 24$	$4(0) + 6(4) = 24$
Machine B:	$8X + 4Y \leq 16$	no change	$8(0) + 4(4) = 16$

MINIMIZATION AND OTHER FORMS OF CONSTRAINTS

The simplex procedure can also be used to solve cost minimization problems which have objective functions of the form $\text{Min } Z = AX_1 + BX_2 + \cdots + MX_n$. Constraints in minimization problems are often of a $\geq$ type rather than the $\leq$ type we just encountered. In these types of constraints we must subtract a *surplus* variable (instead of adding a slack variable). To handle both $=$ and $\geq$ types of constraints, artificial variables are also used (in addition to the S variables). The artificial variables serve only to state the equations in a form suitable for the simplex table and have no other meaning. They are typically assigned very large coefficients (M's), which will quickly drive them out of solution. Problem 8.5 illustrates a graphic solution to a minimization problem, and Prob. 8.11 illustrates the simplex formulation for a different minimization problem.

Questions and Solved Problems

PRODUCT RELIABILITY AND FAILURE RATES

8.1 The reliabilities of system components are shown in Fig. 8-11. What is the reliability of the system AB?

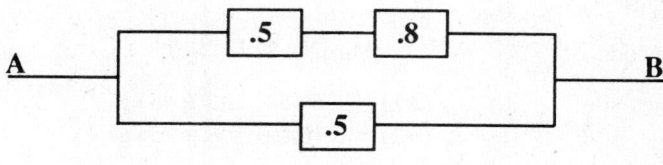

Fig. 8-11

$$R_{s1} = (R_1)(R_2) = (.5)(.8) = .40 \qquad R_{s2} = .5$$
$$R_s = 1 - (1 - R_{s1})(1 - R_{s2}) = 1 - (1 - .4)(1 - .5) = .70$$

8.2 Twenty shipping containers were subject to a pressure test for 300 hours, and 4 developed leaks. What would be the failure rate in number of failures per unit-year?

$$\text{FR}_n = \frac{F}{\text{TT} - \text{NOT}} = \frac{4}{(20)(300)\ \text{hr} - 4(300/2)\ \text{hr}} = .00074\ \text{failures/unit-hr}$$

$$= (.00074\ \text{failure/unit-hr})(24\ \text{hr/day})(365\ \text{day/yr}) = 6.5\ \text{failures/unit-yr}$$

8.3 A manufacturer of computer disk drives has tested 20 units continuously for 4,000 hours and found that 4 units stopped working during the test. What is the MTBF for the disk drives?

$$\text{MTBF} = \frac{\text{TT} - \text{NOT}}{F} = \frac{(20)(4,000)\ \text{hr} - 4(4,000/2)\ \text{hr}}{4\ \text{failures}} = 18,000\ \text{unit-hr/failure}$$

LINEAR PROGRAMMING (GRAPHIC)

8.4 Find the slope of the objective function: Max $Z = 10X + 15Y$.

The slope form is $Y = mX + b$, where $m = $ slope.

Rearranging: $\qquad\qquad\qquad\qquad 15Y = -10X + Z$

$$Y = -\frac{10}{15}X + \frac{Z}{15}$$

Therefore, slope is $-10/15$, or $-2/3$.

8.5 A textile mill has received an order for fabric specified to contain at least 45 kilograms of wool and 25 kilograms of nylon. The fabric can be woven out of any suitable mix of two yarns (A and B). Material A costs $2 per kilogram, and B costs $3 per kilogram. They contain the proportions of wool, nylon, and cotton (by weight) shown in Table 8-14.

Table 8-14

Yarn	Wool (%)	Nylon (%)	Cotton (%)
Material A	60	10	30
Material B	30	50	20

What quantities (kilograms) of A and B yarns should be used to minimize the cost of this order?

(1) Objective function is: $\qquad\qquad$ Min $C = \$2A + \$3B$

The constraints are: $\qquad$ (#1) $\quad$ $.60A + .30B \geq 45$ kg

$\qquad\qquad\qquad\qquad\qquad\quad$ (#2) $\quad$ $.10A + .50B \geq 25$ kg

(2) See Fig. 8-12 for a graph of the two constraints.
For constraint #1: When $A = 0$, then $B = 150$; and when $B = 0$, then $A = 75$.
For constraint #2: When $A = 0$, then $B = 50$; and when $B = 0$, then $A = 250$.
Note: The feasible region is shown shaded in Fig. 8-12. Because this is a minimization problem, the feasible region is above and to the right of the constraints.

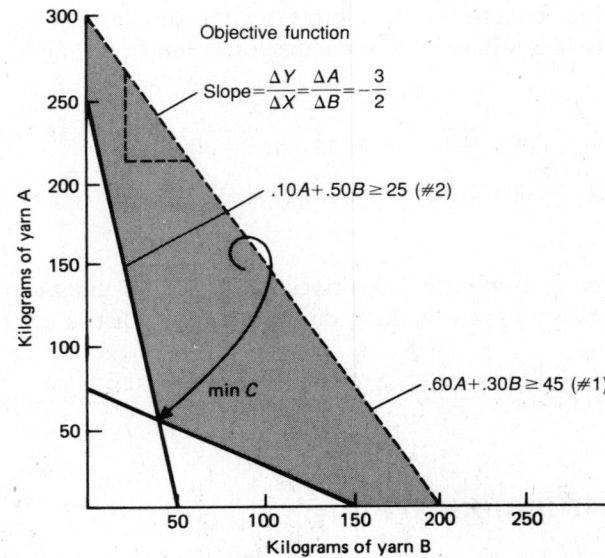

Fig. 8-12

(3) The slope of the objective is: $\qquad\qquad$ $2A = -3B + C$

$$A = -\frac{3}{2}B + \frac{C}{2}$$

Therefore, $\qquad\qquad\qquad\qquad\qquad$ Slope $= -\frac{3}{2}$

This slope of $-3/2$ is plotted as a dashed line in Fig. 8-12 by marking off 300 units (negative) in A for each 200 units (positive) in B.

(4) Move objective function to optimize. The slope of the objective is moved toward the origin until constrained by the wool and nylon supply constraints.

(5) Read the solution values. From the figure it appears that we optimize at approximately $A = 55$ kg and

$B = 40$ kg. Simultaneous solution of the two (binding) constraint equations yields the more precise answer of $A = 55$ kg and $B = 39$ kg.

LINEAR PROGRAMMING (SIMPLEX)

8.6 Mobile Phone Company makes a profit of \$5 on each Model X and \$20 on each Model Y. Each phone requires the following time (in minutes) on the cleaning and testing machines as shown in Table 8-15.

Table 8-15

	X Requirements	Y Requirements	Time Available
Cleaning	2	4	10
Testing	6	3	12

(*a*)   State the objective function and constraints. (*b*) Arrange the equations in a simplex format.

(*a*)   Objective function: Max $Z = 5X + 20Y$

   Constraints:

   Cleaning $2X + 4Y \le 10$
   Testing $6X + 3Y \le 12$

(*b*)   See Table 8-16.

Table 8-16

C →		5	20	0	0	
↓	Variables in Solution	X	Y	S_1	S_2	RHS
0	S_1	2	4	1	0	10
0	S_2	6	3	0	1	12
	Z	0	0	0	0	0
	$C - Z$	5	20	0	0	

8.7 The initial matrix of a maximization linear programming problem was as shown in Table 8-17, where the decision variables are designated A, B, etc. (*a*) State the original constraint equations. (*b*) How many decision variables are there? (*c*) State the objective function. (*d*) What is the value of the first pivot?

Table 8-17

C →		4	8	6	0	0	0	
↓	Variables in Solution							RHS
		5	9	0	1	0	0	36
		0	8	5	0	1	0	24
		2	0	5	0	0	1	7
		0	0	0	0	0	0	0
		4	8	6	0	0	0	

(*a*) $5A + 9B \le 36$, $8B + 5C \le 24$, and $2A + 5C \le 7$

(b)   Three

(c)   Max $Z = 4A + 8B + 6C$

(d)   Ratios are $36/9 = 4$ and $24/8 = 3$. Therefore, pivot is the 8.

8.8   A commercial fertilizer manufacturer produces three grades (W, X, and Y) which net the firm $40, $50, and $60 in profits per ton, respectively. The products require the labor and materials per batch that are shown in Table 8-18.

Table 8-18

	W	X	Y	Total Available
Labor hours	4	4	5	80 hr
Raw material A (lb)	200	300	300	6,000 lb
Raw material B (lb)	600	400	500	5,000 lb

What mix of products would yield maximum profits?

The objective function is: Max $Z = 40W + 50X + 60Y$

The constraints are:

Labor	$4W + 4X + 5Y \leq 80$
Material A	$200W + 300X + 300Y \leq 6,000$
Material B	$600W + 400X + 500Y \leq 5,000$

(1)   Using the simplex method, we set up Table 8-19 as follows:

Table 8-19

$C \rightarrow$ $\downarrow$	Variables in Solution	40	50	60 Decision Variables	0	0	0	Solution Values (RHS)
		W	X	Y	S_1	S_2	S_3	
0	S_1	4	4	5	1	0	0	80
0	S_2	200	300	300	0	1	0	6,000
0	S_3	600	400	(500)	0	0	1	5,000
	Z	0	0	0	0	0	0	0
	C − Z	40	50	60 ↑	0	0	0	

(2)   The pivot column has the largest positive number (60) in the bottom row—column Y. (Therefore, introduce Y into solution.) *Note:* This is because Y contributes the most ($60) to profits.

(3)   The pivot row has the smallest result of:

$$\frac{80}{5} = 16 \qquad \frac{6,000}{300} = 20 \qquad \frac{5,000}{500} = 10$$

Therefore row 3 is the pivot row, and S_3 should be removed from the solution.

Note: S_3 is removed because raw material B is the most restrictive constraint on Y. As computed above, the 80 labor hours divided by 5 hours per unit of Y will permit production of 16 units of Y, and raw material A will permit 6,000 lb ÷ 300 lb per unit = 20 units of Y, but there is only enough raw material B for $5,000 \div 500 = 10$ units.

(4)   The pivot is circled.

(5)   Dividing values in the pivot row by 500, we obtain the new Y row.

	W	X	Y	S_1	S_2	S_3	RHS
Y	$\frac{6}{5}$	$\frac{4}{5}$	1	0	0	$\frac{1}{500}$	10

(6) Convert other values in the pivot column to zero. For S_1 (row 1), multiply new Y row by -5 and add to old S_1 row. For S_2 (row 2), multiply new Y row by -300 and add to old S_2 row.

(7) Compute Z and $C - Z$ values (Table 8-20), and check for optimality.

Table 8-20

$C \rightarrow$ $\downarrow$	Variables in Solution	40 W	50 X	60 Decision Variables Y	0 S_1	0 S_2	0 S_3	Solution Values (RHS)
0	S_1	-2	0	0	1	0	$-\frac{1}{100}$	30
0	S_2	-160	60	0	0	1	$-\frac{3}{5}$	$3{,}000$
60	Y	$\frac{6}{5}$	$\textcircled{$\frac{4}{5}$}$	1	0	0	$\frac{1}{500}$	10
	Z	72	48	60	0	0	$\frac{3}{25}$	600
	$C - Z$	-32	2 $\uparrow$	0	0	0	$-\frac{3}{25}$	

(8) Since the X column has a positive value in the bottom row, repeat.

(9) The pivot column is column X; therefore, introduce X into solution.

(10) Pivot row:

$$\frac{30}{0} = \infty \qquad \frac{3{,}000}{60} = 50 \qquad \frac{10}{\frac{4}{5}} = 12.5$$

Therefore, remove Y from solution, and introduce X.

(11) Dividing values in the pivot row by $4/5$, we get a new pivot row (X-row).

	W	X	Y	S_1	S_2	S_3	RHS
X	$\frac{3}{2}$	1	$\frac{5}{4}$	0	0	$\frac{1}{400}$	$\frac{25}{2}$

(12) By converting other values in the pivot column into zero, we get new values:
S_1 (row 1): The column value is already zero. Leave as is.
S_2 (row 2): Multiply the new X row by -60 and add to the old S_2 row.

(13) Compute the Z and $C - Z$ values (Table 8-21), and check for optimality.

(14) Since no values are >0 in the bottom row, the solution is complete. The only variable in solution is X, and $\frac{25}{2} = 12.5$ units are produced. The profit is \$625.

Table 8-21

$C \rightarrow$ $\downarrow$	Variables in Solution	40 W	50 X	60 Decision Variables Y	0 S_1	0 S_2	0 S_3	Solution Values (RHS)
0	S_1	-2	0	0	1	0	$-\frac{1}{100}$	30
0	S_2	-250	0	-75	0	1	$-\frac{9}{20}$	$2{,}250$
50	X	$\frac{3}{2}$	1	$\frac{5}{4}$	0	0	$\frac{1}{400}$	$\frac{25}{2}$
	Z	75	50	$\frac{125}{2}$	0	0	$\frac{1}{8}$	625
	$C - Z$	-35	0	$-\frac{5}{2}$	0	0	$-\frac{1}{8}$	

Comment: The initial matrix (Table 8-19) was a feasible solution at the W, X, Y origin, where no product was produced and the profit was zero. The \$60 in the bottom row indicated that for each unit of Y introduced, the objective function would be increased by \$60. The next matrix (Table 8-20) called for production of 10 units of Y (only) for a profit coefficient of \$600. However, the positive 2 in the bottom row under the X variable column indicated that for every unit of X introduced, the objective function would be increased by \$2. The final solution (Table 8-21) called for 12.5 units of X, which raised the profit an additional \$25 to a total of \$625. This is the best that can be obtained.

8.9 Table 8-22 shows a simplex solution. Find the upper sensitivity limit of constraint 3.

Table 8-22 Final Solution

$$\text{Sensitivity limits} = \frac{\text{RHS}}{-\text{exch. coef.}}$$

$C \rightarrow$		20	60	0	0	0		
$\downarrow$	Sol	X	Y	S_1	S_2	S_3	RHS	
0	S_1	0	0	1	0	-2	20	$\rightarrow 20 \div (-(-2)) = +10$
0	S_2	4	0	0	1	0	80	$\rightarrow 80 \div (-0) =$
60	Y	.5	1	0	0	.5	20	$\rightarrow 20 \div (-.5) = -40$
	Z	30	60	0	0	30	1200	
	$C - Z$	-10	0	0	0	-30		

Smallest positive value is 10, so the constraint could be increased by 10.

8.10 The initial and final simplex solutions to an LP problem are as shown in Tables 8-23 and 8-24. One of the constraint equations was $6X + 4Y \leq 12$. By how many units could the RHS value of 12 *be increased* until another constraint comes into effect to limit the solution?

Table 8-23 Initial Solution

$C \rightarrow$		4	2	0	0	
$\downarrow$	Sol	X	Y	S_1	S_2	RHS
0	S_1	6	4	1	0	12
0	S_2	2	8	0	1	16
	Z	0	0	0	0	0
	$C - Z$	4	2	0	0	

Table 8-24 Final Solution

$C \rightarrow$		4	2	0	0	
$\downarrow$	Sol	X	Y	S_1	S_2	RHS
4	X	1	2/3	1/6	0	2
0	S_2	0	20/3	$-1/3$	1	12
	Z	4	8/3	2/3	0	8
	$C - Z$	0	$-2/3$	$-2/3$	0	

The constraint referred to is that associated with S_1 (from initial solution). From the final solution, $2 \div (-1/6) = -12$, and $12 \div (1/3) = 36$. Therefore, the RHS value of 12 can be increased by 36 units.

MINIMIZATION

8.11 A livestock supplement is to be mixed to contain exactly 25 pounds of vitamin A, at least 15 pounds of B, and at least 40 pounds of C. The supplement is made from two commercial feeds. Each pound of feed #1 contains 2 ounces of A, 6 ounces of B, and 4 ounces of C, and costs \$5. A pound of feed #2 contains 4 ounces of A, 1 ounce of B, and 3 ounces of C, and costs \$3. Let X_1 be the pounds of feed #1 and X_2 be the pounds of feed #2. (*a*) Formulate the objective functions and constraints for a problem that will minimize the cost of the food supplement while satisfying the vitamin content requirements. (*b*) Arrange the problem in an initial simplex format.

(*a*)
$$\text{Min } Z = 5X_1 + 3X_2$$

Vitamin A: $2X_1 + 4X_2 = 400$ (i.e., 25 lb @ 16 oz/lb = 400 oz)

Vitamin B: $6X_1 + 1X_2 \geq 240$

Vitamin C: $4X_1 + 3X_2 \geq 640$

(b) The initial simplex table (Table 8-25) is shown below. Note that the = constraint (vitamin A requirements) requires one artificial variable (A_1) to ensure its equality. The two $\geq$ constraints each require a slack variable and an artificial variable. The slack variables in $\geq$ constraints represent amounts that must be subtracted from the constraint value; hence, they must have a negative sign. All artificial variables are assigned an extremely large cost M to ensure that they are driven out of solution by the simplex iterative procedure.

Table 8-25

| $C \rightarrow$ | | 5 | 3 | M | 0 | M | 0 | M | Solution |
| $\downarrow$ | | | | Decision Variables | | | | | Values |
	Variables in Solution	X_1	X_2	A_1	S_2	A_2	S_3	A_3	RHS
M	A_1	2	4	1	0	0	0	0	320
M	A_2	6	1	0	-1	1	0	0	240
M	A_3	4	3	0	0	0	-1	1	640
	Z	$12M$	$8M$	M	$-M$	M	$-M$	M	$1{,}200M$
	$C - Z$	$5 - 12M$	$3 - 8M$	0	M	0	M	0	

The solution procedure is the same as in maximization problems except that the variable with the most *negative* value in the bottom ($C - Z$) row is always the one introduced. Problems such as this, or others that involve more than two or three variables or constraints, are most easily solved on a computer. The solution is to produce 136 units of X_1 and 32 units of X_2 for a minimum total cost of $776.

Supplementary Questions and Problems

PRODUCT LIABILITY AND FAILURE RATES

8.12 In response to a customer request for failure-rate data, an instrument manufacturer tested a group of 30 instruments over a 2,000-hour test period and found that 4 failed. Find (a) $FR_\%$ and (b) FR_n (in failures per unit per year). *Ans.* (a) 13.3 percent (b) .625 failure per year

8.13 The purification system in a water treatment plant has three components in series (R_1, R_2, and R_3). The component reliabilities for a 3-month period *remain constant* and are as shown in Fig. 8-13. At the end of each 3-month period all components are replaced regardless of the length of service. In the meantime, each time any component breaks down, the cost of downtime and repair is $300. What is the annual expected cost of downtime and repair? *Ans.* $519.60

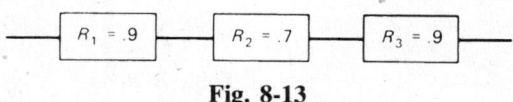

Fig. 8-13

8.14 The maintenance manager for a nationwide trucking firm has found that a substantial savings in tire cost can be gained by contracting with a tire manufacturer to replace tires on the entire fleet of trucks at one time. For safety purposes, the manager feels this should be done at the time 15 percent of the tires are worn out. If tire life is normally distributed with a mean of 30 months and standard deviation of 3 months, when should the replacement take place? *Ans.* 26.9 months

8.15 Pressure relief valves in a chemical plant are known to have a MTBF of 30 years. Assuming the failure rate in constant, what is the likelihood a given will function without failure for 10 years? *Ans.* .72

LINEAR PROGRAMMING

8.16 The objective function for a linear programming problem concerned with product mix is Max $Z = 25X + 5Y$. What is the slope of this objective?   *Ans.* -5

8.17 Use the graphic method of linear programming to solve Prob. 8.6.
Ans.   Slope of objective $= -1/4$, $X = 0$, $Y = 2\frac{1}{2}$, $Z = 50$.

8.18 Sunstroke Paint Co. makes a profit of \$5 per gallon on its oil-based paint and \$7 per gallon on its latex paint. Both paints contain two ingredients, A and B. The oil-based paint contains 80 percent A and 20 percent B, whereas the latex paint contains 40 percent A and 60 percent B. Sunstroke currently has 20,000 gallons of A and 8,000 gallons of B in inventory and cannot obtain more at this time. The company wishes to use linear programming to determine the appropriate mix of oil-based and latex paint to produce to maximize its total profit. (*a*) Let A be X_1 and B be X_2. State the objective function and constraints. (*b*) What is the slope of the objective function?
Ans.   (*a*) Max $5X_1 + 7X_2$, $.8X_1 + .4X_2 \leq 20,000$, $.2X_1 + .6X_2 \leq 8,000$ (*b*) $-5/7$ or $-7/5$ (depending upon choice of variables)

8.19 Max $Z = 6X + 9Y$ subject to $2X + 6Y \leq 30$, $4X + 3Y \leq 36$.   *Ans.* $X = 7$, $Y = 8/3$, $Z = 66$

8.20 Precast Co. can produce grade A material, which yields a profit of \$1 per unit, and grade B material, which yields a profit of \$2 per unit. Each unit of A requires 2 hours of machining and 1 hour of finishing. Each unit of B requires 1 hour of machining and 3 hours of finishing. If 200 hours of machining capacity and 300 hours of finishing capacity are available, (*a*) what amounts of A and B should be produced to maximize profits, and (*b*) what is the profit?   *Ans.*   (*a*) $A = 60$, $B = 80$ (*b*) \$220

8.21 A data processing manager wishes to formulate a linear programming model to help him decide how to use his personnel as programmers (X_1) or systems analysts (X_2) in such a way as to maximize revenues (Z). Each programmer earns \$40 per hour, and each systems analyst earns \$50 per hour. Programming work during the coming week is limited to 50 hours, maximum. The production scheduler has also specified that the total programming time plus two times the systems analysis time be limited to 80 hours or less. State the objective function and constraints.
Ans.   Max $Z = 40X_1 + 50X_2$, $X_1 \leq 50$, $X_1 + 2X_2 \leq 80$

8.22 Solve Prob. 8.21 via the graphic method, using X_1 for the horizontal axis and X_2 for the vertical axis.
Ans. $X_1 = 50$, $X_2 = 15$, $Z = \$2,750$

8.23 Set up the initial simplex table (Table 8-26) for Prob. 8.21.
Ans.

Table 8-26

$C \rightarrow$ $\downarrow$ Variables in Solution	40 X_1	50 Decision Variables X_2	0 S_1	0 S_2	RHS
0 S_1	1	0	1	0	50
0 S_2	1	2	0	1	80
Z	0	0	0	0	0
$C - Z$	40	50	0	0	

8.24   Solve Prob. 8.21 via the simplex method.   *Ans.* $X_1 = 50$, $X_2 = 15$, $Z = \$2,750$

8.25   A manufacturer makes \$5 profit on each unit of X and \$10 on each unit of Y. Each product requires different amounts of time on each of two machines as shown (in hours) in Table 8-27. Use (*a*) the graphic and (*b*) the simplex methods to determine what quantity of X and Y should be produced to maximize profits.
Ans. $X = 0$, $Y = 4$, $Z = 40$ (*Note:* Slope of objective = slope of one of the constraints, so multiple solutions are possible down to $X = 4/5$, $Y = 18/5$.)

Table 8-27

Machine	Requirements for		Total Available
	X	Y	
A	2	4	16 hr
B	6	2	12 hr

8.26   Southern Oak Furniture Association (SOFA) has a plant in Arkansas that produces three models of chairs. The profit contributions per chair are as follows:

$$C = \text{Contemporary} = \$10$$
$$D = \text{Danish} = 15$$
$$E = \text{Early American} = 25$$

The firm's dry-kiln capacity for green lumber limits the total production of any mix of chairs to 1,000 per day. If all production went into contemporary chairs and the dry kilns did not limit production, the firm could produce 1,500 chairs, but the Danish models take 1.5 times as long, and the Early American models take twice as long as the contemporary models. Also, the Danish models require special inlaid backs, which come from a single supplier that cannot supply more than 500 per day. Assuming that the firm's retailers would accept any mix of models, use the simplex method to determine the optimal selection to maximize profits.   *Ans.*   Produce 750 Early American chairs for a profit of \$18,750.

8.27   A company producing a standard and a deluxe line of electric clothes dryers has the time requirements (in minutes) shown in Table 8-28 in departments where either model can be processed:

Table 8-28

	Standard	Deluxe
Metal frame stamping	3	6
Electric motor installation	10	10
Wiring	10	15

The standard models contribute \$30 each and the deluxe \$50 each to profits. The motor installation production line has a full 60 minutes available each hour, but the stamping machine is available only 30 minutes per hour. There are two lines for wiring, so the time availability is 120 minutes per hour. What is the optimal combination of output in units per hour? (Solve graphically.)
Ans.   standard = 2, deluxe = 4

8.28   Solve Prob. 8.27 via the simplex method.   *Ans.*   standard = 2, deluxe = 4

8.29   A perfume maker realizes a $4 profit on each bottle of (X) and $2 on each ($Y$). The two products require different hours of time on each of three mixing containers as in Table 8-29.

Table 8-29

	X Req'ts	Y Req'ts	Hrs avail.
Container 1	1	2	60
Container 2	1		40
Container 3		1	20

(*a*)   State the objective function and constraints to maximize profits.

(*b*)   Graph the constraints, and show your optimal solution.

Table 8-30

C →	Sol	4 X	2 Y	0 S_1	0 S_2	0 S_3	RHS
2	Y	0	1	1/2	−1/2	0	10
4	X	1	0	0	1	0	40
0	S_3	0	0	−1/2	1/2	1	10
	Z	4	2	1	3	0	180
	$C - Z$	0	0	−1	−3	0	

Suppose the optimal solution of the simplex is as shown in Table 8-30:
(*c*) How many bottles should be produced of X and of Y? (*d*) What amount of profit will be realized from the optimal use of resources? (*e*) Suppose container 1 had to be taken out of service for 1 hour for cleaning. What would be the net effect ($ amount) on profits of this reduction?

Ans.   (*a*) Maximize $Z = 4X + 2Y$, subject to $1X + 2Y \leq 60$ (container 1), $1X + 0Y \leq 40$ (container 2), $0X + 1Y \leq 20$ (container 3). (*b*) Graph should show constraints at $Y = 20$, $X = 40$, and a line from $X = 0$, $Y = 30$ down to $X = 60$, $Y = 0$. Slope of objective function is -2, and solution is at $X = 40$, $Y = 10$. (*c*) $X = 40$, $Y = 10$, (*d*) $180, (*e*) $1 decrease to $179.

Chapter 9

Process Planning and Analysis

Simulation

PROCESS SELECTION

Process planning consists of designing and implementing a work system to produce the desired goods or services in the required quantities at the appropriate times and within acceptable costs. The work systems range from those with very general purpose equipment that typically produce a low volume of customized products to those with highly specialized equipment that produce a high volume of more standardized products. These relationships are sometimes depicted in matrix form.

Question: What is the product-process matrix?

The product-process matrix is a synthesis of relationships associated with the four phases in the life cycle of a product (see Chap. 8). It shows that an increase in production volume over time is frequently accompanied by a shift from customized to highly standardized products. This transition from low to high volume also tends to coincide with a shift from job shop and batch processing to repetitive and continuous flow processes that support shorter cycle times. Figure 9-1 is an adaptation from the work of Robert Hayes and Stephen Wheelwright.

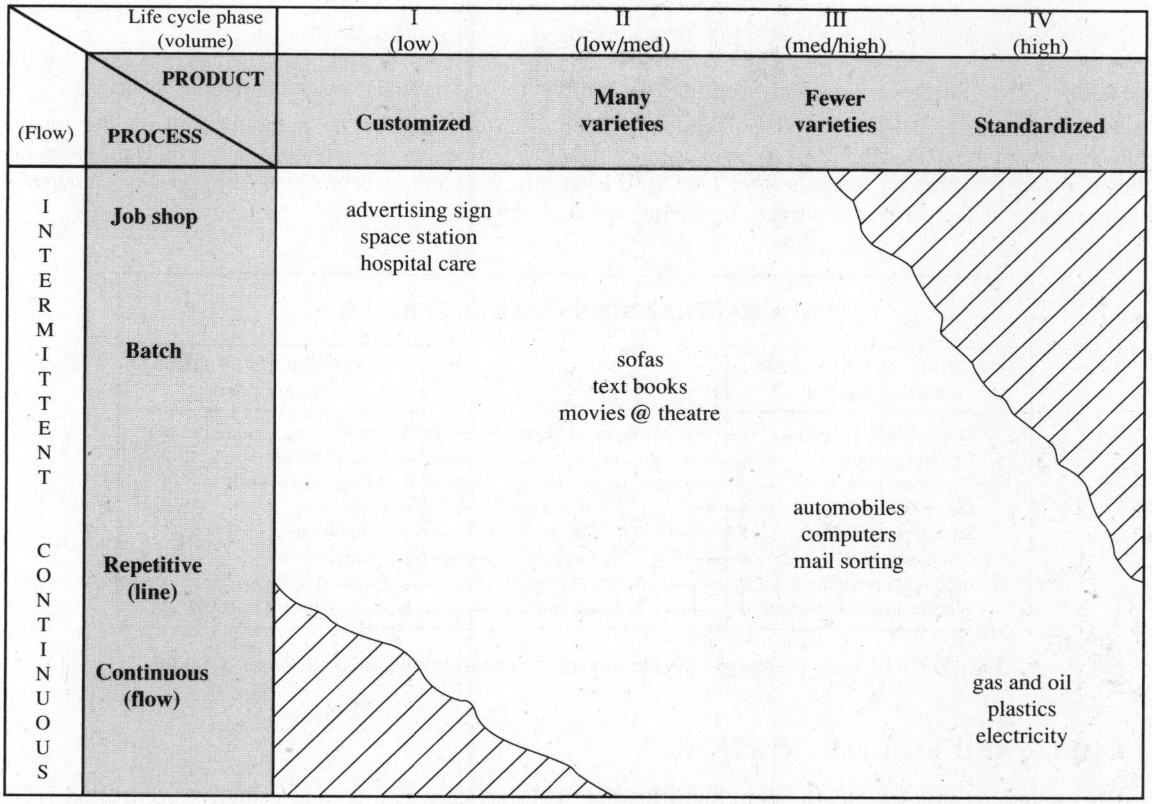

(Flow)	Life cycle phase (volume)	I (low)	II (low/med)	III (med/high)	IV (high)
	PRODUCT / PROCESS	Customized	Many varieties	Fewer varieties	Standardized
I N T E R M I T T E N T	**Job shop**	advertising sign space station hospital care			
	Batch		sofas text books movies @ theatre		
C O N T I N U O U S	**Repetitive (line)**			automobiles computers mail sorting	
	Continuous (flow)				gas and oil plastics electricity

Fig. 9-1 Product-process matrix

Question: The production process dictates how a product will be produced. What are the key considerations involved in selecting a production process?

The process is a function of (*a*) *the type of work flow* (i.e., the layout) and (*b*) *the technology employed in the work center.*

Question: What are the four major types of work flow arrangements?

As depicted in Fig. 9-1, work flow can be designed for (*a*) job shop, (*b*) batch, (*c*) repetitive assembly line, or (*d*) continuous-flow operations, plus a number of hybrid, or mixed, arrangements.

The most effective processing technology will very likely differ from one type of layout to the next. Sometimes the type of layout limits the technology that can be used, and at other times the technology available dictates the layout. Over time, the tendency is to move from job shop toward continuous-flow-type operations.

INTERMITTENT AND CONTINUOUS PRODUCTION SYSTEMS

Question: Distinguish between (*a*) intermittent and (*b*) continuous production systems.

(*a*) *Intermittent systems* are used to produce small quantities (or batches) of many different items on relatively general-purpose equipment. Processing equipment and personnel are located according to function, and products flow through the facilities on irregular paths. Jobs are individually routed, scheduled, and controlled on a job or shop *order-control* system. The goods or services are often customized or *made to order.*

(*b*) *Continuous systems* are used to produce large volumes of a single item (or relatively few items) on specialized equipment following a fixed path. Items follow a similar production sequence, which can range from an assembly line (for TVs) to a pipeline (for oil). Routing and scheduling focus on *flow controls* that govern the rate of flow of raw materials and finished goods. High-volume repetitive and continuous-flow products are often of standardized design and *made for stock.*

Question: How do process planning considerations differ for intermittent versus continuous systems?

In intermittent (job shop and batch) systems, planning efforts tend to be more focused on the *function* (individual equipment capabilities and worker-machine ratios), whereas in assembly-line and continuous-flow systems, major concerns are with the *flow* (and with balancing the large investment in line capacity). Figure 9-2 identifies some of the different concerns in intermittent and continuous systems.

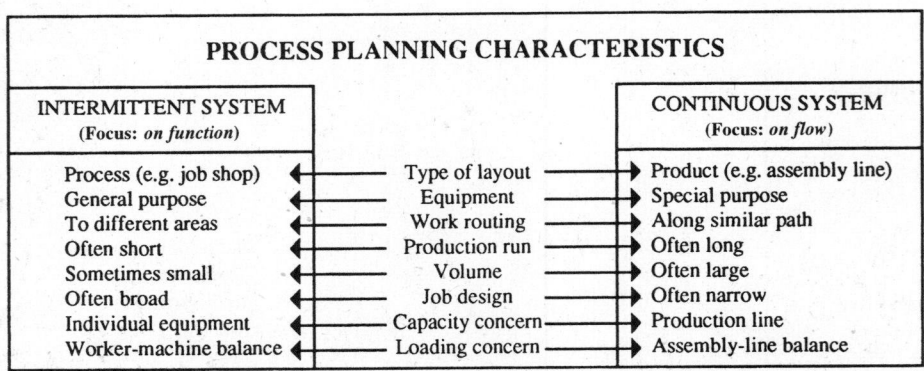

Fig. 9-2 Process planning considerations in intermittent and continuous systems

FLEXIBLE AND ROBOTIC SYSTEMS

By using computer-aided manufacturing (CAM) systems and numerically controlled (NC) machines, firms can obtain some of the flexibility benefits of intermittent layouts along with the volume advantages of continuous-type layouts.

Question: What are flexible manufacturing systems?

Flexible manufacturing systems (see Chap. 7) are computer-enhanced batch or repetitive processes that facilitate the production of high volumes of customized products on highly automated equipment that is responsive to software instructions.

Question: Explain the role of robots in production processes?

Robots are computerized manipulators that can perform a variety of tasks in response to commands or sensory input. The simplest robots do manual manipulations or fixed-sequence activities, and are particularly useful for repetitive and dirty or unsafe tasks. Intelligent robots have microprocessors that can respond to sensory input (vision, sound, or touch) and store, manipulate, and react to information concerning materials, times, storage locations, and manufacturing activities. The quality of robotic work is often higher than that of human efforts.

ASSEMBLY AND FLOW-PROCESS CHARTS

Assembly and flow-process charts are useful aids for planning and managing transformation processes. *Assembly charts* show the material requirements and assembly sequence of components that make up a mechanical assembly. They use standard symbols of $\bigcirc$ for operations and $\square$ for inspections. When the chart also provides complete instructions on how to produce an item, including specifications for the component parts plus operating and inspection times, it is referred to as an *operations-process chart*.

Example 9.1   An electric heater assembly consists of the following parts, as shown in Fig. 9-3:

(1)	Body	(4)	Switch	(7)	Resistor wire	(10)	End plates
(2)	Bracket	(5)	Internal wiring	(8)	Porcelain rod	(11)	Plastic end caps
(3)	Insulator	(6)	Cord	(9)	Copper end caps	(12)	Screen

Parts 2 and 3 are subassembled, as are 7 and 8 plus 7, 8, and 9. A test follows the heater element installation (i.e., after 7, 8, and 9), and a final inspection terminates the assembly. Draw an assembly chart.

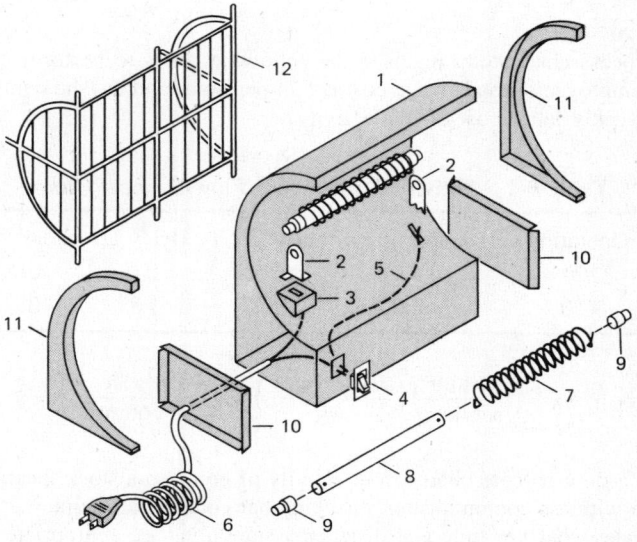

Fig. 9-3   Electric heater parts

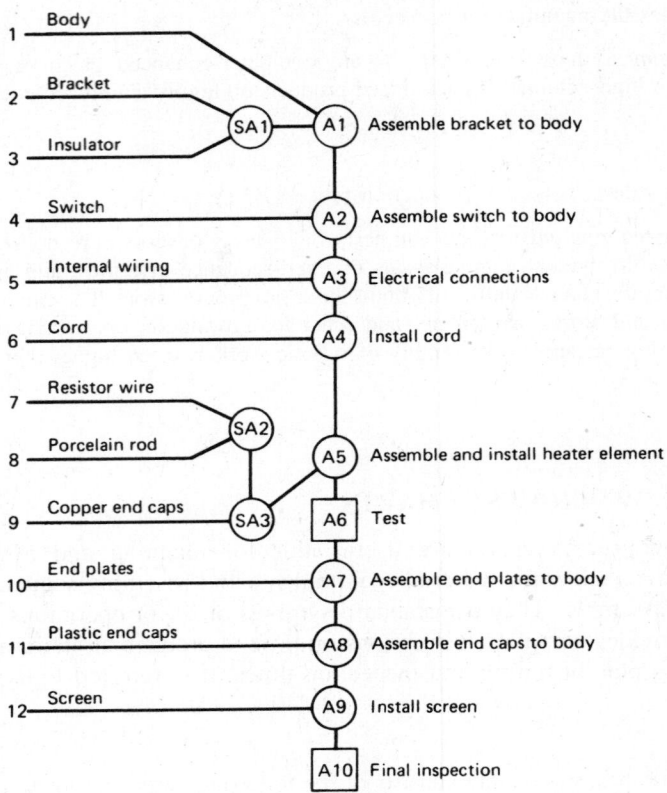

Fig. 9-4 Assembly chart for electric heater

Flow-process charts use symbols similar to assembly charts, except that the nonproductive activities of transport ($\Rightarrow$), delay (D), and storage (∇) are also included. They also have room for time, distance moved, and other relevant information, which permits cost and other analysis. Some flow-process charts are also designed to facilitate analysis by questioning why each activity is done and whether it can be improved upon by eliminating a task, combining tasks, changing the sequence, etc.

Example 9.2 A flow-process chart study revealed the following times to perform a certain type of telephone installation service. If the employee's (total) wage cost is $24 per hour, and he/she completes 10 such services per week, what is the actual weekly labor cost of this activity?

Table 9-1 Activity Times from Flow-Process Chart

Activity	Operation $\bigcirc$	Transport $\Rightarrow$	Inspect $\square$	Delay D	Store ∇	
Number steps	24	4	2	1	0	
Time (min)	20.4	6.0	3.8	1.8	0	Total = 32.0

$$\text{Labor cost/wk} = \left(\frac{32 \text{ min}}{\text{service}}\right)\left(\frac{10 \text{ services}}{\text{wk}}\right)\left(\frac{\$24}{\text{hr}}\right)\left(\frac{\text{hr}}{60 \text{ min}}\right) = \$128/\text{wk}$$

Example 9.3 Construct a flow-process chart for an activity of going to a stock location, counting the stock and entering information into a wireless communicator, checking bar codes and accuracy of the records, and returning to the original location. Make whatever time and distance assumptions are appropriate. See Fig. 9-5 for solution.

FLOW-PROCESS CHART					
Job Inventory Count × Existing method Proposed method			Date 7/6 Charted by B. Roe Chart No. 231		
Details of Method	Activity		Time (min)	Distance (ft)	Notes and Analysis
1. Walk to storage location	○ ➡ □ D ▽		1.25	110	
2. Visually locate item	● ⇨ □ D ▽		.30		
3. Inspect and confirm stock number	○ ⇨ ■ D ▽		.10		
4. Count number in stock	● ⇨ □ D ▽		.10–5.0		
5. Enter value in computer	● ⇨ □ D ▽		.25		
6. Update status card	● ⇨ □ D ▽		.40		
7. Wait for computer analysis	○ ⇨ □ ● ▽		1.25		
8. Check safety stock bar codes	● ⇨ □ D ▽		.70		
9. Check job stock bar codes	● ⇨ □ D ▽		.50		
10. Send computer printout to office	● ⇨ □ D ▽		.10		
11. Wait for office acknowledgment	○ ⇨ □ ● ▽		1.50		
12. Inspect record for accuracy	○ ⇨ ■ D ▽		.30		
13. File inventory printout locally	○ ⇨ □ D ▼		.50		
14. Return to office location	○ ➡ □ D ▽		1.25	110	
15. File inventory record disks	○ ⇨ □ D ▼		.45		

Fig. 9-5   Flow-process chart

MULTIACTIVITY CHARTS

Multiactivity charts are graphic devices for modeling the simultaneous activities of workers and/or the machines they operate. When only one worker is analyzed, the chart is sometimes called a *worker-machine chart*. Multiactivity charts help analysts identify idle time and costs of both workers and machines, and are useful for determining optimal worker-machine combinations.

Worker-machine charts show the time required to complete tasks that make up a work cycle. A *cycle* is the length of time required to progress through one complete combination of work activities. Many worker-machine activities are characterized by a load-run-unload sequence. The chart must be continued long enough past the start-up time to reach an equilibrium cycle time.

Example 9.4   An operator at Roadway Rubber Co. is expected to take 2 minutes to load and 1 minute to unload a molding machine. There are several machines of this type, all doing the same thing, and the automatic run time on each is 4 minutes. Respective costs are $18 per hour for the operator and $20 per hour for each machine. (*a*) Construct a worker-machine chart for the most efficient one-worker two-machine situation. Find the (*b*) cycle time, (*c*) worker's idle time per cycle, (*d*) total idle time per cycle for both machines, (*e*) total cost per hour, (*f*) total cost per cycle, and (*g*) idle time cost per hour.

(*a*)   If the operator begins by loading machine 1, the cycle does not reach an efficient steady state until the ninth minute, as shown in Fig. 9-6.

(*b*)   CT = 7 minutes

(*c*)   The worker is idle 1 minute per cycle.

(*d*)   The machines are not idle (at steady-state operation).

(e)   Cost = worker cost +2 (cost for each machine) = \$18 + 2(\$20) = \$58/hr = \$58 per 60 min

(f) $\text{Cost/cycle} = \dfrac{\$58}{60 \text{ min}} \left(\dfrac{7 \text{ min}}{\text{cycle}} \right) = \$6.77/\text{cycle}$

(g) $\text{Idle time cost/hr} = \dfrac{1 \text{ min}}{\text{cycle}} \left(\dfrac{60 \text{ min/hr}}{7 \text{ min/cycle}} \right) \dfrac{\$18}{60 \text{ min}} = \$2.57/\text{hr}$

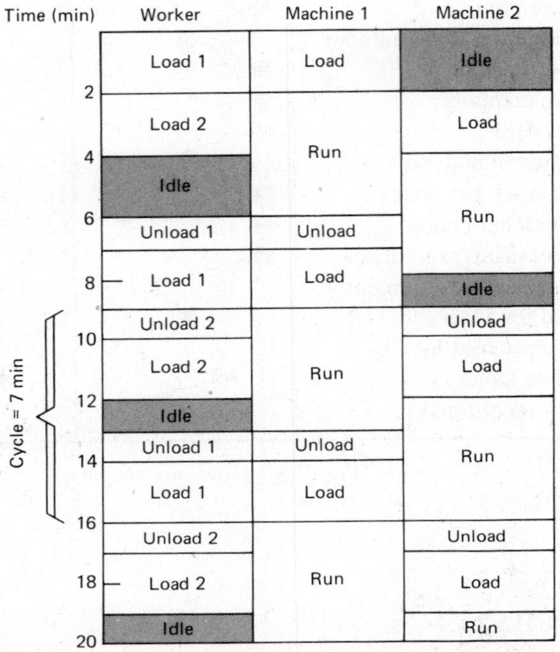

Time (min)	Worker	Machine 1	Machine 2
	Load 1	Load	Idle
2	Load 2	Run	Load
4	Idle		Run
6	Unload 1	Unload	
8	Load 1	Load	Idle
10	Unload 2		Unload
	Load 2	Run	Load
12	Idle		Run
14	Unload 1	Unload	
	Load 1	Load	
16	Unload 2		Unload
18	Load 2	Run	Load
20	Idle		Run

Cycle = 7 min

Summary:

	Worker	Machine 1	Machine 2
Idle time/cycle	1 min	0 min	0 min
Working time/cycle	6 min	7 min	7 min
Cycle time	7 min	7 min	7 min
% utilization	$6 \div 7 = 86\%$	$7 \div 7 = 100\%$	$7 \div 7 = 100\%$

Fig. 9-6   Worker-machine chart

EQUIPMENT SELECTION (MACHINE BREAKPOINTS)

Process planning decisions often concern the selection of equipment capacities required to produce a specified level of output. When the processing costs of alternative ways of doing a job can be allocated into their fixed- and variable-cost components, the most economical alternative is the one with the lowest costs at the expected volume. A graph of the respective costs will reveal the machine breakpoints.

EXAMPLE 9.5   Brackets can be processed on any of three machines with costs as shown in Table 9-2.

Table 9-2

	Machine X	Machine Y	Machine Z
Fixed cost per setup	$100	$200	$600
Variable cost per unit	3	2	1

Which machines should be used for production runs of up to 500 units?

$$TC = FC + VC(Q)$$

At 500 units:

$$TC_X = 100 + 3(500) = \$1600 \qquad TC_Y = 200 + 2(500) = \$1200 \qquad TC_Z = 600 + 1(500) = \$1100$$

For each machine, graph the FC at zero volume and the TC at 500 units as a linear function as shown in Fig. 9-7. For $0 \le 100$ units use X, for 100 to 400 units use Y, and for over 400 units use Z.

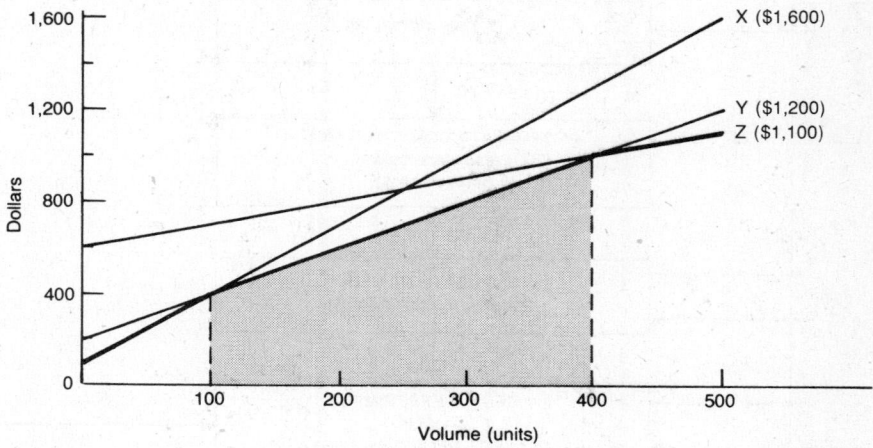

Fig. 9-7 Machine breakpoints

SIMULATION MODELING OF OPERATIONS

Some problems are too complex to solve with pure mathematics, or they involve random elements or risk situations that defy a practical mathematical solution. In such situations, analysts sometimes construct a model of the real-world problem and use a trial-and-error approach to arrive at a reasonable solution to the problem.

Simulation is a means of modeling the essence of an activity or system so that experiments can be conducted to evaluate the system's behavior or response over time. It is not an optimizing technique (like linear programming), but it does permit the decision maker to attack problems that are too complex or unsuitable for ordinary mathematics. Simulations may be done manually or physically, but most realistic business problems are done on computer. No attempt is made to duplicate reality in all respects; only the relevant variables of the problem under study are included.

Question: What are the advantages and disadvantages of using simulation modeling?

Advantages and Disadvantages of Using Simulation

Advantages	Disadvantages
1. Helps in understanding of complex problems/systems	1. May hide critical assumptions that invalidate model
2. Applies to problems that defy mathematical solution	2. Does not apply to deterministic problems
3. Permits study of alternatives under controlled conditions	3. Does not necessarily suggest a solution methodology
4. Avoids risk and disruptive experiments with actual system	4. Does not necessarily yield an optimal solution
5. Compresses time to reveal long-range effects	5. Requires expertise to construct sophisticated models
6. Is less costly than experiments with real-world system	6. Can be costly for data collection, modeling, and analysis

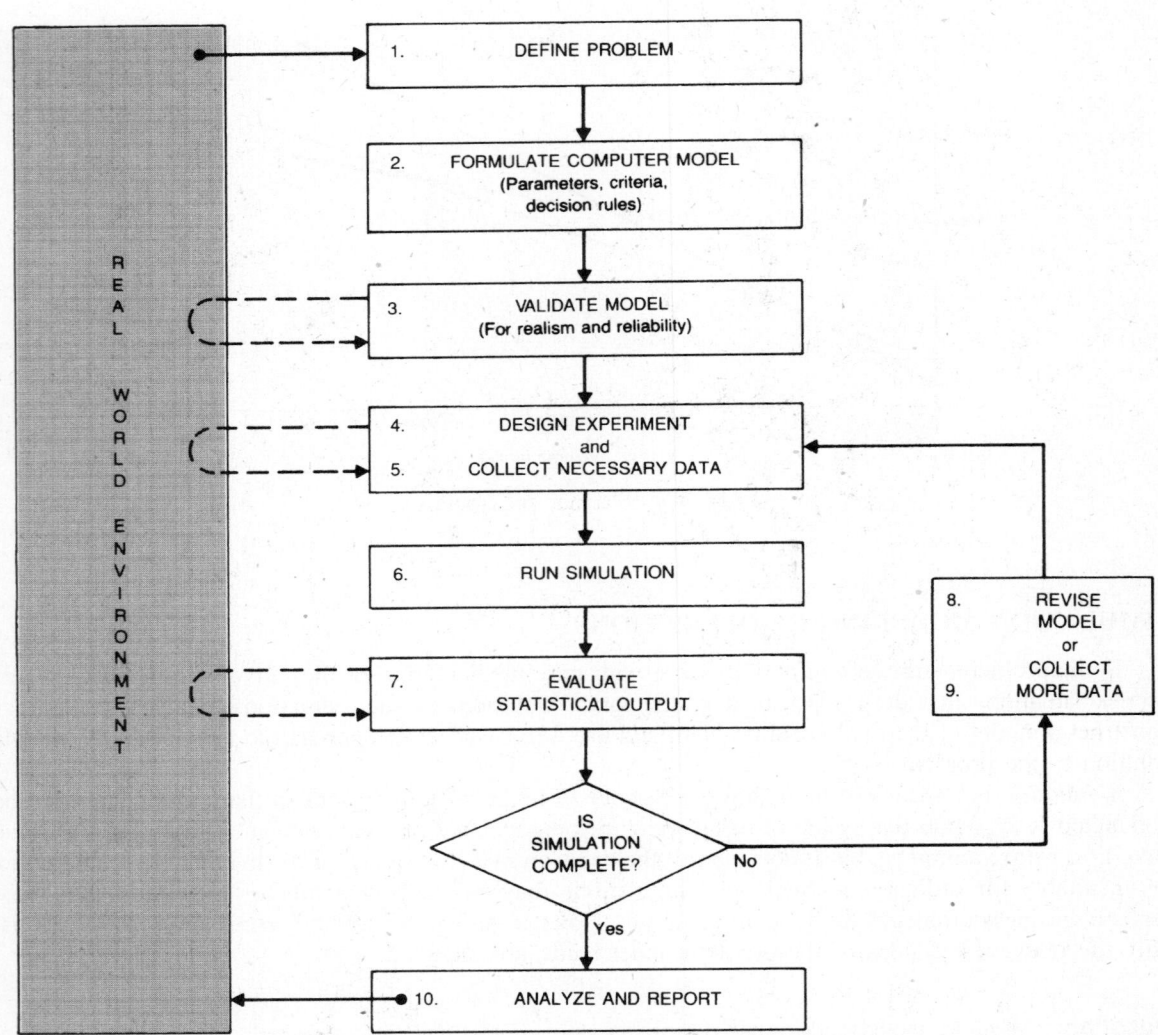

Fig. 9-8 Flowchart showing steps in the simulation process

Figure 9-8 describes the steps in a simulation process. Several commercial simulation languages are available for the model formulation (step 2). Also note that steps 8 and 9 parallel steps 2 and 5, respectively.

Question: What types of activities lend themselves to the use of simulation?

In addition to process planning, the applications are numerous, e.g., steel mill scheduling, office staffing, mining activities, hospital operations, air traffic control, and bank teller activities.

Question: Identify some simulation programs/languages.

Among the software in use are SIMSCRIPT, GPSS, GASP, DYNAMO, SLAM, SIMAN, PC-MODEL, GEMS, MAP/1, and RESQ. Specific application software programs (e.g., SIMFACTORY) are also available. Users can specify the arrival rate of jobs, processing times, batch sizes, number of work centers, etc., and view animated operations of product flowing through the system. Data can then be collected for analysis of product flow, cycle times, queues, delay and idle time, etc.

MONTE CARLO SIMULATION USING EMPIRICAL DATA

Simulations permit the modeling of activities that involve uncertainties, e.g., in demand and work times. When the models incorporate random sampling observations from a probability distribution, they are sometimes referred to as *Monte Carlo simulations*. The following steps simulate an assembly activity.

(1) Collect empirical data on assembly times (or estimate from a pilot activity).

(2) Develop a probability distribution and a cumulative probability distribution.

(3) Assign an interval of random numbers to each class of the distribution. (Optionally, the cumulative distribution may be plotted, showing relative frequency on the vertical axis.)

(4) Using random numbers (RNs), derive simulated assembly times.

(5) Interpret the results (e.g., determine the proportion of actual times that exceed estimated times or the effect of one workstation on the next).

The random-number assignment (step 3) is arranged so that the probability of obtaining a random number in the specified interval corresponds exactly with the empirical frequency reflected in the probability distribution (step 2). Thus, if 10 observations (from 100 total) lie in the first class, the random-number range assigned to that class would be 00 to 09. This represents 10 percent of the possible two-digit numbers from 00 to 99. (*Note:* The upper end of each interval will always be one less than the cumulative probability value.) The actual random numbers used (step 4) can come from either a random number table such as Appendix A (which is useful for small hand calculations) or from computer-generated random numbers (for large studies).

Example 9.6 A process planner is working on plans for producing a new detergent. She wishes to simulate a raw-material demand in order to plan for adequate materials-handling and storage facilities. On the basis of usage for a similar product introduced previously, she has developed a frequency distribution of demand in tons per day for a 2-month period. Use this data (shown in Table 9-3) to simulate the raw material usage requirements for 7 periods (days).

Table 9-3 Demand Pattern for 2-Month Period

Demand, X (tons/day)	10	11	12	13	14	15	
Frequency (no. days)	6	18	15	12	6	3	Total = 60

(1) Data are given in frequencies.

(2) To formulate a probability distribution, divide each frequency by the total (60); for example, $6 \div 60 = .10$, and $18 \div 60 = .30$. Then formulate a cumulative-probability distribution by successively summing the probability values (see Table 9-4 and Fig. 9-9).

Table 9-4

Demand (tons/day)	Frequency (no. days)	Probability $P(X)$	Cumulative Probability
10	6	.10	.10
11	18	.30	.40
12	15	.25	.65
13	12	.20	.85
14	6	.10	.95
15	3 / 60	.05	1.00

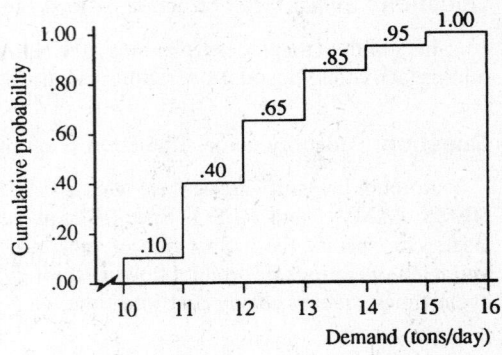

Fig. 9-9 Cumulative distribution

(3) Next, assign random-number intervals so that the number of values available to each class corresponds with the probability (Table 9-5 and Fig. 9-10). Using 100 two-digit numbers (00 to 99) we assign 10 percent (00 to 09) to the first class, 30 percent (10 to 39) to the second class, and so on.

Table 9-5

Demand (tons/day)	Probability $P(X)$	Corresponding Random Numbers
10	.10	00–09
11	.30	10–39
12	.25	40–64
13	.20	65–84
14	.10	85–94
15	.05	95–99
	1.00	RN = 27

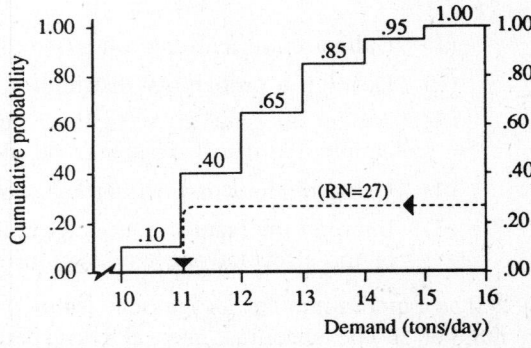

Fig. 9-10 Simulated demand

(4) We obtain random numbers (RN) from a random number table. Assume the RNs are:

    27   13   80   10   54   60   49

The first RN, 27, falls into the second class of the distribution and corresponds to a demand of 11 tons per day.

Table 9-6

Random number	27	13	80	10	54	60	49
Simulated demand	11	11	13	11	12	12	12

(5) This extremely small simulation yields a mean of $\bar{X} = 11.7$ tons (Table 9-6) and a standard deviation of $s = .76$ ton. The expected value from the empirical probability distribution is $E(X) = \Sigma[XP(X)] = 12.05$ tons, suggesting that the small sample size of only 7 periods has resulted in some error. A much larger sample should be simulated before the simulation results are used for making decisions.

Note that in Example 9.6 the width of the random number "target" in each class corresponds *exactly* with the relative frequency of the class. This helps to ensure that the simulated results have the same type of distribution as the original data. This is more apparent from the graph (Fig. 9-10) where the vertical distances correspond to the relative frequencies of the respective classes.

SIMULATIONS USING KNOWN STATISTICAL DISTRIBUTIONS

When values to be used in a simulation follow a known statistical distribution, the computations can be simplified. For example, simulated values can be obtained from uniform and normal distributions as follows:

Uniform Distribution

$$\text{Simulated value} = a + (b - a)(\text{RN}_\%) \qquad (9.1)$$

where a = minimum value

b = maximum value

$\text{RN}_\%$ = random number (as a percent) from a table of uniformly distributed random numbers (see Appendix H)

Normal Distribution

$$\text{Simulated value} = \mu + \sigma(\text{RN}_{\text{ND}}) \qquad (9.2)$$

where μ = mean of database being simulated

σ = standard deviation of database being simulated

RN_{ND} = random number (as a Z score) from a table of normally distributed random numbers (see Appendix I)

Poisson Distribution

$$\text{Simulated value} = \text{value of } c \text{ that corresponds to the cumulative } P(\leq c \,|\, \lambda)$$

where c = cumulative number of occurrences of an event (and the upper limit of the class)

λ = average number of occurrences (and mean of the Poisson distribution)

The simulated class (and value of c) is determined probabilistically by expressing a random number RN_{ND} as a decimal and assigning it to the appropriate class. (See Solved Prob. 9.7 for example.)

Example 9.7 (*Uniform Distribution*)   The diameters of trees arriving at a lumber mill vary uniformly from 2 feet to 3 feet. The time required to saw a 2-foot log is 5 seconds and to saw a 3-foot log is 8 seconds; within that range it varies directly with the diameter. Simulate the time required to saw five logs selected at random in the 2- to 3-foot range.

$$\text{Simulated value} = a + (b - a)(\text{RN}_\%)$$

where $a = 5$ seconds, $b = 8$ seconds, and $b - a = 3$ seconds. $\text{RN}_\%$ will be taken from the first five values of three-digit numbers from column 2 of the Random Number Table (Appendix A). See Table 9-7.

Table 9-7

RN	$a + (b - a)(\text{RN}_\%)$	=	Simulated Value
435	$5 + 3(.435)$	=	6.31 sec
143	$5 + 3(.143)$	=	5.43 sec
362	$5 + 3(.362)$	=	6.09 sec
620	$5 + 3(.620)$	=	6.86 sec
573	$5 + 3(.573)$	=	6.72 sec

Example 9.8 (*Normal Distribution*)   A patient-care service in a hospital has normally distributed times with a mean of 15 minutes and standard deviation of 2 minutes. Simulate four values of the times required to perform this service, using the table of normally distributed random numbers.

$$\text{Simulated value} = \mu + \sigma(\text{RN}_{\text{ND}})$$

where $\mu = 15$, $\sigma = 2$, and RN_{ND} are from column 1 of Appendix I. See Table 9-8.

Table 9-8

RN_{ND}	$\mu + \sigma(\text{RN}_{\text{ND}})$	$=$	Simulated Value
.34	$15 + 2(.34)$	$=$	15.68 min
-1.09	$15 + 2(-1.09)$	$=$	12.82 min
-1.87	$15 + 2(-1.87)$	$=$	11.26 min
1.57	$15 + 2(1.57)$	$=$	18.14 min

Questions and Solved Problems

ASSEMBLY AND FLOW-PROCESS CHARTS

9.1   A production analyst estimated the times for activities associated with a new casting process and came up with the information in Table 9-9. Show the activities in the form of a flow-process chart.
See Fig. 9-11.

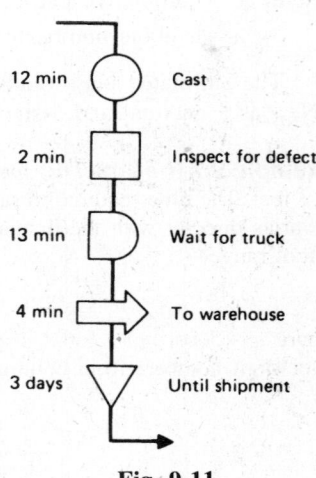

Table 9-9

Number	Classification	Time
1	Perform casting operation	12 min
2	Inspect casting	2 min
3	Wait for lift truck	13 min
4	Transport to warehouse	4 min
5	Store: await shipment	3 days

12 min — Cast

2 min — Inspect for defects

13 min — Wait for truck

4 min — To warehouse

3 days — Until shipment

Fig. 9-11

9.2   A flow-process chart is to be constructed for the activity of testing a temperature controller during a shutdown. The controller is to be transferred from a furnace, tested, and reinstalled. Make whatever assumptions are required to construct the flow-process chart.
See Fig. 9-12.

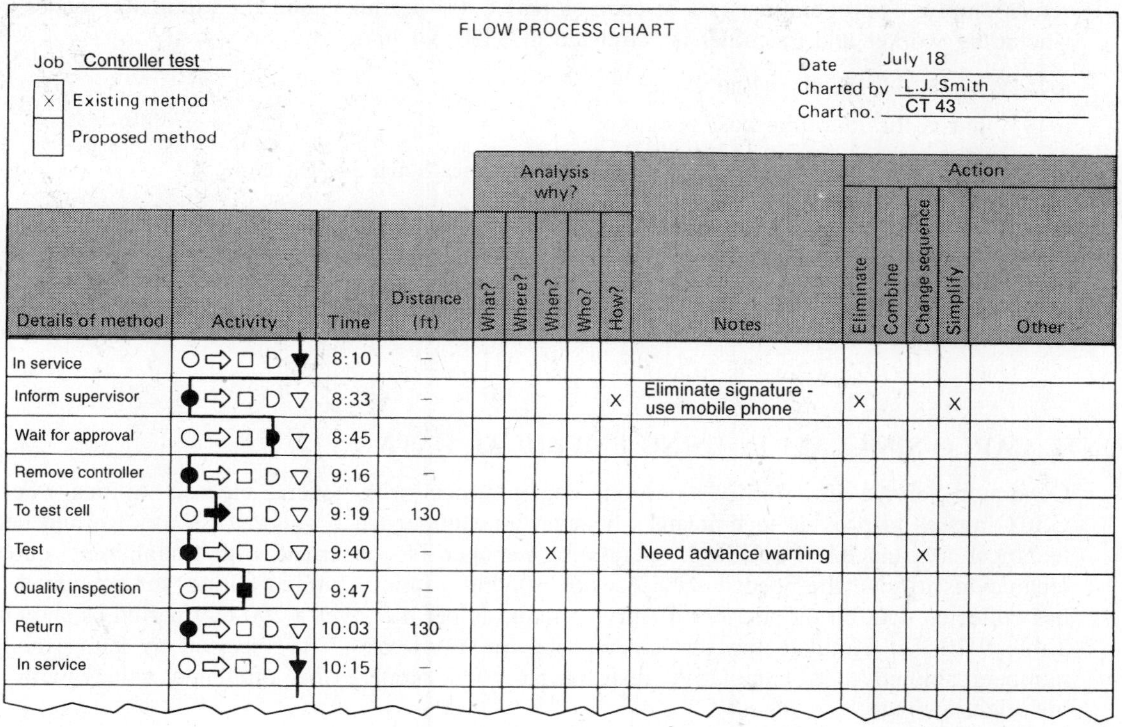

FLOW PROCESS CHART

Job **Controller test**

[X] Existing method
[ ] Proposed method

Date **July 18**
Charted by **L.J. Smith**
Chart no. **CT 43**

Details of method	Activity	Time	Distance (ft)	What?	Where?	When?	Who?	How?	Notes	Eliminate	Combine	Change sequence	Simplify	Other
In service	○⇨□ D ▼	8:10	—											
Inform supervisor	●⇨□ D ▽	8:33	—					X	Eliminate signature - use mobile phone	X			X	
Wait for approval	○⇨□ ■ ▽	8:45	—											
Remove controller	●⇨□ D ▽	9:16	—											
To test cell	○➡□ D ▽	9:19	130											
Test	●⇨□ D ▽	9:40	—			X			Need advance warning			X		
Quality inspection	○⇨■ D ▽	9:47	—											
Return	●⇨□ D ▽	10:03	130											
In service	○⇨□ D ▼	10:15	—											

Fig. 9-12

MULTIACTIVITY CHARTS

9.3 The accompanying portion of an activity chart (Fig. 9-13) is for an automatic-loader mining operation. The loader, in the mine, requires 8 minutes to load a skip car. There are three skips, and they take 9 minutes to travel loaded to the ore dump, 2 minutes to dump, and 7 minutes to

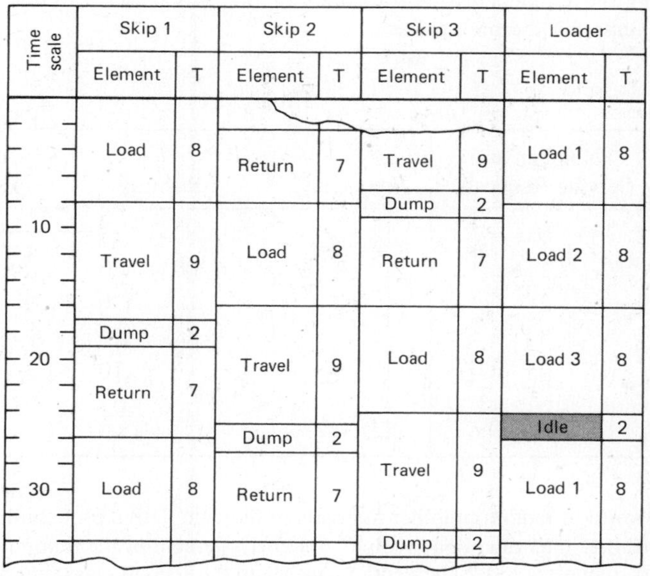

Time scale	Skip 1 Element	T	Skip 2 Element	T	Skip 3 Element	T	Loader Element	T
	Load	8	Return	7	Travel	9	Load 1	8
					Dump	2		
10	Travel	9	Load	8	Return	7	Load 2	8
20	Dump	2	Travel	9	Load	8	Load 3	8
	Return	7						
			Dump	2			Idle	2
30	Load	8	Return	7	Travel	9	Load 1	8
					Dump	2		

Fig. 9-13

return empty. The operating cost of each skip is $200 per hour, and the automatic-loader cost (including worker and machine) is estimated at $350 per hour.

(*a*)   What is the length of the cycle?

(*b*)   What is the idle-time cost per hour?

(*a*)   *Cycle length:* The system is in a similar state at times 8 and 34. Therefore,

$$\text{Cycle length} = 34 - 8 = 26 \text{ min}$$

(*b*)   *Idle-time cost:* The loader is idle 2 minutes per cycle = 2/26 = 1/13 of each hour. Therefore,

$$\text{Cost/hr} = 1/13(\$350) = \$26.92/\text{hr}$$

MONTE CARLO SIMULATION USING EMPIRICAL DATA

9.4   Generator Service Co. (GSCO) has ongoing contracts with several electric utilities wherein GSCO agrees to provide technicians whenever a customer has a generator shutdown and needs technical assistance. The GSCO operations manager is concerned with maintaining enough technicians to give the needed service while staying within a limited budget for personnel. He has collected data on the number of service requests per day over a 200-day period as shown in Table 9-10. (*a*) Simulate the service requests for a 1-week (7-day) period by using random numbers applied to a cumulative distribution. (*b*) Compare the simulated values with the historical average.

Table 9-10

Number of service requests	0	1	2	3	4	5	6
Frequency	30	40	60	44	20	6	0

(*a*)   Data are given in frequencies. Develop a relative cumulative probability distribution (Table 9-11) by first converting the frequencies into probabilities (i.e., divide each by the total frequency) and then successively summing the probabilities.

Table 9-11

Number of Service Requests	Frequency	Probability	Cumulative Probability
0	30	30/200 = .15	.15
1	40	40/200 = .20	.15 + .20 = .35
2	60	.30	.35 + .30 = .65
3	44	.22	.87
4	20	.10	.97
5	6	.03	1.00
	200	1.00	

Now assign two-digit random number intervals to the cumulative probabilities so that they correspond with the probability intervals. For example, assign 15 percent of the random numbers (00 to 14) to the first class (zero requests) and 20 percent (15 to 34) to the second class (one request) as shown in Table 9-12.

Table 9-12

Request Class	Frequency	Probability	Cumulative Probability	Random No. Assigned
0	30	.15	.15	00–14
1	40	.20	.35	15–34
2	60	.30	.65	35–64
3	44	.22	.87	65–86
4	20	.10	.97	87–96
5	6	.03	1.00	97–99

Finally, select 7 two-digit random numbers from a random number table, determine which request class they fall into, and record the corresponding number of service requests as shown in Table 9-13.

Table 9-13

Day	1	2	3	4	5	6	7	
Random number	85	68	99	21	17	56	12	Total
Corresponding number of service requests	3	3	5	1	1	2	0	15

(b) Note that the mean number of requests from the 7-day simulation is $15/7 = 2.14$ service requests. This compares with the mean of the historical data of

$$\mu = 0(.15) + 1(.20) + 2(.30) + 3(.22) + 4(.10) + 5(.03) = 2.01$$

9.5 Empirical data collected on the time required to weld a transformer bracket were recorded to the nearest quarter minute, as shown in Table 9-14.

Table 9-14

Weld Time (min)	Number of Observations
< .25	0
.25 < .75	24
.75 < 1.25	42
1.25 < 1.75	72
1.75 < 2.25	38
2.25 < 2.75	14
2.75 < 3.25	10

(a) Formulate a cumulative distribution in percentage terms.

(b) Graph the frequency and cumulative distributions.

(c) A simulation is to be conducted using random numbers. What simulated weld times (to the nearest .25 minute) would result from the random numbers 25, 90, and 59?

(*d*)   What proportion of the times exceed 2.0 minutes?

(*a*)   Cumulative distributions are usually formulated on a scale where the cumulative percentage is more than or less than a corresponding *x* axis amount. We shall use a "less than" percentage and so will need to identify the upper-class boundaries (UCB) as the *Y* coordinates for the cumulative distribution (Table 9-15).

Table 9-15

Weld Time (min)	Frequency in Numbers	Upper-Class Boundary (UCB)	Cumulative Number of Times <UCB	Cumulative Percentage of Times <UCB
< .25	0	.25	0	0
.25 < .75	24	.75	24	12
.75 < 1.25	42	1.25	66	33
1.25 < 1.75	72	1.75	138	69
1.75 < 2.25	38	2.25	176	88
2.25 < 2.75	14	2.75	190	95
2.75 < 3.25	10	3.25	200	100

(*b*)   The frequency distribution is constructed by extending vertical lines from the class boundaries to the appropriate frequency level for the class. For the cumulative distribution, values of the cumulative percentage of time <UCB are plotted at weld times corresponding to the UCB. For example, the frequency (12 percent) is plotted at UCB = .75 (as illustrated in Fig. 9-14).

(*c*)   The simulated time for random number (RN) 25 is determined by entering the cumulative graph at 25 (as shown by the arrow) and proceeding horizontally to the curve and then down to the weld time. The result is a reading of 1.0 minute (rounded to the nearest .25 minute). Times for random numbers 90 and 59 are 2.5 and 1.5 minutes, respectively. (A larger graph would lend more accuracy.)

(*d*)   From the cumulative distribution, about 12 percent of the times exceed 2.0 minutes.

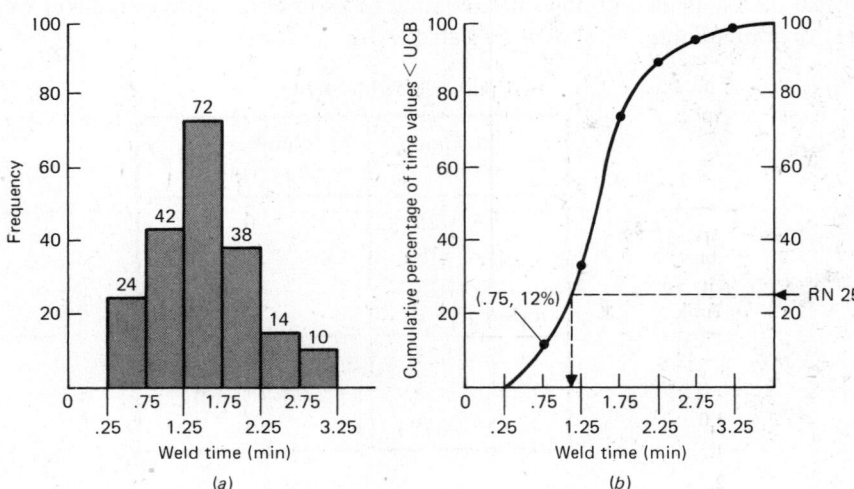

Fig. 9-14   Frequency and cumulative distributions

9.6   In an aircraft assembly operation, activity A precedes activity B, and inventory may accumulate between the two activities. With the use of random numbers, a simulated sample of performance times yielded the values shown (minutes) in Table 9-16.

Table 9-16

Activity A		Activity B	
Random Number	Time (min)	Random Number	Time (min)
07	.3	63	.5
90	.8	44	.4
02	.2	30	.4
50	.5	98	.9
76	.6	30	.4
47	.5	72	.6
13	.3	58	.5
06	.3	96	.9
79	.7	37	.4

(a) Simulate the assembly of six parts, showing idle time in activity B, waiting time of each part, and number of parts waiting. *Note:* Omit the first random number of A so that activity B begins at time zero.

(b) What was the average length of the waiting line ahead of B (in number of units)?

(c) What was the average output per hour of the assembly line?

(a) Our interest lies in activity B, so we can set up Table 9-17 to show when parts arrive at B, how long it takes B to work on them, and the resultant idle and waiting times:

Activity B begins at 0, and it takes .5 minute to complete the first part. B is then idle for .3 minute until part 2 arrives from A at .8 minute. Part 2 takes .4 minute, so the ending time is $.8 + .4 = 1.2$ minutes. By this time part 3 has been waiting .2 minute because it became available at $.8 + .2 = 1.0$ minute, but work could not begin on it until 1.2 minutes. However, before activity B is finished on part 3 at 1.6 minutes, part 4 has arrived (at $1.0 + .5 = 1.5$ minutes), and so one part is waiting. We continue systematically in this manner through part 6, noting that when it is finished at time 3.5 minutes, there are two parts waiting, for their availability times were 2.9 minutes and 3.2 minutes, respectively.

Table 9-17

Part Number	Part Available for Activity B at Time	Activity B Beginning Time	Activity B Ending Time	Activity B Idle Time	Waiting Time of Part	Number Parts Waiting at B End Time
1	—	0	.5	0	0	0
2	.8	.8	1.2	.3	0	1
3	1.0	1.2	1.6	0	.2	1
4	1.5	1.6	2.5	0	.1	1
5	2.1	2.5	2.9	0	.4	2
6	2.6	2.9	3.5*	0	.3	2
7	2.9				1.0†	
8	3.2					

*Total run time.
†Total waiting time.

(b) The average length of the waiting line (that is, average inventory) ahead of B can be expressed in equation form as follows:

$$\text{Average inventory} = \frac{\text{total waiting time}}{\text{total run time}}$$

$$= \frac{1.0 \text{ assembly minute}}{3.5 \text{ minutes}} = .29 \text{ assembly} \qquad (9.3)$$

(c) Average output per hour:

$$\text{Units/hr} = \frac{6 \text{ units}}{3.5 \text{ min}} \left(\frac{60 \text{ min}}{\text{hr}} \right) = 102.9 \text{ units/hr}$$

SIMULATIONS USING KNOWN STATISTICAL DISTRIBUTIONS

9.7 *Simulated Poisson Arrivals.* The pattern of arrivals of patients to the emergency room of a large metropolitan hospital can be described by a Poisson distribution with a mean of four per hour. Simulate the number of patient arrivals for an 8-hour shift. (*Note:* For random numbers, use the first three digits of column 3 of the Random Number Table, Appendix A.)

First, obtain the cumulative probability distribution from the Poisson Distribution Values, Appendix D. For a mean of $\lambda = 4$, the cumulative probabilities can be taken directly from the table as $P(X \leq c | \lambda)$. Thus $P(X \leq 0 | \lambda = 4) = .018$ and $P(X \leq 1 | \lambda = 4) = .092$, etc., as seen in Table 9-18.

Table 9-18

No. patients c	0	1	2	3	4	5	6	7	8	9	10	11	12
Cumulative $P(c)$	.018	.092	.238	.433	.629	.785	.889	.949	.979	.992	.997	.999	1.000

Next, obtain eight random numbers (one to represent the number of patients arriving during each hour of the shift).

    853   540   985   903   266   373   920   164

Next, convert each random number to a number of patients per hour by assigning each to the appropriate cumulative probability class. For example, the first RN 853 falls within the .785 to .889 class (in Table 9-18), which belongs to the interval classified as 6 patients (or anything between 5 and 6) as shown in Table 9-19. Note also that because we are using only three random digits in our random number scale of 000 to 999, we must advance any random numbers that happen to lie on the upper boundary into the next larger class. Thus, the first class should represent 18/1000 of the numbers, and this is satisfied by the digits 000 to 017. So, a random number exactly 018 would be interpreted as one patient arrival rather than as zero.

Table 9-19   Simulated Number of Patient Arrivals

RN	853	540	985	903	260	373	920	164
No. arrivals	6	4	9	7	3	3	7	2

Supplementary Questions and Problems

9.8 A toy car is to be assembled in the following order:

(1) Start with base frame (5) Assemble wheels onto axles

(2) Add upper body (6) Snap on wheel and axle assembly

(3) Install motor (7) Snap on bumper

(4) Snap on hood (8) Inspect

Draw an assembly chart. *Ans.* See Example 9.1 and Fig. 9-4 for a comparable chart.

9.9 Drexron Furniture Co. maintains a constant stock of chair backs for a standard line of dining room chairs. The activities and associated times for producing one lot are as shown in Table 9-20. Illustrate the activities in the form of a simplified flow-process chart.

Ans. See Prob. 9.2 for comparable chart. Symbols, in sequence, are $\bigcirc$, $\Rightarrow$, $\bigcirc$, $\square$, D, $\Rightarrow$, $\bigcirc$, D, $\Rightarrow$, ∇

Table 9-20

Number	Activity	Time
1	Rough cut—saw shop	20 min
2	Move to lathe shop	10 min
3	Turn in lathes	60 min
4	Inspect for flaws	5 min
5	Wait for lift truck	15 min
6	Move to gluing area	10 min
7	Glue rods to backplate	30 min
8	Wait for glue to dry	180 min
9	Move to warehouse	10 min
10	Store for later use	

9.10 The operations shown in Table 9-21 must be performed on a housing that is part of a motor-mounting bracket. The housings are then inspected before going on to the next assembly operation. This takes 6 seconds for each.

Table 9-21

Operation	Machine	Output, parts/hr
Shear	Shear X-100	80
Form	Main press	400
Clean	Ultrasonic tank	150

(*a*) Construct an operations process chart showing the activities, appropriate symbols, and times in minutes.

(*b*) How many of each type of machine would be required for a production rate of 300 parts per hour, assuming 80 percent utilization of the X-100 shear machine and 100 percent utilization of the others?
  Ans. (*a*) shear = .75 min/part, form = .15 min/part, clean = .40 min/part, inspect = .10 min/part
  (*b*) shear = 5 machines, form = 1 machine, clean = 2 machines

9.11 Rework Prob. 9.3, except with load times of 8 minutes, travel times of 6 minutes, and return times of 4 minutes. Assume 2-minute dump times. Find the (*a*) length of the cycle, and (*b*) the idle-time cost per hour.   *Ans.*   (*a*) 24 minutes (*b*) $100 per hour

9.12 A construction firm uses dump trucks to haul asphalt to a distant location, where a paving machine applys a 4-inch layer to a new roadway. Trucks require 3 minutes for loading at the asphalt plant, 7 minutes to travel loaded to the new roadway, 10 minutes to dump the asphalt into the paving machine, and 5 minutes to return empty. The firm has only one paving machine, and it paves only while it is being fed (and pulled) by a truck during its dumping activity. (*a*) How many trucks are required to pave the roadway as quickly as possible? (*b*) Construct an activity chart for a two-truck, one-paving-machine arrangement. (*c*) If the paving machine cost is $80 per hour, and the truck cost is $34 per hour, how many trucks should be used to minimize the idle equipment cost?

Ans.   (*a*) 2.5 trucks are required, so use 3 trucks. (*b*) The cycle time is 25 min. (*c*) With 2 trucks, idle-machine cost = $16 per hour; and with 3 trucks, idle-truck-time cost = $17 per hour; so use 2 trucks.

9.13 A production analyst is planning for the manufacture of valve fittings. Each fitting must be milled on any one of three milling machines, *X*, *Y*, or *Z*. The set-up and operating costs for each are as shown in Table 9-22.

Table 9-22

	Setup	Operating
X	$10	$.30/unit
Y	30	.10/unit
Z	40	.05/unit

(*a*) Graph the cost structure for the three alternatives for volumes up to 250 units. (*b*) For what *range* of outputs should the analyst specify the use of machine *Y*?

Ans.   (*a*) Graph should show $ on *Y* axis, units on *X*. (*b*) $0 \le 100$ use *X*, $100 \le 200$ use *Y*, >200 use *Z*

9.14 A large grocery distribution center in Denver is computerizing its order service department. Orders from customers in Colorado (and adjacent states) will go via phone line directly into the company's main computer, which will generate "pick lists" with standard times to fill each order. To assist in planning for the proper number of delivery trucks, planners have collected the data shown in Table 9-23 from the past year (300 days).

Table 9-23

No. trucks required, *X*	20	21	22	23	24	25	26	27	28	29	30	31	32
No. days, frequency	0	4	18	25	28	41	68	56	31	22	5	2	0

(*a*) Simulate the demand for trucks over a 10-day period. (*Note:* Use the first three digits from column 6, Appendix A, for your random numbers.) Compute (*b*) the mean and (*c*) the standard deviation of your simulated sample.   *Ans.*   (*a*) See Table 9-24. (*b*) $\bar{X} = 26.2$ trucks (*c*) $s = 2.0$ trucks

Table 9-24

RN	697	667	248	063	887	432	732	970	449	425
No. trucks	27	27	24	22	28	26	27	29	26	26

9.15 Data were collected on the assembly times for 1,000 water valves (size 2 inches, 150 pounds) at the Drain Company, as shown in Table 9-25.

Table 9-25

Time		Number of Valves
LCB	UCB	
1.0	Under 1.5 min	0
1.5	Under 2.0 min	20
2.0	Under 2.5 min	120
2.5	Under 3.0 min	280
3.0	Under 3.5 min	430
3.5	Under 4.0 min	120
4.0	Under 4.5 min	30
4.5	Under 5.0 min	0
		1,000

(a) Graph the data as a cumulative distribution. (b) What percentage of the assembly times exceed 4.0 minutes? (c) What would be the simulated assembly time for a random number of 44? (Estimate to the nearest half minute.)

Ans. (a) Graph should show assembly times (in minutes) on X axis and cumulative percentages on Y axis
(b) 3 percent (c) approximately 3 minutes.

9.16 Operation B follows operation A. In a simulated operating time of 440 minutes, operator B was idle 15.2 minutes, and the total waiting time of assemblies before B was 1,980 minutes. If B completed 92 parts during the production run, what was the average inventory of parts waiting for B in terms of number of assemblies?   *Ans.*   4.5 assemblies

9.17 An assembly operation has been simulated using random numbers with the results as shown in Tables 9-26 and 9-27. Operator B started at time zero.

A ⟶ B ⟶

Table 9-26   Operator A

Unit	1	2	3	4	5	6	7	8	9	10
Random no.	76	60	07	22	14	94	87	11	37	88
Time (sec)	8	6	2	4	4	11	10	3	5	10

Table 9-27   Operator B

Unit	1	2	3	4	5	6	7	8	9	10
Random no.	69	46	17	13	38	12	42	29	36	44
Time (sec)	7	6	4	4	5	3	6	5	5	6

A simulation of the assembly of 10 units has begun in Table 9-28. Complete the remaining portion of the table, and utilize the data from the first 8 units to determine the average length of the waiting line ahead of B.

Ans.   For 8 parts: B ending time = 48, total waiting time = 19 seconds, average length of line = .40 units.

Table 9-28

Unit No.	Unit Available for B at Time	Activity B Data			Unit Data	
		Beginning Time	Ending Time	Idle Time	Waiting Time	Number Waiting
1	0	0	7	0	0	1
2	6	7	13	0	1	2
3	8	13	17	0	5	2
4	12	17	21	0	5	1
5						
6						
7						
8						
9						
10						

9.18   In the meal preparation kitchens of New York International Airlines, dinners are prepared on an assembly line where there is limited space for an inventory of partially filled plates. A simulation of two adjacent workers (where Y is dependent upon X) developed the random numbers and times shown (in seconds) in Table 9-29.

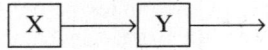

Table 9-29

Activity X		Activity Y	
Random Number	Time	Random Number	Time
72	22	84	32
18	10	26	12
77	23	13	8
84	27	60	24
5	7	53	22
20	11	22	12
46	27	90	36

(*a*) Simulate the preparation of five meals, and determine the idle time for activity Y, the waiting time of each meal, and the number of meals waiting (omitting the first random number of X). (*b*) What was the average length of the waiting line upstream from Y? (*c*) What was the average output per minute of the production line?

Ans.   (*a*) Follow tabular format for simulation (*b*) .47 meal (*c*) 2.83 meals per minute.

9.19   A promotional campaign is being planned where winners have an equal (uniform) chance of winning either 1, 2, 3 . . . up to 10 hours of professional instruction in small computer operations. Simulate the amount of time the firm will have to provide for the first five winners of the contest. Use the first two-digit random numbers from column 7, Appendix A, and round your simulated times to the nearest whole hour.

Ans.   9, 8, 7, 2, 3, for an average of 5.8 hours per winner

9.20   The time required to service a customer at the Pacific Airlines Counter is normally distributed with a mean of 2 minutes and standard deviation of .5 minute. Simulate the service times for three customers using a table of normally distributed random numbers. (Assume the numbers selected from the table at random are −.48, 1.54, and −.22.)   *Ans.*   1.76, 2.77, and 1.89

9.21   A process planner must plan for maintenance facilities capable of repairing an average of three motors per day. Breakdowns are assumed to follow a Poisson distribution. Using the first 3 digits of column 4 of the Random Number Table, Appendix A, simulate the number of motor breakdowns over a 10-day period.

Ans.   5, 1, 2, 3, 1, 3, 7, 6, 0, 7

9.22   A governmental agency must plan for the staffing of a social services office. Service times for one worker can be described by a Poisson distribution with a mean of 2.6 people per hour. Using the first 3 digits of column 5 of the Random Number Table, Appendix A, simulate the number of people served per hour by one social worker over an 8-hour day.   *Ans.*   2, 2, 1, 3, 1, 2, 4, 2

Chapter 10

Job Design and Work Measurement
Statistical Sample Size

OVERVIEW

Employees are the most valuable asset of an organization. They have an intrinsic value that no equipment can match and a diversity of skills, emotions, and levels of performance that cannot be found in any machine.

Jobs are the work activities performed by employees to meet organizational goals. Job designs typically specify *what* task is to be done, *how* to do it, and (if necessary) *when* and *where* to do it. The work goals themselves should be clear and specific, moderately difficult, and well accepted.

A sound human resource management policy rests upon the two cornerstones of maintaining a work environment that:

(1) Respects the human dignity and equality of all employees

(2) Provides employees with jobs that encourage self-actualization (via development of their knowledge, values, and job-related skills)

Question: How is job design related to work methods, work measurement, and job satisfaction?

As depicted in Fig. 10-1, the jobs an organization elects to do stem from the *priorities* it receives from the (social/market) environment and the *capacities* (resources/technology) it has available to satisfy those needs. The resulting job design then dictates work methods that, in turn, require some form of work measurement and yield some degree of job satisfaction.

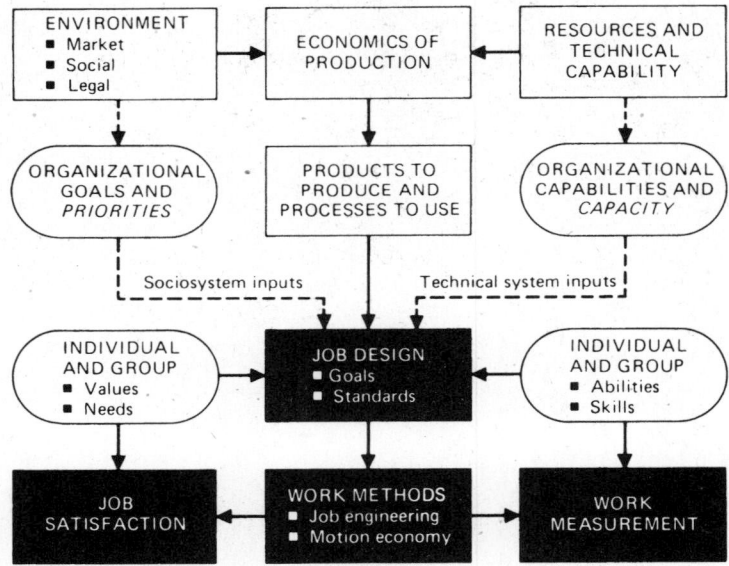

Fig. 10-1 Elements of job design and work methods

SPECIALIZATION AND AUTOMATION

As a worker's efforts become more focused, he or she tends to benefit from experience, become more proficient, and do a better job. However, as the scope of a job narrows and specialization occurs,

176

the work can also become dissatisfying. One of the most fundamental concerns of job design relates to the degree of specialization that should be designed into a job.

Question: Identify some advantages and disadvantages of specialization of labor.

Advantages and Disadvantages of Specialization of Jobs

Advantages	Disadvantages
1. Lower education requirements	1. Repetitive, monotonous, boring work
2. Fewer overall skills required	2. Lower level of employee satisfaction
3. Readily available labor supply	3. Higher costs of absenteeism, turnover
4. Easier/faster training	4. Reduced opportunity for improvement
5. Lower wage cost (for firm)	5. More imbalance of work tasks
6. High output per time period	6. More difficult to control quality

Question: How has automation affected the design of jobs?

Automation is the substitution of mechanical (or electrical/chemical) equipment for human effort. It has resulted in the shifting of many repetitive and less desirable tasks to machines—where the rate of output is generally higher and the level of quality generally better. However, automation often involves a high capital investment in specialized equipment that reduces the firm's flexibility and requires a higher volume of (more standardized) output to maintain its profitability.

APPROACHES TO JOB DESIGN

Question: What is job design, and what approach is taken to the design of jobs?

Job design is the conscious structuring of the content and methods of work effort. Past approaches to job design have emphasized (1) the *objective efficiency* of getting the job done or (2) the *behavioral satisfactions* of the employees or (3) a *combination* of both, such as in the sociotechnical system.

Question: Distinguish between the (*a*) objective and (*b*) behavioral approaches to job design.

(*a*) The *objective*, or *efficiency*, *approach* stems from Taylor's scientific management concepts and has given us quantitative measures such as time study, work sampling, and methods improvement.

(*b*) The *behavioral approach* has developed from the Hawthorn studies, plus the work of Herzberg, Hackman and Oldham, and others, and has been exemplified by some Japanese management systems. It has laid claim to productivity and quality improvements as a result of having more broadly trained and highly motivated employees. Many firms have successfully blended elements of both systems.

A Comparison of Some Job Design Characteristics

Highly objective ←——————— Job Design ————————→ Highly Behavioral		
On *job* to be done	Emphasis	On *individual* hired
Written in detail	Job description	*Unwritten*
Highly *specialized*	Job assignments	Widely *diversified*
Specific—and *limited*	Job training	General—and *continuous*
Highly *specified*—no discretion	Job methods	Highly *unspecified*—much freedom
Immediate objective measure	Performance	Measured over *long run* only

Question: What is the essence of the sociotechnical approach to job design?

This approach, exemplified by the sociosystem and technical system inputs shown in Fig. 10-1, stresses the balancing of the sociological concerns of the worker (or work group) with modern technology (e.g., of robots and computer-controlled machines). It emphasizes the need for individual and group autonomy to do well-defined and responsible tasks that evidence a tangible contribution to the end product or service.

BEHAVIORAL CONSIDERATIONS IN JOB DESIGN

On-the-job studies have extended sociotechnical research to some practical behavioral guidelines. Considerable emphasis has been focused upon motivational techniques and has shown that motivation is enhanced by arranging a variety of tasks into a meaningful work cycle. Figure 10-2 describes some motivational techniques and cites the work of contributors to this field.

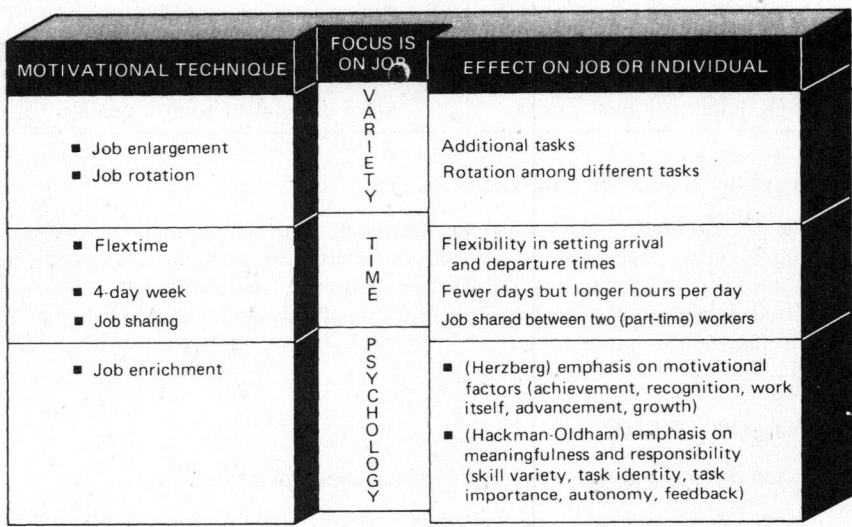

Fig. 10-2   Motivational techniques used in job design

Question: Explain what is meant by (*a*) job rotation, (*b*) job enlargement, and (*c*) job enrichment.

(*a*)   *Job rotation* involves the periodic exchange of jobs within workers to relieve monotony (and possibly help the workers to develop additional skills).

(*b*)   *Job enlargement* involves increasing the horizontal scope of a job by adding more tasks or different tasks of the same skill and responsibility level.

(*c*)   *Job enrichment* involves increasing the vertical scope of a job by requiring additional skills or responsibility, such as planning, organizing, or inspecting one's own work.

Question: What elements are stressed in the Hackman-Oldham approach to job design?

This approach emphasizes the need for meaningful work, responsibility for outcomes, and knowledge of actual results. Five characteristics have emerged as job design principles:

(1)   *Skill variety* (jobs requiring different skill levels generate respect and are more satisfying)

(2)   *Task identity* (clearly defined tasks, with visible outcomes, are best)

(3)   *Task importance* (the perceived importance of a task to others makes a difference)

(4)   *Autonomy* (jobs that give employees discretion and some control over work are preferred)

(5)   *Feedback* (fast, clear, and direct information about effectiveness of one's work is helpful)

HUMAN CONCERNS IN THE WORK SYSTEM

Although the balance sheets of most organizations are dominated by investment in plant and equipment, inventories, etc., employees are often recognized as the most valuable asset of an organization. As a vital component of a work system, humans generate a number of social, legal, and ethical concerns that operations managers must deal with on a day-to-day basis. These concerns can affect productivity and performance as much as, or even more than, the technical operation of the system. Table 10-1 lists some of the policy issues related to operations.

Table 10-1   Ten Human Resource Policy Issues Affecting Operations

General Area	Specific Issues of Concern	Example Questions
1. *Employment*	Hiring, layoff, use of temporaries Plant closures and foreign operations	Should firm close plant and move operations to Mexico?
2. *Wage and salary*	Employee compensation policies Stock plans/employee ownership	Would company do better under employee ownership?
3. *Health care*	Wellness: medical and dental plans Accident, disability, life insurance	Was this a job-related injury if employee was in transit?
4. *Education and training*	Minimal job skill requirement Enhancement of self-actualization	Should company send Joanne to training session in Houston?
5. *Communication and change*	Building and maintaining community Fostering acceptance of change	How can we get employees to "buy into" new total quality program?
6. *Commitment*	Decline in work-force loyalty Office versus home, child care issues	Will this job foster employees' long-term job satisfaction?
7. *Discrimination*	Racial, age, sex, or religious bias Accepting diversity in workplace	Do female production workers have equal chance of promotion?
8. *Ethical behavior*	Fairness, honesty, justice Social responsibility/service	Would discussing the job cost with a competitor be ethical?
9. *Work environment*	Maintaining underlying values Smoking, drug, safety standards	Does this promote cooperation? Are smokers treated fairly?
10. *Community environment*	Asset/liability to community Pollution and waste disposal	Should Brian be given time off for work with United Way?

In many industrialized nations, the traditional (physical-effort-centered) jobs are rapidly giving way to information and knowledge-based work activities. Much of this shift has come about as a result of the electronic revolution, which has spawned computers and telecommunication networks.

Question: What transitions in work systems are likely to continue in the future?

(1)   Continued *automation* of manual work (e.g., via robots, NC machines, CAM, and CIM)

(2)   Enhanced use of *information networks* and wireless links to suppliers and customers (e.g. internet)

(3)   Accelerated trend toward *knowledge-based* activities that require multiskilled effort

(4)   More *decentralization* of work within an organization and to *remote locations*

(5)   More use of *projects* and interdisciplinary self-managed *work teams*

(6)   Improved recognition of the *intrinsic value* of work and dignity of human effort

(7)   Increased emphasis on the *quality* of the process and compressed time-to-market

COMPENSATION

Question: What types of systems do organizations use to set employee compensation?

Two commonly used approaches are (*a*) time-based and (*b*) incentive-based systems.

(*a*)   *Time-based systems* pay employees for the time they spend on the job and result in a fixed hourly or monthly wage. Examples here include job-based systems (where all employees receive the same pay), seniority-based systems (where those with seniority receive a higher pay within a range), and merit-based systems (where better-performing employees are positioned higher within a range).

(b) *Incentive-based systems* pay employees according to a measurement of their performance. Examples here include individual (e.g., piece rate) plans and group (or work group) plans, as well as some organizationwide plans (e.g., profit-sharing bonuses).

Time-based systems are most common, for they are more suitable for jobs that require mental or decision-making activities. Incentive plans motivate a greater output but are more difficult to administer.

Question: Describe (*a*) the Scanlon Plan and (*b*) the Lincoln Electric Plan.

(*a*) *The Scanlon Plan* is a gain-sharing plan (first implemented in the 1940s) that allows labor to share in savings achieved by reducing cost and improving efficiency. A bonus is awarded that is based upon the ratio of the total labor cost to the sales value of production. Versions of this plan are still gaining in popularity.

(*b*) *The Lincoln Electric Plan* incorporates elements of a piecework system, a bonus, and a stock purchase plan. It fosters employee participation in production planning and quality-control evaluation.

LEGAL AND ETHICAL CONSIDERATIONS

Question: What major laws have influenced job design and the work environment in the United States?

(1) *National Labor Relations Act* (1935) protects the rights of workers to organize and bargain.

(2) *Fair Labor Standards Act* (1938) established a minimum wage and governs some working conditions.

(3) *Taft-Hartley Act* (1947) protects management's rights in bargaining and other labor practices.

(4) *Civil Rights Act of 1964* created a commission that protects workers (in hiring, firing, compensation, etc.) against discrimination because of race, color, religion, sex, or national origin.

(5) *Occupational Safety and Health Act* (1971) sets standards for noise, air pollution, and plant safety.

(6) *Worker Adjustment and Retaining Notification Act* (1988) outlines plant closing responsibilities.

(7) *North American Free Trade Agreement* (1994) reduces trade barriers and assists displaced workers.

The above laws have characterized the U.S. political system's response to what was considered unacceptable business behavior. Although they have multifaceted benefits, one result of such laws has been an exponential increase in the amount of compliance investigation, enforcement effort, and paperwork. In the final analysis, no amount of laws can ensure the ethical behavior of free people.

Ethical behavior rests upon honesty, trust, respect, and other God-given virtues. Many observers feel that the United States (and other countries as well) must realign its priorities to these values before real progress can be made in establishing enduring relationships among employees, customers, and society in general. In the midst of technological progress, the turmoil resulting from a focus on self-satisfaction (as opposed to service to others) seems to be increasing. Moreover, the legal system is proving ineffective as a surrogate for a solid moral foundation. The challenge of future sociotechnical progress may well lie in revitalizing the integrity of our culture and embracing value-oriented concepts such as honesty, social responsibility, and justice for all.

WORK METHODS AND MOTION ECONOMY

Work methods are ways of doing work. Both new and existing jobs can be analyzed by the relatively standardized approach described in Table 10-2.

Table 10-2 Steps in a Methods-Improvement Study

> 1. *Select* the job to be studied.
> 2. Document and *analyze* the present method.
> 3. *Develop* an improved method.
> 4. *Implement* the improved method.
> 5. *Maintain* and follow up on the new method.

Jobs that have a high labor content and are done frequently, or are unsafe or tiring, offer the most potential for improvement. Analysis of the present job method can make use of flow-process and worker-machine charts (see Chap. 7) plus a questioning technique that asks of every activity: (*a*) what is its purpose, and (*b*) why is it necessary? In addition, standardized lists of *Principles of Motion Economy* are useful for studying work methods.[1] Micromotion studies using high-speed cameras are also used in job improvement. The improved method of doing a job typically flows from the earlier analysis and must be convincingly applied and monitored.

WORK-MEASUREMENT OBJECTIVES

Question: What are labor standards?

Labor standards are declarations of the amount of time that should reasonably be used to perform a specified activity at a sustainable rate, using established methods under normal working conditions.

Standards satisfy the needs of the worker, provide a measure of performance for the organization, and facilitate scheduling and costing of operations. Methods commonly used to set standards include (1) historical (estimates from experience), (2) time study, (3) predetermined time standards, and (4) work sampling.

TIME-STUDY EQUATIONS

Time-study methods were originally developed by Taylor and are still the most widely used technique for work measurement of short, repetitive tasks. The task is broken down into basic elemental motions, and each element is timed with a stopwatch. Then the average time over several cycles is computed and adjusted for the speed and skill, or performance rating (PR), of the worker studied. Finally, an allowance factor (AF) is applied for personal needs, unavoidable delays, and fatigue. It typically represents a proportion of the normal time (although some analysts use total workday time). Table 10-3 summarizes the calculations; symbols are explained in the examples.

Table 10-3   Steps in Conducting a Time Study

1. Select the job, inform the worker, and define the best method.

2. Time an appropriate number of cycles *n*. Use a sample size chart or graph to determine *n*, or

   If *s* is known:

   $$n = \left(\frac{Zs}{e}\right)^2 \tag{10.1}$$

   If *s* is unknown:

   $$n = \frac{Z^2[n' \Sigma X^2 - (\Sigma X)^2]}{e^2(\Sigma X)^2} \tag{10.2}$$

3. Compute the *cycle time*

   $$CT = \frac{\Sigma \text{ times}}{n \text{ cycles}} \tag{10.3}$$

4. Compute the *normal time*

   $$NT = CT \times PR \tag{10.4}$$

5. Compute the *standard time*

   $$ST = NT \times AF \tag{10.5}$$

   where

   (*a*) If allowances are a percentage of the normal (working) time: (*b*) If allowances are a percentage of the total (workday) time:

   $$AF = (1 + A_{\text{work}}) \qquad\qquad AF = \frac{1}{1 - A_{\text{total}}} \tag{10.6}$$

---

[1] R.M. Barnes, *Motion and Time Study Design and Measurement of Work*, 6th ed., John Wiley, New York, 1968, p. 220.

STATISTICAL SAMPLE SIZE

The required sample size n can be determined from numerous charts, graphs, or equations, as illustrated in Table 10-4 and Fig. 10-3. The chart in Fig. 10-3 is for ± 5 percent accuracy for various coefficient-of-variation values, V.

Table 10-4   Number of Cycles for a Time Study*

When Time per Cycle Is ≥ (minutes):	Minimum Number of Cycles to Be Timed If Annual Activity is:		
	<1,000	1,000–10,000	>10,000
480	1	1	2
120	1	2	4
60	2	3	5
30	3	4	8
12	5	6	12
7.2	6	8	15
4.8	8	10	20
3.0	10	12	25
1.2	15	20	40
.72	20	25	50
.48	25	30	60
.30	30	40	80
.18	40	50	100
.12	50	60	120
<.12	60	80	140

*Benjamin W. Niebel, *Motion and Time Study*, 7th ed., Richard D. Irwin, Inc., Homewood, IL, 1982, p. 337.

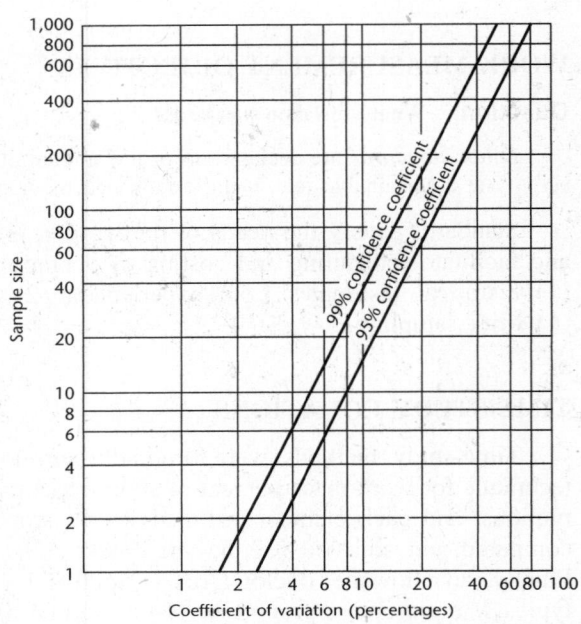

Fig. 10-3 Time-study sample size chart (for $\pm 5\%$ accuracy) (From A. Abruzzi, *Work Measurements*, Columbia University Press, New York, 1952.)

Table 10-4 gives a suggested minimum number of observations to be included as a function of the cycle time and its annual frequency. It is representative of the simplified tables used widely in industry. Figure 10-3 permits us to read the sample size directly from a chart, once a value for the coefficient of variation V has been estimated from a preliminary or partial sample. The coefficient of variation tells how much variability exists in the data relative to the value of the mean.

$$V = \frac{s}{\overline{X}} \qquad\qquad (10.7)$$

Example 10.1   A packaging firm has an order for 8,000 plastic inserts that have a .50-minute assembly time. How large a sample should be taken to set a time standard for the assembly activity?

From Table 10-4, for cycle times between .48 to .72 minute, use $n = 30$ cycles.

Example 10.2   A preliminary sample showed a mean of 1.25 minutes and a standard deviation of 15 seconds. How many cycles should be timed in order to be 95 percent confident that the resultant standard time is accurate within plus or minus 5 percent of the population value?

$$V = \frac{s}{\overline{X}} \qquad \text{where} \qquad s = \frac{15}{60} = .25 \text{ min}$$

$$V = \frac{.25}{1.25} = .20 = 20\%$$

Therefore, from Fig. 10-3, use $n = 60$ cycles.

Whereas tables and graphs always assume some values for the confidence level desired Z and the precision or maximum error e, these values can be individually specified if one uses an equation to calculate the sample size. Then, for example, if one wished to have 92 percent confidence that the error is held to within plus or minus .25 minute, the Z value (from the normal distribution) would be 1.75 and e would be .25. If the standard deviation is known, Eq. 10.1 may be used to find n. Otherwise, data from a preliminary sample of n' is needed and Eq. 10.2 is used.

Example 10.3   How large a sample size is needed to be 92 percent confident that a time-study result is accurate within plus or minus .12 minute, if the standard deviation is $s = .45$ minute?

$$n = \left(\frac{Zs}{e}\right)^2 = \left(\frac{1.75 \times .45}{.12}\right)^2 = 43$$

ADJUSTMENTS, ALLOWANCES, AND STANDARD TIMES

The performance rating (PR), done by an experienced analyst, adjusts the standard so that it is not geared to the skill or effort level of the particular worker being observed. Thus, if the worker under study has a high PR, for instance 120 percent, the cycle time will be multiplied by 1.20, resulting in a longer normal time, which will serve as a fair standard for an average worker.

Allowances take account of unavoidable delays, rest breaks, and personal time. Two methods are in use to compute the allowance factor, so it is *very important* to distinguish whether the percentage (A) applies to the *total available time* (including the allowable time) or to the actual *work* (job) time only. See Benjamin W. Niebel, *Motion and Time Study* (Richard D. Irwin), for a table of allowance percentages. Where the percentages apply to job times—as is often the case and as shown in Eq. 10.6(a)—the $AF = (1 + A)$, so the standard time is simply $ST = NT(1 + A)$. When the allowances are given in minutes and pertain to the total time available in the day, then the other form of AF calculation applies as in Eq. 10.6(b).

Example 10.4   A job is to be time-studied. (a) Compute the allowance factor if the allowances are 18 percent of the normal (job) time. (b) What would the allowance factor have been if the allowances were taken to be 18 percent of the total time?

(a) $AF = (1 + A_{\text{work}}) = (1 + .18) = 1.18$, or 118%

(b) $AF = \dfrac{1}{1 - A_{\text{total}}} = \dfrac{1}{1 - .18} = 1.22$, or 122%

Example 10.5   A time study of a fabrication shop worker yielded an observed (cycle) time of 3.80 minutes and a performance rating of 90 percent. Allowances for this activity are 12 percent of *work* (job) time. Compute the standard time.

$$NT = CT \times PR = 3.80 \times .90 = 3.42 \text{ min}$$

$$ST = NT(1 + A_{\text{work}}) = 3.42(1 + .12) = 3.83 \text{ min}$$

Example 10.6   A time study of an electronics plant worker revealed the actual times shown in Table 10-5. The standard deviation of the sample (with the 10.20-minute cycle omitted) was $s = .21$ minute. The analyst rated the worker at 90 percent PR, and the company allows 20 minutes of personal time and 30 minutes of delay time per 8-hour day. (a) Find the standard time, and (b) determine whether the sample was of adequate size for the analyst to be 99 percent confident that the resultant standard time is within 5 percent of the true value. If it was not, how many cycles should have been time-studied to gain this level of confidence?

Table 10-5 Observations from a Time Study (min/cycle)

Worker time	2.30	1.80	2.00	2.20	1.90	10.20*	2.20	1.80
Machine time	.80	.80	.80	.80	.80	.80	.80	.80

*Unusual, nonrecurring situation.

(a) Cycle time (also called observed time) should omit the unusual situation of taking 10.20 minutes.

$$\text{Worker CT} = \frac{\Sigma \text{ times}}{n \text{ cycles}}$$

$$= \frac{2.30 + 1.80 + 2.00 + 2.20 + 1.90 + 2.20 + 1.80}{7} = 2.03 \text{ min}$$

Machine CT = .80 min

$$\text{NT} = \text{CT} \times \text{PR} = \underset{\text{(worker time)}}{2.03(.90)} + \underset{\text{(machine time)}}{.80(1.00)} = 2.63 \text{ min}$$

$$\text{ST} = \text{NT} \times \text{AF}$$

where

$$A_{\text{total}} = 50 \text{ min as \% of } 480 = \frac{50}{480} = 10.42\%$$

Therefore,

$$\text{AF} = \frac{1}{1 - A_{\text{total}}} = \frac{1}{1 - .1042} = 1.116$$

$$\text{ST} = (2.63)(1.116) = 2.94 \text{ min/cycle}$$

(b) Coefficient of variation:

$$V = \frac{s}{\bar{x}} = \frac{.21}{2.03} = 10.34\%$$

Using Fig. 10-3, $n \cong 40$ cycles would have been required. The 7 cycles were not adequate for 99 percent confidence.

Although time-study methods are widely used, an increasing number of firms with exceptionally good labor relations hesitate to institute them because they fear that definitive standards would damage an already cooperative work environment. Firms that emphasize group activities and participative decision-making are less likely to rely upon published standards to measure or motivate their employees. Also, when the labor cost constitutes a small proportion of the end-product cost (e.g., less than 15 percent), the benefits of enforcing rigid production standards may not outweigh the costs.

PREDETERMINED TIME STANDARDS

Predetermined time standards are job times that are established by defining a job in terms of very small basic elements, using published tables to find the time for each element, and adding the element times to determine a total time for the job. Three measurement systems are *methods time measurement* (MTM), *basic motion time* (BMT), and *work factor*. Advantages of these methods are that (1) the standard can be determined from universally available data, (2) the standard can be completed before a job is done, (3) no performance rating is required, (4) there is no disruption of normal activities, and (5) the methods are widely accepted as fair systems of determining standards.

The MTM system uses times for basic motions (''therbligs'') consisting of activities such as search, select, grasp, and transport loaded. Times are measured in time-measurement units (TMUs) where one TMU equals .0006 minute.

WORK SAMPLING

Work sampling is a work-measurement technique that consists of taking random observations of workers to determine the proportion of time they spend doing various activities. It is particularly useful for analyzing group activities and long-cycle activities. Data are recorded in the form of counts of times working or idle, rather than as stopwatch times. However, once collected, the data can be used for standards purposes, as well as for methods or cost analysis. See Table 10-6.

Table 10-6   Steps in Conducting a Work-Sampling Study

1. Select the job (or group) to be studied, and inform the workers.
2. Delineate the operations, and prepare lists of worker activities.
3. Estimate the number of observations required, n.

$$n = \frac{Z^2 pq}{e^2} \qquad (10.8)$$

where Z = standard normal deviate for desired confidence level
  p = estimated proportion of time of activity of interest
    (use past experience; otherwise let $p = .5$)
  $q = 1 - p$
  e = maximum error percentage for precision level

4. Prepare a schedule of random observation times.
5. Observe, rate, and record worker activities per schedule.
    (*Note:* Sample size is often recomputed as study data become available.)
6. Record starting time, stopping time, and number of acceptable units completed during the period.
7. Compute the *normal time*.

$$\text{NT} = \frac{(\text{total time})(\%\ \text{working})(\text{PR})}{\text{number units completed}} \qquad (10.9)$$

8. Compute the *standard time*.

$$\text{ST} = \text{NT} \times \text{AF} \qquad (10.5)$$

where $\text{AF} = (1 + A_{\text{work}})$ or $\text{AF} = \dfrac{1}{1 - A_{\text{total}}}$ $\qquad (10.6)$

SAMPLE SIZE FOR WORK SAMPLING

Sample size is based upon the same statistical theory as used for time study, except we are dealing with a distribution of proportions rather than a distribution of means. (See Chap. 3, Discrete and Continuous Data.)

Example 10.7   A data processing manager estimates that a computer operator is idle 20 percent of the time and would like to do a work-sampling study that would be accurate within ±4 percentage points. The manager wishes to have 85 percent confidence in the resulting study. How many observations should be made?

$$n = \frac{Z^2 pq}{e^2} \quad \text{where } Z = 1.96 \text{ for 95\% confidence}$$
$$p = \text{idle time estimate} = .20$$
$$q = 1 - p = 1 - .20 = .80$$
$$e = \text{maximum error} = .04$$

$$n = \frac{(1.96)^2 (.20)(.80)}{(.04)^2} = 384 \text{ observations}$$

Note that we have used the estimate of idle time (20 percent) to calculate n. If early study results indicate that p will be outside the range of 20 percent ± 4 percent, then the number of observations may have to be adjusted as the study progresses.

Making Random Observations.   A random number table (Appendix A) may be used to ensure that observations are made at random intervals.

Example 10.8   The manager in Example 10-7 would like to set up a random schedule for the 384 observations over a 1-week period of five 8-hour workdays. (*a*) Illustrate the process by using a random number table to select 8 of the 384 observation times. (*b*) Show how a tally of 100 observations might be recorded if the operator were idle on 14 occasions and on another filing assignment during 20 of the observations.

(*a*)   Minutes   available $= 60$ min/hr $\times 8$ hr/day $= 480$ min/day $= 2{,}400$ min/week. (Thus day 1 includes minutes 1–480, day 2 includes 481–960, etc.) This requires a four-digit column of random numbers between 0001 and 2400 (eliminate numbers >2400). From the first column of Appendix A, the first number is 2776 and is discarded because it is >2400. The next number is 1302, which is the $1302 - 0960$, or the 342d minute of day 3. See Table 10-7. For an 8:00 a.m. to 5:00 p.m. workday, this is 1:42 p.m. After the 384 times are obtained, they would be chronologically arranged.

Table 10-7

Day 1	Day 2	Day 3	Day 4	Day 5
0001–0480	0481–0960	0961–1440	1441–1920	1921–2400
	0810	1302 (1:42 p.m.) 1087 1212	1771 1547	2230 2130

(*b*)   See Table 10-8.

Table 10-8

	Tally of All Observations (Total Time)		Working and Idle Time Only	
	Tally	Number	Number	Percent
Working	ͰͰͰ ͰͰͰ ͰͰͰ ͰͰͰ ͰͰͰ ͰͰͰ ͰͰͰ ͰͰͰ ͰͰͰ ͰͰͰ ͰͰͰ ͰͰͰ ͰͰͰ I	66	66	82.5
Idle	ͰͰͰ ͰͰͰ IIII	14	14	17.5
Other assignment	ͰͰͰ ͰͰͰ ͰͰͰ ͰͰͰ	20		
Total		100	80	100

Work-sampling studies can be done by part-time observers and need not disturb the workers. A quick glance can identify a worker's activity. If the performance is not rated, a 100 percent PR is assumed. Work sampling does not focus on the work methods (as much as time study) and is not as useful for short, repetitive tasks. It is more appropriate for long cycle activities.

Example 10.9   A work-sampling study of customer service representatives in a telephone company office showed that a receptionist was working 80 percent of the time at 100 percent PR. This receptionist handled 200 customers during the 8-hour study period. Company policy is to give allowances of 10 percent of total on-the-job time. Find the normal time and the standard time per customer.

$$NT = \frac{(\text{total time})(\% \text{ working})(PR)}{\text{number units completed}} = \frac{(480 \text{ min})(.80)(1.00)}{200} = 1.92 \text{ min/customer}$$

$$ST = (NT)(AF) \quad \text{where} \quad AF = 1/(1 - .10) = 1.11$$

$$= (1.92)(1.11) = 2.13 \text{ min/customer}$$

Questions and Solved Problems

APPROACHES TO JOB DESIGN: BEHAVIORAL AND HUMAN CONCERNS

10.1   Which is more important, the sociological aspect or the technical aspect of job design?

The needs of companies differ, so no prescription is right for all. While the benefits from technological innovation are frequently more immediate and tangible, the Operations Management Association has identified corporate leaders who feel we have not focused sufficiently on the human side of management (e.g., teamwork and group decision support).

10.2   Do standards exist for the skill requirements of high-tech industry workers?

Yes. For example, the American Electronics Association (in conjunction with Motorola, AT&T, IBM, and Hewlett-Packard) has developed guidelines for the education, training, and assessment of skills commonly used by high-tech firms. Credentials are recognized by firms nationwide. Similar guidelines exist for some administrative and information services jobs.

10.3   Are the (Japanese) job design methods of consensus building and participative decision-making always more successful than those that flow from more traditional (top-down) managerial styles?

Not necessarily. With increasing competition, the slower pace of consensus-reached decisions has forced even some Japanese companies (e.g., Mitsubishi Electric) to act more autocratically. Lifetime job security and compensation based on seniority can be highly inefficient. Some firms (e.g., Pharmacia Biotec K.K.) have moved away from compensation based upon group results toward more personal accountability.

10.4   Is technology having a significant effect on blue- and white-collar employment?

Yes, a profound effect. Many major industrial firms have "downsized" to a more lean production system. For example, General Motors eliminated a quarter million jobs in the 5 years preceding 1993 and planned on many more cuts before the year 2,000. Employment cuts in Britain (e.g., in banks with ATMs), Germany (e.g., Siemens), and elsewhere are largely attributable to technology, or the preparation for advanced technology.

10.5   How are educational requirements of workers changing?

Mindless tasks in factories are being automated with humans shifted to jobs that require technical competency. New workers at General Motors are thoroughly tested (manual dexterity *and mathematics*). At a Chrysler plant, over one-quarter of the newly hired employees for a *third-shift* operation possessed college degrees. Displaced workers with outdated training or inadequate skills are gravitating to service industries, or lower-paying nonunion jobs. In Scotland's shipbuilding industry, for example, the high-cost (unionized) jobs have been recast as lower-paying but higher-productivity (nonunion) jobs. Similar shifts are occurring elsewhere.

STATISTICAL SAMPLE SIZE

10.6   An analyst wants to obtain a cycle time estimate that is within ± 5 percent of the true value. A preliminary run of 20 cycles took 40 minutes to complete and had a calculated standard deviation of .3 minute. What is the value of the coefficient of variation to be used for computing the sample size for the forthcoming time study?

$$V = \frac{s}{\overline{X}} \quad \text{where} \quad s = \text{standard deviation of sample} = .3 \text{ min/cycle}$$
$$\overline{X} = \text{mean of sample} = \frac{\Sigma X}{n} = \frac{40 \text{ min}}{20 \text{ cycles}} = 2 \text{ min/cycle}$$
$$= \frac{.3}{2} = .15$$

10.7   How large a sample should be taken to provide 99 percent confidence that a sample value is within ± 5 percent of the true value if the coefficient of variation is estimated to be 15 percent?

From Fig. 10-3, for $V = 15$ percent, $n \cong 80$.

10.8   Past records of a certain work activity show that it has a mean time of 60 seconds and a standard deviation of 9 seconds. How many time-study observations should be made to be 95 percent confident that the sample mean is within 3 seconds (± 3) of the true population value?

$$V = \frac{s}{\overline{X}} \quad \text{where} \quad s = 9 \text{ sec}$$
$$\overline{X} = 60 \text{ sec}$$
$$V = \frac{9}{60} = .15$$

Figure 10-3 can be used because the 3-second accuracy required corresponds to $3/60 = 5$ percent accuracy. Therefore, $n = 35$ observations.

10.9   Suppose we make a preliminary estimate that the standard deviation of an activity is 9 seconds. How many time-study observations should be made to be 95 percent confident that the sample mean is within 3 seconds (± 3) of the true population value?

Note the similarity between this and the previous problem. In this case we have no mean value available to estimate the coefficient of variation, so we must calculate the sample size instead of using Fig. 10-3. Our method is similar to that followed for the work sampling in Example 10.7, except in this case we are dealing with means ($\bar{x}$'s) rather than sample proportions (p's). Both situations rely on the fact that the sample means and proportions are normally distributed (Fig. 10-4) about the population parameters (that is, μ and π, respectively) if the sample size is sufficiently large (say 30 or more for means and 100 or more for proportions). In solving this problem we wish to set one-half the accuracy interval width ($Zs_{\bar{x}}$) equal to 3 seconds.

$$e = Zs_{\bar{x}} = Z \frac{s}{\sqrt{n}}$$

$$\therefore n = \left(\frac{Zs}{e}\right)^2 \quad \text{where} \quad e = 3$$
$$Z = 1.96$$
$$s = 9$$

$$n = \left[\frac{(1.96)(9)}{3}\right]^2 \cong 35 \text{ observations}$$

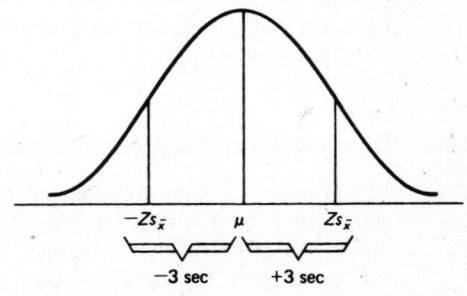

Fig. 10-4

Note that the chart method (previous problem) and the calculation method (this problem) are essentially equivalent, but the chart is perhaps a little easier to use if V can be estimated.

10.10 A time-study analyst wishes to estimate the cycle time for an assembly operation within $\pm.03$ minute at a confidence level of 95.5 percent. If the cycle time standard deviation, σ, is known to be .08 minute, how many observations are required?

$$n = \left(\frac{Z\sigma}{e}\right)^2 \quad \text{(Since } \sigma \text{ is known, we use it instead of } s.)$$

$$= \left[\frac{2(.08)}{.03}\right]^2 = 28.4, \text{ use 29 observations}$$

Note that as the sample size gets below 30, the t is a more appropriate distribution than the normal. However, the normal approximation should be adequate in this case.

10.11 An assembly activity in an ammunition factory is to be analyzed by a time study. A preliminary sample of $n' = 30$ cycles reveals the following:

$$\Sigma X = 670 \text{ sec} \qquad \Sigma X^2 = 16,400 \qquad s = 6.6 \text{ sec}$$

The firm wishes to be 95 percent confident that the resultant time is accurate within ± 5 percent. Find the required sample size (a) by use of the coefficient of variation using Fig. 10-3 and (b) by use of Equation 10.2.

(a)
$$V = \frac{s}{\overline{X}} \quad \text{where} \quad \overline{X} = \frac{\Sigma X}{n} = \frac{670}{30} = 22.3 \text{ sec}$$

$$V = \frac{6.6}{22.3} = .30 = 30\%$$

From Fig. 10-3, $n \cong 150$.

(b)
$$n = \frac{Z^2[n' \Sigma X^2 - (\Sigma X)^2]}{e^2(\Sigma X)^2} = \frac{(1.96)^2[30(16,400) - (670)^2]}{(.05)^2(670)^2} = 147$$

ADJUSTMENTS, ALLOWANCES, AND STANDARD TIMES

10.12 A job being time-studied has allowances of 20 minutes per 8-hour day for rest breaks and 28 minutes per day for unavoidable delays. Compute the allowance factor if allowances are computed as a percentage of (a) total available time and (b) working time only.

(a)   Allowances as a percentage of total time:

$$A_{\text{total}} = \frac{20 + 28}{(8 \text{ hr})(60 \text{ min/hr})} = \frac{48}{480} = .10 = 10\% \qquad \text{AF} = \frac{1}{1 - A_{\text{total}}} = \frac{1}{1 - .10} = 1.11$$

(b)   Allowances as a percentage of working time:

$$A_{\text{work}} = \frac{20 + 28}{480 - (20 + 28)} = \frac{48}{432} = .111 = 11.1\%$$

$$\text{AF} = (1 + A_{\text{work}}) = 1 + .11 = 1.11 \text{ (same)}$$

10.13 A time study of a nursing home activity yielded a cycle time of 4.00 minutes, and the nurse was rated at PR = 95 percent. The nursing home uses a 20 percent of normal work time allowance factor. Find the standard time.

$$ST = NT \times AF = NT(1 + A_{work})$$
$$= 3.80(1 + .20) = 4.57 \text{ min}$$

10.14 An operator in a fruit-packing plant was clocked by a stopwatch with results as shown in Table 10-9. The allowance for this type of work is 15 percent of total time. Find (*a*) the normal time per cycle and (*b*) the standard time per cycle.

Table 10-9

Element	Time for Cycle (min)					Performance Rating
	1	2	3	4	5	
1. Obtain 2 boxes	.82	—	.80	—	.85	130
2. Pack 4 items/box	.44	.42	.46	.40	.41	110
3. Set box aside	.71	.67	.69	.71	.68	115

(*a*) $$NT = CT \times PR$$

For element 1, each box suffices for 2 cycles.

$$NT = \frac{.82 + .80 + .85}{6}(1.30) = .535$$

For element 2:

$$NT = \frac{.44 + .42 + .46 + .40 + .41}{5}(1.10) = .469$$

For element 3:

$$NT = \frac{.71 + .67 + .69 + .71 + .68}{5}(1.15) = \underline{.796}$$

$$\text{Total} = 1.800$$

Thus, NT = 1.80 min/cycle.

(*b*) $$ST = NT \times AF$$

where $$AF = \frac{1}{1 - .15} = 1.18$$

$$ST = (1.80)(1.18) = 2.12 \text{ min/cycle}$$

WORK SAMPLING

10.15 The State Mental Health Division has a health-care activity that has a normal time of 8.3 minutes, but the activity seems to have been prolonged recently by an increasing number of unavoidable delays. D. R. Mix, a management analyst called in to determine a new standard, conducted a work-sampling study and obtained the results shown in Table 10-10:

Table 10-10

Activity	Number of Observations	Percentage of Working Time
Working	650	100%
Unavoidable delay	78	12%
Personal time	52	8%
Total	780	

The Mental Health Division grants its workers a personal-time allowance of 8 percent of working time, and Mr. Mix wishes to retain that allowance in the new standard.

(*a*)   Incorporate the unavoidable-delay time, and determine a standard time for this activity.

(*b*)   Determine how precise the estimate is of unavoidable time, assuming the analyst wishes to have 95 percent confidence in the estimate.

(*c*)   State whether the same precision applies to the estimate of personal time.

(*a*)   Allowances should now consist of 8% personal time + 12% unavoidable delay = 20%

$$ST = NT \times AF = NT(1 + A_{work})$$

$$= 8.3(1 + .20) = 9.96 \text{ (round to 10.0 min)}$$

(*b*)   For 95% confidence interval, $Z = 1.96$. Half the interval width h is

$$h = e = Zs_p$$

where
$$s_p = \sqrt{\frac{pq}{n}} = \sqrt{\frac{(.12)(.88)}{780}} = .011$$

$$\therefore h = 1.96(.012) = .023$$

The interval is $\pm 2.3\%$—that is, the analyst could be 95% confident that the true unavoidable delay time is from $12.0\% - 2.3\% = 9.7\%$ to $12.0\% + 2.3\% = 14.3\%$ of the work time.

(*c*)   The precision interval for the personal-time estimate would be slightly smaller (better) due to the use of 8% instead of 12% for the value of p. In general, for a given level of precision, the sample size required for various activities is governed by the activities with p values closest to .5.

Supplementary Questions and Problems

10.16   A customer-service activity in a bank takes an estimated 3 minutes and must be performed 25 times per day and 200 days per year. Estimate the sample size required to set a time standard for this activity.
Ans.   From Table 10-4, $n = 12$

10.17   A textile workers' union in New York City has requested that a new time study be made of a skirt-sewing activity. Previous data indicate the activity has a mean time of $\bar{X} = 1.80$ minutes and a standard deviation of .40 minute. What is the best preliminary estimate of the sample size required in order to have 95 percent confidence in the result. (*Hint:* Use Fig. 10-3, and assume an accuracy of ± 5 percent.)
Ans.   From Fig. 10-3, $n \cong 65$

10.18 Sixty samples of an electronic-assembly operation revealed an average time of 3.20 minutes per unit. The performance rating was estimated at 105 percent, and allowances are set at 20 percent of the work time available. What is the standard time in minutes per piece?   *Ans.*   4.03 minutes

10.19 A time study of an Iowa City grain-elevator loading activity revealed a cycle time of 8.57 minutes for a worker rated at 107 percent. The allowances are based upon total (480 minutes per day) time and are as follows: personal time = 25 minutes per day, fatigue = 84 minutes per day, delay = 35 minutes per day. Determine the standard time for an 8-hour-per-day operation.   *Ans.*   13.10 minutes

10.20 An activity has a cycle time of 2.20 minutes per cycle and a calculated normal time of 2.64 minutes per cycle. Allowances are 10 percent of work (normal) time. What was (*a*) the performance rating factor of the worker studied and (*b*) the resultant standard time?   *Ans.*   (*a*) 120 percent (*b*) 2.90 minutes per cycle

10.21 A time study of 40 cycles of a worker-machine operation revealed an operator time of .60 minute per cycle and a machine time of 1.40 minutes per cycle. The worker was rated at 115 percent, and allowances for the operation, based on an 8-hour workday, are as follows: personal = 30 minutes per day, fatigue = 20 minutes per day, delay = 30 minutes per day. Calculate the standard time per cycle for the (combined) worker-machine operation.   *Ans.*   2.51 minutes per cycle

10.22 Time-study data taken for a bulk-filling activity in a cannery in Baltimore were recorded on a *continuous* basis, as shown in Table 10-11.
(*Note:* The times given are cumulative amounts, so one must subtract the previous amount to obtain the time for each element. For example, in cycle 1, the time for locating for fill is 16 − 4 = 12 seconds, and for machine-fill it is 26 − 16 = 10 seconds, etc.)

Table 10-11

	Cycle Time (sec)					
	1	2	3	4	5	RF
Grasp bag	4	37	74	105	338	120
Locate for fill	16	51	84	117	352	120
Machine-fill	26	61	94	127	362	
Set on conveyor	34	68	102	334*	369	110

*Bag broke open due to presence of a foreign object on the conveyor.

The firm's labor contract requires a 15 percent allowance based on total time for all workers on the bulk-filling line. Compute the standard time for this activity.   *Ans.*   44.62 ($\cong$45) seconds per cycle

10.23 A work-sampling study is to be made of an airline ticket counter in a major airport. The operations manager feels the ticket agents are idle 30% of the time and wishes to have 95.5% confidence that the accuracy is within 4 percentage points. How many observations should be made?   *Ans.*   525

10.24 An analyst wishes to develop a labor-cost standard for a manual card-sorting activity. The elements consist of (1) collecting the cards, (2) sorting them, and (3) filing the sorted deck. For element 2, the standard deviation is estimated to be $\sigma = 2.25$. To determine the sorting time to an accuracy of within ±.5 minute with 95.5% confidence, how large a sample should be taken?   *Ans.*   81

10.25 A work-sampling study is to be made of food-service activities in a major hotel chain. Analysts have defined the activities as shown in Table 10-12. They wish to be 95.5% confident that the true proportions of time of the various elements are accurate within ±3%. How many samples should be taken to be sure the 95.5% confidence level holds for all elements?   *Ans.*   1,100

Table 10-12

Element	Estimated Time (%)
Taking orders	20
Filling and serving orders	45
Table set-up	10
Billing	10
Delays	15
Total	100

10.26 A work-sampling study was made of a cargo-loading operation for the purpose of developing a standard time. During the total 120 minutes of observation the employee was working 80% of the time and loaded 60 pieces of cargo. The analyst rated the performance at 90%. The firm wishes to incorporate an allowance factor for fatigue, delays, and personal time of 10% of normal (work) time. What is the standard time for this operation in minutes per piece? *Ans.* 1.58 min per piece

Chapter 11

Aggregate Planning and Master Scheduling

PLANNING AND SCHEDULING STRATEGIES

Production planning is a complex problem! The capacity of most firms is relatively stable, whereas the demand for goods and services is often quite variable. Demand cannot always be met. The objective of aggregate planning is to respond to irregular market demands by utilizing the organization's equipment, personnel, and other resources in the most effective manner possible.

Question: What is aggregate planning?

Aggregate planning is the process of planning the *quantity* and *timing* of output over the intermediate range (often 3 to 18 months) by adjusting the production rate, employment, inventory, and other controllable variables.

Figure 11-1 illustrates how aggregate planning links long-range and short-range planning activities. It is "aggregate" in the sense that the planning activities at this early stage are concerned with homogeneous categories (families) such as gross volumes of products or number of customers served, rather than specific models of individual goods or categories of services.

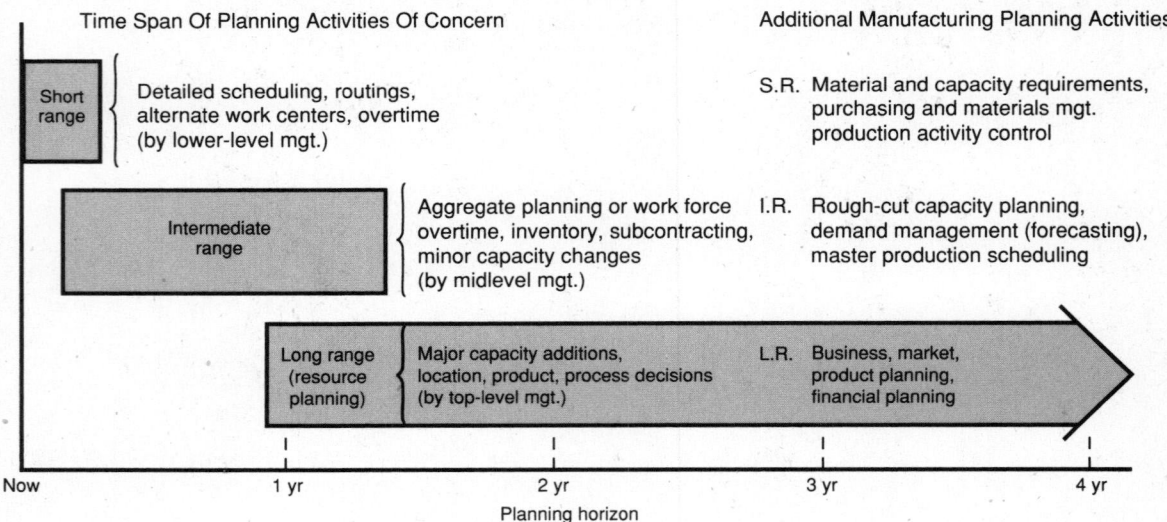

Fig. 11-1 Planning levels and activities

Question: What is master scheduling, and does it differ from aggregate planning?

Master scheduling follows aggregate planning and expresses the overall plan in terms of the amounts of specific end items to produce and dates to produce them. It uses information from both forecasts and orders on hand, and it is the major control (driver) of all production activities. Figure 11-2 illustrates a simplified aggregate plan and master schedule.

Aggregate Plan

Month	J	F	M	A	M	J	J	A	S
Number of motors	40	25	55	30	30	50	30	60	40

Master Schedule

Month	J	F	M	A	M	J	J	A	S
AC motors:									
5 hp	15	—	30	—	—	30	—	—	10
25 hp	20	25	25	15	15	15	20	30	20
DC motors:									
20 hp	—	—	—	—	—	—	10	10	—
WR motors:									
10 hp	5	—	—	15	15	5	—	20	10

Fig. 11-2 Aggregate plan and master schedule for electric motors

VARIABLES USED IN AGGREGATE PLANNING

Aggregate planning is a complex problem largely because of the need to coordinate interacting variables in order for the firm to respond to the (uncertain) demand in an effective way. Table 11-1 identifies some of the key variables available to planners and the costs associated with them.

Table 11-1   Some Decision Variables and Costs in Aggregate Planning

Decision Variable	Associated Cost
1. Varying size of work force	1. Hiring, training, and layoff costs
2. Using overtime or accepting idle time	2. Wage premiums and nonproductive time costs
3. Varying inventory levels	3. Carrying and storage costs
4. Accepting back orders	4. Stockout costs of lost orders
5. Subcontracting work to others	5. Higher labor and material costs
6. Changing the use of existing capacity	6. Delayed response and higher fixed costs

To best understand the effect of changes in these variables, it is useful to first focus upon the impact of a change in only one variable at a time, with other variables held constant. The examples that follow show the effect on production costs of (isolated) changes in the decision variables. They are presented in a simplified format in order to best convey the underlying concept; more realistic examples follow in later sections.

Example 11.1 Paris Candy Company has estimated its quarterly demand (cases) as shown in Table 11-2 and Fig. 11-3. It expects the next demand cycle to be similar to this one and wishes to restore ending inventory, employment, etc., to beginning levels accordingly.

Table 11-2   Demand

Quarter	Units
1st	500
2nd	900
3rd	700
4th	300

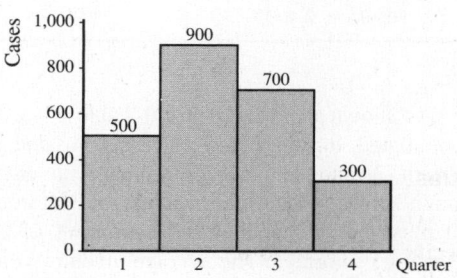

Fig. 11-3 Histogram of demand by quarter

(Example 11.1: *Vary work-force size*)   Each quarterly change of 200 units output has an incremental labor cost of $2,000, and ending levels must be restored to initial levels. What is the cost associated with changing the work-force size?

As shown in Fig. 11-4, six changes of 200 are required.

$$\text{Employment change cost} = 6\ (\$2,000)$$
$$= \boxed{\$12,000}$$

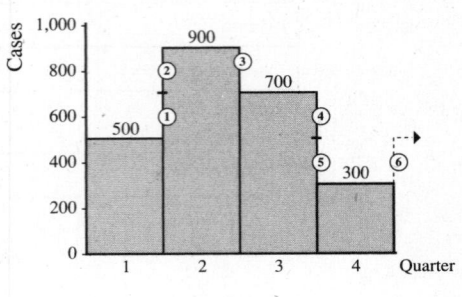

Fig. 11-4

Example 11.2 (*Overtime and idle time*)   Maintain a stable work force capable of producing 600 units per quarter, and use OT (at $5 per unit) and IT (at $20 per unit).

As shown in Fig. 11-5, 400 units will be produced on overtime, and workers will be idle when 400 units could have been produced.

$$\text{OT + IT cost} = (400)(\$5) + (400)(\$20)$$
$$= \boxed{\$10,000}$$

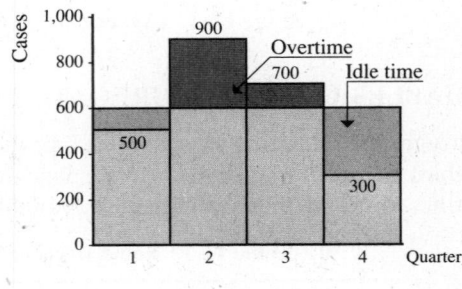

Fig. 11-5

Example 11.3 (*Vary inventories*)   Vary inventory levels, but maintain a stable work force producing at an average requirement rate (of 2,400 units ÷ 4 quarters = 600 units per quarter) with no OT or IT. The carrying cost (based on average inventory) is $32 per unit per year, and the firm can arrange to have whatever inventory level is required before period 1 at no additional cost. Annual storage cost (based on maximum inventory) is $5 per unit.

Table 11-3   Inventory Level Balances

Qtr.	Fcst.	Rate of Production	Change in Inventory	Prelim. End Bal.	End Balance w/300 on 1/1
1	500	600	100	100	400
2	900	600	−300	−200	100
3	700	600	−100	−300	0
4	300	600	300	0	300
	Totals	2,400			800

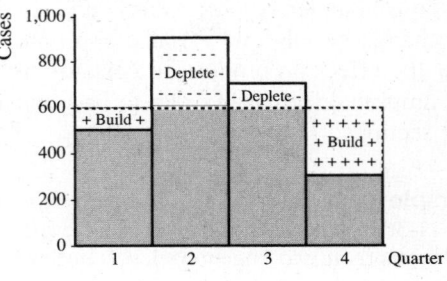

Fig. 11-6

As shown in Fig. 11-6 and Table 11-3, inventory is accumulated during quarters 1 and 4, and depleted in quarters 2 and 3. The preliminary inventory balance column shows a negative inventory of 300 in quarter 3, so 300 must be on hand at the beginning of quarter 1 to prevent any shortage. The average inventory on hand is the ending balance total of 800 units divided by 4 quarters = 200 units.

Carrying cost, C_c (on avg. inventory)

$$C_c = (\$32/\text{unit-yr})(200\ \text{units}) = \$6,400$$

Storage cost, C_s (on max. inventory)

$$C_s = (\$5/\text{unit})(400\ \text{units}) = \$2,000$$

Total inventory cost $(C_c + C_s)$ $\boxed{\$8,400}$

Example 11.4 (*Back orders*)   Produce at a steady rate of 500 units per period, and accept a limited number of back orders when demand exceeds 500. The stockout cost of lost sales is $20 per unit.

A *back order* is an arrangement to fill a current order during a later period. Stockout costs occur when some sales (or customers) are lost because products are not immediately available. In this example, 200 units of the excess demand in period 2 are placed on back order for delivery in period 4, as shown in Fig. 11-7. The other 200 units demanded in period 2 are lost, along with the 200 in period 3.

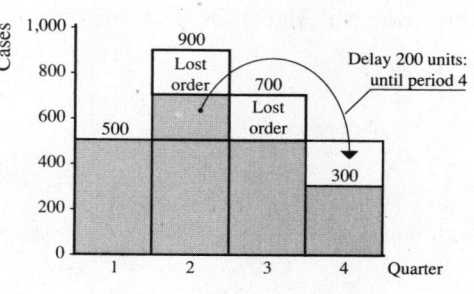

Fig. 11-7

Stockout cost = (200 + 200)($20/unit) = $8,000

Example 11.5 (*Subcontract*)   Produce at a steady rate of 300 units per period, and subcontract for excess requirements at a marginal cost of $8 per unit.

The firm must subcontract 200 units in period 1, 600 in period 2, and 400 in period 3, as shown in Fig. 11-8.

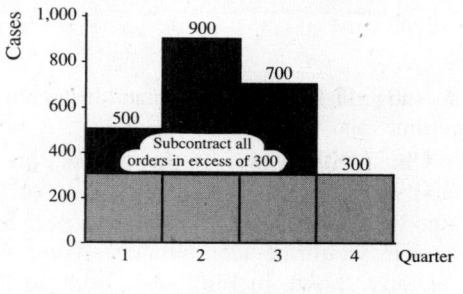

Fig. 11-8

Subcontract cost = (1200)($8/unit) = $9,600

Note: Of the five decision variables considered above, *accepting back orders* results in the least cost ($8,000).

FOCUSED AGGREGATE PLANNING STRATEGIES

Several different strategies have been employed to assist in aggregate planning. Three so-called "pure" strategies are recognized, and a myriad of other "mixed" strategies are possible. The pure strategies stem from early models that depicted production results when only one of the decision variables (see previous examples) was permitted to vary—all others being held constant. The concept of a "pure" strategy has, however, largely shifted to one that has a single focus (focused strategy), such as level employment, but does not necessarily preclude the use of other variables, e.g., overtime. Mixed strategies can incorporate any mix of variables and can be unique to a firm.

Question: What focused strategies are employed by production planners to meet nonuniform demands?

Three focused strategies (apart from numerous mixed strategies) are:

(1) *Vary production to match demand*—by changes in employment. (*Chase demand strategy:* This strategy permits hiring and layoff of workers, use of overtime, and subcontracting as required in each period. However, inventory buildup is not used.)

(2) *Produce at a constant rate*—and use inventories. (*Level production strategy:* This strategy retains a stable work force producing at a constant output rate. Inventory can be accumulated to satisfy peak demands. In addition, subcontracting is allowed and back orders can be accepted. Promotional programs may also be used to shift demand.)

(3) *Produce with stable work force*—but vary the utilization rate. (*Stable work-force strategy:* This strategy retains a stable work force but permits overtime, part-time, and idle time. Some versions of this strategy permit back orders, subcontracting, and use of inventories. Although this strategy uses overtime, it avoids the detrimental effects of layoff.)

We can use the data in Figs. 11-9 and 11-10 to illustrate the three focused strategies described above. These figures display a histogram and cumulative graph of a 9-month forecast for motors. The total requirement for the 9 months is 360 motors. This works out to an average (mean) of 40 motors per month, which is shown as a dotted line in Fig. 11-9.

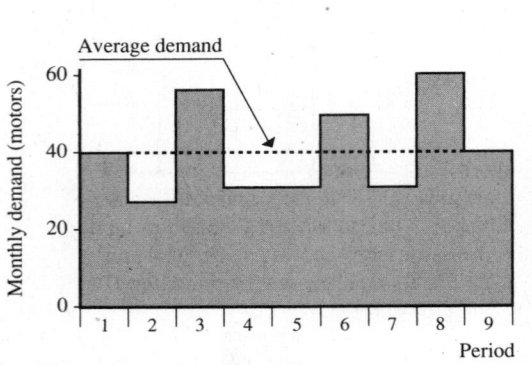

Fig. 11-9 Monthly demand histogram

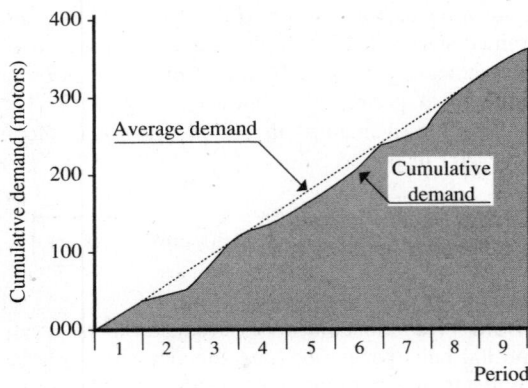

Fig. 11-10 Cumulative demand graph

(#1) Chase Strategy. If the production planner designed a plan to exactly match the forecast of demand shown in Fig. 11-9, by adding or laying off employees to change the level of production, the planner would be using a *chase strategy*. Some overtime or subcontracting might also be used, but no inventories would be accumulated. (*Note:* The resulting aggregate plan would coincide exactly with the one already shown in Fig. 11-2, because the forecast values shown in Fig. 11-9 are the same as the values shown in the aggregate plan of Fig. 11-2.)

(#2) Level Production Strategy. The graph in Fig. 11-9 shows (visually) that the demand exceeds the average requirement in some months and is below average in others. A production plan could be developed to produce at the constant rate of 40 motors per month, accumulating inventory in months 2, 4, 5, and 7, and using that inventory to meet the above average demands in months 3, 6, and 8. Figure 11-10 shows that the cumulative demand (forecast) never exceeds the cumulative averages (production), so no initial inventory is needed to prevent shortages. However, if there were shortages, some back orders could be allowed under a *level production, or inventory, strategy*.

(#3) Stable Work-Force Strategy. Referring to Fig. 11-9, suppose the firm has a stable work force capable of producing 36 motors per month on regular time. Production might go as high as 60 motors per month by using overtime, but if demand falls to less than 36 motors per month, some workers would be idle. Using overtime and idle time to meet demand would be employing a *stable work-force strategy*. As part of this strategy, however, it seems likely that planners would build up some inventory during what might otherwise be idle time periods.

Formats for Presentation and Comparison. Numerous formats have been developed to display the comparative data from an analysis of different plans. Both tabular and graphic presentations are useful; this type of analysis also lends itself to computer modeling and the use of spreadsheets. The solved problems at the end of the chapter illustrate some additional ways of structuring the analysis. Note that when making a comparison among different plans, only the relevant production and cost information (i.e., variables that change from one alternative to another) need be included.

Service activities do not have the inventory strategy available to them, so they tend to rely more upon shifting or managing demand (by fixed schedules, appointments, etc.).

Example 11.6 (*Vary production to match demand*) An aggregate plan is to be developed for the forecast of demand covering nine periods shown in Table 11-4. Other relevant production and cost information is also provided. (Note that since this plan does not allow for any inventory buildup, a decision has been made to carry 10 units of safety stock, but no overtime or subcontract labor is used.) Find the cost associated with an aggregate plan that involves varying the size of the work force in order to have a production rate that matches demand.

Table 11-4   Demand, Production, and Cost Information

Month	Jan.	Feb.	Mar.	April	May	June	July	Aug.	Sept.	Total
Forecast	40	25	55	30	30	50	30	60	40	360

Production information		Cost information	
Current number workers	10	Hiring cost	$600/employee
Worker time/mo	160 hr/mo	Layoff cost	$500/employee
Time to produce one unit	40 hr/unit	Regular-time cost	$ 30/hr
Individual worker output:		Overtime cost	$ 45/hr
(160 hr/mo ÷ 40 hr/unit)	4 units/mo	Subcontract labor cost	$ 50/hr
Safety stock of inventory req'd	10 units	Inventory carrying cost	$ 35/period

The cost associated with changing the employment level is calculated in Table 11-5. The number of workers required is first determined by dividing the forecast amount by the worker output of 4 units per month. Fractional values have been rounded up. Beginning with the current level of 10 workers, the number that must be hired, or laid off, is then determined. Costs are then computed for (1) regular-time hours, (2) hiring and layoff, and (3) carrying safety stock. These are added to get the total plan cost of $470,450.

Table 11-5   Cost Calculation for Varying Work Force to Match Demand

Period	1	2	3	4	5	6	7	8	9
Production forecast (units/mo)	40	25	55	30	30	50	30	60	40
WORK-FORCE SIZE DATA									
No. workers required	10	7	14	8	8	13	8	15	10
(fcst. ÷ output of 4 units/wkr-mo)									
(a) No. hired @ beg. of mo	0	0	7	0	0	5	0	7	0
(b) No. laid off @ beg. of mo	0	3	0	6	0	0	5	0	5
COSTS									
Regular time cost, $	48,000	33,600	67,200	38,400	38,400	62,400	38,400	72,000	48,000
(No. wkr)($30/wkr-hr)(160 hr/wkr-mo)									
Hiring or layoff cost, $: @	0	1,500	4,200	3,000	0	3,000	2,500	4,200	2,500
(a) ($600) or (b) ($500)									
Inventory carrying cost	350	350	350	350	350	350	350	350	350
(10 units) ($35/unit-period)									

Total cost for plan = Σ Regular-time employment + Σ Hiring and layoff + Σ Inventory carrying cost
= $446,400 + $20,900 + $3,150
= $470,450

Example 11.7 (*Produce at a constant rate*)   Using the demand shown in Table 11-4 (plus 10 more units in periods 8 and 9), develop an aggregate plan based upon the use of the 10 regular-time production workers at a constant rate, with *inventories used to satisfy peak demand*. The inventory carrying cost is $35 per unit per period. Some subcontracting can be used at a labor cost of $50 per hour if necessary. Assume a constant output rate of 40 units per period. No safety stock is required, but total demand of 380 units must be met.

The costs associated with producing at a constant rate and using inventories to help meet nonuniform demands are shown in Table 11-6. Note that the constant production rate of 40 units per period yields 360 units, which is 20 units short of total demand. Insofar as the additional demand is in periods 8 and 9—when demand already consumes all production—the additional demand will be subcontracted out in these two periods. The labor cost for each subcontracted unit is (40 hours per unit)($50 per hour) = $2,000 per unit, so for the 10 units in periods 8 and 9, the subcontracting costs are $20,000 in each period.

Table 11-6   Cost Calculation for Using Inventories to Meet Demand

Period	1	2	3	4	5	6	7	8	9
Production forecast	40	25	55	30	30	50	30	70	50
PRODUCTION DATA									
Output: regular time	40	40	40	40	40	40	40	40	40
subcontract	—	—	—	—	—	—	—	10	10
Output − forecast	0	15	−15	10	10	−10	10	−20	0
Inventory:									
Beginning-of-period	0	0	15	0	10	20	10	20	0
End-of-period	0	15	0	10	20	10	20	0	0
Average inventory	0	7.5	7.5	5	15	15	15	10	0
COSTS									
Regular-time cost, $ (10)($30/hr)(160 hr)	48,000	48,000	48,000	48,000	48,000	48,000	48,000	48,000	48,000
Subcontract cost, $: (@ $2,000/unit)	—	—	—	—	—	—	—	20,000	20,000
Inventory carrying cost (avg. inv.) ($35/period)	0	263	263	175	525	525	525	350	0

Total cost for plan = Σ Regular-time employment + Σ Subcontract cost + Σ Inventory carrying cost
= $432,000 + $40,000 + $2,826
= $474,826

Inventory costs under this format are computed by first determining how many units go into (or out of) inventory. This amount (i.e., the output minus production forecast) is shown in the table. For period 1, where forecast and output are both 40 units, it is zero. For period 2, when 40 units are produced and only 25 are needed, 15 go into inventory. The beginning- and end-of-period inventory rows in the table show how the inventory balance fluctuates. Average inventory is the sum of beginning- plus end-of-period inventory divided by two. The inventory carrying cost is then this average amount multiplied by the $35 per period carrying charge.

MIXED STRATEGIES

The number of mixed strategy alternative production plans is almost limitless. However, the realities of the situation will most likely limit the number of practical solutions. These can be evaluated on a trial-and-error basis to find which plan best satisfies the requirements, taking cost, employment policies, etc., into account.

Example 11.8 (*Vary work force, and use inventory*)   Custom Furniture Co. currently has 100 employees and has forecast quarterly demand as shown in Table 11-7. The historical average production rate is 40 units per employee per quarter, and the firm has a beginning (safety stock) inventory of 1,000 units. The hiring and training cost is $400 per employee, and the layoff cost is $600 per employee. Inventory is carried at a cost of $8 per unit per quarter. Use the data to develop an aggregate plan that uses variable employment and inventory to meet demand.

Table 11-7   Quarterly Demand Forecast for Furniture Manufacture

Quarter	1	2	3	4	Total
Demand	3,500	5,000	4,000	3,450	15,950

One alternative plan is shown in Table 11-8. (Many others, including better ones, are possible.) The planner has chosen to build some extra inventory in quarter 1 with the workers already on the payroll. Producing at a rate of 40 units per employee, the first quarter production of 4,000 units is 500 more than demand (3,500), so the ending inventory equals the beginning 1,000 plus 500, or 1,500 units. This results in a carrying cost of $8 (1,500) = $12,000. Twenty employees are hired at the beginning of quarter 2 to help meet the larger demand during the quarter. This results in a hiring cost of $400 (20) = $8,000. Employment is cut back again at the beginning of quarter 3, as the firm dips into safety stock, and employment is restored to its original level at the beginning of quarter 4. (*Note:* This firm bases inventory cost on ending inventory balance.)

Table 11-8   Aggregate Plan for Vaying Work Force and Inventory Levels

(1) Qtr.	(2) Fcst. of Demand	(3) No. of Empl.	(4) Change in Empl.	(5) Total Prod'n	(6) Cumul. Prod'n	(7) Cumul. Demand	(8) Ending Inv.	(9) Inv. Cost @ $8	(10) Empl. Chg Cost @ $400 or $600
1	3,500	100	—	4,000	4,000	3,500	1,500	12,000	—
2	5,000	120	+20	4,800	8,800	8,500	1,300	10,400	8,000
3	4,000	80	−40	3,200	12,000	12,500	500	4,000	24,000
4	3,450	100	+20	4,000	16,000	15,950	1,050	8,400	8,000
							Totals	34,800	40,000

The total (comparative) cost for this is $34,800 + $40,000 = $74,800. Note that inventory and employment costs are not well balanced, and employment is the lowest during quarter 3 when demand is relatively high. With some additional trials, the planner could undoubtedly develop a plan that would result in a lower total cost. Large fluctuations in production often result in more problems (and higher costs) than more steady-state operations.

MATHEMATICAL PLANNING MODELS

Mathematical models attempt to refine or improve upon the trial-and-error approaches. Table 11-9 identifies four mathematical approaches. The value from some of these models is more theoretical than practical. The LDR is not easily understood, nor are the outputs always realistic. The management coefficients model is nonoptimal and not easily transferable, whereas the computer search models do not necessarily yield a "global" minimum cost.

Table 11-9   A Summary of Some Mathematical Aggregate Planning Models

Approach	Linear Programming	Linear Decision Rule (LDR)	Management Coefficients	Computer Search Models
Application	Minimizes costs of employment, over-time, and inven-tories subject to meeting demand	Uses quadratic cost functions to derive rules for work-force size and num-ber of units	Develops regres-sion model that in-corporates mana-gers' past decisions to predict capacity needs	Computer routine searches numerous combinations of capacity and selects the one of least cost

A useful version of the linear-programming model (the transportation algorithm) views the aggregate planning problem as one of allocating capacity (supply) to meet forecast requirements (demand) where supply consists of the inventory on hand and units that can be produced using regular time (RT), overtime (OT), and subcontracting (SC), etc. Demand consists of individual-period requirements plus any desired ending inventory. Costs associated with producing units in the given period or producing them and carrying them in inventory until a later period are entered in the small boxes inside the cells in the matrix, as is done in the standard transportation linear-programming format.

Example 11.9 Given the accompanying supply, demand, cost, and inventory data (Tables 11-10, and 11-11) for a firm that has a constant work force and wishes to meet all demand (that is, with no back orders), allocate production capacity to satisfy demand at minimum cost.

Table 11-10   Supply Capacity (units)

Period	Regular Time ($100/unit)*	Overtime ($125/unit)	Subcontract ($130/unit)
1	60	18	1,000
2	50	15	1,000
3	60	18	1,000
4	65	20	1,000

*50 percent of cost is labor

Table 11-11   Demand and Inventory

Demand:				
Period	1	2	3	4
Units	100	50	70	80

Inventory:
  Initial = 20, Final = 25
  Carrying cost = $2/unit-period

The initial linear-programming matrix in units of capacity is shown in Fig. 11-11, with entries determined as explained below. Because total capacity exceeds demand, a *slack* demand of unused capacity is added to achieve the required balance in supply versus demand.

Supply, units from		Demand, units for				Capacity	
		Period 1	Period 2	Period 3	Period 4 and Final	Unused	Total Available
Initial inventory		0	2	4	6	8	20
Period 1	Regular	100	102	104	106	50	60
	Overtime	125	127	129	131	0	18
	Subcontract	130				0	1,000
Period 2	Regular		100	102	104	50	50
	Overtime		125	127	129	0	15
	Subcontract		130			0	1,000
Period 3	Regular			100	102	50	60
	Overtime			125	127	0	18
	Subcontract			130		0	1,000
Period 4	Regular				100	50	65
	Overtime				125	0	20
	Subcontract				130	0	1,000
Demand		100	50	70	105	4,001	4,326

Fig. 11-11   Linear programming format for scheduling

Initial inventory. There are 20 units available at no additional cost if used in period 1. Carrying cost is $2 per unit per period if units are retained until period 2, $4 per unit until period 3, and so on. If the units are unused during any of the four periods, the result is a $6-per-unit cost, plus $2 per unit to carry it forward to the next planning horizon, for $8 total if unused.

Regular time. Cost per unit-month is $100 if units are used in the month produced; otherwise, a carrying cost of $2 per unit-month is added on for each month the units are retained. Unused regular time costs the firm 50 percent of $100 = $50.

Overtime. Cost per unit is $125 if the units are used in the month produced; otherwise, a carrying cost of $2 per unit-month is incurred, as in the regular-time situation. Unused overtime has zero cost.

Subcontracting. Cost per unit is $130 plus any costs for units carried forward. This latter situation is unlikely, however, because any reasonable demand can be obtained when needed, as indicated by the arbitrarily high number (1,000) assigned to subcontracting capacity. There is no cost for unused capacity here.

Note: If the initial allocations are made so as to use regular time as fully as possible, the solution procedure is often simplified. Overtime and subcontracting amounts can also be allocated on a minimum-cost basis.

Final inventory. The final-inventory requirement (25 units) must be available at the end of period 4 and has been added to the period 4 demand of 80 units to obtain a total of 105 units.

Since no back orders are permitted, production in subsequent months to fill demand in a current month is not allowed. These unavailable cells, along with the cells associated with carrying forward any subcontracted units, may therefore be blanked out, since they are infeasible. The final solution, following normal methods of distribution linear programming, is shown in Fig. 11-12. This result flows from a least-cost allocation.

Supply, units from		Demand, units for — Period 1	Period 2	Period 3	Period 4 and Final	Capacity — Unused	Total Available
Initial inventory		0 / 20	2	4	6	8	20
Period 1	Regular	100 / 60	102	104	106	50	60
Period 1	Overtime	125 / 18	127	129	131	0	18
Period 1	Subcontract	130 / 2				0 / 998	1,000
Period 2	Regular		100 / 50	102	104	50	50
Period 2	Overtime		125	127	129 / 12	0 / 3	15
Period 2	Subcontract		130			0 / 1,000	1,000
Period 3	Regular			100 / 60	102	50	60
Period 3	Overtime			125 / 10	127 / 8	0	18
Period 3	Subcontract			130		0 / 1,000	1,000
Period 4	Regular				100 / 65	50	65
Period 4	Overtime				125 / 20	0	20
Period 4	Subcontract				130 / 0	0 / 1,000	1,000
Demand		100	50	70	105	4,001	4,326

Fig. 11-12 Matrix for planning decision

The optimal solution values can be taken directly from the cells. Thus in period 2, for example, the planners will schedule the full 50 units to be produced on regular time plus 12 units on overtime to be carried forward to period 4. This leaves 3 units of unused overtime capacity and no subcontracting during that period. Because of the similar carrying cost for units produced on regular time or overtime, it does not matter which physical units are carried forward, once overtime production is required. Thus, different optimal solutions (but with identical costs) may be obtained.

MASTER SCHEDULING OBJECTIVES, INPUTS, AND PLANNING HORIZON

The master production schedule (MPS) formalizes the production plan and translates it into specific end-item requirements over a short to intermediate planning horizon. The end items are then exploded into specific material and capacity requirements by the Material Requirements Planning (MRP) and Capacity Requirements Planning (CRP) systems. Thus the MPS essentially drives the entire production and inventory system.

Planners frequently "trial fit" the MPS on the MRP and CRP systems to confirm that a tentative schedule can be met before it is considered firm. A good MPS system should also incorporate feedback from operations to ensure that the order priorities and capacity use data in the system remain valid as the schedule is actually carried out.

Question: What are the major inputs to the master production schedule?

(1) *Forecasts* of demand, e.g., of end items and service parts

(2) *Customer orders*, i.e., including any warehouse and interplant needs

(3) *Inventory* on-hand from the previous period

Forecasts of demand are the major input for make-to-stock items. However, to be competitive, many make-to-order firms must anticipate orders by using forecasts for long lead-time items and by matching the forecasts with customer orders as the orders become available.

Question: What determines the planning horizon length (time span) of a master schedule?

The time horizon depends upon the type of product, volume of production, and component lead times. It can be weeks, months, or some combination, but the schedule must normally extend far enough into the future so that the lead times for all purchased and assembled components are adequately encompassed. Figure 11-13 illustrates a 10-week lead time for an item assembled from three component parts. The master scheduler should allow 10 weeks to produce this item, unless raw materials (or components) are stocked, or unless the machining or assembly times can be shortened (e.g., via use of overtime, subcontracting, etc.).

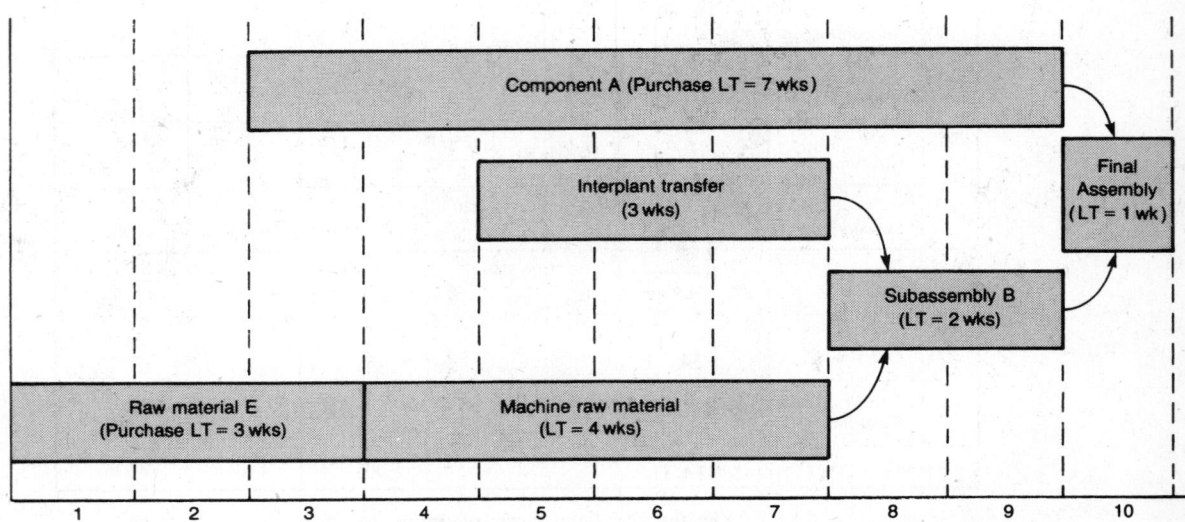

Fig. 11-13 Assembly with 10-week cumulative lead time

Question: How do firms accommodate changes in their master schedule?

Master schedules frequently have both firm and flexible (or tentative) portions. Table 11-12 illustrates an MPS for a furniture company that has one such schedule—where the firm and flexible portions have been marked. The firm portion encompasses the minimum lead time necessary and is not open to change.

Table 11-12   Master Schedule for a Furniture Company

Item	Week																	
	1	2	3	4	5	6	7	8	9	10	11	12	13	14	15	16	17	18
R28 table	50		50	50		40			40	40		40	40			40		
R30 table		80		20	60		80	80			60				80			
L7 lamp	20		20		10	20			20	20	10	20	20	20				

←———— Firm ————→   ←—————— Flexible ——————→   ←————— Open ————→
(emergency changes only)   (capacity firm and material ordered)   (additions and changes OK)

Question: What is meant by the terms (*a*) demand time fence and (*b*) planning time fence?

(*a*)   A *demand time fence* is the firm or "frozen" portion of the master schedule (beginning with the current period) during which no changes can be made to the schedule without management approval.

(*b*)   A *planning time fence* is the portion of the master schedule (also beginning with the current period) during which changes will not automatically be made (i.e., via computer) to accommodate demand. This gives the master scheduler a manageable base to work from and still allows some discretion in overriding the constraint.

Question: How does master scheduling differ under manufacturing strategies of (*a*) make-to-stock, (*b*) assemble-to-order, and (*c*) make-to-order?

See Fig. 11-14 where the shorter line segments represent fewer items. In (*a*) make-to-stock operations, the (fewer) end items are stocked to support customer service, and the master production schedule (MPS) is structured around those end items. In (*b*) assemble-to-order plants, such as in automobile manufacturing, the master scheduling is done for the major subassembly-level items. In (*c*) make-to-order products, such as customized furniture, there are fewer raw materials than end items; the end items may even be one of a kind. Here, the master schedule is typically structured around the raw material usage.

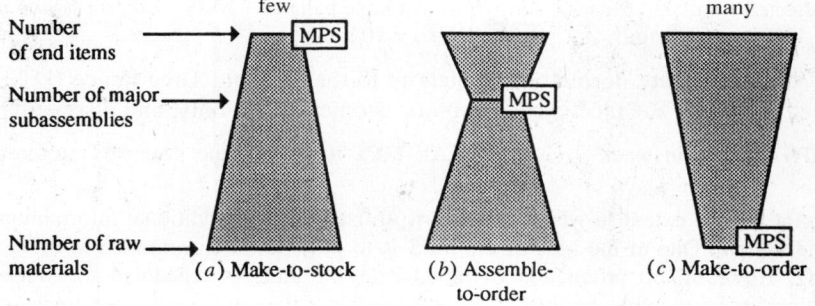

Fig. 11-14   Location of master scheduling activity

MASTER SCHEDULING FORMAT

Planning production that integrates a forecast of demand, incoming customer orders, and current inventory levels is difficult—especially when it must be done over a multiweek period. These difficulties are amplified as hundreds or thousands of items become involved. Numerous computer programs have been designed to assist in the scheduling and to provide detailed reports and graphs for analyzing and testing proposed master schedules. Although the extras differ from one program to another, much of the logic is similar. It consists of (1) incorporating the forecast and customer orders, (2) determining whether the inventory balance is sufficient to satisfy the *larger* of either the forecast or the orders on hand for the period, and (3) scheduling the production of a predetermined lot size in periods whenever the inventory balance is inadequate.

Example 11.10 Use the master schedule shown in Table 11-13 to answer the following:

(a) Does this product appear to be made primarily for stock, or is it made-to-order?

(b) How long is the planning period, and how many production runs will be scheduled in response to demand?

(c) Why is no production run scheduled for week 1, and how is the projected available balance determined?

(d) How many end items will be "exploded" into component parts in the MRP system as a result of the MPS requirements during week 3?

(e) Does the capacity appear to be fully utilized during the 6-week planning period?

Table 11-13   Master Schedule for TR28 Blood Analyzer Unit

Lead Time 0 On-Hand 30		Lot Size 25 Safety Stock 0					Demand Time Fence 0 Planning Time Fence 6			
Period	1	2	3	4	5	6	7	8	9
Forecast	20	20	20	20	20	20			
Customer orders (booked)									
Projected available balance	10	15	20	0	5	10			
Master production schedule		25	25		25	25			

(a) Product appears to be made for stock in response to a forecast; no customer orders are shown.

(b) The planning time fence is 6 weeks, and the schedule calls for four production runs (i.e., of 25 units each in weeks 2, 3, 5, and 6).

(c) No production is needed in week 1 because the beginning inventory of 30 is more than enough to meet the forecast demand of 20 needed in week 1.

$$\text{Projected Available Balance} = \text{Previous available balance} + \text{MPS} - \text{Current period requirements} \quad (11.1)$$
$$(\text{@ end of period}) \qquad = 30 + 0 - 20 = 10 \text{ units}$$

(Note: No changes are normally accepted up to the Demand Time Fence (DTF). Prior to the DTF the Projected Available Balance is based upon customer orders only, and disregards the forecast.)

(d) The MPS amount in week 3 (25 units). All MPS items become projected requirements in the MRP system.

(e) We cannot tell the extent to which capacity is utilized without additional information from the capacity planning system. One of the uses of the MPS is to provide the information to drive rough-cut capacity planning. However, no production of blood analyzer units is scheduled for weeks 1 and 4, so the production facilities might be idle at that time—unless they are being used for another product.

AVAILABLE-TO-PROMISE QUANTITIES

In make-to-order operations, as actual customer orders are received, they essentially take the place of an equivalent amount in the forecast, or *consume the forecast*. For this reason, the scheduled production of a lot is initiated by the *larger* demand of either the forecast amount or the actual (booked) customer orders.

As new orders are evaluated (and received), it is important to provide marketing with realistic promises of when shipments can be made. In well-designed master scheduling systems, this information is provided by a simple calculation that yields an available-to-promise inventory.

Question: What is available-to-promise inventory, and how is it determined?

Available-To-Promise (ATP) inventory is that portion of the on-hand inventory plus scheduled production that is not already committed to customer orders. For the first (current) period, the ATP includes the beginning inventory plus any MPS amount in that period, minus the total of booked orders up to the time when the next MPS amount is available. In subsequent periods, the ATP inventory consists of the MPS amount in that period, minus the actual customer orders already received for that period and all other periods until the next MPS amount is available.

Example 11.11   Find the ATP inventory values for the master schedule shown in Table 11-14.

Table 11-14   Master Schedule for Tractor Levelers

On-Hand 23			Lot Size 25			Planning Time Fence 6		
Period	1	2	3	4	5	6	7	8 9
Forecast	10	10	10	10	20	20		
Customer orders (booked)	13	5	3	1				
Projected available balance	10	0	15	5	10	15		
Master production schedule			25		25	25		
Available-to-promise								

Available-To-Promise values are computed for the current period (1) and for other periods when the MPS shows that a lot will be produced.

For period 1: $ATP_1 = \text{(On-hand Inv.)} - \text{(orders in periods 1 and 2)} = 23 - (13 + 5) = 5$

  3: $ATP_3 = \text{(MPS amount in 3)} - \text{(orders in periods 3 and 4)} = 25 - (3 + 1) = 21$

  5: $ATP_5 = \text{(MPS amount in 5)} - \text{(orders in period 5)} = 25 - 0 = 25$

  6: $ATP_6 = \text{(MPS amount in 6)} - \text{(orders in period 6)} = 25 - 0 = 25$

The last row should have values of 5, 21, 25, and 25 in the columns for periods 1, 3, 5, and 6, respectively.

Questions and Solved Problems

VARIABLES AND STRATEGIES USED IN AGGREGATE PLANNING

11.1 High Point Furniture Co. maintains a constant work force (no overtime, back orders, or subcontracting) that can produce 3,000 tables per quarter. The annual demand is 12,000 units and is distributed seasonally in accordance with the quarterly indexes: $Q_1 = .8$, $Q_2 = 1.40$, $Q_3 = 1.00$, $Q_4 = .80$. Inventories are accumulated when demand is less than capacity and are used up during periods of strong demand. To supply the total annual demand: (*a*) How many tables must be accumulated during each quarter? (*b*) What inventory must be on hand at the beginning of the first quarter?

See Table 11-15. (*a*) The inventory accumulation is given in column 3. (*b*) From column 4, the largest negative inventory is 600 units; therefore, 600 must be on hand on January 1. Column 5 shows the resulting balance at the end of each quarter.

Table 11-15

Quarter	(1) Production at 3,000/Q	(2) Seasonal Demand (SI)$Y_c = Y_{sz}$	(3) Inventory Change	(4) Inventory Balance	(5) Balance with 600 on Jan. 1
1st	3,000	(.8)(3,000) = 2,400	600	600	1,200
2d	3,000	(1.4)(3,000) = 4,200	−1,200	−600	0
3d	3,000	(1.0)(3,000) = 3,000	0	−600	0
4th	3,000	(.8)(3,000) = 2,400	600	0	600

Note: The next two examples add precision to the aggregate planning process by taking account of the actual production days available per month. They also demonstrate a different format for determining how much inventory must be on hand at the beginning of the planning period and include an inventory storage cost.

11.2 (*Plotting the requirements*) A firm has developed the following forecast in units (Table 11-16) for an item that has a demand influenced by seasonal factors.

Table 11-16

Jan.	220	Apr.	396	July	378	Oct.	115
Feb.	90	May	616	Aug.	220	Nov.	95
Mar.	210	June	700	Sept.	200	Dec.	260

(*a*) Prepare a chart showing the daily demand requirements. (*Note:* Available workdays per month are given below.) (*b*) Plot the demand as a histogram and as a cumulative requirement over time. (*c*) Determine the production rate required to meet average demand, and plot this as a dotted line on the graphs.

(*a*)   See Table 11-17.

Table 11-17   Chart of Production Requirements

Month	(1) Forecast Demand	(2) Production Days	(3) Demand/Day (1) ÷ (2)	(4) Cumulative Production Days	(5) Cumulative Demand
January	220	22	10	22	220
February	90	18	5	40	310
March	210	21	10	61	520
April	396	22	18	83	916
May	616	22	28	105	1,532
June	700	20	35	125	2,332
July	378	21	18	146	2,610
August	220	22	10	168	2,830
September	200	20	10	188	3,030
October	115	23	5	211	3,145
November	95	19	5	230	3,240
December	260	20	13	250	3,500
	3,500	250			

(*b*)   See Fig. 11-15.

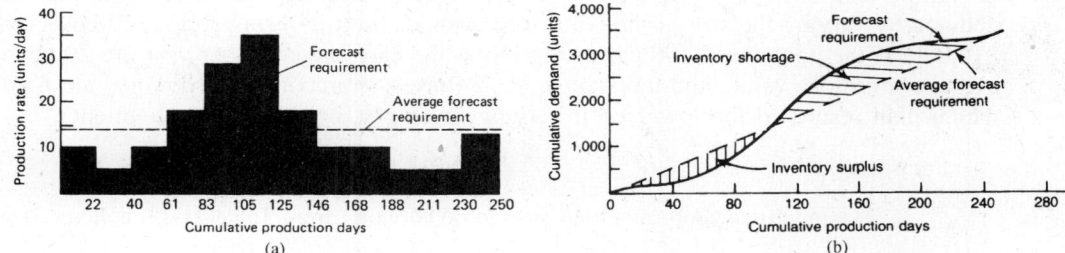

Fig. 11-15   Histogram and cumulative graph of forecast

(*c*) $$\text{Average requirement} = \frac{\text{total demand}}{\text{total production days}} = \frac{3,500}{250} = 14 \text{ units/day}$$

11.3   Use the data from Prob. 11.2 to determine the monthly inventory balances required to follow a plan of letting the inventory absorb all fluctuations in demand. In this case we have a constant work force, no idle time or overtime, no back orders, no use of subcontractors, and no capacity adjustment. Assume that the firm does not use safety stock or cushion inventory to meet the demand.

The firm can satisfy demand by producing at an average requirement (14 units per day) and by accumulating inventory during periods of slack demand (periods below the dashed line in Fig. 11-15*b*) and depleting it during periods of strong demand. Disregarding any safety stock, the inventory balance is:

$$\text{Inventory balance} = \Sigma \, (\text{production} - \text{demand})$$

See Table 11-18 for the solution. The pattern of demand is such that column 4 reveals a maximum negative balance of 566 units at the *end* of July, so 566 additional units must be carried in stock initially if demand is to be met. Column 5 shows the resulting inventory balances required.

Table 11-18

Month	(1) Production at 14 Units/Day	(2) Forecast Demand	(3) Inventory Change	(4) Ending Inventory Balance	(5) Ending Balance with 566 on Jan. 1
January	308	220	+88	88	654
February	252	90	+162	250	816
March	294	210	+84	334	900
April	308	396	−88	246	812
May	308	616	−308	−62	504
June	280	700	−420	−482	84
July	294	378	−84	−566	0
August	308	220	+88	−478	88
September	280	200	+80	−398	168
October	322	115	+207	−191	375
November	266	95	+171	−20	546
December	280	260	+20	0	566
		3,500			

11.4   Given the data of Prob. 11.2, suppose the firm has determined that to follow a plan of meeting demand by varying the size of the work force would result in hiring and layoff costs estimated at $12,000. If the units cost $100 each to produce, the carrying costs per year are 20 percent of the average inventory value, and the storage costs (based on maximum inventory) are $.90 per unit, which plan results in the lower cost: varying inventory, or varying employment?

From Prob. 11.3,

Maximum inventory requiring storage = 900 units (from Table 11-18, column 5)

$$\text{Average inventory balance} \cong \frac{654 + 816 + 900 + \cdots + 566}{12} \cong 460 \text{ units}$$

Plan 1 (varying inventory):

Inventory cost = carrying cost + storage cost

$$= (.20)(460)(\$100) + (\$.90)(900) = \$10,010$$

Plan 2 (varying employment): $12,000
Therefore, varying inventory is the strategy with the lower cost.

11.5   In comparing the costs of various aggregate plans, how can analysts incorporate the costs associated with not being able to meet the customer demand; i.e., what mechanism is available to take account of these costs?

Assuming that the costs of being out of stock can be estimated, they can be incorporated into the analysis by means of the back-order costs. The higher costs assigned to units that cannot be supplied in the time period desired by the customer can include some cost for the delay as well as some cost for being totally unable to supply the item.

11.6   Michigan Manufacturing produces a product that has a 6-month demand cycle, as shown in Table 11-19 and Fig. 11-16. Each unit requires 10 worker-hours to produce, at a labor cost of $6 per hour regular rate (or $9 per hour overtime). The total cost per unit is estimated at $200, but units can be subcontracted at a cost of $208 per unit. There are currently 20 workers employed in the subject department, and hiring and training costs for additional workers are $300 per person, whereas layoff costs are $400 per person. Company policy is to retain a safety stock equal to 20 percent of the monthly forecast, and each month's safety stock becomes the beginning inventory for the next month. There are currently 50 units in stock carried at a cost of $2 per unit-month. Unit shortage, or stockouts, have been assigned a cost of $20 per unit-month.

Table 11-19

	January	February	March	April	May	June
Forecast demand	300	500	400	100	200	300
Workdays	22	19	21	21	22	20
Work hr at 8/day	176	152	168	168	176	160

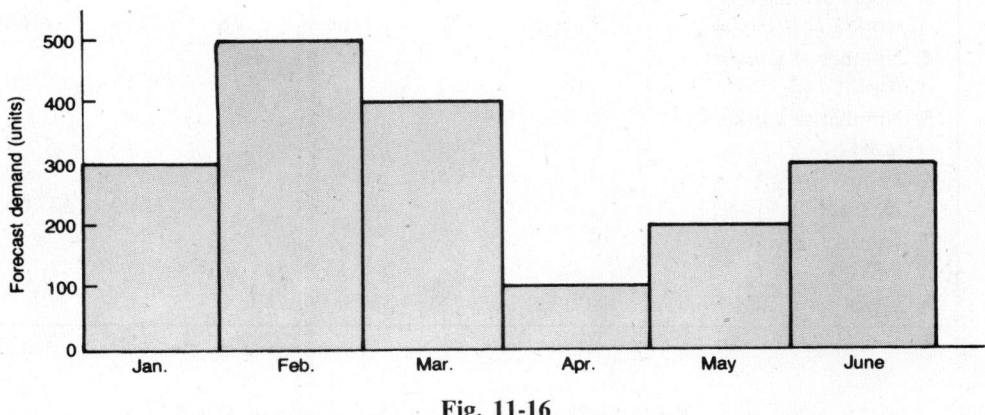

Fig. 11-16

Three aggregate plans are proposed.

Plan 1:   Vary work-force size to accommodate demand.

Plan 2:   Maintain constant work force of 20, and use overtime and idle time to meet demand.

Plan 3:   Maintain constant work force of 20, and build inventory or incur stockout cost. The firm must begin January with the 50-unit inventory on hand.

Compare the costs of the three plans in table form.

We must first determine what the production requirements are, as adjusted to include a safety stock of 20 percent of next month's forecast. Beginning with a January inventory of 50, each subsequent month's inventory reflects the difference between the forecast demand and the production requirement of the previous month. See Table 11-20. The costs of the three plans are shown in Tables 11-21, 11-22, and 11-23.

Table 11-20

	Forecast Demand	Cumulative Demand	Safety Stock at 20 Percent Forecast	Beginning Inventory	Production Requirement (fcst. + SS − beg. inv.)
January	300	300	60	50	300 + 60 − 50 = 310
February	500	800	100	60	500 + 100 − 60 = 540
March	400	1,200	80	100	400 + 80 − 100 = 380
April	100	1,300	20	80	100 + 20 − 80 = 40
May	200	1,500	40	20	200 + 40 − 20 = 220
June	300	1,800	60	40	300 + 60 − 40 = 320

Table 11-21   Plan 1 (Vary Work-Force Size)

	January	February	March	April	May	June	Total
1. Production required	310	540	380	40	220	320	
2. Production hours required (**1** × 10)	3,100	5,400	3,800	400	2,200	3,200	
3. Hours available per worker at 8/day	176	152	168	168	176	160	
4. Number of workers required (**2** ÷ **3**)	18	36	23	3	13	20	
5. Number of workers hired		18			10	7	
6. Hiring cost (**5** × $300)		$5,400			$3,000	$2,100	$10,500
7. Number of workers laid off	2		13	20			
8. Layoff cost (**7** × $400)	$800		$5,200	$8,000			$14,000

Table 11-22   Plan 2 (Use Overtime and Idle Time)

	January	February	March	April	May	June	Total
1. Production required	310	540	380	40	220	320	
2. Production hours required (**1** × 10)	3,100	5,400	3,800	400	2,200	3,200	
3. Hours available per worker at 8/day	176	152	168	168	176	160	
4. Total hours available (**3** × 20)	3,520	3,040	3,360	3,360	3,520	3,200	
5. Number of OT hours required (**2** − **4**)		2,360	440			0	
6. OT prem.* (**5** × $3)		$7,080	$1,320			0	$8,400
7. Number IT hours (**4** − **2**)	420			2,960	1,320		
8. IT cost (**7** × $6)	$2,520			$17,760	$7,920		$28,200

*Incremental cost of OT = overtime cost − regular time cost = $9 − $6 = $3.

Table 11-23   Plan 3 (Use Inventory and Stockout Based on Constant 20-Worker Force)

	January	February	March	April	May	June	Total
1. Production required	310	540	380	40	220	320	
2. Cumulative production required	310	850	1,230	1,270	1,490	1,810	
3. Total hours available at 20 workers	3,520	3,040	3,360	3,360	3,520	3,200	
4. Units produced ($3 \div 10$)	352	304	336	336	352	320	
5. Cumulative production	352	656	992	1,328	1,680	2,000	
6. Units short ($2 - 5$)		194	238				
7. Shortage cost ($6 \times \$20$)		\$3,880	\$4,760				\$8,640
8. Excess units ($5 - 2$)	42			58	190	190	
9. Inventory cost ($8 \times \$2$)	\$84			\$116	\$380	\$380	\$960

Note that plan 3 assumes that a stockout cost is incurred if safety stock is not maintained at prescribed levels of 20 percent of forecast. The firm is in effect managing the safety-stock level to yield a specific degree of protection by absorbing the cost of carrying the safety stock as a policy decision.

Summary:

Plan 1: \$10,500 hiring + \$14,000 layoff = \$24,500

Plan 2: \$8,400 overtime + \$28,200 idle time = \$36,600

Plan 3: \$8,640 stockout + \$960 inventory = \$9,600 (least-cost plan)

11.7   *(Mixed strategy: constant minimal work force with subcontracting)*   Use the data from Example 11.6 except modify as follows: Monthly demand and number of workdays per month are as shown below, employees work 8 hours per day, and time to produce one unit is 40 hours. Regular-time cost is (\$30 per hour) (40 hours per unit) = \$1,200 per unit, and subcontract time cost is (\$50 per hour) (40 hours per unit) = \$2,000 per unit. Produce with a (minimal) constant work force of six workers on regular time and subcontract to meet additional requirements.

See Table 11-24 for solution. The total cost of this plan is \$268,800 + \$270,720 = \$539,520.

Table 11-24   Plan for Mixed Strategy: Constant Work Force and Subcontracting

[1] Month	Jan.	Feb.	Mar.	April	May	June	July	Aug.	Sept.	Total
[2] Forecast	40	25	55	30	30	50	30	60	40	360
[3] Workdays/mo	22	18	21	22	22	20	21	22	20	
[4] Prod. hr avail. [3] (6 wkrs) (8 hr)	1056	864	1008	1056	1056	960	1008	1056	960	
[5] Reg.-time prod. [4] $\div$ 40 hr/unit	26.4	21.6	25.2	26.4	26.4	24.0	25.2	26.4	24.0	
[6] Units subcon't [2]−[5]	13.6	3.4	29.8	3.6	3.6	26.0	4.8	33.6	16.0	
[7] Subcon't cost [6] (\$2,000)	27,200	6,800	59,600	7,200	7,200	52,000	9,600	67,200	32.000	268,800
[8] Reg.-time cost [5] (\$1,200)	31,680	25,920	30,240	31,680	31,680	28,800	30,240	31,680	28,800	270,720

MATHEMATICAL PLANNING MODELS

11.8 Idaho Instrument Co. produces calculators in its Lewiston plant and has forecast demand over the next 12 periods, as shown in Table 11-25. Each period is 20 working days (approximately 1 month). The company maintains a constant work force of 40 employees, and there are no subcontractors available who can meet its quality standards. The company can, however, go on overtime if necessary and encourage customers to back-order calculators. Production and cost data follow.

Table 11-25

Period	Units	Period	Units	Period	Units
1	800	5	400	9	1,000
2	500	6	300	10	700
3	700	7	400	11	900
4	900	8	600	12	1,200

Production capacity:

Initial inventory: 100 units (final included in period 12 demand)

RT hours: (40 employees)(20 days/period)(8 hr/day) = 6,400 hr/period

OT hours: (40 employees)(20 days/period)(4 hr/day) = 3,200 hr/period

Standard labor hours per unit: 10 hr

Costs:

Labor: RT = \$6/hr   OT = \$9/hr

Material and overhead: \$100/unit produced

Back-order costs: apportioned at \$5/unit-period (and increasing in reverse)

Inventory carrying cost: \$2/unit-period

Option A. Assume that five periods constitute a full demand cycle, and use the transportation linear programming approach to develop an aggregate plan based on the first five periods only. (*Note:* A planning length of five periods is useful for purposes of methodology, but in reality the planning horizon should cover a complete cycle, or else the plan should make inventory, personnel, and other such allowances for the whole cycle.)

Option B. Determine the optimal production plan for the 12-period cycle using a transportation linear-programming format. (*Note:* This more realistic option involves a substantial amount of calculation and should be done on a computer, using a transportation LP code.)

Option *A*

RT capacity, avail./period = 6,400 hr ÷ 10 hr/unit = 640 units

OT capacity, avail./period = 3,200 hr ÷ 10 hr/unit = 320 units

RT cost = (10 hr/unit)(\$6/hr) + \$100 mat'l. and OH = \$160/unit

OT cost = (10 hr/unit)(\$9/hr) + \$100 mat'l. and OH = \$190/unit

See Fig. 11-17. Note that the back orders are shown in the lower left portion of the matrix.
The solution of *Option B* is left as an exercise. See Prob. 11.16.

Supply, units for		Demand, units for					Capacity	
		Period 1	Period 2	Period 3	Period 4	Period 5	Unused	Total available
Initial inventory		0 — 100	2	4	6	8	10	100
Period 1	Regular	160 — 640	162	164	166	168	60	640
	Overtime	190	192	194	196	198	0 — 320	320
Period 2	Regular	165 — 60	160 — 500	162 — 60	164 — 20	166	60	640
	Overtime	195	190	192	194	196	0 — 320	320
Period 3	Regular	170	165	160 — 640	162	164	60	640
	Overtime	200	195	190	192	194	0 — 320	320
Period 4	Regular	175	170	165	160 — 640	162	60	640
	Overtime	205	200	195	190	192	0 — 320	320
Period 5	Regular	180	175	170	165 — 240	160 — 400	60	640
	Overtime	210	205	200	195	190	0 — 320	320
Demand		800	500	700	900	400	1,600	4,900

Fig. 11-17

MASTER SCHEDULING

11.9 Taiwan Shoe Company schedules running shoe production in lot sizes of 40 units (each of which consists of a carton of pairs). They have a beginning inventory of 45 units and have developed a forecast of demand as shown in Table 11-26. The company has received orders for 22 units in week 1, 9 units in week 2, 4 units in week 3, 15 units in week 4, and 5 units in week 5. Set up a master production schedule, and find the ATP inventory values for weeks 1 through 8.

Table 11-26 Master Schedule for Running Shoe Production

On-Hand 45			Lot Size 40				Planning Time Fence 8	
Period	1	2	3	4	5	6	7	8
Forecast	20	20	30	20	20	13	15	20
Customer orders (booked)	22	9	4	15	5			
Projected inventory balance	23	3	13	33	13	0	25	5
MPS			40	40			40	
Available-To-Promise	14		36	20			40	

See Table 11-26. The period 1 balance is $45 - 22$ (i.e., 22, because orders are *larger* than forecast) $= 23$, and period 2 balance is $23 - 20 = 3$. Period 3 balance would be a negative $(3 - 30)$, so a lot size of 40 goes into the MPS in week 3, resulting in a balance of $(3 + 40) - 30 = 13$.

For period 1, $ATP_1 = $ (On-hand Inv.) $-$ (Orders in periods 1 and 2) $= 45 - (22 + 9) = 14$; and $ATP_3 = $ (MPS amount in 3) $-$ (Orders for period 3) $= 40 - (4) = 36$. For week 4, $ATP_4 = $ (MPS amount in 4) $-$ (Orders for periods 4, 5, and 6) $= 40 - (15 + 5 = 0) = 20$.

Supplementary Questions and Problems

11.10 A manufacturer of prefabricated bathrooms has forecast the demand shown in Fig. 11-18. Ending inventory, employment, etc., levels are to be restored to beginning levels. Find the cost associated with each of the following.

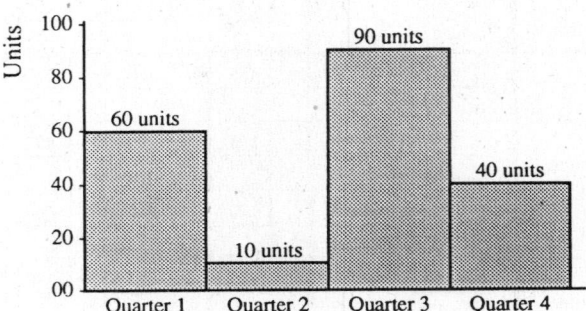

Fig. 11-18   Demand histogram

(a) Vary the work force to meet demand. Each quarterly change of 10 units in output has an incremental labor cost of $500.

(b) Maintain a stable work force capable of producing 50 units per quarter, and use overtime (at $100 per unit) and idle time (at $200 per unit).

(c) Maintain a steady work force on regular hours capable of producing 50 units per quarter. The carrying cost is $240 per unit per year based on average inventory, and storage cost is $10 per unit based on maximum inventory. Assume that an adequate inventory is on hand at the start of quarter 1 and is to be the same at the end of quarter 4.

(d) Produce at a steady rate of 40 units per quarter, and accept back orders. The stockout cost of lost sales is $200 per unit. (*Note:* Assume continuing operations, with units already on back order.)

(e) Produce at a steady rate of 10 units per quarter, and subcontract excess requirements at a marginal cost of $35 per unit.

Ans. (*a*) (20 changes)($500/change) = $10,000, (*b*) [(50 units)($100) + (50 units)($200)] = $15,000,
(*c*) requires 10 units on hand, $C_c + C_s = 40$ units ($10/unit) + 12.5 units ($240/unit-yr) = $3,400,
(*d*) [200 units $- 40(4)$] ($200/unit) = $8,000, (*e*) [200 units $- 10(4)$] ($35/unit) = $5,600

11.11 Rainwear Manufacturing, Inc., produces outdoor apparel that has a demand projected to be as shown in Table 11-27. The plant has a 2-week vacation shutdown in July, so the available production days per month are 22, 19, 21, 21, 22, 20, 12, 22, 20, 23, 19, and 21, respectively.

Table 11-27

January	4,400	April	6,300	July	1,200	October	9,200
February	4,750	May	4,400	August	3,300	November	7,600
March	6,300	June	2,000	September	5,000	December	7,350

(*a*) Prepare a chart showing the daily production requirements. (*b*) Plot the demand as a histogram and as a cumulative requirement over time. (*c*) Determine the production rate required to meet average demand, and plot this as a dotted line on your graph.

Ans.   (*a*) Chart should show January through December daily demands of 200, 250, 300, 300, 200, 100, 100, 150, 250, 400, 400, 350. (*b*) Histogram should show cumulative production days on *x* axis, production rate (units per day) on *y* axis. Cumulative requirement should show cumulative production days on *x* axis and cumulative demand (units) on *y* axis. (*c*) 255.4 units

11.12   The Speedee Bicycle Co. makes 3-speed bikes that sell for $100 each. This year's demand forecast is shown in Table 11-28. Units not sold are carried in stock at a cost of 20 percent of the average inventory value per year, and storage costs are $2 per bike per year, based upon maximum inventory.

Table 11-28   Bike Demand Forecast

Quarter	First	Second	Third	Fourth
Units	30	120	60	70

(*a*) Plot the demand as a histogram on a quarterly basis, and show the average requirement as a dotted line on your graph. (*b*) Assume Speedee wishes to maintain a steady work force and to produce at a uniform rate (that is, with no overtime, back orders, subcontracting, or capacity changes) by letting inventories absorb all fluctuations. How many bikes must they have on hand on January 1 in order to meet the forecast demand throughout the year? (*c*) For an incremental amount of $400 in labor costs (total), Speedee can vary its work-force size so as to produce exactly to demand. Compare the costs of producing at a uniform versus variable rate, indicate which plan is less costly, and show the net difference in cost.

Ans.   (*a*) Histogram should show quarters on *x* axis and production rate (units per quarter) on *y* axis. (*b*) 10
(*c*) Variable rate is $50 per year less costly.

11.13   An aggregate planner at Duotronix has estimated the demand requirements (Table 11-29) for forthcoming work periods, which represent one complete demand cycle for them. The company is a ''going concern'' and expects the next demand cycle to be similar to this one. Five plans are being considered.

Table 11-29

Period	Forecast	Period	Forecast
1	400	6	1,200
2	400	7	600
3	600	8	200
4	800	9	200
5	1,200	10	400

Plan 1: Vary the labor force from an initial capability of 400 units to whatever is required to meet demand. See Table 11-30.

Table 11-30

Amount of Change	Incremental Cost to Change Labor Force	
	Increase	Decrease
200 units	$ 9,000	$ 9,000
400 units	15,000	18,000
600 units	18,000	30,000

Plan 2: Maintain a stable work force capable of producing 600 units per period, and meet demand by overtime at a premium of $40 per unit. Idle-time costs are equivalent to $60 per unit.

Plan 3: Vary inventory levels, but maintain a stable work force producing at an average requirement rate with no overtime or idle time. The carrying cost per unit per period is $20. (The company can arrange to have whatever inventory level is required before period 1 at no additional cost.)

Plan 4: Produce at a steady rate of 400 units per period and accept a limited number of back orders during periods when demand exceeds 400 units. The stockout cost (profit, goodwill, and so on) of lost sales is $110 per unit.

Plan 5: Produce at a steady rate of 200 units per period, and subcontract for excess requirements at a marginal cost of $40 per unit.

Graph the forecast in the form of a histogram, and analyze and relevant costs of the various plans. You may assume the initial (period 1) work force can be set at a desired level without incurring additional cost. Summarize your answer in the form of a table showing the comparative costs of each plan.
Ans.   Graph shows period on x axis and demand level on y axis. Plan costs are as follows: plan 1 = $90,000, plan 2 = $140,000, plan 3 = $160,000, plan 4 = $220,000, plan 5 = $160,000.

11.14   Two mixed-strategy plans have been proposed for the Duotronix situation in Prob. 11.13. Assume that the pattern inherent in the demand cycle given will be repeated in the next demand cycle.

Plan 6 (back orders and limited inventory): Produce at a steady rate of 500 units per period, and carry inventory at $20 per unit-period. Assume that the 700 units of excess demand can be satisfied by back orders placed in period 5 and filled in periods 8, 9, and 10. No inventory is available at the beginning of period 1, and none should be available at the beginning of the next cycle.

Plan 7 (subcontracting and limited inventory): Produce at a steady rate of 300 units per period, and subcontract for excess requirements at a marginal cost of $40 per unit. A 200-unit inventory is available at the beginning of period 1 and should also be available at the beginning of the next cycle. Carry inventory at $20 per unit-period.

Determine the comparative costs of the two plans.
Ans.   Plan 6: carrying cost = $8,000, back-order cost = $110,000, total = $118,000. Plan 7: carrying cost = $8,000, subcontract cost = $120,000, total = $128,000.

11.15   Sun Valley Ski Co., producers of the famous *Sun-Ski*, has a production cost of $60 per pair during regular time and $70 per pair on overtime. The firm's production capacity and forecast quarterly demands are shown in Table 11-31. Beginning inventory is 200 pairs, and stock is carried at a cost of $5 per pair-quarter. Demand is to be met without any hiring, layoff, subcontracting, or back orders. Unused regular time has a $20-per-pair cost.

(*a*)   Develop the preferred plan, and present it in the form of a solved matrix.

(*b*)   What is the minimum total cost of the plan?

Table 11-31

Supply, Units from		Demand, Units for					Total Capacity Available
		First Quarter	Second Quarter	Third Quarter	Fourth Quarter and Final	Unused Capacity	
Initial inventory							200
1	Regular						700
	Overtime						300
2	Regular						700
	Overtime						300
3	Regular						700
	Overtime						300
4	Regular						700
	Overtime						300
Forecast demand		900	500	200	1,900	700	4,200

Ans.   (*a*) One optimal solution is:

Initial inventory: use in 1st Q

1 RT: use in 1st Q	3 OT: use 200 in 4th Q
2 RT: use 500 in 2nd Q, 200 in 4th Q	4 RT: use 700 in 4th Q
3 RT: use 200 in 3rd Q, 500 in 4th Q	4 OT: use 300 in 4th Q

   (*b*) $208,500

11.16   Complete option (*b*) of Prob. 11.8.
   Ans.   The solution should have the following entries in the row-column (r, c) matrix locations: (Initial inventory, 1) = 100, (1 RT, 1) = 640, (2 RT, 1) = 60, (2 RT, 2) = 500, (2 RT, 3) = 60, (2 RT, 4) = 20, (3 RT, 3) = 640, (4 RT, 4) = 640, (5 RT, 4) = 240, (5 RT, 5) = 400, (6 RT, 6) = 300, (6 RT, 9) = 80, (6 RT, 10) = 60, (6 RT, 11) = 200, (7 RT, 7) = 400, (7 RT, 9) = 240, (8 RT, 8) = 600, (8 RT, 9) = 40, (9 RT, 9) = 640, (10 RT, 10) = 640, (11 RT, 11) = 640, (11 OT, 11) = 60, (11 OT, 12) = 240, (12 RT, 12) = 640, (12 OT, 12) = 320.

11.17   Rework Prob. 11.9 assuming the company has received orders for 18 units in week 1 and 23 units in week 2.
   Ans.   Projected balances for weeks 1 through 9 are 25, 2, 12, 32, 12, 39, 24, and 4. ATP values are 4 in week 1, and 40 in weeks 3, 4, and 6.

11.18   Shown in Table 11-32 is the expected demand for an end item *X*, which has a beginning inventory of 30 units. The production lot size is 70 units, and the firm maintains a safety stock of 5 units.

Table 11-32

	Week Number							
	1	2	3	4	5	6	7	8
Customer forecast	5	5	5	30	15	5	25	25
Service forecast	—	—	20	—	—	20	—	—
International orders	—	—	—	30	—	25	—	40
Warehouse orders	—	5	—	10	20	—	30	—

Complete a master schedule by determining the projected inventory balance, MPS entries, and the ATP amounts. (*Hint:* Sum the forecast amounts; enter them in the forecast row, and do the same for orders. Also, the safety stock requirement means that a minimum of 5 units should always be on hand.)
Ans.   See Table 11-33.

Table 11-33   Master Schedule for Product X

On-Hand 30		Safety Stock 5			Lot Size 70		Planning Time Fence 8	
Period	1	2	3	4	5	6	7	8
Forecast	5	5	25	30	15	25	25	25
Customer orders (booked)		5		40	20	25	30	40
Projected inventory balance	25	20	65	25	5	50	20	50
MPS			70			70		70
Available-To-Promise	25		10			15		30

Chapter 12

Materials Management:
Purchasing, Inventory, and JIT Systems

Calculus

SCOPE AND STATUS OF MATERIALS MANAGEMENT

Materials are the raw materials, components, subassemblies, and supplies used to produce a good or service. Most materials are transformed into finished products, whereas supplies are typically consumed in daily operations.

Question: What is materials management?

Materials management is the planning, organizing, and controlling of the flow of materials from their initial purchase, through internal operations, to the distribution of finished goods.

Question: What activities does materials management encompass?

It includes (1) purchasing, (2) transportation (incoming and outgoing), (3) control through production-and-inventory management (i.e., receiving, storage, shipping, materials handling, and inventory counting), and (4) warehousing and distribution. Although some firms have *centralized* these responsibilities (under a materials manager), others rely more upon the coordination of decentralized activities.

Materials management is a strategically important activity. The investment in materials constitutes one of the largest uses of cash of a manufacturing company; materials account for close to two-thirds of the cost of an average product. However, materials handling *per se* adds little (or nothing) to the value of the product; and even holding materials in storage can be costly. Insofar as there are numerous opportunities here for improvement in most firms, materials management has become an increasingly competitive tactic in today's international marketplace.

Question: What trends have reinforced the importance of materials management?

1. *Globalization* (sources of supply/distribution are increasingly more international)
2. *Reduced time-to-market* (products must be delivered to customers in shorter time periods)
3. *Just-in-time inventories* (must rely on supplier coordination rather than large inventories)
4. *Variety/life cycle* (must accommodate more product varieties with shorter life cycles)
5. *Responsibility* (must accept more product warranty and environmental liabilities)

PURCHASING PROCESS

The materials used in large organizations are normally obtained through the purchasing process, which is the major link between an organization and its suppliers, or vendors.

Question: What is purchasing, and who does it?

Purchasing is the process of acquiring goods or services in exchange for funds. In large organizations, the negotiations are done by professional buyers who have specialized knowledge about selected product lines and who are familiar with specifications, contract law, shipping regulations, and a myriad of related factors. However, buyers also work closely with the engineering unit (which prepares specifications), operations (which receives and uses the materials), accounting (which processes invoices), and other units of the organization.

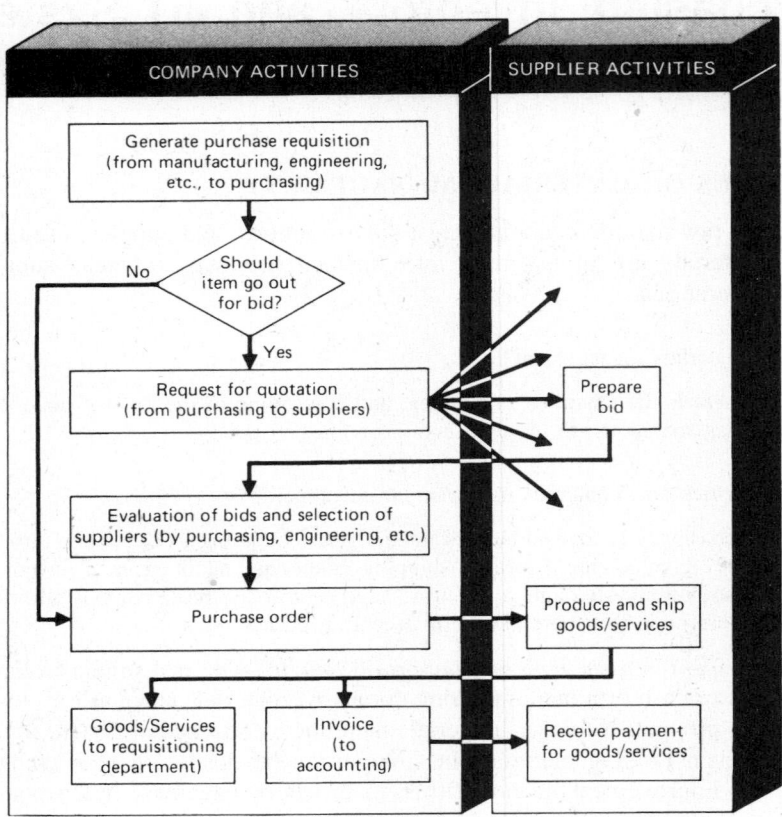

Fig. 12-1 The purchasing process

Figure 12-1 illustrates the purchasing process, which differs depending upon the type of item. High-value items (e.g., turbine generators) typically require bids, whereas low-value items (e.g., stationery) are often obtained under *open purchase orders*, with little day-to-day involvement of the purchasing department. High-volume items are sometimes supplied under a *blanket purchase order*, which establishes a firm price but enables the buyer to use extended delivery dates.

Question: What are the major responsibilities of a purchasing department?

1. Delineate company needs for goods/services (specify quantity, quality, timing, etc.).
2. Identify sources of supply, and maintain supply system database.
3. Select suppliers, and negotiate contracts (that satisfy price, delivery, and other needs).
4. Maintain good relations with suppliers, and monitor their performance.
5. Learn about new goods/services, and channel appropriate information to others.
6. Initiate value analysis, make-or-buy, and other studies as appropriate.

In the past, domestic firms have tended to use more suppliers than some foreign competitors, but that is changing. Alternative sources of supply sometimes force more competitive prices and reduce the risk of materials shortages. However, close relations with a few (certified) suppliers can give the firm more consistent, high-quality supplies and better coordination of deliveries—which is essential to just-in-time operations. The result is fewer rejects and lower inventory levels.

Value Analysis.   Buyers employ various strategies to obtain the most value they can get for the money they spend. *Value analysis* entails a close examination of the function and cost of a purchased item in an attempt to find a lower-cost item that will perform the same function. The analysis includes inquiries such as whether the need is genuine, whether all features are really needed, whether quality specifications might be relaxed, whether standard components might replace custom parts, etc.

VENDOR SELECTION CRITERIA

Important variables to consider in selecting suppliers include (1) quantity, (2) quality, (3) price, (4) delivery, (5) service, (6) maintenance, (7) technical support, (8) financial stability, and (9) terms of purchase. Trade discounts (based upon whether the buyer is a manufacturer, distributor, or user), cash discounts (for prompt payment), and shipping terms (such as F.O.B. shipping point) all affect the ultimate cost of the item.

Choosing one vendor from among several can be difficult—especially when many variables must be taken into account. One technique for arriving at a quantitative ranking of suppliers is to assign importance ratings to the various criteria and base the selection on an expected score.

$$\text{Expected score} = \Sigma \ (\text{importance rating} \times \text{criteria value}) \qquad (12.1)$$

Example 12.1   A municipal utility in Texas has four suppliers for watt-hour meters. The company's computer has recognized a low-stock situation and must issue a purchase recommendation to the buyer on the basis of the criteria shown in Table 12-1. What rank will the computer give to the respective vendors?

Table 12-1

		Vendor			
Criteria	Importance Rating (1–10)	(A) Western Supply	(B) General Selectric	(C) Roundy Corp.	(D) Ohio Meters
1. Price	6	.4	.4	.6	.7
2. Field service	3	.7	.7	.3	.2
3. Delivery reliability	4	.8	.9	.2	.3
4. Delivery time	1	.5	.3	.2	.2
5. Ease of maintenance	8	.6	.4	.3	.3
6. Adaptability to computer	2	.5	.6	.0	.9
7. Product life	3	.5	.4	.3	.3

The computer will sum the weighted scores for each potential vendor and print out a list of ranks, with "1" being the highest and "4" being the lowest.

$$\text{Supplier A} = 6(.4) + 3(.7) + 4(.8) + 1(.5) + 8(.6) + 2(.5) + 3(.5) = 15.50$$
$$\text{Supplier B} = 6(.4) + 3(.7) + 4(.9) + 1(.3) + 8(.4) + 2(.6) + 3(.4) = 14.00$$

$$\text{Supplier C} = 6(.6) + 3(.3) + 4(.2) + 1(.2) + 8(.3) + 2(.0) + 3(.3) = 8.80$$
$$\text{Supplier D} = 6(.7) + 3(.2) + 4(.3) + 1(.2) + 8(.3) + 2(.9) + 3(.3) = 11.30$$

Ranks are 1 = supplier A, 2 = supplier B, 3 = supplier D, 4 = supplier C.

MAKE-OR-BUY DECISIONS

Decisions concerning whether to make or buy components involve both economic and noneconomic considerations. Economically, an item is a candidate for in-house production if the firm has sufficient capacity and if the component's value is high enough to cover all the variable costs of production plus make some contribution to fixed costs. Low volumes of usage favor buying, which takes little or no fixed costs. Figure 12-2 illustrates.

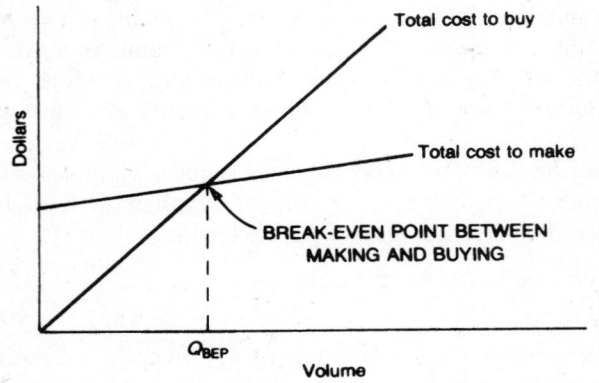

Inputs
- Availability of funds and skilled personnel
- Availability and volume of supply from others
- Desire for alternative sources of supply

Processing
- Employee preferences and stability concerns
- Desire to develop R&D facilities
- Need to control trade secrets
- Desire to expand into new product line
- Need to control delivery lead times
- Impact upon production flexibility

Outputs
- Need to control quality or reliability
- Goodwill and reciprocity impact on customers

Fig. 12-2 Economic and less economic factors influencing make-or-buy decisions

Example 12.2 Auburn Machine Co. produces parts that are shipped nationwide. It has an opportunity to produce plastic packaging cases, which are currently purchased at $.70 each. Annual demand depends largely on economic conditions, but long-run estimates are as shown in Table 12-2.

If the company produces the cases itself, it must renovate an existing work area and purchase a molding machine, which will result in annual fixed costs of $8,000. Variable costs for labor, materials, and overhead are estimated at $.50 per case. (a) Should Auburn Machine make or buy the cases? (b) At what volume of production is it more profitable to produce in-house rather than purchase from an outside supplier?

Table 12-2

Demand	Chance, %
20,000	10
30,000	30
40,000	40
50,000	15
60,000	5

(a) First, determine the expected volume by treating the percentage of chance as an empirical probability. Thus, $E(D) = 37,500$ units as seen in Table 12-3.

Table 12-3

Demand D	Chance P(D)	D · P(D)
20,000	.10	2,000
30,000	.30	9,000
40,000	.40	16,000
50,000	.15	7,500
60,000	.05	3,000
		37,500

Next, Auburn Machine should produce the cases if the expected cost to produce is less than the expected cost to purchase.

Expected cost to produce:

$$TC = FC + VC(V) = \$8,000 + (\$.50/\text{unit})(37,500 \text{ units}) = \$26,750$$

Expected cost to purchase:

$$TC = (\text{Price})(V) = (\$.70/\text{unit})(37,500 \text{ units}) = \$26,250$$

Conclusion: Continue to purchase the cases.

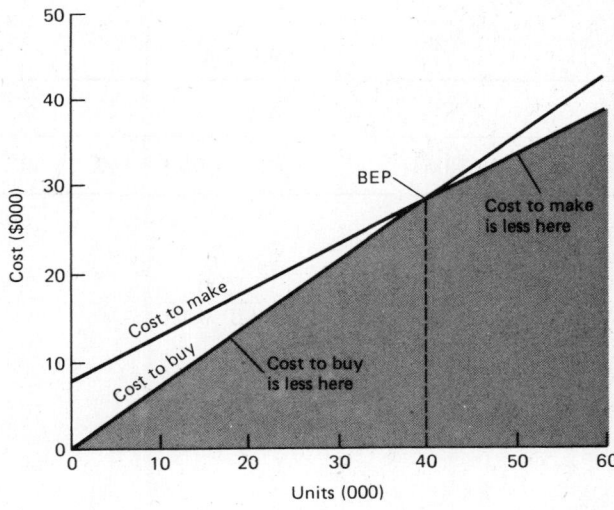

Fig. 12-3

(b) The break-even point (Fig. 12-3) is the volume of production where the total costs to make equal the total costs to buy:

$$TC \text{ to make} = TC \text{ to buy}$$
$$FC + VC(V) = P(V)$$
$$\$8,000 + (\$.50)V = (\$.70)V$$
$$\$.20V = \$8,000$$
$$V = 40,000 \text{ units}$$

For volumes above 40,000 units, it is more economical to make the cases in-house.

PURCHASE-QUANTITY DECISIONS: SINGLE-PERIOD MODEL

Marginal Analysis. In an economic (theoretical) sense, firms should purchase and hold materials until the marginal profit (MP) from acquiring and holding the last unit is equal to or greater than the marginal loss (ML) incurred if that unit is unsold. But the MP from an item during a single period is a function of the probability that the item will be demanded $P(D)^*$ during that period, and the ML is a function of the probability that it will not be demanded, $1 - P(D)^*$. In equation form, the decision rule would suggest that additional stock be carried up to the point where:

$$P(D)^* \, (MP) \geq [1 - P(D)^*](ML)$$

Solving for $P(D)^*$ we obtain the critical probability of sale for the last unit that should be stocked.

$$P(D)^* \, (MP) \geq ML - P(D)^* \, (ML)$$
$$P(D)^* \, (MP) + P(D)^* \, (ML) \geq ML$$
$$P(D)^* \geq \frac{ML}{MP + ML} \tag{12.2}$$

where $P(D)^*$ represents the cumulative probability of needing the *next* unit. If the probability of sale of the next unit is less than the $P(D)^*$ amount, the item should not be added to inventory. Note that in this expression, the values of $P(D)^*$ are cumulated on a declining or $\geq$ basis, so that each additional unit added to stock brings with it a lower probability of sale.

Example 12.3 An item costs $6 to produce, sells for $10, and has an estimated cumulative probability distribution of demand during the next period as shown in Table 12-4. No salvage values apply.

Table 12-4

Unit number	1	2	3	4	5	6	7	8	9
P(selling $\geq$ this unit)	1.00	.92	.82	.75	.62	.40	.15	.10	0

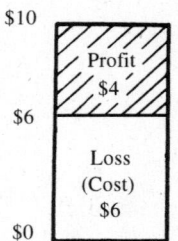

How many units should be ordered?

$$P(D)^* \geq \frac{ML}{MP + ML}$$

where   ML = $6
       MP = $10 − $6 = $4

$$P(D)^* \geq \frac{\$6}{\$4 + \$6} = .60$$

Therefore, order 5 units where P(selling $\geq$ 5 units) of .62 is just greater than the equating probability of .60. It seems intuitively correct that since the cost ($6) represents 60 percent of the value of the item, the likelihood of sale, $P(D)^*$, should also equal 60 percent. If the cost were a higher proportion (for example, 80 percent), then fewer units would be justified (3 units) because the marginal profit would be less (i.e., $2).

Single-Period Model. Marginal, or incremental, analysis can be applied to single-period inventories in an attempt to balance the costs of carrying excess stock (C_{os} = overstocking) against the opportunity costs of having too little stock (C_{us} = understocking). The *single-period model* applies to situations where unused items are not normally carried forward from one period to the next, or a penalty exists for doing so. The assumptions for this model are that (1) demand can be estimated, (2) purchase quantity is limited and may not be increased beyond the initial amount, and (3) costs exist for overstocking and understocking. The (optimal) balance point is that where the expected cost of understocking $C_{us}P(D)^*$ equals the expected cost of overstocking $C_{os}[1 - P(D)^*]$. In this expression, $P(D)^*$ is the cumulative probability of needing the *next* unit and takes on a value that makes the two expected costs equal.

$$\text{cost of understocking} = \text{cost of overstocking}$$
$$C_{us}P(D)^* = C_{os}[1 - P(D)^*]$$

Single period
$$P(D)^* = \frac{C_{os}}{C_{os} + C_{us}} \qquad\qquad (12.3)$$

Purchase quantities appropriate for multiple-period situations involve consideration of stockout costs and reorder frequency. They are discussed in Chap. 13.

Example 12.4   An operations manager of Nationwide Car Rentals must decide on the number of vehicles of a certain model to allocate to her agency in the Nashville area. The cars are obtained from an auto leasing firm at a cost of $40 per day. Nationwide rents the cars to its customers for $60 per day. If a car is not used, the auto leasing firm gives Nationwide a $16 rebate.

Records of past demand have yielded the empirical probability distribution shown in Table 12-5. How many units of this model should Nationwide stock if it seeks to balance the costs of overstocking and understocking?

Table 12-5   Demand for Rental Cars

Demand for cars, d	≤6	7	8	9	10	11	12	13	14	≥15
Probability $P(d)$	0	.03	.07	.15	.20	.23	.15	.12	.10	0
Cumul. prob. $P(D^* \geq d)$	1.00	1.00	.97	.90	.75	.55	.32	.17	.05	0

The cumulative probability $P(D)^*$ that demand will be at least the amount d is shown in the last row of the table. It is computed by recognizing that 100 percent of the time demand was greater than or equal to 7 cars (therefore $P(D)^* = 1.0$). Because the $P(d = 7 \text{ cars}) = .03$, the cumulative $P(D)^* \geq 8$ cars is $1.00 - .03 = .97$. Other values follow by subtraction in the same manner. The equating probability is:

$$P(D)^* = \frac{C_{os}}{C_{os} + C_{us}}$$

where $C_{os} = \$40 - \$16 = \$24$
$C_{us} = \$60 - \$40 = \$20$

∴ $P(D)^* = \dfrac{24}{24 + 20} = .545$

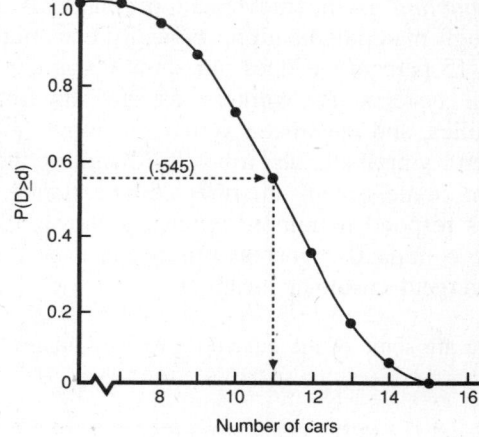

Fig. 12-4

Stock 11 cars, the amount closest to the $P(D^* \geq d)$ value. See Fig. 12-4. *Note:* For computed values midway between the cumulative values, stock less rather than more because the cost of overstocking ($24) is greater than the cost of understocking ($20).

Insofar as $P(D)^*$ is the cumulative probability that demand will be exceeded, it also represents the stockout risk (SOR) associated with the corresponding level of stock.

Single period
$$\text{SOR} = \frac{C_{os}}{C_{os} + C_{us}} \qquad\qquad (12.4)$$

Service Level Version of Single-Period Model.   In Example 12.4, the value $P(D)^* = .545$ was a cumulative probability identifying the stocking level where the likelihood of renting more cars was getting so low that it did not justify holding more units in stock. $P(D)^*$ represented the (decreasing) cumulative probability that a given level of demand (e.g., for 12 cars) would be realized. On the other hand, the decline in cumulative probabilities of renting 12, 13, or 14 cars (i.e., to .32, .17, and .05, respectively) also represented a corresponding decline in the risk of running out of stock. If the rental company stocked 14 cars, and limited the chance of running out of stock to .05, it would be offering its customers a .95 chance of having cars in stock when they are needed. Service level is a complementary (and more positive) way of viewing stockout probability. Suppliers frequently judge their performance on the basis of the service level they can offer their customers.

Question: What is service level?

Service level of an inventory is a number that represents the percentage of units or percentage of order cycles in which all demand requests can be supplied from stock. Service level is $1.00 -$ stockout risk.

Because service level is the complement of the SOR, we can restate Eq. 12.4 in a more commonly encountered format as:

Single period $$\text{SL} = 1.00 - \text{SOR} = \frac{C_{us}}{C_{os} + C_{us}}$$ *(12.5)*

Example 12.5   Use Eq. 12.5 to find the service level that balances the overstocking and understocking costs of Example 12.4.

$$\text{SL} = \frac{C_{us}}{C_{os} + C_{us}} = \frac{20}{24 + 20} = .455$$

Note: This is the same value that would result from letting $\text{SL} = 1.00 - \text{SOR}$, or $1.00 - .545 = .455$.

MATERIALS HANDLING, STORAGE, AND RETRIEVAL

Materials handling is the movement of materials from receiving, through operations, to final shipment. Although materials handling typically constitutes a significant portion of the cost-of-goods-sold (e.g., 10 to 15 percent), it does not show up as any tangible value in a finished good.

Conventional systems use workers, assisted by trucks, forklifts, cranes and hoists, conveyors, pipelines, hydraulics, and pneumatic systems to move goods. Modern facilities make extensive use of bar codes for identifying materials, robots for handling them, and automatic guided vehicles (AGV) for transporting them. Some systems are directed by software commands emitted from wires (e.g., buried in the floor); others respond to remote wireless controls. Some of the more highly automated systems respond to voice commands, whereas others can even respond directly to orders entered remotely on computers at divergent customer locations.

Question: What are some of the guidelines for designing conventional materials-handling systems?

> 1. Plan handling as a complete system.
> 2. Minimize handling volume and frequency.
> 3. Optimize load size and weight.
> 4. Use direct, rapid, steady flows of materials.
> 5. Minimize idle time of equipment and operators.
> 6. Allow for breakdowns, changes, and maintenance.

Automated storage and retrieval systems (AS/AR) are computer-controlled materials-handling systems which receive, store, and deliver inventory to high cube-storage locations in quantities specified by the computer. Highly automated systems are integrated with production so that material requirements are automatically identified from the bill-of-material database. Items are selected by part-retrieval robots; computer-controlled stackers and conveyors move them to appropriate kitting locations or workstations. In addition, inventory, work-in-process, and material-requirements-planning (MRP) records are automatically updated.

PURPOSE OF INVENTORIES

Question: Define the term inventory, and identify the types of manufacturing inventories.

An *inventory* is a stock of an item or idle resource held for future use. For example, a refinery may have an inventory of oil, whereas an airline may have an inventory of available seating. Manufacturing inventories are frequently classified as either (1) raw materials, (2) work in process, (3) finished goods, or (4) supplies.

Inventories represent investments designed to assist in production activities and/or serve customers. However, holding inventories consumes working capital, which may not be earning a return on investment and may be urgently needed elsewhere. Hence the problem of inventory management is to maintain adequate, but not excessive, levels of inventories. Poor inventory management leads to liquidity problems, which have been the cause of failure for many firms.

Question: Why do firms carry inventories?

Major Reasons for Carrying Inventories

1. *Service customers* with variable (immediate and seasonal) demands.
2. *Protect against* supply errors, shortages, and stockouts.
3. *Help level production activities*, stabilize employment, and improve labor relations.
4. *Decouple successive stages* in operations so that breakdowns do not stop the entire system.
5. *Facilitate the production of different products* on the same facilities.
6. Provide a means of obtaining and handling materials in *economic lot sizes* and of *gaining quantity discounts*.
7. Provide a means of *hedging against future price and delivery uncertainties*, such as strikes, price increases, and inflation.

DEPENDENT AND INDEPENDENT DEMAND

Dependent demand inventory consists of the raw materials, components, and subassemblies that are used in the production of parent or end items. For example, the demand for computer chips depends on the demand for the parent item, computers. Manufacturing inventory demand is largely dependent and predictable. The requirements for all components vis-à-vis other components are fixed by design, and production quantities are dictated by the firm's master schedule.

Independent demand inventory consists of the finished products, service parts, and other items whose demand arises more directly from the uncertain market environment. Thus, distribution inventories often have an independent and highly uncertain demand. Dependent demands can usually be calculated, whereas independent demands usually require some kind of forecasting.

INVENTORY COSTS AND THE EOQ EQUATION

The major costs associated with procuring and holding inventories are as follows:

(1) *Ordering and set-up costs* for placing orders, expediting, inspection, and changing or setting up facilities to produce in-house

(2) *Carrying costs* on invested capital, handling, storage, security, insurance, taxes, obsolescence, spoilage, and data-processing costs

(3) *Purchase costs* including the price paid, or the labor, material, and overhead charges necessary to produce the item

The total cost (TC) of stocking inventory is the sum of the cost of ordering, plus the cost of carrying, plus the purchase cost. If D equals demand in units on an annual basis, S equals the cost to prepare (or set up) for an order, H equals the cost to carry (or "hold") a unit in stock for a given time period, P equals the purchase cost, Q equals lot size, and $Q/2$ equals average inventory, then the relationship can be expressed mathematically:

$$\text{Total cost} = \text{ordering cost} + \text{carrying cost} + \text{purchase cost}$$

where $\text{Ordering cost} = \left(\dfrac{D \text{ units}}{\text{yr}}\right)\left(\dfrac{\text{order}}{Q \text{ units}}\right)\left(\dfrac{S \text{ \$}}{\text{order}}\right) = \left(\dfrac{D}{Q}\right)S$

$\text{Carrying cost} = \left(\dfrac{Q \text{ units}}{2}\right)\left(\dfrac{H \text{ \$}}{\text{unit-yr}}\right) = \left(\dfrac{Q}{2}\right)H$

$\text{Purchase cost} = \left(\dfrac{P \text{ \$}}{\text{unit}}\right)\left(\dfrac{D \text{ units}}{\text{yr}}\right) = PD$

Thus,
$$\text{TC} = \frac{D}{Q}S + \frac{Q}{2}H + PD \tag{12.6}$$

Differentiating with respect to the order quantity Q yields the slope of the TC curve. (Refer to Prob. 12.8 for an explanation of differentiation.)

$$\frac{d\text{TC}}{dQ} = -DS\,Q^{-2} + \frac{Q^{0}H}{2} + 0$$

Setting this first derivative equal to zero identifies the point where the TC is a minimum.

$$0 = -\frac{DS}{Q^{2}} + \frac{Q^{0}H}{2} + 0$$

Solving for Q (*Note:* $Q^{0} = 1$) $Q = \text{EOQ} = \sqrt{\dfrac{2DS}{H}} \tag{12.7}$

Equation 12.7 is known as the economic order quantity equation. It yields the order quantity that will satisfy the specified demand at the lowest total cost provided that certain conditions are satisfied.

Question: What assumptions underlie the EOQ model?

Assumptions of the EOQ Model

1. Demand and lead time are known and constant.
2. Replenishment is instantaneous at the expiration of the lead time.
3. Purchase costs do not vary with the quantity ordered.
4. Ordering and carrying-cost expressions include all relevant costs, and these costs are constant.

Figure 12-5 describes the relationship between the relevant ordering and carrying costs.

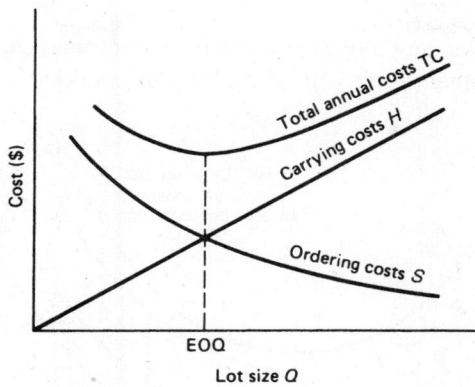

Fig. 12-5   Economic order quantity (EOQ)

Example 12.6   Overland Motors uses 25,000 gear assemblies each year and purchases them at $3.40 each. It costs $50 to process and receive an order, and inventory can be carried at a cost of $.78 per unit-year. (*a*) How many assemblies should be ordered at a time? (*b*) How many orders per year should be placed?

(*a*) $$\text{EOQ} = \sqrt{\frac{2DS}{H}} = \sqrt{\frac{2(25,000)(50)}{.78}} = 1,790 \text{ assemblies}$$

(*b*) $$\text{Orders/yr} = \frac{D}{Q} = \frac{25,000}{1,790} = 14 \text{ orders/yr}$$

GRADUAL REPLENISHMENT/USAGE

The EOQ equation assumes instantaneous replenishment. When a firm takes time to produce its own inventory and uses some of it as it is produced, only a portion of the production goes into inventory. If the proportion used is represented by the ratio of demand rate d over production rate p, the proportion going into inventory is $[1 - (d/p)]$. The economic production run length is thus:

$$Q = \text{ERL} = \sqrt{\frac{2DS}{H\left(1 - \dfrac{d}{p}\right)}} = \sqrt{\frac{2DS}{H}\left(\frac{p}{p-d}\right)} \qquad (12.8)$$

where S = set-up cost in $/set-up
D = annual demand in units/yr
H = holding cost in $/unit-yr
d = demand rate in units/period
p = production rate in units/period

Example 12.7   A plastics molding firm produces and uses 24,000 teflon bearing inserts annually. The cost of setting up for production is $85, and the weekly production rate is 1,000 units. If the production cost is $2.50 per unit and the annual storage and carrying cost is $.50 per unit, how many units should the firm produce during each production run?

The demand and production rates must be in the same units, so we arbitrarily put both into annual terms, assuming a 52-week year,

$$\text{ERL} = \sqrt{\frac{2DS}{H}\left(\frac{p}{p-d}\right)} = \sqrt{\frac{2(24,000)(85)}{.50}\left(\frac{52,000}{28,000}\right)} = 3,893 \text{ inserts}$$

At a production rate of 1,000 units per week, each production run will last about 1 month, so the firm will be producing inserts about every other month.

QUANTITY DISCOUNTS

Quantity discounts are lower unit prices offered to buyers who purchase in large volumes. They make price P a function of lot quantity Q with a resulting discontinuous total-cost function (Fig. 12-6).

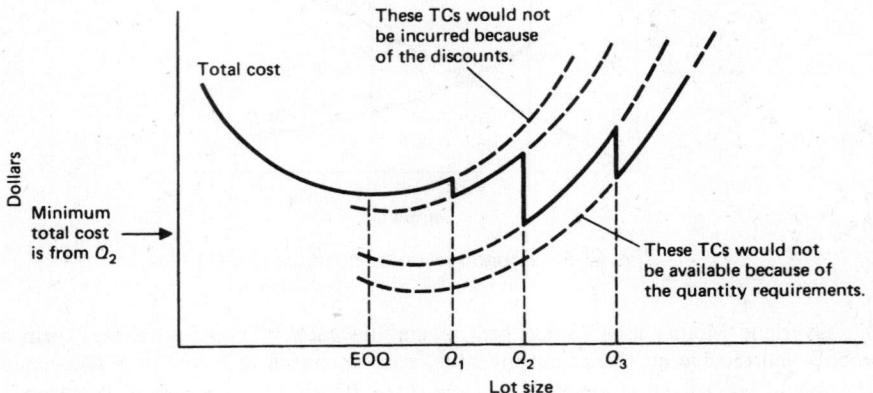

Fig. 12-6 Quantity discount situation

One approach to finding an optimal order quantity in a discount situation is to:

(1) Determine the EOQ on the basis of the nondiscounted base price.

(2) Compare the total cost at this EOQ point with that for price breakpoints at higher volumes.

(3) If the EOQ happens to fall in a quantity discount range, recalculate it using the quantity discount price and recheck to see if the revised EOQ or a price breakpoint (to the right) has the lower total cost.

Example 12.8   A producer of photo equipment buys lenses from a supplier at $100 each. The producer requires 125 lenses per year, and the ordering cost is $18 per order. Carrying costs per unit-year (based on average inventory) are estimated to be $20 each. The supplier offers a 6 percent discount for purchases of 50 lenses and an 8 percent discount for purchases of 100 or more lenses at one time. What is the most economical amount to order at a time?

Disregarding quantity discounts, the EOQ amount would be:

$$\text{EOQ} = \sqrt{\frac{2DS}{H}} = \sqrt{\frac{2(125)(18)}{20}} = 15 \text{ lenses}$$

And the total annual cost associated with this EOQ is:

$$\text{TC} = \text{ordering} + \text{carrying} + \text{purchase}$$

$$= \frac{D}{Q}S + \frac{Q}{2}H + PD$$

$$= \left(\frac{125}{15}\right)18 + \left(\frac{15}{2}\right)20 + 100(125) = \$12,800$$

For a 50-unit order, the purchase cost is reduced by 6 percent of $100, or $6. Assuming that the ordering and carrying costs remain constant, the total annual cost associated with a 50-unit order is:

$$\text{TC} = \left(\frac{125}{50}\right)18 + \left(\frac{50}{2}\right)20 + (100 - 6)125 = \$12,295$$

Similarly, the total annual cost associated with a 100-unit order is:

$$TC = \left(\frac{125}{100}\right)18 + \left(\frac{100}{2}\right)20 + (100 - 8)125 = \$12,522$$

The 50-unit lot size results in the lowest total annual cost. Although the purchase price per unit is less with the 100-unit order, the carrying costs begin to outweigh such savings. The costs and direction of change up ($\uparrow$) or down ($\downarrow$) are shown in Table 12-6.

Table 12-6   Cost Comparison

Order Quantity	Ordering Cost	+ Carrying Cost	+ Purchase Cost	= Total
15-unit order	$150	$ 150	$12,500	$12,800
50-unit order	45$\downarrow$	500$\uparrow$	11,750$\downarrow$	12,295$\downarrow$
100-unit order	22$\downarrow$	1,000$\uparrow$	11,500$\downarrow$	12,522

Example 12.8 assumed a constant carrying cost of $20. If the carrying cost is given as a percentage of the price ($\%P$), it will also be reduced by a small but proportionate amount as larger quantity discounts come into effect. See Prob. 12.20.

ABC CLASSIFICATION AND BAR CODING

The ABC classification system is a widely used method of categorizing inventories according to quantity and value. Table 12-7 summarizes the key characteristics of this system.

Table 12-7   Characteristics of ABC Classification System

Groups	Quantity (% of items)	Value (% of $)	Degree of Control	Types of Records	Safety Stock	Ordering Procedures
A items	10–20%	70–80%	Tight	Complete, accurate	Low	Careful, accurate; frequent reviews
B items	30–40%	15–20%	Normal	Complete, accurate	Moderate	Normal ordering; some expediting
C items	40–50%	5–10%	Simple	Simplified	Large	Order periodically: 1- to 2-year supply

Bar codes (the alternating vertical wide and narrow lines that label items) have done much to simplify the accounting associated with inventories, as well as other production-control activities. Bar codes contain digitally encoded information that is read by optical scanners linked to a computerized database. At the retail level, the computer can then identify the item, display its cost, and update inventory record quantities. At the factory level, bar codes direct and monitor production processes by conveying information on the location of parts, assemblies, and finished goods. Computers can then monitor what work has been done, who did it, what should be done next, etc.

INVENTORY COUNTING

Inventory records must be highly accurate (98 to 99 + percent) to facilitate automated production systems. Two methods of auditing inventory records are by (1) periodic physical counting (e.g., once a year) and (2) cycle counting. *Cycle counting* is a continuous physical counting of inventory so that all items are counted at a specified frequency, and inventory records are periodically reconciled with actual data. A *cycle* is the time required to count all items in inventory at least once.

Example 12.9   Empire Electronics has 6,400 items in stock; 400 are class A items, 1,000 are B items, and 5,000 are C items. The company operates 250 days per year and wishes to count A, B, and C items with a relative frequency of 5, 2, and 1 times a year. How many items should the company count per day, on the average?

Item		Count Frequency	Total Counts
Type	Number		
A	400	5	2,000
B	1,000	2	2,000
C	5,000	1	5,000
			9,000

$$\text{Number of items counted/day} = \frac{\Sigma \text{ total counts}}{\text{number of days}} = \frac{9,000 \text{ counts}}{250 \text{ days}} = 36 \text{ items counted/day}$$

Note: In a well-disciplined inventory management system, a cycle counter may be able to count and reconcile as many as 40 items per day, so this *may* be within the capacity of one cycle counter at Empire Electronics.

INVENTORY MEASURES: TURNOVER

Cycle Inventory and Safety Stock.   Insofar as inventory exists in many forms and at different stages of production, it is sometimes categorized by function. Some common classifications are:

(a)   *Anticipation inventory*, which is held to offset uneven demand peaks, e.g., seasonal variation

(b)   *Transit* or *pipeline inventory*, which is in the system, i.e., inbound, work in process, and outbound

(c)   *Cycle inventory*, which represents the average order quantity amount on hand, i.e., $Q \div 2$

(d)   *Safety stock inventory*, which is used to protect against uncertainties, e.g., 2-week supply

Example 12.10   Plumbing Supply Company orders a certain fitting from a supplier in lot sizes of 180 units per order and receives the shipment 2 weeks later. Demand for the fittings averages 60 units per week, and the company holds an extra 10 fittings to guard against stockout. What is (a) the cycle inventory and (b) the safety stock.

(a)   Cycle inventory $= Q \div 2 = 180$ units $\div 2 = 90$ units

(b)   Safety stock $= 10$ units

Turnover.   *Inventory turnover* is the number of times the total inventory value "turns," or is sold, in a given time period (year). It is found by dividing the annual sales (at cost) by the average inventory value. The number of turns is a commonly used measure of comparison for firms in a given industry and may vary from a low near 1 to over 20. A relatively high number of turns signifies high sales for a given amount of investment and a correspondingly low carrying cost per unit.

Example 12.11   Medco Appliances realized sales worth $4.6 million (at cost) last year (52 weeks) from an average inventory of $.6 million. Find the (a) weeks of supply on hand and (b) inventory turnover rate.

(a)   Weeks of supply $= \dfrac{\$600,000}{\$4.6 \text{ million/yr} \div 52 \text{ wks/yr}} = 6.7$ weeks of supply

(b)   Turnover $= \dfrac{\$4.6 \text{ million/yr}}{\$600,000/\text{turn}} = 7.7$ turns/yr

JUST-IN-TIME SYSTEMS

In the never-ending drive to improve productivity, few concepts have had such a widespread impact as the Just-In-Time (JIT) philosophy. Beginning with Japan, JIT thinking has influenced not only inventory systems but the entire culture of corporations worldwide.

Question: What is Just-In-Time?

Just-In-Time is a managerial philosophy that fosters continuous improvement by reducing in-plant inventories and developing the supplier and system capabilities to produce quality goods in relatively small lots when needed, i.e., just in time.

Some authors illustrate the JIT with a boat crossing a body of water that has dangerous rocks below the surface. Lowering the water (inventory) level reveals the large boulders (problems) that, once removed (solved), allow the boat (system) to function more effectively. Profits stem not only from the lower inventory cost but also from the numerous other benefits of enhanced coordination.

With JIT, work is "pulled" through the system in response to control from the following (downstream) work center. It, in turn, is responding to the next work center, which is ultimately responding to the master schedule and timely, but perhaps fragmented, customer demand. This means that lot sizes are typically smaller, variety is greater (i.e., requiring mixed-model assembly), and more set-ups are required (to produce different models). It also necessitates that the work force be more highly trained (i.e., and flexible) to function in a cooperative way as a team, taking more responsibility for the quality level of their own work, production methods, preventive maintenance of equipment, etc. With low inventories, there is little or no cushion of buffer stock to keep production going if a machine breaks down or if deliveries from a supplier are late. Defect-free supplies must arrive exactly when needed, in the correct quantities. And production problems must be solved as they occur—not hidden under the protection of inventory.

Question: What are some of the key elements of JIT systems?

Key Elements of JIT Systems

1. Close ties with *few reliable suppliers*
2. *Low inventories* of raw materials, work in process, and finished goods
3. *Pull-type* movement of work through the system
4. *Small lot sizes* with shorter lead times
5. Short, *low-cost set-up times*
6. *Product layout* tendency in smaller, more focused facilities
7. Smoothed (*mixed-model*) master schedule
8. Multiskilled, flexible, *responsible work force*
9. Supportive, team-oriented, *problem-solving environment*
10. *Preventive maintenance* system to minimize breakdowns
11. Strong commitment of everyone to *continuous improvement*

KANBAN SYSTEMS

Many firms have adopted a form of the Japanese kanban system to control inventory in a series of workstations.

Question: What are kanbans?

Kanbans are visible cards affixed to containers (or containers themselves) that authorize the (*a*) movement or (*b*) production of materials in response to downstream demand. Kanbans limit inventory to that needed to support current production and are an effective mechanism for identifying production constraints.

Production kanbans authorize the production of the number of parts needed to fill a limited size container. *Move, or withdrawal, kanbans* authorize the movement of containers from the output of one work center to the input of the next. Inventory levels are controlled by adding or removing card sets from the system.

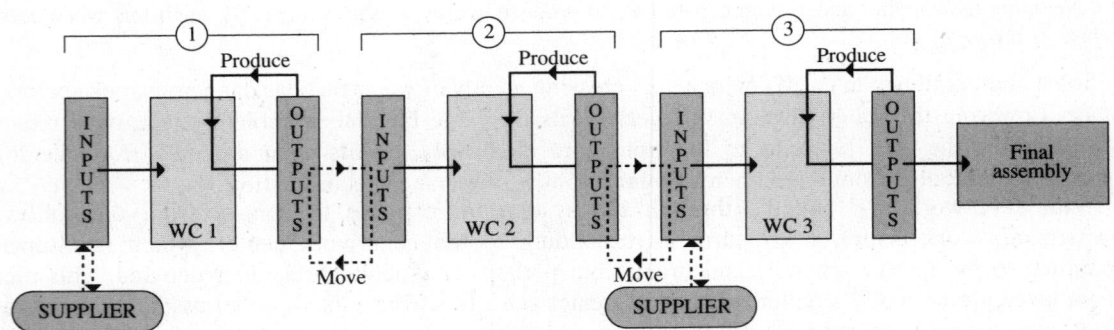

Fig. 12-7   Kanban system

Figure 12-7 illustrates the inventory flow in a portion of an assembly line with three work centers. The daily production schedule is given to the final assembly line (on the right) and to outside suppliers. Assume it authorizes production at WC 3 and, in doing so, uses up a full container of parts. The empty container and a move Kanban are returned to the output area of WC 2, which has a full container (with a production Kanban attached). The production Kanban in the full container is removed and placed on a receiving post at WC 2 (signaling the authorization to produce another container of parts). The move Kanban from the empty container is then transferred to the full container, which is moved back to the input area of WC 3. As WC 2 needs more inputs, its supply requirements are in turn transmitted back to WC 1 (and possibly to outside suppliers).

Kanban systems are deceptively simple yet effective. Although they do seem to require unnecessary worker involvement in materials handling, they foster cooperation and avoid unnecessary paperwork. As improvements in set-up and production times are realized, inventory levels can be reduced by removing some of the available containers.

Questions and Solved Problems

MAKE-OR-BUY DECISIONS

12.1   The Evergreen Garden Tractor Co. has extra capacity that can be used to produce gears that the company is now buying for $10 each. If Evergreen makes the gears, it will incur materials costs of $3 per unit, labor costs of $4 per unit, and variable overhead costs of $1 per unit. The annual fixed cost associated with the unused capacity is $8,000. Demand over the next year is estimated at 4,000 units. Would it be profitable for the company to make the gears?

We assume that the unused capacity has no alternative use.

Cost to make:

$$VC/unit = materials + labor + overhead$$
$$= \$3 + \$4 + \$1 = \$8/unit$$
$$TVC = (4{,}000 \text{ units})(\$8/unit) = \$32{,}000$$

Add: FC	+8,000
Total costs	$40,000

Cost to buy:

$$\text{Purchase cost} = (4{,}000 \text{ units})(\$10/unit) = \$40{,}000$$

Add: FC	+8,000
Total costs	$48,000

Making the gears is advantageous because it can be done for $32,000 in variable costs versus the $40,000 cost of purchasing the gears. The $40,000 total cost to make the gears covers both fixed and variable costs, whereas the $40,000 purchase cost does not help cover any fixed cost, yet the fixed cost must be covered.

12.2 A sporting goods company can purchase warm-ups for $15 each or produce them (where FC = $45,000 and VC = $10 per unit). Management figures a 70 percent chance that demand will be 4,000 units and a 30 percent chance that it will be 6,000 units. What is the economic advantage of buying the warm-ups rather than making them?

$$\text{Expected demand} = E(D) = 4{,}000\,(.70) + 6{,}000\,(.30) = 4{,}600 \text{ units}$$
$$TC_{\text{make}} = FC + VC(V) = \$45{,}000 + \$10(4{,}600) = \$91{,}000$$
$$TC_{\text{buy}} = P(V) = \$15(4{,}600) = \$69{,}000$$
$$\text{Advantage of buying} = \$91{,}000 - \$69{,}000 = \$22{,}000$$

12.3 A telephone equipment company manager must decide whether to make or buy a relay and a switch. Capacity is available, and no other uses of the capacity are apparent. See Table 12-8.

Table 12-8

	Relay	Switch
Quantity required (units/yr)	30,000	18,000
If make:		
Material cost/unit	$.085	$.025
Direct labor hours	400	220
Overhead (variable)	$12/labor hr	$12/labor hr
If buy:		
Purchase price/unit	$.48	$.22

The direct labor cost is $16 per hour, and the variable portion of overhead is $12 per labor hour. The fixed overhead of the available capacity is estimated at $2,600 per year.

Table 12-9   Make-or-Buy Comparison

	Relay Only	Switch Only	Relay and Switch
Cost to make:			
Materials cost [units × cost/unit]	$ 2,550	$ 450	$ 3,000
Direct labor [labor hr × $16/hr]	6,400	3,520	9,920
Overhead (variable) [labor hr × $12/hr]	4,800	2,640	7,440
Total variable cost	$13,750	$6,610	$20,360
Add: Fixed cost	2,600	2,600	2,600
Total cost to make	$16,350	$9,210	$22,960
Cost to buy:			
Purchase price [price/unit]	$ .48	$ .22	
Total purchase price [units × price/unit]	$14,400	$3,960	$18,360
Add: Unavoidable fixed cost	2,600	2,600	2,600
Total cost to buy	$17,000	$6,560	$20,960
Summary:			
Purchase price [price/unit]	$ .48	$ .22	
Variable cost [Total VC ÷ units]	$ .4583	$.3672	
Cost to make	$16,350	$9,210	$22,960
Cost to buy (including FC)	17,000	6,560	20,960
Advantage	$ 650	$2,650	$ 2,000
	(make)	(buy)	(buy)

Note that the fixed cost of $2,600 must be paid regardless of whether the company makes the relay only, the switch only, or both. The total cost to buy both ($20,960) is equal to the cost to make ($22,960), so on first glance it may appear advantageous to buy both. However, on closer examination we see that making the relay is profitable, but making the switch is not. The variable cost of the relay ($.4583) is less than the purchase price ($.48), so there will be enough contribution to cover the fixed cost of $2,600, plus a profit of $650. (This is not highly profitable, however, so if something better comes along, it should be investigated.)

The variable cost of producing the switch ($.3672) exceeds the purchase price ($.22), so the firm would be losing money on every switch it produced, with no contribution to the fixed overhead. The switch should be bought.

PURCHASE-QUANTITY DECISIONS: SINGLE-PERIOD MODEL

12.4   A product costing $70 sells for $120. Unsold units can be returned for $30. What is the probability value that represents the point to which units should be stocked (i.e., the optimal probability of the last unit being sold)?

Note: Salvage reduces the ML to $70 − $30 = $40.

MP = $120 − $70 = $50.

$$P(D)^* \geq \frac{\text{ML}}{\text{MP} + \text{ML}} = \frac{\$40}{\$50 + \$40} = .44$$

Continue stocking units until the probability of selling the next unit to be added drops below .44.

12.5   [*Using P(D)**] Suppose the product demand of the previous problem is uniformly distributed between 800 and 900 units. How many units should be stocked?

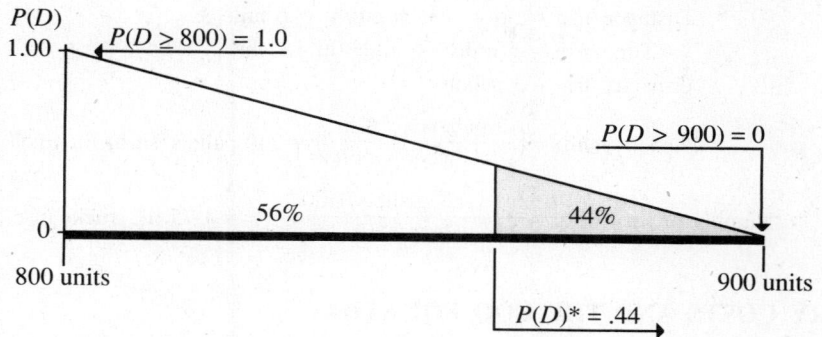

With uniform demand, the P(selling ≥ 800) is 1.0; and the P(selling ≥ 900) drops to zero at a uniform rate. Thus the probability value of .44 is 44 percent of the distance away from the zero point at 900 (i.e., or 56 percent of the distance beyond 800). Therefore the amount to stock is:

$$\text{Stocking level} = 800 + .56 \ (900 - 800) = 856 \text{ units}$$

12.6 (*Using service level*) Demand for videotapes of a financial TV program varies uniformly between 400 and 800 copies. The distributor pays \$3.50 per tape and realizes \$17.00 income on each sale. Unsold tapes cannot be returned and have no salvage value. Find (*a*) the optimal stocking level and (*b*) the stockout risk associated with that quantity.

(*a*) The optimal stocking level is that service level corresponding to the ratio of the cost associated with understocking (or shortage) to the sum of the overstock (excess) and understocking (shortage) costs. Using Eq. 12.5 for service level we have:

$$\text{Service level} = \frac{C_{\text{us}}}{C_{\text{os}} + C_{\text{us}}} = \frac{C_{\text{shortage}}}{C_{\text{excess}} + C_{\text{shortage}}} = \frac{(\$17.00 - \$3.50)}{\$3.50 + (\$17.00 - \$3.50)} = .79$$

Enough tapes should be stocked to meet demand 79 percent of the time. We can compute the optimal quantity to stock directly by starting with the minimal amount of 400 and adding 79 percent of the difference between the highest and lowest demands.

$$\text{Stocking level} = 400 + .79 \ (800 - 400) = 716 \text{ tapes}$$

(*b*) SOR = 1.00 − SL = 1.00 − .79 = .21

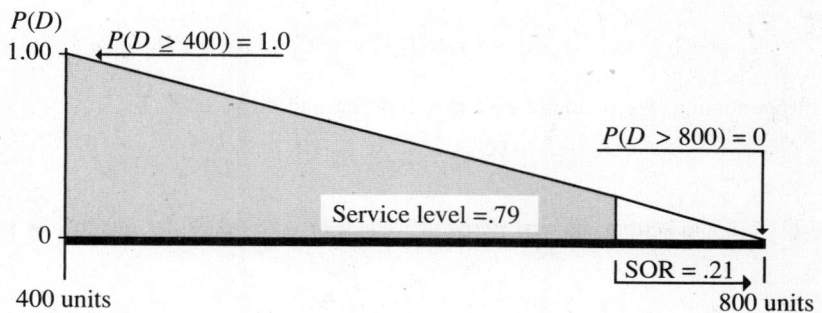

MATERIALS HANDLING, STORAGE, AND RETRIEVAL

12.7 Lakeview Lumber uses forklift trucks to transport lumber from the mill to a storage warehouse .3 mile away. The lift trucks can move three loaded pallets per trip and travel at an average speed of 6 miles per hour (allowing for loading, delays, and travel). If 420 pallet loads must be moved during each 8-hour shift, how many lift trucks are required?

$$\text{Distance/trip} = .3(\text{over}) + .3(\text{return}) = .6 \text{ mile}$$

$$\text{Time/trip} = .6 \text{ mile} \div 6 \text{ miles/hr} = .1 \text{ hr}$$

$$\text{Capacity/trip} = 3 \text{ pallets}$$

$$\text{Capacity/shift} = \left(\frac{3 \text{ pallets}}{.1 \text{ hr}}\right)\left(\frac{8 \text{ hr}}{\text{shift}}\right) = 240 \text{ pallets/shift–lift truck}$$

$$\text{Number of lift trucks} = \frac{420 \text{ pallets/shift}}{240 \text{ pallets/shift–lift truck}} = 1.75 \text{ lift trucks (use 2 lift trucks)}$$

INVENTORY COSTS AND THE EOQ EQUATION

12.8   (*Calculus explanation and derivation*)   If carrying costs consist of two components—(1) C_i = interest cost per unit-year on the average inventory investment and (2) C_s = storage space cost per unit-year to accommodate Q units—set up an equation for total costs and derive an expression for the EOQ that includes both these terms.

$$\text{TC} = \text{ordering} + \text{interest} + \text{storage} + \text{purchase}$$

$$= \frac{D}{Q} S + \frac{Q}{2} C_i + QC_s + PD$$

The TC equation can be differentiated by standard calculus methods, where the differential of $Y = X^n$ is

$$\frac{dY}{dX} = nX^{n-1}$$

and when a constant a is included, the differential of Y with respect to X is

$$Y = aX^n$$

$$\frac{dY}{dX} = naX^{n-1}$$

The differential of a constant (by itself) is, of course, equal to zero; so, for example, if $Y = 4 + 5X^3$,

$$\frac{dY}{dX} = 0 + 15X^2$$

The differential of TC with respect to Q can be obtained most easily if we first move the Q's into the numerator (by adjusting to a negative exponent) so that

$$\text{TC} = DSQ^{-1} + \frac{Q}{2} C_i + QC_s + PD$$

Upon differentiating, the purchase cost is a constant and drops out:

$$\frac{d\text{TC}}{dQ} = -DSQ^{-2} + \frac{Q^o}{2} C_i + Q^o C_s + 0$$

Letting $Q^o = 1$ and setting the first derivative equal to zero, the order quantity is now

$$0 = -\frac{DS}{Q^2} + \frac{C_i}{2} + C_s$$

$$\frac{DS}{Q^2} = \frac{C_i + 2C_s}{2}$$

$$Q = \sqrt{\frac{2DS}{C_i + 2C_s}} \qquad\qquad (12.9)$$

12.9   (*Calculus application*)   An inventory cost function is

$$Y = 5X^2 - 400X + 20,000$$

where Y = total cost (\$) and X = units of inventory. Find (*a*) the amount of inventory that will result in the lowest total cost and (*b*) the (lowest) total cost.

(*a*)   This represents a total cost curve like the one shown in Fig. 12-5. To find the point of minimum cost, we find the first derivative of the equation and set it equal to zero.

$$Y = 5X^2 - 400X + 20,000$$

$$\frac{dy}{dx} = 10X - 400$$

Setting this equal to zero we have: $10X = 400$

$$X = 40 \text{ units}$$

(*b*) $Y = 5X^2 - 400X + 20,000 = 5(40)^2 - 400(40) + 20,000 = \$5,600$
   [*Note:* Values on either side of 40 (e.g., 30 and 50 units) all result in higher costs.]

12.10 (*EOQ model*)   A firm with an annual demand of 900 units per year estimates its ordering costs at \$15.00 per order and its carrying costs at \$.30 per unit per year. Assuming that all the conditions of the EOQ model are met, what is the most economical quantity to order?

$$Q = \sqrt{\frac{2DS}{H}} = \sqrt{\frac{2(900)(15)}{.30}} = 300 \text{ units}$$

12.11 (*EOQ model with carrying costs in percentage*)   A distributor pays \$80 each for an electrical switch, for which it has an annual demand of 4,000 units. Its ordering cost is estimated at only \$20 per order, and it estimates carrying charges at 20 percent. What is the EOQ for this item?

This is a standard EOQ problem except that the carrying charge (H) is expressed as an interest percentage (I) of the cost of the item (C). This cost can be calculated beforehand as $H = IC$, which in this problem would be $H = (.20)(\$80) = \$16$. However, this modification is frequently incorporated into the EOQ and ERL equations resulting in the reformulations of:

(*basic EOQ model*) $\text{EOQ} = \sqrt{\dfrac{2DS}{IC}}$ (*12.10*)

(*EOQ with gradual replenishment*) $\text{ERL} = \sqrt{\dfrac{2DS}{IC}\left(\dfrac{p}{p-d}\right)}$ (*12.11*)

Applying Equation 12.10 to this problem we have:

$$Q = \sqrt{\frac{2DS}{IC}} = \sqrt{\frac{2(4,000)(20)}{(.20)(80)}} = 100 \text{ units/order}$$

12.12 An inventory manager is reviewing some annual ordering data for 3 years ago, when the firm used only 2,000 cases and had carrying charges of only 6 percent of the \$20 per case purchase price. At that time, it cost the firm only \$10 to write up an order. The manager has come across the following equation:

$$Q = \sqrt{\frac{2(2,000)(10)}{1.20}} = \sqrt{\frac{2(\quad)2,000(\quad)10(\quad)}{1.2(\quad)}}$$

Identify the units associated with the numbers used in the equation, and show what units Q results in.

$$Q = \sqrt{\frac{2\binom{\text{pure}}{\text{number}}2,000(\text{cases/yr})10(\$)}{1.20(\$/\text{case-yr})}} = 183 \text{ cases}$$

Note that the \$ and year units cancel, leaving cases^2, so the answer is in cases.

12.13 A San Antonio stockyard uses about 200 bales of hay per month and pays a broker $80 per order to locate a supplier that handles the ordering and delivery arrangements. Its own storage and handling costs are estimated at 30 percent per year. If each bale costs $3, what is the most economical order quantity?

The purchase price is relevant for computing carrying charges (only), and they must be in the same units as demand. We will (arbitrarily) use months, and let the carrying charge $H = IC$.

$$\text{EOQ} = \sqrt{\frac{2DS}{H}} = \sqrt{\frac{2DS}{IC}}$$

where $IC = (\%)(\text{cost})$
$\qquad\quad = (.30/\text{yr})(\$3/\text{unit}) = \$.90/\text{unit-yr}$
$\qquad\quad = \$.90/12 = \$.075/\text{unit-mo}$

$$= \sqrt{\frac{2(200)(80)}{.075}} = 653 \text{ bales}$$

12.14 Far West Freeze Dry purchases 1,200 tins of tea annually in economic order quantity lots of 100 tins and pays $9.85 per tin. If processing costs for each order are $10, what are the implied carrying costs of this policy?

$$Q = \sqrt{\frac{2DS}{H}}$$

Solving for H we have:

$$H = \frac{2DS}{Q^2} = \frac{2(1,200)(10)}{(100)^2} = \$2.40/\text{tin-year}$$

12.15 What is the most economic order quantity for a firm selling containers at $280 each if its annual demand is 600 units and its ordering cost is $53 per order? Include consideration of carrying costs (at $40 per container per year based on average inventory) and storage costs (at $8 per container based on maximum inventory).

We could set up a total cost expression and derive the EOQ. However, the cost elements of ordering, interest on average inventory, storage on maximum inventory, and purchasing suggest that the use of Eq. 12.9 is appropriate here.

$$Q = \sqrt{\frac{2DS}{C_i + 2C_s}} = \sqrt{\frac{2(600)(53)}{40 + 2(8)}} = 33.7, \text{ say 34 containers}$$

12.16 A hospital uses 250 gallons of a certain gel per year at a purchase price of $2.50 per gallon. The cost associated with placing an order is $8 and with holding stock (on an annual basis) is 20 percent of the purchase cost per unit. Assuming that the basic conditions of the EOQ model apply, (a) what is the most economic order quantity, and (b) what is the total annual cost of ordering, carrying, and purchasing the gel?

(a)

$$Q = \sqrt{\frac{2DS}{H}} = \sqrt{\frac{2(250)(8)}{(.20)(2.50)}} = 89.4, \text{ say 90 units}$$

(b)

$$TC = \frac{D}{Q}S + \frac{Q}{2}H + PD$$

$$= \left(\frac{250}{90}\right)(8) + \left(\frac{90}{2}\right)(.20)(2.50) + (2.50)(250) = \$670/\text{yr}$$

GRADUAL REPLENISHMENT/USAGE

12.17 An electronics firm operates 52 weeks per year with an annual requirement for 156,000 chips, which it produces in-house. The cost of setting up production is $1,600, and the rate at which it produces the chips is 5,000 per week. If the carrying cost is $30 per unit per year, what is the most economical number of chips to make on a production run?

This is a problem of gradual replenishment/usage where the firm produces the product on an intermittent basis, but uses it continuously. We can use Eq. 12.8.

$$\text{ERL} = \sqrt{\frac{2DS}{H}\left(\frac{p}{p-d}\right)} = \sqrt{\frac{2(156,000)(1600)}{30}\left(\frac{5,000}{5,000 - 3,000}\right)} = 6,450 \text{ chips/run}$$

12.18 The Finish Creamery Co. produces ice cream bars for vending machines and has an annual demand for 72,000 bars. The company has the capacity to produce 400 bars per day. It takes only a few minutes to adjust the production set-up (cost estimated at $7.50 per set-up) for the bars, and the firm is reluctant to produce too many at one time because the storage cost (refrigeration) is relatively high at $1.50 per bar-year. The firm supplies vending machines with its "Fin-Barrs" on 360 days of the year. (*a*) What is the most economical number of bars to produce during any one production run? (*b*) What is the optimal length of the production run in days?

(*a*)
$$\text{ERL} = \sqrt{\frac{2DS}{H[1 - (d/p)]}}$$

where S = setup cost = $7.50
D = annual demand = 72,000 bars/yr
H = carrying cost = $1.50/bar-yr

d = daily demand rate = $\dfrac{72,000}{360}$ = 200 bars/day

p = daily production rate = 400 bars/day

$$\text{ERL} = \sqrt{\frac{2(72,000)(7.50)}{1.50[1 - (200/400)]}}$$
$$= 1,200 \text{ bars/run}$$

(*b*)   Optimal number of days of the run is

$$\text{Number of days} = \frac{1,200 \text{ bars}}{400 \text{ bars/day}} = 3 \text{ days}$$

12.19 A firm has a yearly demand for 52,000 units of a product that it produces. The cost of setting up for production is $80, and the weekly production rate is 1,000 units. The carrying cost is $3.50 per unit-year. How many units should the firm produce on each production run?

$$Q = \sqrt{\frac{2DS}{H[1 - (d/p)]}} = \sqrt{\frac{2(52,000)(80)}{3.50[1 - (1,000/1,000)]}} = \sqrt{\infty} = \infty$$

Note that the demand and production rates in this problem are equal, and the equation (rightly) suggests that they should have an infinite (continuous) run.

QUANTITY DISCOUNTS

12.20 Magic Mountain Packers use 80,000 containers annually and can purchase any quantity up to 10,000 at $.10 per container. Costs of ordering are $30.00 per order; interest costs are 20 percent of the price per container and apply to the average inventory. Storage costs are $.02 per container per year and are based on the maximum inventory. Quantity discount costs are available as shown.

No. containers ordered	20,000	40,000	60,000	80,000
Unit cost per container, $	.07	.05	.048	.046

Assuming the firm purchased in quantities of $Q = 40,000$, what is the total annual cost associated with ordering, carrying, storage, and purchase of this inventory?

Letting S = ordering, H = carrying, W = storage, and P = purchase costs, we have:

$$TC = \frac{D}{Q}S + \frac{Q}{2}H + QW + PD$$

$$= \left(\frac{80,000}{40,000}\right)(30) + \left(\frac{40,000}{2}\right)(.20)(.05) + (40,000)(.02) + (.05)(80,000)$$

$$= \$60/yr + \$200/yr + \$800/yr + \$4,000/yr$$

$$= \$5,060$$

INCREMENTAL APPROACH: OPTIONAL MATERIAL

12.21 *Incremental Approach: Uniformly Distributed, Single-Period Demand.* The *City Chronicle* has a daily newspaper demand that varies uniformly between 20,000 and 24,000 copies per day. The paper costs 8 cents per issue to produce and generates a revenue of 20 cents per issue. Unsold papers have no value. (*a*) What is the optimal level of papers to stock, and (*b*) what service level would that optimal level correspond with?

(*a*) This is a problem of balancing the cost of understocking inventory C_{us} with that of overstocking C_{os}. From Eq. 12.3, the balance point is where the cumulative probability of demand $P(D)$ establishes the equality of

$$C_{us}P(D)^* = C_{os}[1 - P(D)]$$

Thus
$$P(D)^* = \frac{C_{os}}{C_{os} + C_{us}}$$

where C_{os} = cost/unit − salvage value = $.08 − 0 = $.08$
C_{us} = revenue/unit − cost/unit = $.20 − $.08 = $.12$

Therefore,
$$P(D)^* = \frac{.08}{.08 + .12} = .40$$

Because demand is uniform, we can depict it with a straight (linear) line from the minimum demand (20,000) to the maximum demand (24,000). Going into the curve from .4 and down to the horizontal axis yields an optimal value of 22,400.

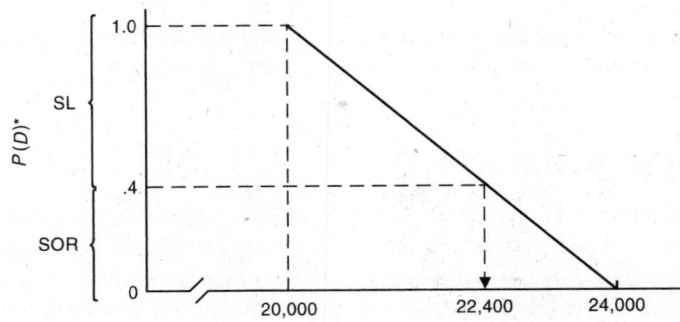

(b) The .40 is the probability demand will be exceeded, and represents a stockout risk established on the basis of the costs given. Thus the corresponding service level is:

$$SL = 1 - SOR = 1 - .4 = .6$$

Note: Algebraically, the point representing a 60 percent service level is an inventory of:

$$I_{OPT} = D_{min} + \%SL(\Delta \text{ inventory}) = 20,000 + .60(24,000 - 20,000) = 22,400 \text{ units}$$

12.22 *Incremental Approach: Normally Distributed Demand.* Demand for a chemical product is normally distributed with μ equals 80 gallons per week and σ equals 5 gallons per week. If C_{os} is \$.15 per gallon and C_{us} is \$.50 per gallon, what is the optimal level to stock?

$$P(D)^* = \frac{C_{os}}{C_{os} + C_{us}} = \frac{.15}{.15 + .50} = .231$$

$$I_{OPT} = \mu + Z\sigma$$

where Z is for a probability area of $.500 - .231 = .269$. Therefore $Z = .73$.

$$I_{OPT} = 80 + .73(5) = 83.65 \text{ gal}$$

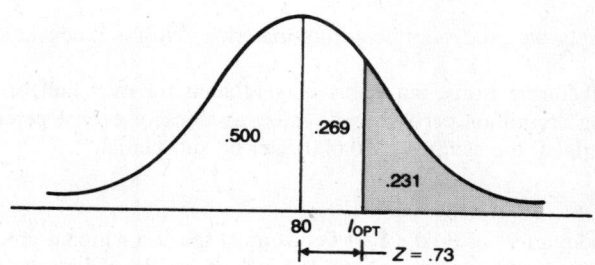

Note: The solution above corresponds to an SL of $1 - SOR = 1.00 - .231 = .769 = 76.9$ percent. Some analysts compute the SL percentage directly by reversing the equation for $P(D)$ to:

$$SL\% = \frac{C_{us}}{C_{os} + C_{us}} = \frac{.50}{.15 + .50} = .769$$

12.23 *Incremental Approach: Poisson Demand.* A large city hospital has determined that the demand for ambulances can be approximated by a Poisson distribution with a mean of 6 per day. The cost for having an ambulance available is \$460 per day, and its life-support value, when used, is placed at \$2,000 per day. If unused, of course, its service value is zero. What is the optimum number of ambulances to maintain?

The two expected costs are equal for the cumulative probability:

$$P(D)^* = \frac{C_{os}}{C_{os} + C_{us}}$$

where C_{os} = cost/unit − salvage = \$460 − 0 = \$460
  C_{us} = value/unit − cost/unit = \$2,000 − \$460 = \$1,540

$$P(D)^* = \frac{\$460}{\$460 + \$1540} = .23$$

Thus, $SL = 1.00 - .23 = .77$.

Cumulative probabilities for a Poisson distribution with a mean of 6.0 are then obtained from Appendix D.

Demand (ambulances per day)	0	1	2	3	4	5	6	7	8	9
Cumulative probability	.002	.017	.062	.151	.285	.446	.606	.744	.847	...

Seven ambulances would closely approximate the service level which balances the expected costs and benefits (i.e., 77 percent), but eight would be required to equal or exceed the 77 percent.

Supplementary Questions and Problems

12.24 Purchasing is simply the process of acquiring materials. Why is it considered such an important activity in an organization?

Ans. For manufacturing firms, purchases may account for over half of the funds spent by a firm. If a company spending $6 million per year can realize a saving of even 1 percent (e.g., from getting a quantity discount on materials), the savings ($60,000) can be substantial.

12.25 The Madison headquarters of Fred's Fast Foods must select a national grocery chain to supply its fast-food chains with fresh vegetables on a nationwide basis. It has developed the following criteria and scores for prospective vendors (Table 12-10). Which supplier has the highest expected score, and what is its value?

Table 12-10

	Rating (1–10)	Potential Suppliers of Vegetables		
		Food Fair	A & R	SaleWay
Price	8	.7	.3	.5
Quality	3	.2	.8	.5
Delivery	5	.5	.4	.4
Location	2	.1	.7	.6

Ans. Food Fair, 8.9

12.26 Banktel, Inc., produces an automatic cash machine in leased facilities in an industrial park. The product uses special-purpose punch keys, which the firm currently purchases at $4.20 each. It is considering leasing additional space and producing the keys itself. The long-term lease would result in annual costs of $4,800. The labor, materials, and variable overheads are estimated at $2.80 per key, and fixed overhead would be an additional $1,080 per year. Demand is estimated as shown in Table 12-11.

Table 12-11

Demand D	Probability P(d)
2,000	.05
3,000	.10
4,000	.30
5,000	.40
6,000	.15

(a) Should Banktel produce the keys? (b) What is the break-even volume where it becomes profitable to produce them rather than buy from a supplier?   *Ans.*   (a) Cost to produce is $420 less. (b) 4,200 units

12.27 New England Grocery Supply Co. has collected the historical data shown in Table 12-12 of weekly demand for a line of breakfast cereals. An operations analyst estimates that it costs the company $1.50 per case per week to be overstocked because of spoilage and carrying costs. Being understocked results in lost profits of $2.40 per case.

Table 12-12

Weekly Demand (no. of cases)	Probability of Demand P(d)
500	.10
600	.20
700	.30
800	.30
900	.05
1,000	.05

(a) Compute the cumulative probability of demand that equates the cost of understocking with the cost of overstocking. (b) Assuming that the company must purchase in lots of 100, how many lots should it stock?   *Ans.*   (a) .385 (b) 800 cases

12.28 One of the four assumptions underlying the EOQ model is that:
(a) The demand pattern follows a normal distribution over the order cycle.
(b) The purchase price per unit varies depending upon the quantity ordered.
(c) Replenishment is instantaneous at the expiration of the lead time.
(d) The model makes allowance for stockout by including an understocking cost.
Ans.   (c)

12.29 The operations manager at St. Mary's Hospital has asked the stores department to begin ordering inventory on an economic lot-size basis. The hospital uses 500 electrocardiogram tape rolls per year, has an ordering cost of $10 per order, and estimates carrying costs at $.25 per unit per year. How many tapes should be ordered each time?   *Ans.*   200

12.30 If the EOQ is 100 units when demand is $D = 1,200$ units per year and the ordering cost is $S = \$10$, what is the implied carrying cost, H?   *Ans.* $2.40

12.31 A furniture company uses rubber moldings at a rate of 200 per week (50 wks/yr) and can produce them in its own shop at a rate of 500 per week. The cost to set up for a production run is $34.56 each time. Carrying costs are $.80 per molding per year. What is the most economical production run length?
Ans.   1,200 moldings

12.32 An office uses 120 boxes of photocopy paper per year. Their cost is $25 per box, but the office gets a 10 percent discount for ordering quantities of 10 boxes or more and 20 percent for orders of 30 or more. Carrying costs are always taken to be $5 per box per year based upon average inventory, and storage costs are $2 per box per year based upon the maximum inventory received at one time. If ordering costs are estimated at $30 per order, what is the total annual cost associated with ordering in quantities of 30 boxes per order?
Ans.   TC = order + carry + storage + purchase = (120/30) ($30) + (30/2) (5) + 30 (2) + (25 − 5) (120) = 120 + 75 + 60 + 2,400 = $2,655

12.33 Factory Built Homes (FBH), Inc. purchases paneling components from a nearby western New York mill for $5 per unit. It expects to use about 4,000 units during the coming year. FBH estimates that it costs $30 to place an order and $1.50 per unit-year for carrying and storage costs. The mill can provide FBH with immediate delivery of any reasonable quantity. (*a*) What is the most economical quantity for FBH to order? (*b*) How many orders per year should be placed? (*c*) What is the total yearly cost associated with ordering, carrying, and purchasing the EOQ amount?   *Ans.*   (*a*) 400 units per order (*b*) 10 (*c*) $20,600

12.34 In the ERL model, if the demand rate is 40 units per month, and the production rate is 200 units per month, what percentage of the production goes into inventory?   *Ans.*   80 percent

12.35 Spokane Public Power Co. purchases transformers at a cost of $330 and uses an ordering cost of $45 per order. Inventory is carried at a cost of 10 percent of the per-unit price (based on average inventory). Storage costs, based upon adequate space for maximum inventory, are $6 per transformer. Annual demand is 800 units. (*a*) Compute the total yearly cost if the firm orders in EOQ amounts. (*b*) The supplier has offered the power company a 10 percent discount for purchasing in quantities of 200. Assuming this affects all costs except ordering and per-unit storage costs, compute the total yearly cost for this quantity discount situation.   *Ans.*   (*a*) TC = $900 + $660 + $240 + $264,000 = $265,800 (*b*) $241,950

12.36 A distributor of fabric has an annual demand for 5,400 yards of silk. The cost to place an order is $50, and the carrying cost is 20 percent of the unit cost. (*a*) What is the most economical order quantity if the unit price is $30 per yard? (*b*) Suppose the supplier offers a 10 percent discount if the distributor purchases a lot of 1,000 yards. What would be the ordering, carrying, and purchasing costs ($ per year) if the firm takes advantage of this discount?
Ans.   (*a*) $Q = 300$ units, (*b*) $(D/Q)S = \$270/\text{yr}$, $(Q/2)H = \$2,700/\text{yr}$, $P(D) = \$145,800$

12.37 Golden Valley Cannery uses 64,000 size 7X cans annually and can purchase any quantity up to 10,000 cans at $.040 per can. At 10,000 cans the unit cost drops to $.032 per can, and for purchases of 30,000 it is $.030 per can. The costs of ordering are $24 per order, and interest costs are 20 percent of the price per can and apply to the average inventory. Storage costs are $.02 per can-year and are based upon maximum inventory. (Disregard safety stock costs.) (*a*) What is the EOQ, disregarding the quantity discounts? (*b*) What is the most economical order quantity, considering the quantity discounts?
Ans.   (*a*) 8,000 cans (*b*) TC for 10,000 unit purchase is least at $2,433.

12.38 A governmental supply warehouse is implementing a cycle counting system whereby class A items are counted monthly, B items quarterly, and C items annually. Of the 6,300 items in inventory, 900 are in class A, 2,100 are in class B, and the remainder are in class C. Assuming that there are 250 workdays per year, how many items should be counted per day?   *Ans.*   90 items per day

12.39 In the ABC system, the class of inventory items that contains a high percentage of items and a low percentage of dollar volume is (*a*) A, (*b*) B, (*c*) C, or (*d*) D?   *Ans.*   (*c*)

12.40 Paris Lighting Co. manufactured and sold components having a cost-of-goods value of $2.6 million during the past year. The average value of inventory on hand, at cost, was $840,000. What was the inventory turnover rate for this company?   *Ans.*   3.1 turns

12.41 (*Uniformly Distributed Demand*) The publisher of a weekly news magazine has a demand that varies uniformly between 10,000 and 12,000 copies per week. The magazine costs 12 cents per issue to produce and generates a revenue of 50 cents per issue. Unsold magazines have no value. (*a*) What is the appropriate value for $P(D)^*$? (*b*) With what service level does this value of $P(D)^*$ correspond? (*c*) What is the optimal number of copies to publish?   *Ans.*   (*a*) .24, (*b*) .76, (*c*) 11,520 copies

12.42 (*Normally Distributed Demand*) A seafood restaurant has a normally distributed demand for a certain shellfish with a mean of 60 pounds per week and standard deviation of 12 pounds per week. Overstocking costs are estimated at $4.00 per pound and understocking costs at $7.00 per pound. What is the optimal level to stock?   *Ans.*   62 pounds

12.43 (*Poisson Distributed Demand*) The demand for intensive care units (ICU's) at a large hospital is Poisson distributed with a mean of 4 beds per day. The cost, including overhead, of having the facilities available is estimated at $500 per day and life support value has been placed at $4,000 per day. Recognizing that unused facilities have no value, what is the optimum number of ICU beds to have available?
Ans.   6 beds

Chapter 13

Inventory Control:
Order Points, Safety Stocks,
and Service Levels

Statistical Methods

INVENTORY CONTROL SYSTEMS

Inventory control systems are the ordering and monitoring techniques used to control the *quantity* and *timing* of inventory transactions. The traditional inventory control systems have been classified as (1) *continuous*, or fixed order-quantity systems, and (2) *periodic*, or fixed order-period systems. However, a number of (3) *hybrid systems* exist, e.g., base stock system, including the use of some relatively straightforward measures, such as inventory turns. In addition, many broad-based production control systems incorporate inventory control as an integral part of their larger systems. Chief among these are the (4) *MRP systems*, which lend themselves to use of (5) *just-in-time* (JIT) as a means of inventory control, and (6) *kanban* to control the material movement subsystems. Inventory turns, just-in-time, and kanban are reviewed in Chap. 12, and MRP is covered in Chap. 14. This chapter covers order points, safety stocks, and service levels associated with traditional inventory systems.

Question: How does the inventory control approach of the (*a*) more traditional inventory systems differ from that of (*b*) MRP and JIT-type systems?

(*a*) *Continuous and periodic monitoring systems*, by themselves, are essentially order launching techniques that look back at historical usage. They call for the release of new orders whenever the quantity on hand or the elapsed time reaches a predetermined point. This is often satisfactory for distribution inventories where extra stock (safety stock) is carried to allow for uncertainties.

(*b*) *MRP and JIT philosophy* is to plan ahead to a forecast, or to orders already received. Under MRP, material requirements are ordered so as to arrive precisely when needed. This time-phased sequencing is most useful in a manufacturing environment where the inventory consists largely of components that are being transformed into planned quantities of finished goods in an orderly manner.

Constant Demand and Constant Lead Time. Inventory control systems monitor both the pattern of demand and the time required to deliver an order (lead time). If both demand and lead time are known and constant (as assumed in the EOQ model), the required inventory levels and inventory costs can be readily computed.

Example 13.1 Green Bluff Warehouse has a contract to ship 100 boxes per day of Washington apples to a distributor. They order apple boxes in quantities of $Q = 3,000$ per order from a supplier who always delivers them in 10 days. (*a*) What is the maximum (I_{max}) and average (I_{ave}) amount of inventory on hand? (*b*) How much inventory is still on hand when they place an order for more boxes? (*c*) What annual carrying cost is incurred if the boxes cost \$2 each and the carrying cost is 25 percent per year?

(*a*) See Fig. 13-1. Since there is no uncertainty about either demand or lead time, Green Bluff should arrange for the boxes to arrive just as the previous supply is used up. Therefore the most boxes in stock at any time is the order quantity, so $I_{max} = 3,000$ boxes. Because demand is constant, the average inventory on hand is $I_{max} \div 2 = 1,500$ boxes.

(*b*) Insofar as there is a 10-day delivery time, Green Bluff must have a 10-day supply of boxes on hand when they reorder. With a demand of 100 boxes per day, the inventory on hand (reorder point) must be:

Reorder point = (demand rate) (lead time) = (100 boxes/day) (10 days) = 1,000 boxes

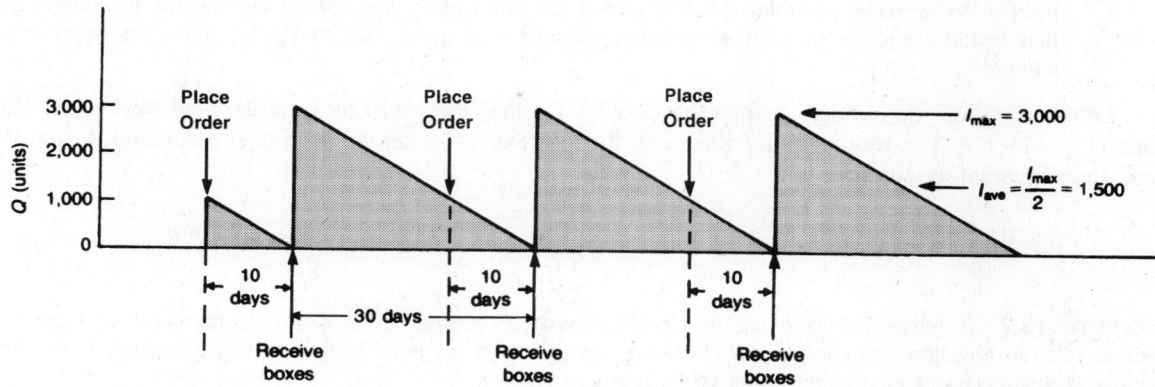

Fig. 13-1　Inventory levels under constant demand and lead time

(c)　　　　　　　　Carrying cost $= (\% \text{ rate})(\text{unit cost})(I_{ave})$
　　　　　　　　　　$= (.25/\text{yr})(\$2/\text{box})(1{,}500 \text{ boxes}) = \$750/\text{yr}$

Variable Demand and/or Variable Lead Time.　The major uncertainties associated with managing inventories are the variability of demand and the variability of lead time. A common technique for handling these uncertainties is to hold an extra amount of inventory (safety stock) in excess of regular usage quantities. Safety stocks add an important controllable variable to inventory systems.

Question:　The variables associated with an inventory system are the (a) order quantity, (b) lead time, (c) safety stock, and (d) reorder point. Define these terms, and (e) show their relationship.

(a)　*Order quantity* (Q) is the replenishment amount ordered.

(b)　*Lead time* (LT) is the time period between placing and receiving an order.

(c)　*Safety stock* (SS) is the extra inventory needed for cycles when demand is stronger than average.

(d)　*Reorder point* (ROP or OP) is the inventory level at which a replenishment order is placed.

(e)　Figure 13-2 illustrates the relationship. Note that both demand and lead time are variable, and this sometimes necessitates the use of safety stock. A general guideline is to have enough safety stock to

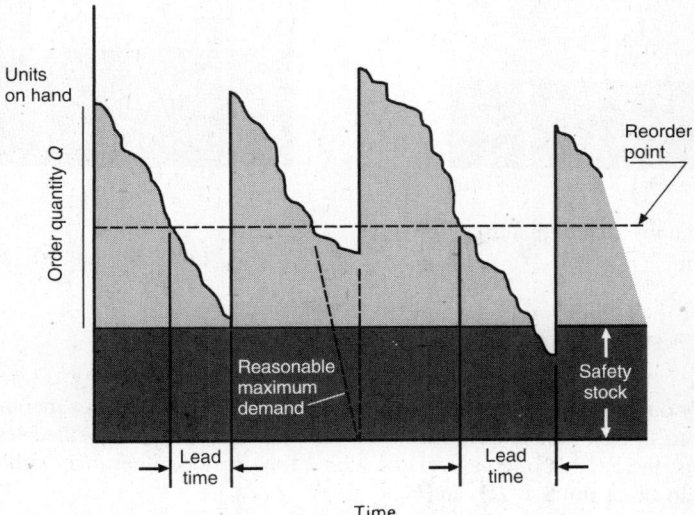

Fig. 13-2　Inventory variables: Q, LT, SS, and ROP

meet a "reasonable maximum demand during the lead time." The reorder point is the inventory level that includes enough to meet the average demand during the lead time, D_{LT}, plus the safety stock amount.

Letting $\bar{d}$ equal the average demand rate and LT equal the lead time, the demand during the lead time $D_{LT} = \bar{d} \cdot LT$. The reorder point ROP can then be expressed as the sum of the demand during the lead time plus safety stock:

$$ROP = \bar{d} \cdot LT + SS$$
$$= D_{LT} + SS \qquad\qquad (13.1)$$

Example 13.2 A nationwide trucking firm has an average demand of 10 new tires per week and receives deliveries from an Ohio tire company 20 business days (5 days per week) after placing an order. If the firm maintains a safety stock of 15 tires, what is the reorder point?

$$= \bar{d} \cdot LT + SS$$
$$= (10 \text{ tires/wk})(4 \text{ wks}) + 15 \text{ tires} = 55 \text{ tires}$$

Continuous and Periodic Review Systems. The continuous and periodic review (traditional) systems of inventory control both use safety stock to protect against uncertainties.

Question: Describe a continuous review (fixed-quantity) system.

Continuous review (Q) systems maintain a current (perpetual) record of the amount of inventory in stock at all times. This record is frequently on computer with entries from bar code readers or other computers.

Figure 13-3 illustrates a continuous system, where Q is the order quantity, ROP is the reorder point, SS is the safety stock, and LT is the lead time. Note that demand is variable, sometimes requiring use of the safety stock.

Whenever the amount on hand drops to the ROP, a fixed quantity, Q, is ordered—hence the name fixed-quantity system. This can be an EOQ or ERL amount, although that is not a necessity.

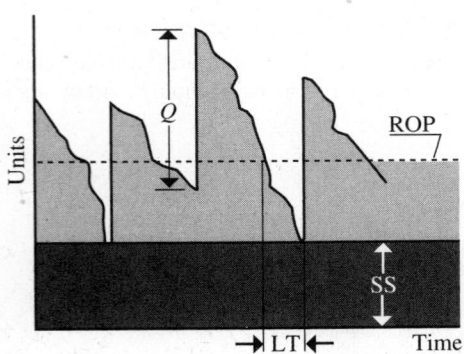

Fig. 13-3 Continuous (fixed-quantity) system

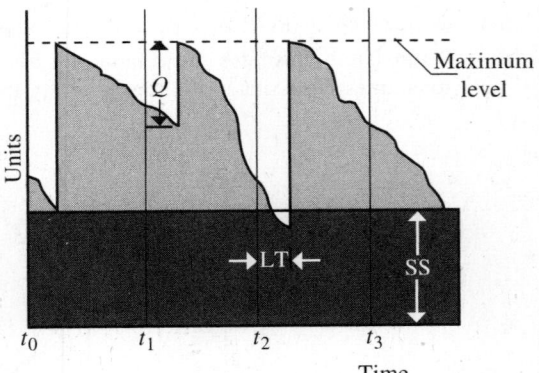

Fig. 13-4 Periodic (fixed-period) system

Question: Describe a periodic review (fixed-period) system.

Periodic review (P) systems are referred to as fixed-interval systems. See Fig. 13-4. Under periodic systems, the amount of inventory on hand is assessed at periodic intervals, such as weekly or monthly. A variable quantity is then ordered—enough to bring the inventory on hand and on order up to a specified level.

In periodic systems, the ROP is often associated with a time period rather than with a quantity on hand. Thus the reorder points would be at times t_1, t_2, and t_3.

Periodic systems lend themselves to situations in which periodic physical counting of on-hand materials is more practical than maintaining a perpetual count.

Because continuous systems are constantly being monitored, they are especially useful for inventories that have large, unexpected fluctuations in demand. Periodic systems, on the other hand, are more useful for processes that call for a consistent use of material and for conditions where a single review period can be used to identify several items that can be combined into one order. Insofar as the safety stock in a periodic system must provide protection over the entire cycle (of t_1 to t_2), the safety stock level is typically larger than would be required under a continuous system, where the safety stock protects against large demands in the lead time only.

Question: What is the base stock system?

The base stock system is a combination of the continuous (fixed-quantity) and periodic (fixed-interval) systems. Inventory levels are reviewed periodically, but orders are placed only when the stock is below some specified level. This yields some of the control aspects of periodic systems, but can result in the placement of fewer orders and orders of a more economic lot size.

METHODS OF HANDLING INVENTORY UNCERTAINTIES

Several approaches exist for handling inventory uncertainties, and they include different techniques for determining the amount of safety stock. Table 13-1 summarizes some of the more widely recognized methods, and examples follow.

Table 13-1   Methods of Protecting against Shortages and Stockouts

Method	Description
Informal decision rules	(1) *Ratios:* Order on basis of ratio (e.g., on-hand inventory/expected usage during LT) (2) *Ultra conservative:* Provide stock for largest daily demand × longest LT (3) *Safety stock percentage:* Let SS = D_{LT} plus a 25 to 40 percent safety factor (4) *Square root of D_{LT}:* Let SS = $\sqrt{D_{LT}}$
Expected-value approach	Construct payoff and expected-value tables where alternatives are amounts of inventory to stock and uncontrollable variable is D or LT (especially suitable for handling perishable inventories).
Incremental approach	Add inventory to point where incremental cost = incremental gain × P(gain). (1) *Single-period model:* Use understocking cost per unit and overstocking cost per unit to set stock level for one-time demand situation. (See Chap. 12.) (2) *Multiple-period model:* Use stockout cost per unit and carrying cost per unit to set stock level for multiple-period demands. (See Chap. 13.)
Statistical distribution methods	(1) *Empirical:* Use empirical data to formulate a probability distribution of D or LT, and compute required SS for specified service level. (2) *Known distribution:* Use known (or assumed) statistical distribution of D or LT, and compute required SS for specified service level. (*a*) Service level based upon *order cycle* stockouts (*b*) Service level based upon *individual* unit shortages

INFORMAL DECISION RULES

Although ratios and percentages are quick and easy to apply, they do not always make the best use of available information about variability of demand and lead time.

Example 13.3   Demand for a product ranges from 50 to 140 units over each 4-week period, with a mean of 60 units. Lead time averages 3 weeks, but has been as long as 4 weeks. How much safety stock should be maintained under a strategy of carrying (a) 20 percent of D_{LT} and (b) the square root of D_{LT}?

$$(a) \quad D_{LT} = (60 \text{ units}/4 \text{ wks})(3 \text{ wks}) = 45 \text{ units}$$
$$\text{SS} = (\%)D_{LT} = (.20)(45 \text{ units}) = 9 \text{ units}$$

$$(b) \quad \text{SS} = \sqrt{D_{LT}} = \sqrt{45} = 6.7, \text{ say } 7 \text{ units}$$

EXPECTED-VALUE APPROACH

Although somewhat theoretical, expected value is a useful method of incorporating uncertainty by assigning probabilities to the different possible values of demand (States of Nature). Once the outcomes (payoffs) are estimated, the optimal level of stock is the level that yields the highest expected profit. For each stocking option, the $E(X) = \Sigma[X \cdot P(X)]$.

Table 13-2   Payoff Matrix

	State of Nature (demand)	
	25 units	60 units
Stock 40	250	1,000
Stock 60	(750)	2,400

Example 13.4   The payoff matrix shows the $ profit resulting from stocking either 40 or 60 units of a product depending upon whether demand turns out to be 25 or 60 units. Assuming the probabilities have been estimated at .40 for a demand of 25 and .60 for a demand of 60, what level of stock will maximize profits?

$$\text{For stocking 40: } E(X) = 250(.40) + 1,000(.60) = \$700$$
$$\text{For stocking 60: } E(X) = (-750)(.40) + 2,400(.60) = \$1,140$$

Therefore stock 60 units.

INCREMENTAL APPROACH: MULTIPLE-PERIOD MODEL

(The *single-period model* concerns a one-time order quantity and is discussed with other order quantity models in Chap. 12.)

Multiple-period models offer users a means of incorporating stockout costs into inventory decisions that apply to continuous operations. The overstocking and understocking costs of the singe-period model are replaced by per unit carrying costs, H, and per unit stockout costs C_{so}. In addition, carrying costs per unit are known (not probabilistic), and stockout costs per unit must be multiplied by the number of opportunities for stockout, or order cycles D/Q. The multiple-period stock out risk, SOR, is thus:

$$\begin{pmatrix} \text{Expected stockout} \\ \text{cost/unit} \end{pmatrix} \begin{pmatrix} \text{number of order} \\ \text{cycles this occurs} \end{pmatrix} = \begin{pmatrix} \text{carrying cost} \\ \text{of the unit} \end{pmatrix}$$

$$\underbrace{\begin{pmatrix} \text{Probability of} \\ \text{stockout} \end{pmatrix}}\begin{pmatrix} \text{stockout} \\ \text{cost/unit} \end{pmatrix} \left(\frac{D \text{ units/yr}}{Q \text{ units/order}} \right) = \begin{pmatrix} \text{carrying cost} \\ \text{per unit-yr} \end{pmatrix}$$

$$(SOR)C_{so}\left(\frac{D}{Q}\right) = H$$

Therefore, $$(\text{Multiple-period}) \ SOR = \frac{H}{C_{so}}\left(\frac{Q}{D}\right) \qquad\qquad (13.2)$$

Note that Eq. 13.2 applies to situations where individual stockout costs per unit are known (or estimated).

Example 13.5 A construction equipment dealer experiences an annual demand of about 300 electric generators and orders in quantities of 50 units per order. Carrying costs are \$900 per unit-year, and stockout costs are estimated at \$2,000 per unit. What optimum probability of stockout should be used to determine the appropriate inventory-stocking level?

$$SOR = \frac{H}{C_{so}}\left(\frac{Q}{D}\right) = \left(\frac{\$900/\text{unit-yr}}{\$2,000/\text{unit-order}}\right)\left(\frac{50 \ \text{units/order}}{300 \ \text{units/yr}}\right) = .075$$

The stock should be enough that demand is exceeded only 7.5 percent of the time.

STATISTICAL DISTRIBUTION METHODS: USE OF EMPIRICAL DATA

Empirical data describing past demand, and lead-time variations, may be used to establish safety-stock levels if the service level and carrying cost of inventory is specified. First, assume that demand is variable and that lead times are constant. Once the stockout risk is established, and data collected on past demand, a cumulative distribution of demand can be formulated, and the maximum demand for a given stockout risk (D_{SOR}) can be obtained directly from the cumulative distribution. The required safety stock is then the difference between this maximum demand (D_{SOR}) and the average demand (D_{LT}).

$$SS = D_{SOR} - D_{LT} \qquad\qquad (13.3)$$

In the example that follows, the service level (SL) represents the *percentage of order cycles* that demand is met from stock on hand.

Example 13.6 (*Variable demand, constant lead time*) The data shown in Table 13-3 represent weekly demand on a \$250 item that has a constant lead time of 1 week. The firm has a 20 percent per year cost for carrying inventory. Determine the safety-stock level and carrying cost for providing a service level of (*a*) 90 percent and (*b*) 95 percent.

Table 13-3   Frequency Distribution of Demand During Lead Time

Weekly Demand (number of units)	Frequency (number of weeks this demand occurred)	Cumulative Frequency (number of weeks demand exceeded lower-class boundary)	Cumulative Percentage (percentage of weeks demand exceeded lower-class boundary)
< 50	1	104	100.0
50 < 100	7	103	99.0
100 < 150	11	96	92.3
150 < 200	16	85	81.7
200 < 250	19	69	66.3
250 < 300	20	50	48.1
300 < 350	14	30	28.8
350 < 400	9	16	15.4
400 < 450	5	7	6.7
450 < 500	2	2	1.9
	104		

The 90 percent and 95 percent service levels represent stockout risks of 10 percent and 5 percent, respectively (per Eq. 12.5). We can formulate a frequency distribution (histogram) and cumulative distribution (ogive) of demand as shown in Figs. 13-5 and 13-6.

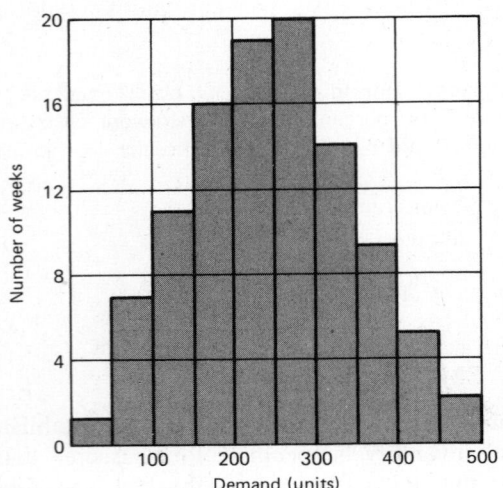

Fig. 13-5 Frequency of demand

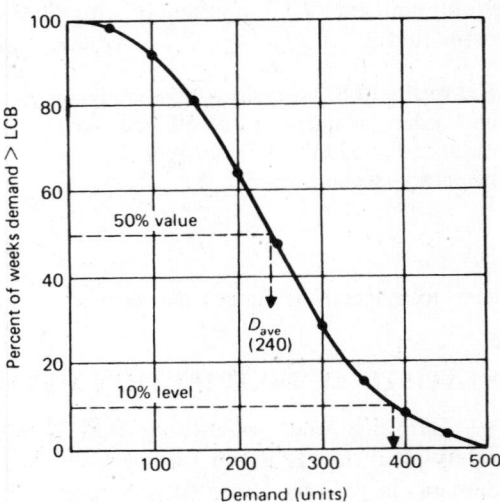

Fig. 13-6 Cumulative distribution of demand

Because it is readily available and sufficiently representative of the central tendency of the data, we use the median (the fiftieth percentile value) of the cumulative distribution as our estimate of average demand.

$$D_{LT} = \text{average demand} \cong 240 \text{ units}$$

(a)   90 percent service level (10 percent SOR):

$$D_{SOR} = \text{demand level corresponding to 10 percent risk of stockout}$$
$$\cong 385 \text{ units (from graph)}$$
$$SS = \text{safety stock level} = D_{SOR} - D_{LT} = 385 - 240 = 145 \text{ units}$$

$$SS \text{ cost} = \left(\frac{\$250}{\text{unit}}\right)\left(\frac{20\%}{\text{yr}}\right)(145 \text{ units}) = \$7,250/\text{yr}$$

(b)   95 percent SL (5 percent SOR):

$$D_{SOR} = \text{demand level corresponding to 5 percent risk of stockout}$$
$$\cong 430 \text{ units (from graph)}$$
$$SS = D_{SOR} - D_{LT} = 430 - 240 = 190 \text{ units}$$

$$SS \text{ cost} = \left(\frac{\$250}{\text{unit}}\right)\left(\frac{20\%}{\text{yr}}\right)(190 \text{ units}) = \$9,500/\text{yr}$$

Variations in lead time are the second major cause of uncertainty in inventory management, although lead times are generally more controllable than demand. Lead time uncertainty can also be handled by developing cumulative probability distributions of lead times for the given inventory item. The safety-stock level (in weeks of usage) required to guard against this uncertainty can be determined by referring to the cumulative distribution. This is done by using the cumulative distribution to find the lead time required to limit the stockout risk to a given level (LT_{ave}), and subtracting the average lead time (LT_{ave}) from LT_{SOR}.

$$SS = LT_{SOR} - LT_{ave} \tag{13.4}$$

Example 13.7 (*Constant demand, variable lead time*)   A product has a constant demand of 40 units per week and an average lead time of 5 weeks. However, lead time varies from 3 weeks to 10 weeks: it exceeds 8 weeks 10 percent of the time and exceeds 9 weeks only 5 percent of the time. How much safety should be carried to provide a 95 percent service level?

The required safety stock, in weeks of usage, is:

$$SS = LT_{SOR} - LT_{ave}$$
$$= 9\,wk - 5\,wk = 4\,wk$$

This is a safety stock of (40 units per week)(4 weeks) = 160 units.

STATISTICAL DISTRIBUTION METHODS: USE OF KNOWN DISTRIBUTIONS

Order Cycle Stockouts and Individual Unit Shortages. If the distribution of demand during lead time is symmetrical and unimodal, the normal distribution may satisfactorily describe it. In such cases, two statistical approaches are available for estimation of safety-stock requirements:

(a) *Order cycle stockouts* (i.e., using probabilities that demand will be exceeded during any order cycle—without regard to the specific number of units of shortage)

(b) *Individual unit shortages* (i.e., using probabilities based upon the possible number of units of shortage during a time period, such as monthly or annually)

We begin with order cycle stockouts (a), which involve relatively straightforward calculations based directly upon areas under the normal curve. Later examples show how calculations are made easier by the use of a table of normal distribution service levels. The individual unit shortage approach (b) necessitates that consideration be given to the probabilities of demand for discrete units. However, calculations are simplified by use of a table of unit normal loss functions.

(a) SERVICE LEVEL BASED UPON ORDER CYCLE STOCKOUTS

With a normally distributed demand during lead time, D_{LT}, an average demand would just use up the cycle (nonsafety stock) inventory during the lead time, as illustrated in Fig. 13-7.

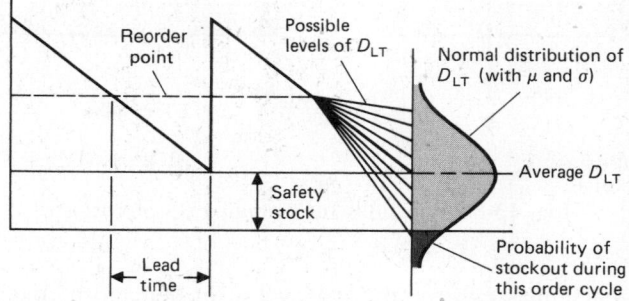

Fig. 13-7 Normal distribution of demand during lead time

If no safety stock were carried, we could expect the firm to run out of stock on about 50 percent of the order cycles (i.e., 50 percent service level). Safety stock increases the service level in proportion to the area under the normal curve that corresponds to the safety stock. We first use the standard normal distribution to illustrate the calculation of safety-stock requirements.

Example 13.8 (*Constant lead time*)   Demand for a product during its (constant) lead time is normally distributed with $\mu = 1,000$ units and $\sigma = 40$ units. How many units of SS are required to provide a service level of 97 percent on any given order cycle?

Using the normal distribution, the Z-value for an area of 97.0 percent $-$ 50 percent $=$ 47.0 percent is 1.88.

$$Z = \frac{x - \mu}{\sigma} \quad \text{or} \quad x = \mu + Z\sigma$$

$$x = 1{,}000 + 1.88(40) = 1{,}075$$

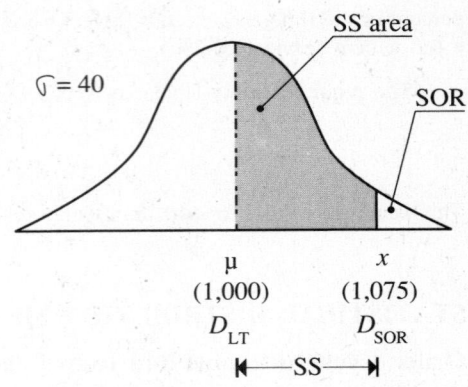

$$\text{SS} = D_{\text{SOR}} - D_{\text{LT}} = 1{,}075 - 1{,}000 = 75 \text{ units}$$

Computation of SS and ROP. In Eq. 13.1, we saw that $\text{ROP} = D_{\text{LT}} + \text{SS}$ where the expected demand during the lead time, D_{LT}, could be viewed as the product of the average demand per period, $\bar{d}$, multiplied by the length of the lead time, LT, so that $D_{\text{LT}} = \bar{d} \cdot \text{LT}$. From the normal distribution pattern (and as computed in Example 13.8), the safety stock is:

$$\text{SS} = Z\sigma_L \qquad\qquad (13.5)$$

where σ_L is the standard deviation of demand during lead time. Note that whenever the LT is less than the order cycle—as commonly occurs—σ_L can be computed directly from the demand values during the lead times of the several order cycles. See Fig. 13-8 (left side).

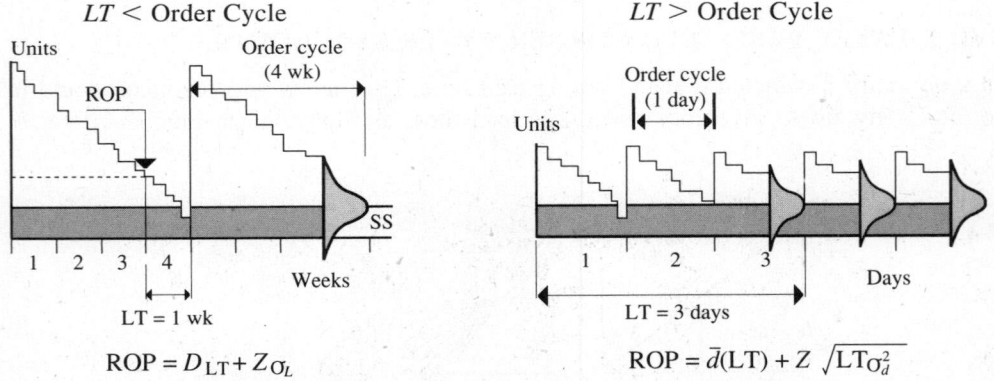

Fig. 13-8 Equations for comuting SS and ROP

When the lead time encompasses several order cycles (or demand periods), the standard deviation value must take account of the combined variability in demand over the several periods. See Fig. 13-8 (right side). Recalling that variances are additive, the standard deviation of demand during lead time, σ_L, can be computed by summing the variances of each demand period, σ_d^2 and taking the square root, i.e.,

$$\sigma_L = \sqrt{\sigma_{d1}^2 + \sigma_{d2}^2 + \cdots + \sigma_{dn}^2} = \sqrt{\Sigma \sigma_d^2}$$

However, given our assumption of constant lead times and assuming that demand in each period is independent of other periods, we can replace the Σ sign with the number of order cycles in the lead time, LT_c, and multiply that by a representative variance σ_d^2. Restating the equation for σ_L we have:

$$\sigma_L = \sqrt{\text{LT}_c \, \sigma_d^2} \qquad\qquad (\text{Also: } \sigma_L = \sqrt{\text{LT}_c} \, \sigma_d)$$

where σ_d represents the standard deviation of the daily demand or of the period demand rate. The ROP can thus be restated in a more general form as:

$$(LT > \text{order cycle}) \qquad \text{ROP} = \bar{d}(\text{LT}) + Z\sqrt{\text{LT}_c\,\sigma_d^2} \qquad\qquad (13.7)$$

Example 13.9 (*Constant lead time with LT < order cycle*)   Demand averages 50 units per week, with a standard deviation of 20 units per week. Lead time is constant at 1 week. Find (*a*) the standard deviation of demand during lead time, σ_L, and (*b*) the ROP for a service level of 98 percent. (*Note:* From normal curve, $Z = 2.05$ for 48 percent of area.)

(*a*) $\qquad \sigma_L = \sqrt{\text{LT}_c\,\sigma_d^2} = \sqrt{(1)(20)^2} = 20$ units

(*b*) $\quad \text{ROP} = D_{\text{LT}} + \text{SS} = D_{\text{LT}} + Z\sigma_L = 50 + (2.05)(20) = 91$ units

Note: Insofar as Eq. 13.7 is general in form, we could also use it to compute the ROP:

$$\text{ROP} = \bar{d}(\text{LT}) + Z\sqrt{\text{LT}_c\,\sigma_d^2} = (50)(1) + 2.05\sqrt{(1)(20)^2} = 91 \text{ units}$$

Example 13.10 (*Constant lead time with LT > order cycle*)   Demand averages 10 units per day, with a standard deviation of 4 units per day. Orders are placed daily and lead time is constant at 3 days. For a service level of 99 percent, find (*a*) the standard deviation of demand during lead time, σ_L, (*b*) the safety stock, and (*c*) the ROP.

(*a*) $\qquad \sigma_L = \sqrt{\text{LT}_c\,\sigma_d^2} = \sqrt{3(4)^2} = 7$ units

(*b*) $\qquad \text{SS} = Z\sqrt{\text{LT}_c\,\sigma_d^2} = 2.33\sqrt{3(4^2)} = 16$ units

(*c*) $\quad \text{ROP} = \bar{d}(\text{LT}) + Z\sqrt{\text{LT}_c\,\sigma_d^2}$

$\qquad\qquad = 10(3) + 2.33\sqrt{3(4^2)} = 30 + 16 = 46$ units

Variable Lead Time.   With constant demand and variable (but normally distributed) lead time, the ROP is:

$$\text{ROP} = d(\overline{\text{LT}}) + Z(d)\sigma_{\text{LT}} \qquad\qquad (13.8)$$

where d = constant demand rate, $\overline{\text{LT}}$ = average lead time, and σ_{LT} = standard deviation of lead time.

Example 13.11 (*Variable lead time, constant demand*)   A plant consumes 800 pounds of acid per day. Delivery time from the supplier is normally distributed with a mean of 14 days and standard deviation of 4 days. What should be the reorder point if the chemical company wishes to ensure a service level of 99 percent?

$$\text{ROP} = d\,\overline{\text{LT}} + Z(d)\sigma_{\text{LT}}$$
$$= 800(14) + 2.33(800)(4) = 11{,}200 + 7{,}456 = 18{,}656 \text{ pounds}$$

Variable Demand and Variable Lead Time.   When the rate of demand and the lead time are both variable, the two variances can be added to obtain a combined standard deviation of demand during lead time. If both demand and lead time are normally distributed, we have:

$$\sigma_{d\text{LT}} = \sqrt{\sum \sigma_d^2 + \sum \sigma_{LT}^2} = \sqrt{\overline{\text{LT}}\sigma_d^2 + \bar{d}^2\sigma_{\text{LT}}^2} \qquad\qquad (13.9)$$

The ROP is then:

$$\text{ROP} = \bar{d}(\overline{\text{LT}}) + Z\sqrt{\overline{\text{LT}}\sigma_d^2 + \bar{d}^2\sigma_{\text{LT}}^2} \qquad\qquad (13.10)$$

Example 13.12 (*Variable demand and variable lead time*) Demand for an industrial bearing is normally distributed with a mean of 40 units per day and $\sigma_d = 6$ units per day. Lead time from a supply shop is also normally distributed with a mean of 10 days and $\sigma_{LT} = 5$ days. Find (*a*) the amount of safety stock required and (*b*) the reorder point for a 95 percent service level.

(*a*) $$\text{SS} = Z\sqrt{\overline{\text{LT}}\sigma_d^2 + \bar{d}^2\sigma_{LT}^2} = 1.645\sqrt{10(6)^2 + (40)^2(5)^2} = 330 \text{ units}$$

(*b*) $$\text{ROP} = \bar{d}(\overline{\text{LT}}) + \text{SS} = 40(10) + 330 = 730 \text{ units}$$

Using Tables to Find Safety-Stock and Service Levels. Computations of safety-stock and service levels can be simplified by using the safety-stock level factors (SF) for normally distributed variables provided in Table 13-4. The factors are simply the number of standard (and mean absolute) deviations required to include the specified percentage of area under the normal curve cumulated in the positive direction (only). Both standard deviation (σ) and mean absolute deviation (MAD) values are provided. Recall that $\sigma \cong 1.25\,\text{MAD}$ and

$$\sigma = \sqrt{\frac{\Sigma(X-\mu)^2}{N}} \qquad \text{and} \qquad \text{MAD} = \frac{\Sigma|X-\mu|}{N}$$

Examples 13.13 and 13.14 illustrate the computation of SS amounts using values from the safety factor table of normally distributed variables. Example 13.13 also provides additional insight on assigning a value to the desired service level.

Although these examples relate to a constant lead time situation, where $\text{LT} \leq$ order cycle, the safety factors can be used with the other equations of the chapter as well—once the appropriate standard deviation has been calculated. The factors simply take the place of looking up values in the normal distribution table. They would, however, not be applicable to demands that follow other statistical patterns. Many demands have a normal distribution at the production-plant level and a Poisson distribution at the retail level. The Poisson distribution also has some applicability in estimating lead time.

Example 13.13 (*Computation of SS using σ*) A firm has a normal distribution of demand during a (constant) lead time, with $\sigma_L = 250$ units. The firm wants to provide 98 percent service. (*a*) How much safety stock should be carried? (*b*) If the demand during the lead time averages 1,200 units, what is the appropriate reorder point?

(*a*) $$\text{SS} = \text{SF}_\sigma(\sigma_L) = (2.05)(250) = 512 \text{ units}$$

(*b*) $$\text{ROP} = D_{LT} + \text{SS} = 1,200 + 512 = 1,712 \text{ units}$$

If the number of stockouts allowed per time period is designated, estimate the total number of reorder cycles. The service level is the number of cycles the stock is adequate as a percentage of the total number of cycles.

Example 13.14 (*Computation of SS using MAD*) A firm has a normally distributed forecast of demand, with $\text{MAD} = 60$ units during the fixed lead time of 1 week. It desires a service level that limits stockouts to one order cycle per year. (*a*) How much safety stock should be carried? (*b*) If D_{LT} averages 500 units, what is the appropriate reorder point?

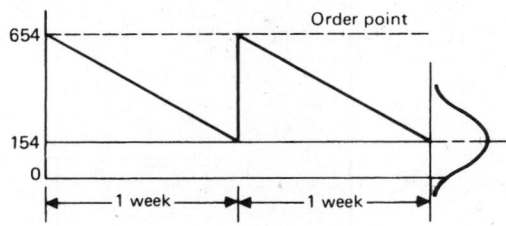

Table 13-4 Safety-Stock Factors for Normally Distributed Variables

$SS = SF_\sigma(\sigma)$ or $SS = SF_{MAD}(MAD)$ $ROP = D_{LT} + SS$		
	Safety Factor Using	
Service Level (percentage supplied without stockout)	Standard Deviation SF_σ	Mean Absolute Deviation SF_{MAD}
50.00	.00	.00
75.00	.67	.84
80.00	.84	1.05
84.13	1.00	1.25
85.00	1.04	1.30
89.44	1.25	1.56
90.00	1.28	1.60
93.32	1.50	1.88
94.00	1.56	1.95
94.52	1.60	2.00
95.00	1.65	2.06
96.00	1.75	2.19
97.00	1.88	2.35
97.72	2.00	2.50
98.00	2.05	2.56
98.61	2.20	2.75
99.00	2.33	2.91
99.18	2.40	3.00
99.38	2.50	3.13
99.50	2.57	3.20
99.60	2.65	3.31
99.70	2.75	3.44
99.80	2.88	3.60
99.86	3.00	3.75
99.90	3.09	3.85
99.93	3.20	4.00
99.99	4.00	5.00

Source: Adapted from G. W. Plossl and O.W. Wight, *Production and Inventory Control: Principles and Techniques*, 1967, p. 108. Reprinted by permission of Prentice-Hall, Englewood Cliffs, NJ.

(*a*) $$SS = SF_{MAD}(MAD)$$

where SF_{MAD} depends upon the service level:

$$1 \text{ week's supply/order} = 52 \text{ orders/yr}$$

$$1 \text{ stockout in } 52 = \frac{51}{52} \text{ in stock} = 98\% \text{ service}$$

Thus, $SF_{MAD} = 2.56$ (from Table 13-4)

$SS = 2.56(60) = 154$ units

(b) $ROP = D_{LT} + SS = 500 + 154 = 654$ units

(b) SERVICE LEVEL BASED UPON INDIVIDUAL UNIT SHORTAGES

Although the service level computations provide useful information about the likelihood of stockout during an order cycle, they do not address the amount of the shortage, i.e., whether 1 unit, 10 units, or 100 units. When the distribution of demand during lead time is normal, the expected *number* of units short during an order cycle, $E(n)$, can be estimated by using the unit normal loss function values contained in Table 13-5.

Table 13-5 Expected Units Short for Standard Deviations of Safety Stock*

Z	E(Z)	Z	E(Z)	Z	E(Z)	Z	E(Z)
−4.00	4.000	−.70	.843	.65	.155	1.50	.029
−3.50	3.500	−.60	.769	.70	.143	1.60	.023
−3.00	3.000	−.50	.698	.75	.131	1.65	.021
−2.50	2.502	−.40	.630	.80	.120	1.70	.018
−2.00	2.008	−.30	.567	.85	.110	1.75	.016
−1.90	1.911	−.20	.507	.90	.100	1.80	.014
−1.80	1.814	−.10	.451	.95	.092	1.85	.013
−1.70	1.718	0.00	.399	1.00	.083	1.90	.011
−1.60	1.623	.10	.351	1.05	.076	2.00	.008
−1.50	1.529	.20	.307	1.10	.069	2.10	.006
−1.40	1.437	.30	.267	1.15	.062	2.20	.005
−1.30	1.346	.35	.248	1.20	.056	2.30	.004
−1.20	1.256	.40	.230	1.25	.051	2.40	.003
−1.10	1.169	.45	.214	1.30	.046	2.50	.002
−1.00	1.083	.50	.198	1.35	.041	2.60	.001
−.90	1.000	.55	.183	1.40	.037	2.80	.001
−.80	.920	.60	.169	1.45	.033	3.00	.000

*This is an abbreviated table for illustrative purposes. Adapted from Robert G. Brown, *Decision Rules for Inventory Management* (New York: Holt Rinehart & Winston, 1967), pp. 95–103.

Computing Number of Units Short. The Z values in Table 13-5 should correspond to a designated service level and can be taken either from a normal distribution table or from the SF_σ or SF_{MAD} values of Table 13-4. The $E(Z)$ values then apply to a standardized (normal) distribution. They are the expected number of units short, if the mean and standard deviation of demand during lead time are 0 and 1, respectively—which is rarely the case. When σ_L is some value other than one, the $E(Z)$ values must be multiplied by σ_L to obtain the expected number of units short for the specific data being analyzed.

$$E(n) = E(Z)\sigma_L \qquad\qquad (13.11)$$

where $E(Z)$ = standardized number of units short

σ_L = standard deviation of demand during lead time

The total number of units short over a year, $E(N_{yr})$, can then be computed as the product of the expected (i.e., or average) number of units short during one order cycle, $E(n)$, times the number of order cycles, which is the annual D divided by the order quantity, Q.

$$E(N_{yr}) = E(n)\left(\frac{D}{Q}\right) \qquad (13.12)$$

Example 13.15 A product has a normally distributed demand during its lead time, with a mean of 300 units and standard deviation of 65 units. Assuming the firm carries enough safety stock to provide 90 percent service, what is the expected number of units of shortage during any given order cycle?

$$E(n) = E(Z)\sigma_L$$

where $E(Z)$ = standardized number for 90 percent service, which from Table 13-4 corresponds to $Z = 1.28$
= .046 (from Table 13-5) (*Note:* Interpolation between Z values 1.25 and 1.30 would be .048)

$$E(n) = E(Z)\sigma_L = .048(65) = 3 \text{ units short}$$

Example 13.16 The product in Example 13.15 has an annual demand of 3,600 units, and the firm orders in quantities of 900 units per order. What is the expected annual shortage for the service level of 90 percent?

$$E(N_{yr}) = E(n)\left(\frac{D}{Q}\right) \quad \text{where } E(n) = 3 \text{ units (from Example 13.15)}$$

$$= 3\left(\frac{3,600}{900}\right) = 12 \text{ units short per year}$$

Finding the Service Level. Previously, when we associated service levels with an order cycle, we assumed that any shortage constituted a stockout–even if most of the demand during the cycle was supplied from stock. The alternative method of defining the service level (on the basis of individual unit shortages) generally results in a higher annual service level.

For this analysis, if we let SL_{ind} represent the service level arising from satisfying a specified percentage of the demand of individual items from stock, then SOR_{ind} is the annual percent of shortage. Multiplying SOR_{ind} by the annual demand, D, yields the total number of items short per year. This is equivalent to the expected number of units of shortage during any given order cycle $[E(Z)\sigma_L]$ times the number of cycles per year $[D \div Q]$. Then $SL_{ind} = (1 - SOR_{ind})$:

$$\begin{pmatrix} \text{Annual \%} \\ \text{shortage} \end{pmatrix} \begin{pmatrix} \text{units} \\ \text{per yr} \end{pmatrix} = \begin{pmatrix} \text{units short} \\ \text{per order} \end{pmatrix} \begin{pmatrix} \text{orders} \\ \text{per yr} \end{pmatrix}$$

$$(SOR_{ind}) \qquad D \quad = \quad E(Z)\sigma_L \quad \left(\frac{D}{Q}\right)$$

Dividing through by annual demand, D, gives:

$$SOR_{ind} = \frac{E(Z)\sigma_L}{Q} \qquad (13.13)$$

Example 13.17 A product with an annual demand of $D = 3,600$ units is ordered in quantities of $Q = 900$ units. The distribution of demand during lead time is normal with $\sigma_L = 65$ units. If the service on an order cycle basis is .90, what is the annual service level, as based upon individual unit shortages, SL_{ind}?

From Table 13-5, for a 90 percent service level (per cycle), where $Z = 1.28$, then $E(Z) = .046$.

$$SOR_{ind} = \frac{E(Z)\sigma_L}{Q} = \frac{(.046)(65)}{900} = .003$$

Therefore $SL_{ind} = 1 - SOR_{ind} = 1.000 - .003 = .997$ (substantially higher than .90).

Finding SS and ROP. Service levels based upon the percentage of units that can be supplied from stocks, SL_{ind}, can be used to establish safety stocks and reorder points by using the equivalent value of the service level based upon the probability of a stockout during an order cycle, SL.

Example 13.18 A product has a normally distributed demand during its (constant) lead time, with a mean of 300 units per month and $\sigma_L = 65$ units and $Q = 900$ units. For an individual shortage service level of $SL_{ind} = .997$, what is (a) the amount of safety stock required and (b) the reorder point?

- (a) $SS = SF_\sigma(\sigma_L)$ where $SF_\sigma = 1.28$ (from Table 13-4, for an equivalent SL = 90.0% per Example 13.17)
$$= 1.28(65) = 83 \text{ units}$$

- (b) $ROP = D_{LT} + SS = 300 + 83 = 383 \text{ units}$

Example 13.19 Suppose an individual shortage service level of $SL_{ind} = 90.0$ percent is deemed adequate for the product of Example 13.18. (Recall that $D_{LT} = 300$ units/mo, $\sigma_L = 65$ units, $Q = 900$ units.) (a) By how much might the safety stock and order point be reduced from that required for $SL_{ind} = 99.7$? (b) Comment on the impact of this difference.

- (a) For $SL_{ind} = 99.7\%$, SOR = .03 $SOR = \dfrac{E(Z)\sigma_L}{Q}$

  $\therefore E(Z) = \dfrac{SOR(Q)}{\sigma_L} = \dfrac{.03(900)}{65} = .415$

  This corresponds to Z value of 0 (per Table 13-5)

  $\therefore ROP = D_{LT} + Z\sigma_L = 300 + 0(65) = 300 \text{ units}$.

  Reduction in SS is from 83 to zero, and so the ROP is also 83 units lower.

- (b) This shows that even with no safety stock, the SL_{ind} can be substantially higher (97%) than the order cycle service level, which would be 50% with no safety stock.

ORDER QUANTITIES UNDER PERIODIC REVIEW SYSTEMS

Recall that with periodic review systems, inventory is counted and orders for variable quantities are generated at fixed intervals. Under conditions of *variable demand and constant lead time*, the order quantity derived from a fixed-period model differs from that of a fixed-quantity model.

In the fixed-period model (i.e., without continuous monitoring) the safety stock must protect against stockout over the time interval between placing orders (TI) plus the lead time (LT) from when an order is placed until it is received. The order quantity must then take account of the demand during this period of exposure to stockout plus the designated safety stock needs less any inventory already on hand or on order (I_{oh}). The order size, Q, for a fixed-period model is thus:

$$\begin{array}{c}\text{Order} \\ \text{quantity}\end{array} = \begin{bmatrix}\text{expected demand} \\ \text{during (TI + LT)}\end{bmatrix} + \begin{pmatrix}\text{safety} \\ \text{stock}\end{pmatrix} - \begin{pmatrix}\text{inventory on hand and} \\ \text{on order at reorder time}\end{pmatrix}$$

$$Q = \bar{d}(TI + LT) + Z\sigma_d\sqrt{TI + LT} - I_{oh} \qquad (13.14)$$

where $\bar{d}$ = average demand per period

$\quad\quad\quad Z$ = value from normal distribution or Table 13-4

$\quad\quad\quad \sigma_d$ = standard deviation of demand during (normally distributed) review and lead time

Example 13.20 (*With SL applied to order cycle*) Demand averages 50 units per day and is normally distributed over the review and lead time with $\sigma_d = 6$ units and LT = 5 days. The time interval between orders is 15 days, and 80 units are on hand at the reorder time. A 98.0 percent service level over the order cycle is desired. How many units should be ordered?

TI = 15 days and Z = 2.05 for 98 percent service level.

$$Q = \bar{d}(\text{TI} + \text{LT}) + Z\sigma_d \sqrt{\text{TI} + \text{LT}} - I_{\text{oh}}$$

$$= 50(15 + 5) + 2.05(6)\sqrt{15 + 5} - 80$$

$$= 1{,}000 + 55 - 80 = 975 \text{ units}$$

Example 13.21 (*With SL_{ind}*)   Use the data of Example 13.20 except let the service level be based upon supplying 98.0 percent of the individual unit demand from stock, SL_{ind} (i.e., rather than the SL that is associated with an order cycle). Find the order quantity.

Note: We can use the same equation (13.14) for calculation of the order quantity, except the Z-value must correspond to the expected units short from 13-5. The expected units short, $E(Z)$, can be computed from the equation:

$$E(Z) = \frac{\bar{d}(\text{TI})(\text{SOR}_{\text{ind}})}{\sigma_{\text{TI+LT}}} \qquad (13.15)$$

where $\text{SOR}_{\text{ind}} = 1 - \text{SL}_{\text{ind}} = 1.00 - .980 = .020$

$\sigma_{\text{TI+LT}} = \sigma_d \sqrt{\text{TI} + \text{LT}} = 6\sqrt{15 + 5} = 26.83$

$$E(Z) = \frac{\bar{d}(\text{TI})(\text{SOR}_{\text{ind}})}{\sigma_{\text{TI+LT}}} = \frac{50(15)(.02)}{26.83} = .559$$

From Table 13-5, for $E(Z) = .56$, we find the value $Z = -.30$.

$$Q = \bar{d}(\text{TI} + \text{LT}) + Z\sigma_d \sqrt{\text{TI} + \text{LT}} - I_{\text{oh}}$$

$$= 50(15 + 5) - .30(26.83) - 80$$

$$= 1{,}000 - 8 - 80 = 912 \text{ units}$$

Note that the order quantity, as calculated on an expected number of shortage basis, is somewhat less than that derived from service level based on the order cycle (review and lead time).

Questions and Solved Problems

METHODS OF HANDLING INVENTORY UNCERTAINTIES

13.1   Why is inventory control such a dominant consideration in production operations?

First, inventories (any stock of materials held for future use) are essential to both manufacturing and customer service activities. *Second*, they represent a large investment—averaging over half of the cost of manufactured products. *Third*, close control takes considerable effort; inventories are part of an interrelated production system. *Fourth*, inventory policies can "make or break" an organization. Poor inventory control has been a major cause of the failure of many organizations.

13.2   How is inventory control accomplished?

Methods vary widely. However, the (*a*) *monitoring* of inventory levels is essential to all methods, along with control over the (*b*) *frequency* and (*c*) *size* of reorders, and a careful determination of (*d*) the amounts (*safety stock*) needed to offset uncertainties.

Some firms employ continuous review (fixed-quantity) and periodic review (fixed-period) systems—especially for managing inventories that must accommodate uncertainties associated with variable demands from customers and lead times from suppliers. Material requirements planning (MRP), just-in-time (JIT), and kanban methods are especially useful in manufacturing environments.

13.3   A producer of SUN-STOP suntan lotion uses 400 gallons per week of a chemical that is ordered in EOQ quantities of 5,000 gallons at a quantity discount cost of \$3.75 per gallon. Procurement lead time is 2 weeks, and a safety stock of 200 gallons is maintained. Storage cost is \$.01 per gallon-week. Find (a) the maximum inventory on hand (on the average), (b) the average inventory maintained, and (c) the order point (in units).

(a)   Maximum inventory: $I_{max} =$ safety stock + EOQ $= 200 + 5{,}000 = 5{,}200$ gal

(b)   Average inventory: $I_{ave} = \dfrac{I_{max} + I_{min}}{2} = \dfrac{5{,}200 + 200}{2} = 2{,}700$ gal

(c)   Order point: $OP = D_{LT} + SS = (400)(2) + 200 = 1{,}000$ units (also called ROP)

13.4   (*Informal decision rules*) Akita Distributors has experienced a demand that averages $\bar{d} = 21$ units per day with $d_{max} = 42$ and $d_{min} = 10$. Lead times for reorders vary from 2 to 8 days. What amount of safety stock would be carried under a decision rule of (a) largest daily usage times longest lead time, (b) square root of average demand during average lead time, and (c) average demand during lead time plus a 40 percent safety factor?

(a) $SS = (d_{max})(LT_{max}) = (40 \text{ units/day})(8 \text{ days}) = 320$ units

(b) $SS = \sqrt{(\bar{d})(LT_{ave})}$ where $LT_{ave} = (2 + 8) \div 2 = 5$ days
              $= \sqrt{(21 \text{ units/day})(5 \text{ days})} = \sqrt{105} = 10.25$ units (Use 11 units.)

(c) $SS = D_{LT} + 40\% = 105 + .40(105) = 147$ units

13.5   (*Multiple-period model*) St. Jude Medical Center has an annual demand for antiseptic gel averaging 6,000 pints and orders in quantities of 2,000 pints per order. Carrying costs are \$18 per pint per year, and the center estimates its costs of running out of stock at \$100 per pint. For these conditions, what is the resulting (a) stockout risk and (b) service level?

(a) $SOR = \dfrac{H}{C_{so}}\left(\dfrac{Q}{D}\right) = \dfrac{18}{100}\left(\dfrac{2{,}000}{6{,}000}\right) = .06$

(b) $SL = 1 - SOR = 1.00 - .06 = .94$

13.6   (*Multiple-period model with SS calculation*) Farm Supply Co. has a demand averaging 800 brooder houses per year and orders from a manufacturer in quantities of 80 units per order. Demand during lead time is normal with $\sigma = 12$. Carrying costs are \$20 per unit and stockout costs are estimated at \$50 per unit. (a) What probability of SOR should be used to determine the optimum amount of safety stock? (b) How much safety stock would be justified by the above costs—with an order cycle based SL?

(a) $SOR = \dfrac{H}{C_{so}}\left(\dfrac{Q}{D}\right) = \dfrac{20}{50}\left(\dfrac{80}{800}\right) = .04$

(b) $SL = 1 - SOR = 1.00 - .04 = .96$

From Table 13-4, $SS = SF_{\sigma}(\sigma) = 1.75(12) = 21$ units.

13.7   (*EOQ with multiple-period model, individual stockout costs, and order cycle stockout*) Supermarket Supply Co. distributes grocery products to customers in Mexico. Demand for canned corn averages 280 cases per month. Lead time to obtain a shipment is about a month, and demand during the lead time is normally distributed with a standard deviation of 60 cases. The company estimates its ordering cost at \$8.00 per order, carrying cost at \$2.40 per case per year, and stockout cost at \$2.00 per case. Find (*a*) the EOQ disregarding stockout costs, (*b*) the number of orders per year if the EOQ is used, (*c*) the optimum level of SOR based upon the 1-month order cycle, and (*d*) the reorder point.

(*a*) $\text{EOQ} = \sqrt{\dfrac{2DS}{H}} = \sqrt{\dfrac{2(280 \times 12)(8)}{2.40}} = 150 \text{ cases/order}$

(*b*) $\dfrac{\text{Orders}}{\text{Year}} = \dfrac{D}{Q} = \dfrac{(280 \times 12) \text{ cases/yr}}{150 \text{ cases/order}} = 22.4 \text{ orders/yr}$

(*c*) $\text{SOR} = \dfrac{H}{C_{so}} \left(\dfrac{Q}{D} \right) = \left(\dfrac{\$2.40/\text{case-yr}}{2.00/\text{case-order}} \right) \left(\dfrac{150 \text{ cases/order}}{3,360 \text{ cases/yr}} \right) = .054, \text{ or } 5.4\% \text{ risk of stockout}$

Thus, $\text{SL} = 1.00 - .054 = .946 = 94.6$ percent.

(*d*) $\text{ROP} = D_{\text{LT}} + \text{SS}$

where $\text{SS} = \text{SF}_\sigma(\sigma_{\text{LT}}) = 1.56(60) = 94 \text{ cases}$

$\text{ROP} = (280 \text{ cases/month})(1 \text{ month LT}) + 94 \text{ cases} = 374 \text{ cases}$

13.8   Using the data of Prob. 13.7, assume the 94.6 percent SL for order cycle stockouts found in part (*c*) was deemed to be sufficient for individual unit shortages (i.e, $\text{SOR}_{\text{ind}} = .054$). What would be the revised ROP?

Knowing SOR_{ind}, we can restate Eq. 13.13 to find $E(Z)$ and use it to obtain the corresponding Z-value from Table 13-5. Then the $\text{SS} = Z(\sigma_L)$

$$E(Z) = \dfrac{\text{SOR}_{\text{ind}}(Q)}{\sigma_L} = \dfrac{(.054)(150)}{60} = .135$$

From Table 13-5, $Z = .73$ (by interpolation).

$$\text{Revised SS} = Z(\sigma_L) = (.73)(60) = 44 \text{ units}$$
$$\text{Revised ROP} = D_{\text{LT}} + \text{SS} = 280 + 44 = 324 \text{ cases}$$

USE OF EMPIRICAL DATA TO SET SAFETY-STOCK LEVELS

13.9   Given the data from Example 13.6, suppose the manager is willing to allocate only \$3,000 per year to the carrying of safety stock for the \$250 item. For what percentage of order cycles can he or she expect to run out of stock?

We can compute how much safety stock the \$3,000 would fund by dividing the \$3,000 by the carrying cost per unit-year.

$$\text{SS} = \dfrac{\$3,000 \text{ allocated/yr}}{\$250/\text{unit } (20\%/\text{yr})} = 60 \text{ units}$$

The stockout risk corresponding to $\text{SS} = 60$ units is:

$$\text{SS} = D_{\text{SOR}} - D_{\text{ave}}$$
$$D_{\text{SOR}} = \text{SS} + D_{\text{ave}}$$

where $D_{\text{ave}} = 240 \text{ units (from Fig. 13-6)}$.

$$D_{\text{SOR}} = 60 + 240 = 300 \text{ units}$$

From the cumulative distribution (Fig. 13-6) a demand of 300 units corresponds to a percentage value of approximately 29 percent. Therefore the manager may expect to run out of stock on approximately 29 percent of the order cycles. That is, if the firm places an order each week, it may run out of stock on $(.29)(52) \cong 15$ occasions. Knowing this, the manager may want to reconsider the $3,000 allocation.

USE OF KNOWN DISTRIBUTIONS TO SET SAFETY-STOCK LEVELS

13.10 (*LT less than order cycle*) A firm has a normally distributed demand, with $D_{LT} = 500$ units, and MAD = 60 units, during the fixed lead time of 1 week, and it orders a 5-week supply. It wishes to limit stockouts to one order cycle per year. (*a*) How much safety stock should be carried, and (*b*) what is the appropriate order point?

Because the uncertainty is confined to the latter phase of a single demand cycle, no special adjustments are required.

(*a*)   A 5-week supply results in $52/5 = 10.4$ orders per year. Being out of stock one time results in $9.4/10.4 = 90$ percent SL. From Table 13-4, $SF_{MAD} = 1.60$. Therefore,

$$SS = SF_{MAD}(MAD) = 1.60(60) = 96 \text{ units}$$

(*b*) $OP = D_{LT} + SS = 500 + 96 = 596$ units

See Fig. 13-9.

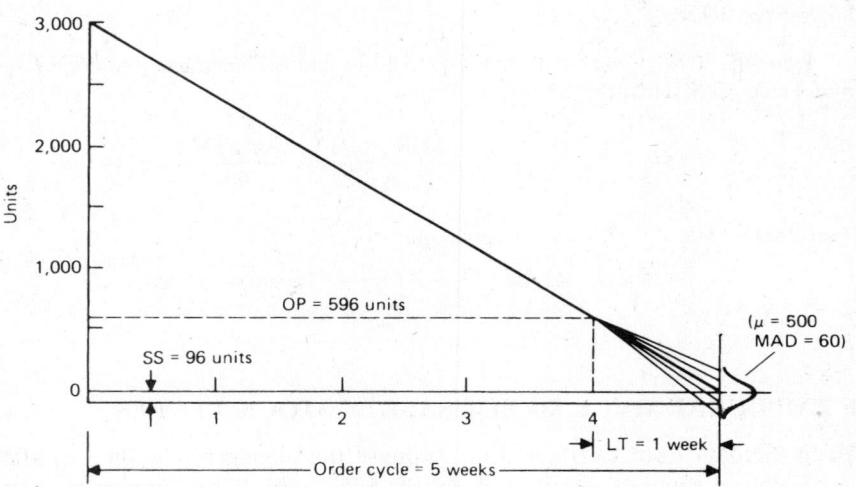

Fig. 13-9   Product with 5-week order cycle and 1-week lead time (90 percent SL)

13.11 (*LT equals order cycle*) Hospital Supply Co. obtains surgical masks from a factory in Chicago and sells them to hospitals on a year-round basis. Demand during the 1-month lead time is normally distributed with a mean of 30 cases per month and MAD equal to 6 cases.

(*a*) How much safety stock should Hospital Supply Co. carry to ensure that demand is met on 90 percent of the order cycles? (*b*) What should be the reorder point for the service level used in (*a*)? (*c*) Suppose the masks cost $100 per case, and the carrying cost is 20 percent per year. What level of service would result from allocating $400 per year to carry safety stock?

(*a*) $SS = SF_{MAD}(MAD) = 1.60(6) = 9.6$, say 10 cases

(*b*) $ROP = D_{LT} + SS = 30 + 10 = 40$ cases

(*c*)   Carrying cost/unit = $(.20)(\$100/\text{unit}) = \$20/\text{unit-yr}$

Therefore, $400 would fund a safety stock of

$$\frac{\$400/yr}{\$20/unit/yr} = 20 \text{ units}$$

From $\qquad\qquad\qquad\qquad\qquad SS = SF_{MAD}(MAD)$

we have $\qquad\qquad\qquad\qquad SF_{MAD} = SS/MAD = 20/6 = 3.33$

From Table 13-4 of safety-stock-level factors for normally distributed variables, the service level is

$$SL = 99.6\%$$

13.12 (*LT greater than order cycle*) A tire distributor has a weekly order cycle and has experienced a normally distributed demand with a mean of 40 units per week and a standard deviation of 5 units per week. Lead time is constant at 6 weeks, and each demand is independent. A 90 percent service level is desired. (*a*) How much safety stock should be carried? (*b*) How much stock is on hand when the reorder is placed?

(*a*) $\qquad\qquad\qquad\qquad\qquad SS = SF_\sigma(\sigma_L)$

where $\quad SF_\sigma$ for 90% = 1.28

$$\sigma_L = \sqrt{LT_c\sigma_d^2} = \sqrt{(6)5^2} = 12.25 \text{ units.}$$

Thus, $SS = (1.28)(12.25) = 15.68$, use 16 units.

(*b*) $\qquad\qquad\qquad ROP = \bar{d}(LT) + SS = 40(1) + 16 = 56 \text{ units}$

Supplementary Questions and Problems

13.13 Using the appropriate symbols, (*a*) identify the interrelated variables of an inventory control system, and (*b*) write an expression for the reorder point that incorporates the standard normal deviate, Z.
Ans. (*a*) Q, LT, SS, and ROP (*b*) $ROP = \bar{d}(LT) + Z\sigma_L$

13.14 A firm with a steady demand orders a raw material in quantities of 5 tons per order from a supplier who always delivers in 30 days. The price is $250 per ton, and the carrying and storage cost is 35 percent per year. Find (*a*) I_{max}, (*b*) I_{ave}, and (*c*) the annual carrying and storage cost.
Ans. (*a*) 5 tons (*b*) 2.5 tons (*c*) $219/year

13.15 Distinguish between the equations (*a*) $ROP = D_{LT} + Z\sigma_L$ and (*b*) $ROP = \bar{d}(LT) + Z\sqrt{LT_c\sigma_d^2}$.
Ans. They are simply two ways of stating the same thing because the demand during lead time, $D_{LT} = \bar{d}(LT)$, and the safety stock can be computed as $Z\sigma_L$ or as $Z\sqrt{LT_c\sigma_d^2}$. The second equation (*b*) is more general, however, for in addition to applying to situations where LT< order cycle, it can be used directly in situations where LT> order cycle.

13.16 Distinguish between (*a*) σ, (*b*) σ_L, (*c*) σ_d, and (*d*) σ_d^2.
Ans. (*a*) σ is the symbol for standard deviation and can apply to whatever variable is being analyzed, such as demand or even demand during lead time. (*b*) σ_L is the one (combined) value of the standard deviation of demand during lead time, whereas (*c*) σ_d is the standard deviation of an individual (e.g., daily) demand period during the lead time. (*d*) σ_d^2 is the variance of an individual demand period.

13.17   Demand for a product averages 180 units per week. The firm receives its deliveries 3 weeks after placing
an order and maintains a safety stock of 200 units. What is the reorder point?   *Ans.*   740 units

13.18   A medical supply product has a demand that varies between 8,000 and 11,500 units per week, with a mean
of 9,400 units per week. Lead time averages 6 weeks, but is occasionally as long as 10 weeks. How much
safety stock should be maintained under an informal decision rule of (*a*) square root of D_{LT}, (*b*) $D_{LT} + 30$
percent, and (*c*) largest weekly demand times longest lead time?
Ans.   (*a*) 238 units (*b*) 73,320 units (*c*) 115,000 units

13.19   A home products firm has a nursery item that they feel could have a demand of 300, 500, or 1,000 units
with probabilities of .2, .3, and .5, respectively. The table below shows the profit they could expect from
stocking 300, 500, or 1,000 units. What is the optimal stocking level based upon an expected value
approach?   *Ans.*   Stock 500 with expected payoff of $920

	Profit from Levels of Demand of		
	300 units	500 units	1,000 units
Stock 300	600	600	600
Stock 500	−200	1,200	1,200
Stock 1,000	−2,200	−1,000	2,000

13.20   What is the (*a*) multiple-period SOR and (*b*) SL associated with a product that has $D = 2,200$ units per
year, $Q = 500$ units per order, $H = \$80$ per unit per year, and $C_{so} = \$600$?   *Ans.*   (*a*) .03 (*b*) .97

13.21   *Multiple-period model with stockout costs.* Muffler Installations, Inc., has an annual demand of about 7,500
mufflers per year. Demand during the 2-week lead time is normally distributed with a mean of 300 and
standard deviation of 50. The firm estimates its ordering cost at $20 per order, holding cost at $10 per
unit-year, and stockout cost at $15 for each lost muffler sale. The operations manager would like to
establish a fixed-quantity inventory system that would carry an optimal amount of safety stock, i.e., where
the cost of adding one more unit of stock just equals the expected gain from adding that unit. (*a*) How
many mufflers should be ordered at one time? (*b*) How much safety stock should be carried? (*c*) What
reorder point should the company use?
Ans.   (*a*) 173 units (*b*) SOR = .0154, SS = 108 (*c*) 408 units

STATISTICAL DISTRIBUTION METHODS: USE OF EMPIRICAL DATA

13.22   A farm implement company manager has plotted the following histogram and cumulative distributions of
demand for a hydraulic lift that they sell. It can be obtained from a supplier in 3 weeks at a cost of $400 per
unit. Assume a 20 percent charge for carrying inventory.

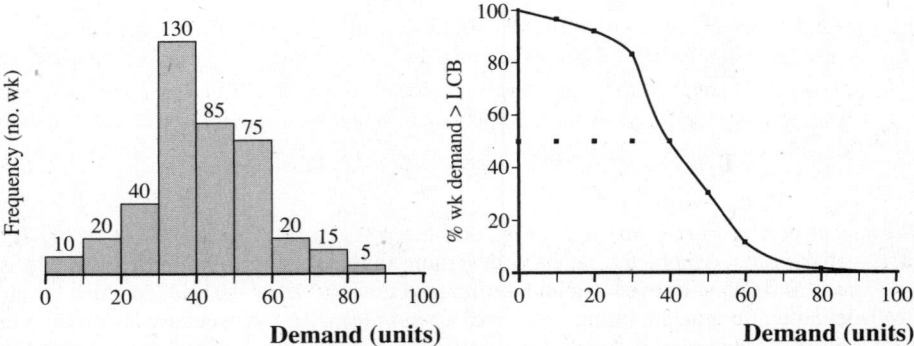

Fig. 13-10   Histogram and cumulative distribution of demand for lift

(a) How much safety stock would be required to satisfy weekly demand during 90 percent of the order cycles? (b) What is the annual cost of carrying that amount of safety stock? (c) Compute the reorder point that would include the required safety stock.   *Ans.*   (a) 20 units (b) $1,600 per year (c) 140 units

13.23   A Kansas City feedlot operator supplies beef to several meat-packing plants. The steers have an average value of $450, and the operator finances them through a local bank, paying 10 percent interest on the borrowed funds. The data given below represents the weekly demand over the past 100 weeks.

Weekly demand (# steers)	0 < 100	100 < 200	200 < 300	300 < 400	400 < 500
Frequency (# wk)	10	35	40	10	5

(a) Prepare a histogram of the frequency distribution of demand. (b) Graph the cumulative distribution of demand. The feedlot operator wishes to keep enough stock in the lot to supply weekly demand 90 percent of the time. Feed costs to maintain a steer at a prescribed weight are $5 per week. (c) What level of safety stock (that is, how many steers) should the operator carry to provide the 90 percent service level? (d) What is the annual cost of carrying this safety stock?

Ans.   (a) and (b) should be of sufficient size and accuracy to yield reasonably accurate values for the 50 percent and 10 percent demand levels. Values may vary, but should be around 130 to 150 steers for (c), and from $39,650 to $45,750 for (d).

13.24   Data on the distribution of lead times for a pump component were collected as shown in Table 13-6. Management would like to set safety-stock levels that will limit the stockout risk to 10 percent. (a) Graph the cumulative distribution. (b) How many weeks of safety stock are required to provide the desired service level?

Table 13-6

Lead Time (weeks)	Frequency of Occurrence
0 < 1	10
1 < 2	20
2 < 3	70
3 < 4	40
4 < 5	30
5 < 6	10
6 < 7	10
7 < 8	10
	200

Ans.   (a) Graph should have weeks on x axis and the percentage of time the lower boundary is exceeded on the y axis (b) 3

SERVICE LEVELS BASED UPON ORDER CYCLE STOCKOUTS

13.25   A firm has adopted a policy of limiting stockouts to "three order cycles per year" and places orders every 4 weeks (assume 52 weeks per year). What is the desired service level?   *Ans.*   77 percent

13.26   Webster Machine orders inventory in EOQs of 600 units every 3 weeks. There is 1 week between ordering and receiving the goods. If usage is constant, and its order point is 450 units, how much safety stock does it keep?   *Ans.* $D_{LT} = 200$ units per week, so SS = 250 units.

13.27 Assuming D_{LT} is normally distributed with mean = 240 units and MAD = 50, what would be the amount of SS required for an order cycle service level of 85 percent?   *Ans.*   65 units

13.28 Demand for an item is normally distributed during its 1-week lead time with $d = 200$ units per week and $\sigma_L = 15$ units. What order point should be used if the firm wishes to provide a 97 percent service level?   *Ans.*   228 units

13.29 Madrid Metal Products Co. orders steel sheets every 3 weeks (16 times per year) in quantities of 50 sheets per order. Demand during its 3-week lead time is normally distributed with MAD = 6 sheets. How much safety stock should it carry if it wishes to limit stockouts to one order cycle per year?
Ans.   SL = 15/16 = .9375. SS = (1.95)(6) = 12 sheets

13.30 San Palo Building Center has an annual demand of approximately 10,000 sheets of 1/2″ plywood. (At 250 working days per year, this is an average demand of 40 sheets per day.) The firm orders in quantities of 500 sheets at a time, and demand during the 10-day lead time is normally distributed with $\sigma_L = 15$ sheets. Management wishes to carry an amount of safety stock that will (theoretically) give them only one stockout per year. (*a*) How many orders per year are placed? (*b*) How much safety stock should be carried? (*c*) What reorder point should be used?
Ans.   (*a*) 20 orders per year (*b*) For SL of 95 percent, SS = 25 sheets (*c*) 425 sheets

13.31 A manufacturer of water filters purchases components in EOQs of 850 units per order. The total need (demand) averages 12,000 components per year, and MAD = 32 units during the lead time. If the manufacturer carries a safety stock of 80 units, what service level does this give the firm? Assume a normal distribution of D_{LT}.   *Ans.*   97.72 percent

13.32 An aluminium firm purchases coke from a foreign supplier who takes 4 weeks to deliver it. The company uses 800 tons of coke per week and places an order every 4 weeks. Demand during the 4-week lead time is normally distributed with a mean absolute deviation of MAD = 600 tons. The company wishes to carry enough safety stock to limit their stockouts to one order cycle over a 2-year period (104 weeks). What order point should they use?   *Ans.*   4,520 tons (from 4,514 to 4,532, depending upon rounding)

13.33 The Hotel-Restaurant Supply Co. orders potatoes in units of 500 bags per order and receives them 10 days later. Its deliveries and usage average 20 bags per day, and it maintains an extra rotating stock of 40 bags to be sure it does not run out of stock. Assume that the demand is normally distributed during the lead time, with $\sigma = 16$ bags. (*a*) What is the lead time? (*b*) What is the order point? (*c*) What service level does the safety stock provide? (*d*) Suppose the firm's management felt that running out of stock during two order cycles of the year were acceptable. By how much could the safety stock be reduced? Assume that there are 250 working days per year.   *Ans.*   (*a*) 10 days (*b*) 240 bags (*c*) 99.38 percent (*d*) 26 or 27 bags

SERVICE LEVELS BASED UPON INDIVIDUAL UNIT SHORTAGES

13.34 Given a normally distributed demand during lead time of 120 units, $\sigma_L = 30$ units, and the firm carries sufficient safety stock to provide 97 percent service. Find the expected number of units of shortage during an order cycle.   *Ans.*   .33 unit

13.35 For the data in the previous problem, the annual demand is 1,440 units and the firm orders in quantities of 240 units per order. Find the expected annual shortage for the service level of 97 percent.
Ans.   2 units short per year

13.36 A product has an annual demand of $D = 1,440$ units and is ordered in quantities of $Q = 240$ units. The distribution of D_{LT} is normal with $\sigma_L = 30$ units. What service level on an individual unit shortage basis, SL_{ind}, would be equivalent to a 97 percent SL on an order cycle basis?
Ans. From Table 13-5, for $Z = 1.88$, $E(Z) = .011$. Then SOR $= .0014$ and $SL_{ind} = .9986$.

13.37 A grocery item has a normally distributed demand during a constant lead time where $D_{LT} = 240$ units per month and $\sigma_L = 90$ units, $Q = 280$ units. Management wishes to have an individual unit shortage service level of 98 percent. Find (*a*) the SS required and (*b*) the ROP.
Ans. (*a*) $E(Z) = \dfrac{SOR\,(Q)}{\sigma_L} = \dfrac{(.02)(280)}{90} = .062$ $\therefore Z = 1.15$ (Table 13-5), $SL_{\sigma_L} = 87.2\%$ (Table 13-4)
and $SS = SF_\sigma \sigma_L = (1.15)(90) = 104$ units. (*b*) ROP $= D_{LT} + SS = 240 + 104 = 344$ units.

ORDER QUANTITIES UNDER PERIODIC REVIEW SYSTEMS

13.38 An electronic product has a normally distributed demand that averages $\bar{d} = 40$ units per day with $\sigma_d = 15$ units per day and LT $= 8$ days. The time interval between orders is 16 days, and 60 units are on hand at the reorder time. Assuming a 99.0 percent service level over the order cycle is desired, how many units should be ordered at one time? *Ans.* $Q = 960 + 171 - 60 = 1071$

Chapter 14

Material Requirements Planning: MRP and CRP

MRP AND CRP OBJECTIVES

The demand for a finished good tends to be independent and relatively stable (see Chap. 13). However, firms typically make more than one product on the same facilities, so production is generally done in lots, e.g., of different end items or models. The quantities and delivery times for the materials needed to make those end items are determined by the production schedule. Chapter 14 concerns those dependent demand items—which constitute most of the items used by a firm.

Question: What is material requirements planning?

Material requirements planning (MRP) is a computer-based technique for determining the quantity and timing for the acquisition of dependent demand items needed to satisfy the master schedule requirements.

By identifying precisely what, how many, and when components are needed, MRP systems are able to reduce inventory costs, improve scheduling effectiveness, and respond quickly to market changes.

Question: What is capacity requirements planning (CRP)?

Capacity requirements planning (CRP) is the process of determining what personnel and equipment capacities (times) are needed to meet the production objectives embodied in the master schedule and the material requirements plan.

Whereas MRP focuses upon the priorities of *materials*, CRP focuses primarily upon *time*. Although both MRP and CRP can be done manually and in isolation, they are typically integrated within a computerized system, and CRP (as well as production activity control) functions are often assumed to be included within the concept of ''an MRP system.'' Computerized MRP systems can effectively manage the flow of thousands of components throughout a manufacturing facility. Figure 14-1 defines some of the terminology used to describe the functioning of MRP systems.

- *MRP.* A technique for determining the quantity and timing dependent demand items.
- *Dependent demand.* Demand for components that is derived from the demand for other items.
- *Parent and component items.* A parent is an assembly made up of basic parts, or *components*. The parent of one subgroup may be a component of a higher-level parent.
- *Bill of materials.* A listing of all components (subassemblies and materials) that go into an assembled item. It frequently includes the part numbers and quantity required per assembly.
- *Level code.* The level on which an item occurs in the structure, or bill-of-materials format.
- *Requirements explosion.* The breaking down (exploding) of parent items into component parts that can be individually planned and scheduled.
- *Time phasing.* Scheduling to produce or receive an appropriate amount (lot) of material so that it will be available in the time periods when needed—not before or after.
- *Time bucket.* The time period used for planning purposes in MRP—usually a week.
- *Lot size.* The quantity of items required for an order. The order may be either purchased from a vendor or produced in-house. *Lot sizing* is the process of specifying the order size.
- *Lead-time offset.* The supply time, or number of time buckets between releasing an order and receiving the materials.

Fig. 14-1   MRP terminology

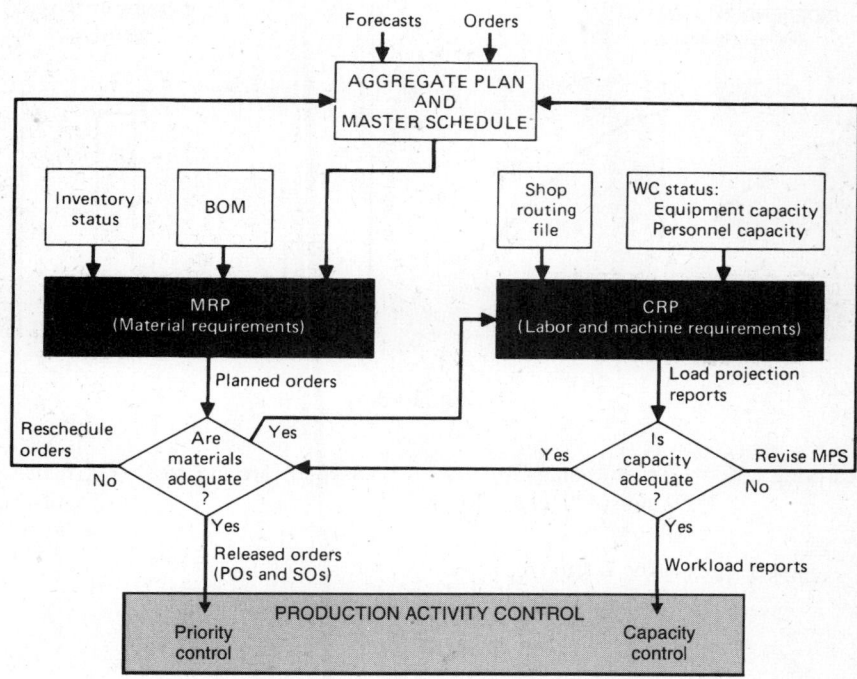

Fig. 14-2   Material and capacity planning flowchart

Figure 14-2 describes MRP and CRP activities in schematic form. Forecasts and orders are combined in the production plan, which is formalized in the master production schedule (MPS). The MPS, along with a bill-of-material (BOM) file and inventory status information, is used to formulate the material-requirements plan. The MRP determines what components are needed and when they should be ordered from an outside vendor or produced in-house. The CRP function translates the MRP decisions into hours of capacity (time) needed. If materials, equipment, and personnel are adequate, orders are released and the workload is assigned to the various work centers.

TIME-PHASING CONCEPTS

End items, such as TV sets, have an *independent demand* that is closely linked to the ongoing needs of consumers. It is random but relatively constant. *Dependent* demand is linked more closely to the production process itself. Many firms use the same facilities to produce different end items because it is economical to produce large lots once the set-up cost is incurred. The components that go into a TV set, such as 24-inch picture tubes, have a dependent demand that is governed by the lot size and model of TV set that happen to be in production at any one time. Dependent demand is predictable, but "lumpy."

MRP systems compute material requirements and specify when orders should be released so that materials arrive exactly when needed. The process of scheduling the receipt of inventory as needed over time is *time phasing*.

Example 14.1   (*a*) Use a sketch to illustrate the difference between on-hand inventory levels under independent and under dependent demand. (*b*) Suppose the use of traditional order point techniques for a component resulted in an average inventory of 80 units. How much carrying cost would be saved by time phasing if the average inventory dropped to 15 units? Assume an item value of $20 and a carrying charge of 30 percent per year. (*c*) What would be the impact of extending this same savings to 2,000 components?

(*a*)   See Fig. 14-3.

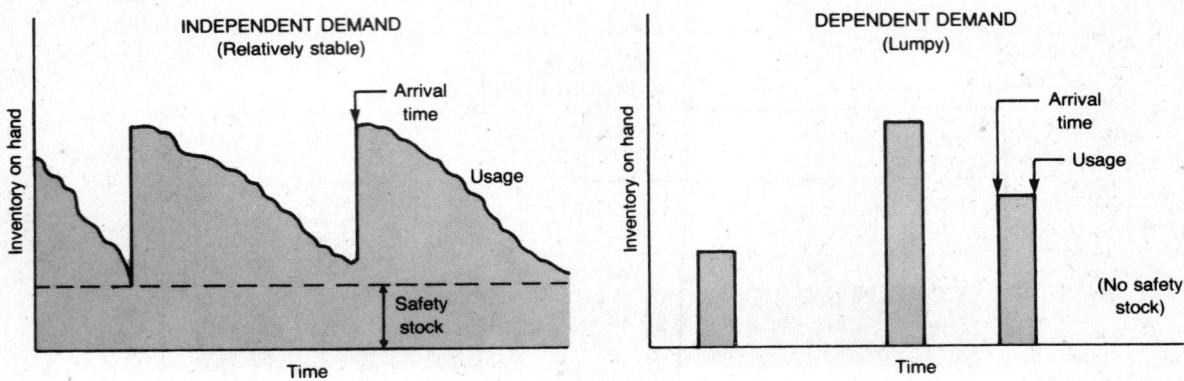

Fig. 14-3

(b)   First carrying cost $= I_{ave}$(item value)(%)   Second carrying cost $= I_{ave}$(item value)(%)
$\qquad\qquad\qquad\quad$ = 80 units($20/unit)(.30/yr) $\qquad\qquad\qquad\qquad$ = 15 units($20/unit)(.30/yr)
$\qquad\qquad\qquad\quad$ = $480/yr $\qquad\qquad\qquad\qquad$ = $90/yr

$\qquad$ Thus, savings = $480 − $90 = $390/yr

(c) $\qquad\qquad\qquad\qquad\qquad$ Savings = 2,000 items($390/item-yr)
$\qquad\qquad\qquad\qquad\qquad\qquad\qquad$ = $780,000/yr

MRP INPUTS AND OUTPUTS

Question: What are the essential inputs and outputs in an MRP system?

Inputs	Outputs
• MPS of end items required • Inventory status file of on-hand and on-order items, lot sizes, lead times, etc. • Product structure (BOM) file of what components and subassemblies go into each end product	• Order release data to CRP for load profiles • Orders to purchasing and in-house production shops • Rescheduling data to MPS • Management reports and inventory updates

BILL OF MATERIALS AND LOW-LEVEL CODING

BOM. A *bill of materials* (BOM) is a listing of all the materials, components, and subassemblies needed to assemble one unit of an end item. Different methods of describing a BOM are in use. Figure 14-4 shows (*a*) a product structure tree and (*b*) an indented BOM. Both are common ways of depicting the *parent-component* relationships on a hierarchical basis. Knowledge of this dependency structure reveals clearly and immediately what components are needed for each higher-level assembly. A third method (*c*) is to use single-level bills of material.

Low-Level Coding. Figure 14-4 also includes *level coding* information. Level 0 is the highest (e.g., the end-item code) and level 3 the lowest for this particular BOM. Note that the four clamps (C20) constitute a subassembly that is combined with base (A10) and two springs (B11) to complete the end-item bracket (Z100). However, the same clamp (C20) is also a component of the base (A10). To facilitate the calculation of net requirements, the product tree has been restructured from where the clamp components might have been (shown dashed) to the lower level consistent with the other (identical) clamp. This *low-level coding* enables the computer to scan the product structure *level by level*, starting at the top, and obtain an accurate and complete count of all components needed at one level before moving on to the next.

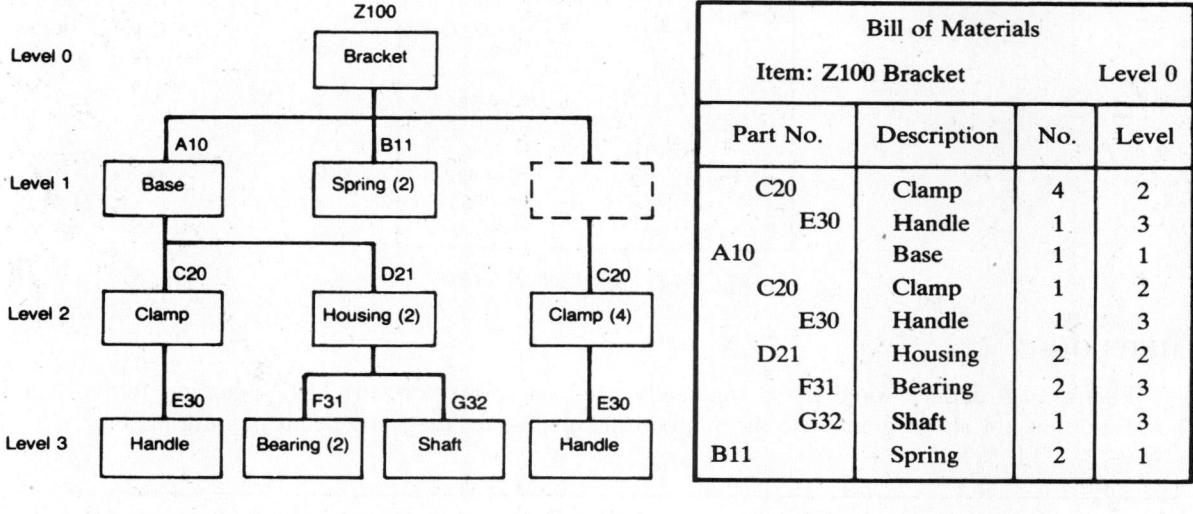

(a) **Product structure tree**

(b) **Indented bill of materials**

Fig. 14-4   Bill of materials for Z100 bracket

Example 14.2   Determine the quantities of A10, B11, C20, D21, E30, F31, and G32 needed to complete 50 of the Z100 brackets depicted in Fig. 14-4. (For simplicity use A, B, . . . , E as part numbers.)

First determine the requirements for one bracket as shown in Table 14-1, and then multiply by 50. Note that parts C and E are used in two different subassemblies, so their separate amounts must be summed. For 50 brackets, each of the requirements column amounts must be multiplied by 50 to obtain the gross requirements.

Table 14-1   Determining BOM Requirements

Component	Dependency Effect	Requirements
A (base)	1A per Z	1
B (spring)	2B's per Z	2
C (clamp)	(1C per A) · (1A per Z) + (4C's per Z)	5
D (housing)	(2D's per A) · (1A per Z)	2
E (handle)	(1E per C) · (1C per A) · (1A per Z) + (1E per C) · (4C's per Z)	5
F (bearing)	(2F's per D) · (2D's per A) · (1A per Z)	4
G (shaft)	(1G per D) · (2D's per A) · (1A per Z)	2

Single-Level BOM.   Figure 14-5 depicts a single-level bill of materials for the same Z100 bracket described in Fig. 14-4. It is a less intuitive but more efficient means of storing the information on computer. Note that in the single-level bill each entry (on the left) contains only an item or part number followed by a list of the part numbers and quantities of components needed to make up the parent item only. This type of listing avoids the searches for duplicate items down through several levels of a tree. On the other hand, it necessitates that the computer search through many single-level bills to find all the components that are included in a product that has several levels of code. Single-level bills frequently contain ''pointers'' to link the records of components with their parents and accommodate the retrieval of a complete bill of materials for an item.

Number	Description
Z100	Bracket
A10 (1)	Base
B11 (2)	Spring
C20 (4)	Clamp
A10	Base
C20 (1)	Clamp
D21 (2)	Housing
C20	Clamp
E30 (1)	Handle
D21	Housing
F31 (2)	Bearing
G32 (1)	Shaft

Fig. 14-5   Single-level BOM

MRP LOGIC

Figure 14-6 defines some terms frequently used on (computerized) MRP planning forms. Note, however, that not all programs use the same terms or provide the same detail of information.

---

1. *Gross requirements*. Projected needs for raw materials, components, subassemblies, or finished goods by the end of the period shown. Gross requirements come from the master schedule (for end items) or from the combined needs of other items.

2. *Scheduled receipts*. Materials *already on order* from a vendor or in-house shop due to be received *at the beginning* of the period. MRP form shows quantity and projected time of receipt. (Note: Some MRP forms include planned receipts here too.)

3. *On hand/Available*. The quantity of an item expected to be available *at the end* of the time period in which it is shown. This includes amount available from previous period plus planned-order receipts and scheduled receipts less gross requirements.

4. *Net requirements*. Net amount needed in the period. This equals the gross requirements less any projected inventory available from the previous period along with any scheduled receipts.

5. *Planned-order receipt*. Materials that *will be ordered* from a vendor or in-house shop to be received at the beginning of the period. Otherwise similar to a scheduled receipt.

6. *Planned-order release*. The planned amount to be ordered in the time period adjusted by the lead-time offset so that materials will be received on schedule. Once the orders are actually released, the planned-order releases are deleted from the form and the planned-order receipts they generated are changed to scheduled receipts.

Fig. 14-6   Terms frequently used on MRP planning forms

The master production schedule dictates *gross or projected requirements* for end items to the MRP system. Gross requirements do not take account of any inventory on hand or on order. The MRP computer program then "explodes" the end-item demands into requirements for components and materials by processing all relevant bills of materials on a level-by-level (or single-level) basis. Net requirements are then calculated by adjusting for existing inventory and items already on order as recorded in the inventory status file.

$$\text{Net requirements} = \text{gross requirements} - (\text{on hand/available} + \text{scheduled receipts}) \qquad (14.1)$$

Order releases are planned for components in a time-phased manner (using lead-time data from the inventory file) so that materials will arrive precisely when needed. At this stage the material is referred to as a planned-order receipt. When the orders are actually issued to vendors or to in-house shops, the planned receipts become scheduled receipts. The inventory on hand at the end of a period is the sum of the previous period on-hand amount plus any receipts (planned or scheduled) less the gross requirements.

On hand/available = on hand at end of previous period + receipts − gross requirements   (14.2)

Example 14.3   A firm producing wheelbarrows is expected to deliver 40 wheelbarrows in week 1, 60 in week 4, 60 in week 6, and 50 in week 8. Among the requirements for each wheelbarrow are two handlebars, a wheel assembly, and one tire for the wheel assembly. Order quantities, lead times, and inventories on hand at the beginning of period 1 are shown in Table 14-2.

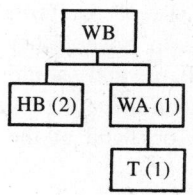

Table 14-2   BOM and Inventory Data for Wheelbarrow Components

Part	Order Quantity	Lead Times	Inventory on Hand
Handlebars	300	2 wk	100
Wheel assemblies*	200	3 wk	220
Tires	400	1 wk	50

*90 wheel assemblies are also needed in period 5 for a garden tractor shipment.

A shipment of 300 handlebars is already scheduled to be received at the beginning of week 2 (i.e., a scheduled receipt). Complete the MRP for the handlebars, wheel assemblies, and tires; and show what quantities or orders must be released and when they must be released to satisfy the master schedule.

Figure 14-7 depicts the master schedule and component part schedules. We shall assume that the customer completes the final assembly, so no time allowance is required there. Note that because each wheelbarrow requires two handlebars, the gross requirements for handlebars are double the number of end products. Thus the gross requirements in period 1 are $40 \times 2 = 80$ units.

Master schedule (Wheelbarrow)	Week number	1	2	3	4	5	6	7	8
	Quantity	40			60		60		50

(×2)

Item ID: HB	Gross requirements	80			120		120		100
Level code: 1	Scheduled receipts		300						
On hand: 100	On hand/Available	20	320	320	200	200	80	80	280
Lot size: 300	Net requirements								20
LT (wk): 2	Planned-order receipts								300
Safety stock: 0	Planned-order releases						300		

Item ID: WA	Gross requirements	40			60	90*	60		50
Level code: 1	Scheduled receipts								
On hand: 220	On hand/Available	180	180	180	120	30	170	170	120
Lot size: 200	Net requirements						30		
LT (wk): 3	Planned-order receipts						200		
Safety stock: 0	Planned-order releases			200					

Item ID: Tire	Gross requirements			200					
Level code: 2	Scheduled receipts								
On hand: 50	On hand/Available	50	50	250	250	250	250	250	250
Lot size: 400	Net requirements			150					
LT (wk): 1	Planned-order receipts			400					
Safety stock: 0	Planned-order releases		400						

* Requirements from another product (garden tractor) that uses the same wheel assembly.

Fig. 14-7   Master schedule and MRP component plans for wheelbarrows

The 100 handlebars on hand at the beginning of period 1 are adequate to supply the gross requirement of 80 handlebars, leaving 20 on hand at the end of period 1. With the (scheduled) receipt of 300 handlebars in period 2, the on-hand inventory remains adequate until the end of week 8, when 80 units are on hand. However, the gross requirement for 100 units in period 9 exceeds the on-hand inventory. This results in net requirements (using Eq. 14.1) of $100 - 80 = 20$ units.

To satisfy this, a planned-order receipt for the standard order quantity (300) is scheduled for the beginning of period 8. Insofar as the handlebars have a 2-week lead time, the planned order for the handlebars must be released 2 weeks earlier (week 6). The planned-order receipt will result in a projected end-of-period on-hand inventory (using Eq. 14.2) of $80 + 300 - 100 = 280$ units.

Different Formats. Insofar as the planned-order receipts are incorporated into the On hand/Available row, some MRP formats simply eliminate the planned-order receipts row. Others eliminate the net requirements row because whenever inventory is needed, the planned-order release will show the amount ordered anyway. Other relatively common formats use the term *On hand* or *Projected on hand* or *Available* or *Available Inventory* for the row we have designated "on hand/available." The logic underlying use of the term *available* is that sometimes inventory may actually be on hand but may be committed to cover other requirements including safety stock. In this situation, the "allocated" inventory is not available and should not be included in the "On hand/Available" amount that is available for normal production activities. (Another variation is to assume the On hand/Available is the inventory expected to be on hand at the *beginning* of each period, rather than at the end of the period.)

SYSTEM REFINEMENTS

Key features of MRP systems are (1) the generation of lower-level requirements, (2) time phasing of those requirements, and (3) the planned-order releases that flow from them. Note particularly that planned-order releases of parent items (such as the wheel assemblies in Example 14.3) generate gross requirements at the component level (e.g., tires).

Example 14.4   In Example 14.3, suppose two tires had been required for each wheel assembly instead of one. What would be the revised gross requirement for tires?

Gross requirement would have been $200 \times 2 = 400$ in week 3 instead of 200.

Example 14.5 (*Planned-order releases, different MRP format*)   Given the MRP format in Fig. 14-8, where 80 units are on hand at the beginning of period 1. What planned-order releases should be scheduled?

Period OQ = 100, LT = 4		1	2	3	4	5	6	7	8
Gross requirements			50			30		60	50
Plan/Sched. receipts									
Projected on hand	80								
Planned-order releases									

Fig. 14-8   MRP format

Note that this MRP format does not include a separate row for net requirements or for planned-order receipts. (For working purposes, you may want to note any planned receipts in the scheduled receipts row—although there is a technical difference in the two.) The on hand drops to zero in period 5, so a planned-order receipt (of 100) is needed in period 7, plus another in period 8. With a 4-period lead time, this means planned orders of 100 must be released in periods 3 and 4.

Additional features of many MRP systems are their capability to handle (*a*) simulations, (*b*) firm-planned orders, (*c*) pegging, and (*d*) the availability of planning bills of material.

Simulation.   The *"what-if" simulation capability* allows planners to "trial fit" a master schedule onto the MRP system before the schedule is actually accepted and released. With this feature, a planner can "try" a potential customer order on the system to see if materials and delivery dates can be met—even before the order is accepted. If lead times, materials, and capacities are sufficient, the order can be accepted; otherwise, changes in quantities or delivery times may have to be negotiated, or the order may even have to be turned down.

Question: Why is it so essential to have accurate delivery quantities, times, etc.?

Aside from the obvious benefits associated with good customer service, keeping dates "honest" and customer priorities current is vital to maintaining system integrity. Invalid priorities undermine the credibility of the system, employees lose faith in it, and eventually it fails to do the job intended.

Firm-Planned Orders.   Sometimes the "normal" manufacturing times are not enough to meet an emergency, secure a special order, or service a valued customer. *Firm-planned order capability* enables planners to instruct the computer to accept certain requirements, even though normal MRP logic would automatically delay or reschedule such orders. By designating certain orders as "firm-planned orders," planners can ensure that the computer will not automatically change the release date, the planned-order receipt date, or the order quantity. In addition, the system can establish a "time fence" around the planned-order release date to preclude the scheduling of other orders near that time so as to ensure that resources are available to do the special job.

Question: What is accomplished by the use of firm-planned orders?

The firm-planned order capability allows planners to create a planned order that does not "fit" into the standard lot size, lead time, or other specifications for the item. The computer will not automatically adjust or reschedule such orders. Firm-planned orders also freeze the timing of the planned-order release.

Pegging.   *Pegging* refers to the ability to work backward from component requirements to identify the parent item, or items that generated those requirements. For example, suppose an automobile manufacturer learned that some of the brake materials (already used in production) were defective. The "where used" pegging file would allow production analysts to trace requirements upward in the product structure tree to determine what end-item models contained the defective components.

Modular and Planning Bills.   *Modular bills of materials* describe the product structure for basic subassemblies of parts that are common to different end items. For example, several models of a manufacturer's automobiles may contain the same transmission, drive train, air-conditioning, and braking systems. By scheduling these items as (common) modules, production can sometimes be more effectively "smoothed" and inventory investment minimized.

Question: For what types of items are bills of material modularized?

Modularization is used for end items that contain many options.

Planning bills of materials are artificial or "pseudo" bills created by production control to facilitate planning. They do not necessarily constitute a "buildable" combination of items but can be very useful for scheduling purposes. For example, a planning bill might consist of the minor assembly items, such as the bolts and nuts, handle grips, and wheel for a wheelbarrow. The "kit" of materials can be treated as an assembly item for purposes of the master schedule.

A major use of planning bills is in the scheduling of options, which present some real challenges when working with components that have long lead times. To provide quick response to customers, options are frequently scheduled on a "percentage of end item" basis. For example, a scheduler may use a planning bill to call for 30 percent of the output to be a "heavy duty option." Then, as orders arrive prior to production, the actual build schedule can be adjusted to accommodate the specific customer demands.

SAFETY STOCK, LOT SIZING, AND SYSTEM UPDATING

Safety Stock. Note that the MRP component plan shown in Fig. 14-7 includes space for an entry of safety stock. Although one of the reasons for using MRP to manage dependent demand inventory items is to avoid the need for safety stock, in reality firms may elect to carry safety stock on some items for a variety of reasons. Safety stock amounts are sometimes deducted before showing the On hand/Available quantities. If not, they can be provided for as shown in Example 14.6.

Example 14.6 (*MRP with different format, LFL ordering, and safety stock requirement*) The master schedule and product structure for an end item "Z" are as shown. Numbers in parentheses represent quantities required, and the lead times (LTs) are also given. Use the space provided to determine the planned-order release dates and quantities for components A, D, and F. Order quantity (OQ) and safety stock (SS) requirements are as shown. On-hand quantities are also shown.

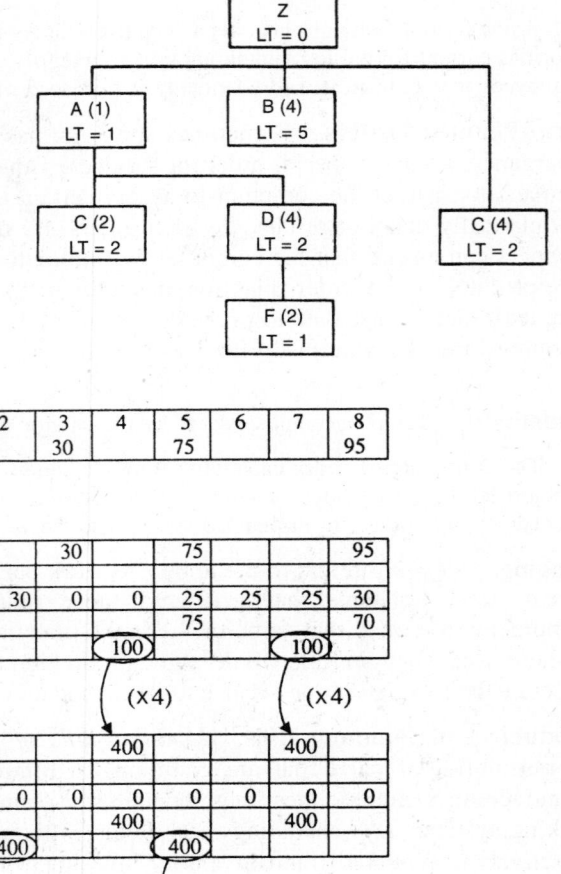

Master Schedule for Z

Week number	1	2	3	4	5	6	7	8
Requirements			30		75			95

A: OQ = 100, LT = 1

Gross requirements				30		75			95
Scheduled receipts									
On hand end-of-period	30	30	30	0	0	25	25	25	30
Net requirements						75			70
Planned-order release					(100)			(100)	

(×4)   (×4)

D: OQ = LFL, LT = 2

Gross requirements					400			400	
Scheduled receipts									
On hand end-of-period	0	0	0	0	0	0	0	0	0
Net requirements					400			400	
Planned-order release			(400)			(400)			

(×2)   (×2)

F: OQ = 1,000, SS = 200 LT = 1

Gross requirements			800			800			
Scheduled receipts									
On hand end-of-period	900	900	1,100	1,100	1,100	300	300	300	300
Net requirements			100						
Planned-order release		1,000							

Fig. 14-9 MRP format for end item Z

Question: List some reasons for carrying safety stock of components on an MRP format?

(1) Not all demand is dependent. Some items (e.g., repair parts) may have a service requirement that has an independent demand component.

(2) Variable lead times from suppliers are a common source of uncertainty to many firms.

(3) Firms may experience machine breakdowns, scrap losses, and last-minute customer changes.

Lot Sizing. Order quantities are not always specified in advance as was done in Example 14.3. Different lot-sizing methods are in use, including (1) fixed-order quantity amounts, e.g., 300 handlebars;

(2) EOQ or ERL amounts; (3) lot for lot, which is ordering the exact amount of the net requirements for each period; (4) fixed-period requirements, e.g., a 2-month supply; and (5) various least-cost approaches, e.g., least-unit cost, least-total cost. The part-period algorithm is a method that uses a ratio of ordering costs to carrying costs per period, which yields a part-period number. Then requirements for current and future periods are cumulated until the cumulative holding cost (in part-period terms) is as close as possible to this number.

System Updating.   MRP system designs typically use one of two methods to process data, update files, and ensure that the system information is valid and conforms with actual: (1) regenerative processing or (2) net change processing.

- *Regenerative MRP systems* use batch processing to replan the whole system (full explosion of all items) on a regular basis (e.g., weekly).
- *Net change MRP systems* are online and react continuously to changes from the master schedule, inventory file, and other transactions.

Early MRP installations were largely of the regenerative type, but then as net change systems became perfected, more firms began installing them. However, being ''activity driven,'' net change systems are sometimes ''nervous'' and tend to overreact to changes. The major disadvantage of regenerative systems is the time lag that exists until updated information is incorporated into the system.

System Application.   Although MRP systems are widely used, they are most beneficial in manufacturing environments where products are manufactured to order, or assembled to order or to stock. MRP does not provide as much advantage in low-volume, highly complex applications or in continuous flow processes, such as refineries. It does, however, enjoy wide application in metals, paper, food, chemical, and other processing applications.

CRP INPUTS AND OUTPUTS

Capacity is a measure of the productive capability of a facility per unit of time. In terms of the relevant time horizon, capacity management decisions are concerned with the following:

(1) *Long range*—resource planning of capital facilities, equipment, and human resources

(2) *Medium range*—requirements planning of labor and equipment to meet MPS needs

(3) *Short range*—control of the flow (input-output) and sequencing of operations

Capacity-requirements planning (CRP) applies primarily to medium-range activities. The CRP system receives planned and released orders from the material-requirements planning system and attempts to develop loads for the firm's work centers that are in good balance with the work-center capacities. Like MRP, CRP is an iterative process that involves planning, revision of capacity (or revision of the master schedule), and replanning until a reasonably good load profile is developed.

Question: What are the essential inputs and outputs in a CRP system?

Inputs	Outputs
• Planned and released orders from the MRP system	• Load reports of planned and released orders on key work centers
• Loading capacities from the work-center status file	• Verification reports to the MRP system
• Routing data from the routing file	• Capacity modification data
• Changes that modify capacity, give alternative routings, or alter planned orders	• Rescheduling data to the MPS

CRP ACTIVITIES: INFINITE AND FINITE LOADING

Planned-order releases (in the MRP system) are converted to standard hours of load on key work centers in the CRP system. Figure 14-10 illustrates this transition for a planned-order release of 300 handlebars in period 6 (from Fig. 14-7) to 6.2 standard hours of work in work center (WC) 4. These hours and the hours for other jobs planned for WC 4 during period 6, plus the hours for orders already released, make up the total expected load of 185 standard hours. Note that 20 hours have been allowed for unplanned or emergency jobs.

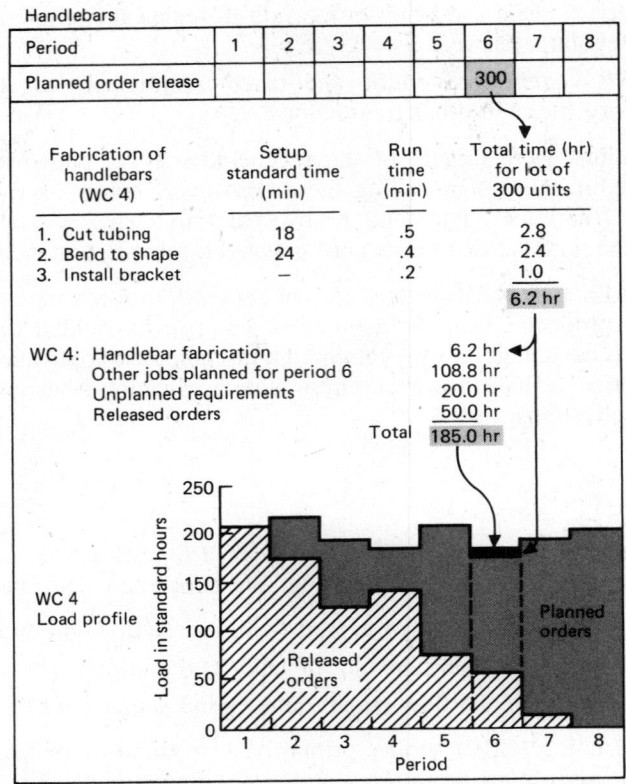

Fig. 14-10   Transition of planned-order release from MRP
to load in CRP system

Example 14.7   Using the data of Fig. 14-10, suppose a change is proposed to increase the planned order for period 6 from 300 to 600 handlebars. Assuming WC 4 has 200 standard hours of capacity available, could it handle the proposed increase in load?

Note that we cannot simply double the 6.2 hours already computed for the handlebar fabrication because the set-up portion of the time will not be increased. Therefore the handlebar fabrication will be less than twice the 6.2 hours (i.e., <12.4 hours). Thus the total will still be less than the 200 standard hours available.

Fabrication time required:

$$\text{Cut} = [18 + (.5)(600)]/60 = 5.3$$
$$\text{Bend} = [24 + (.4)(600)]/60 = 4.4$$
$$\text{Install} = [(.2)(600)]/60 = \underline{\quad 2.0 \quad}$$
$$11.7 \text{ hr}$$

Replacing the 6.2-hour figure with 11.7 hours brings the projected load on WC 4 up to 190.5 hours, which is still well within the 200-hour standard. The WC should be able to absorb the proposed change.

Infinite loading is the practice of loading work centers with all loads when they are required, without regard to the actual capacity of the work centers. CRP often uses infinite loading on an initial load profile to evaluate, or "size up," the proposed load from the MPS and help planners make decisions regarding the use of overtime, alternative routings, etc. *Finite loading* limits the load assigned to a work center to the maximum capacity of the work center. Finite loading is most useful when making the final assignment of work to work centers, after the major adjustments have been made by using infinite loading procedures. Fig. 14-11 illustrates the difference between infinite and finite loading.

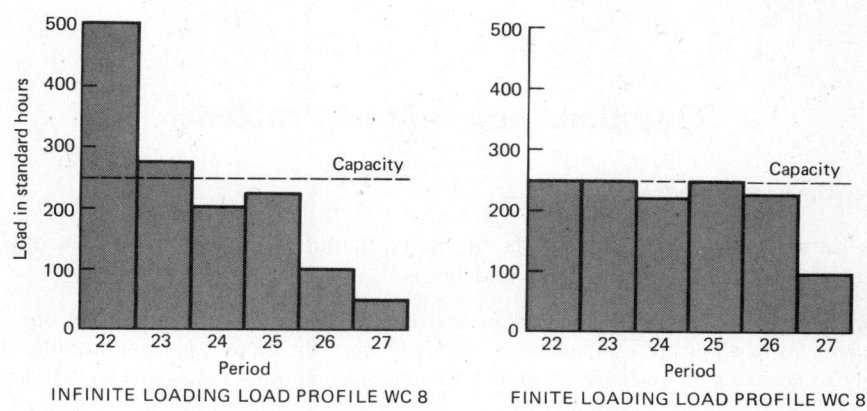

Fig. 14-11   Infinite and finite loading profiles

MRP II: MANUFACTURING RESOURCE PLANNING

The post-World War II industrial expansion, in an environment of increased competition, generated a strong need for closer control of the resources used in manufacturing activities. Early MRP and CRP systems, capitalizing upon the data processing capability of computers, focused upon the management of materials and capacities in a relatively isolated manner. As performance measurement and information feedback systems became more advanced, MRP systems began to resemble the closed-loop configuration, shown earlier in Fig. 14-2.

Question: What is implied in the concept of closed-loop MRP?

Closed-loop MRP implies the existence of a material requirements planning system that includes information from the aggregate plan, the master schedule, and the capacity requirements plan, plus feedback from production activity control functions. The information must be sufficiently complete, accurate, and timely to maintain valid priorities and capacities.

MRP logic has also been extended to plan interrelated orders at various levels of a distribution network (distribution requirements planning, DRP). Also, MRP systems have been expanded to incorporate the influence of other functional activities, such as purchasing, engineering, marketing, and finance. The term most frequently used to define this broader concept (of MRP + associated functions) is *manufacturing resource planning*, or MRP II.

Question: What is implied in the concept of MRP II?

MRP II is a term for describing manufacturing resource planning, which is generally understood to be a planning and scheduling process that incorporates an expanded scope of activities through closed-loop systems that exchange production, marketing, financial, and other information. The use of simulation as a means of anticipating companywide effects is generally assumed to be included in this concept.

Implementation. Not all MRP systems are successful. Implementing an MRP system requires (a) top management commitment, (b) extensive user involvement, and (c) significant education and training. Inventory records, bills of material, and other data must be highly accurate. Numerous software programs are available, but they must be adapted to specific company operations.

Questions and Solved Problems

MRP AND CRP CONCEPTS

14.1 Why is demand for component parts of many manufactured products "lumpy" rather than relatively stable like the demand for end items?

Manufacturing is frequently done on an intermittent basis in lots, or models, of one type or another. The components of a finished product are needed only when the product is being manufactured. Thus there may be large demands on inventory at some times and none at other times, making the demand "lumpy."

14.2 Master schedules typically contain end items, but most of the items in an MRP system are component parts. How are the end items from a master schedule converted into the component parts in an MRP system?

All end items (or major subassemblies) in the master schedule have a bill of material (BOM) associated with them. When the MRP system receives an end-item requirement from the master schedule, it accesses the BOM file and "explodes" the end item into its component requirements in accordance with the product structure as delineated in the BOM file. This explosion process also tells quantities needed, lead times, etc. These component parts all become items in the MRP system.

14.3 What are the primary output reports of an MRP system?

(1) A schedule of planned orders to be released in the future

(2) Order release notices to authorize work on planned orders

(3) Changes (of due dates, quantities, etc.) in planned and open orders

(4) Cancellations or suspensions of open orders due to master schedule changes

(5) Inventory status updates

(6) Various planning, performance, and exception reports

14.4 A firm uses an MRP system in a plant that manufactures a wide assortment of children's furniture. What system features would be used to assist in solving these two problems? (a) Some brackets received from a supplier (last year) have failed because they contained defective welds. The production planner has been asked to identify the crib and playground equipment models that used brackets from this particular supplier. (*b*) Cribs use an assortment of metal and wood screws, inserts, plastic strips, and rollers that are currently included as low-level coded items on numerous different bills of materials. The planner has been asked to reduce the number of BOM levels on these items.

(*a*) Use pegging feature. (*b*) Set up a planning bill of materials for these items.

14.5 Distinguish between the terms utilization and efficiency as used in capacity planning.

Utilization refers to the actual (or percentage of) time a resource is in use. *Efficiency* tells how well the resource is performing during that time, relative to some standard.

BILL OF MATERIALS

14.6 (*a*) What is the major function of the bill of materials? (*b*) What happens if the bill of materials is inaccurate?

(*a*) The BOM provides the product structure hierarchy that guides the explosion process.

(*b*) If inaccurate, the materials needed to manufacture an end item may not be available.

14.7 Distinguish between an indented BOM and a single-level BOM.

An *indented BOM* uses physical offsets to depict the parent-component relationships for each branch of the product structure tree. Insofar as a component of one item might well be the parent of another, the product tree may have several levels.

A *single-level BOM* lists parent items (e.g., A) and only one sublevel of components (e.g., B). If the component (B) is also a parent to some other components (e.g., C and D), then the component (B) is also listed as a parent item.

14.8 Given the product structure tree shown in Fig. 14-12 for wheelbarrow W099, develop an indented bill of materials.

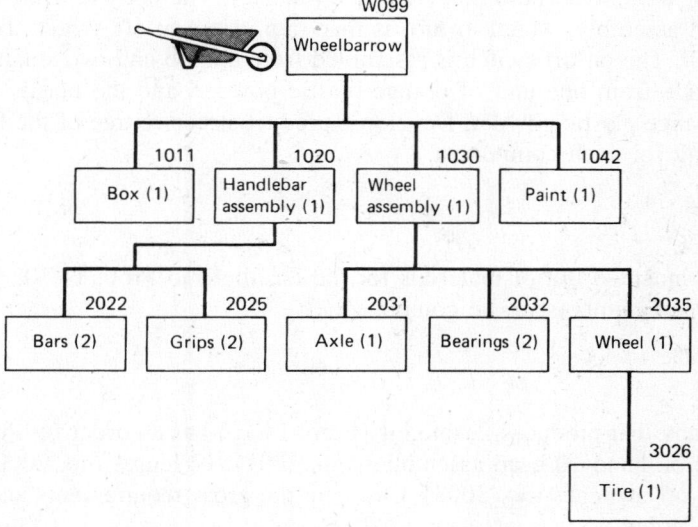

Fig. 14-12

See Table 14-3.

Table 14-3

Bill of Materials				
Part No. W099: Wheelbarrow				Level: 0
Part No.	Description	Quantity/Assembly	Units	Level
1011	Box: deep size, aluminum	1	each	1
1020	Handlebar assembly	1	each	1
2022	Aluminum bars	2	each	2
2025	Grips: neoprene	2	each	2
1030	Wheel assembly	1	each	1
2031	Axle	1	each	2
2032	Bearing: normal-duty	2	each	2
2035	Wheel	1	each	2
3026	Tire: size A	1	each	3
1042	Paint: blue	1	pint	1

14.9 For the wheelbarrow of Prob. 14.8, there are 5 types of boxes available, 3 handlebar options, 2 choices of bearings, 3 sizes of tires, and 5 possible colors of paint. (*a*) If every option were considered a master schedule item, how many items would the master schedule have to accommodate for this product (wheelbarrows) alone? (*b*) What would be the effect of offering the choice of a white stripe on the front?

(*a*) There are $5 \times 3 \times 2 \times 3 \times 5 = 450$ combinations.

(*b*) Offering a stripe would mean that each wheelbarrow could come either with or without a stripe, so the potential number of end items would double to 900. *Note:* To simplify the master scheduling, planners would most likely develop schedules of the level-1 items (rather than the finished product, level zero) and specify the specific end item on a final assembly bill of materials.

14.10 A flashlight is assembled from three major subassemblies: a head assembly, two batteries, and a body assembly. The head assembly consists of a plastic head, a lens, a bulb subassembly (comprising a bulb and bulb holder), and a reflector. The body assembly consists of a coil spring and a shell assembly, which in turn is made up of an on-off switch, two connector bars, and a plastic shell. The on-off switch is assembled from a knob and two small metal slides. The plastic head is made from one unit of orange plastic powder, and the plastic shell is made from three units of orange plastic powder. Develop a product structure tree of the flashlight, and include the level coding for each component.

See Fig. 14-13.

14.11 Design an indented bill of materials for the flashlight in Prob. 14.10. (*Note:* Assign appropriate four-digit part numbers to the components.)

See Table 14-4.

14.12 The company that produces flashlights (Prob. 14.10) has an order for 200 end items (flashlights). They have on hand 10 head assemblies (no. 1001), 12 lenses (no. 2002), 50 springs (no. 2006), and 15 on-off switches (no. 3003). Compute the gross requirements and the net requirements to satisfy the order.

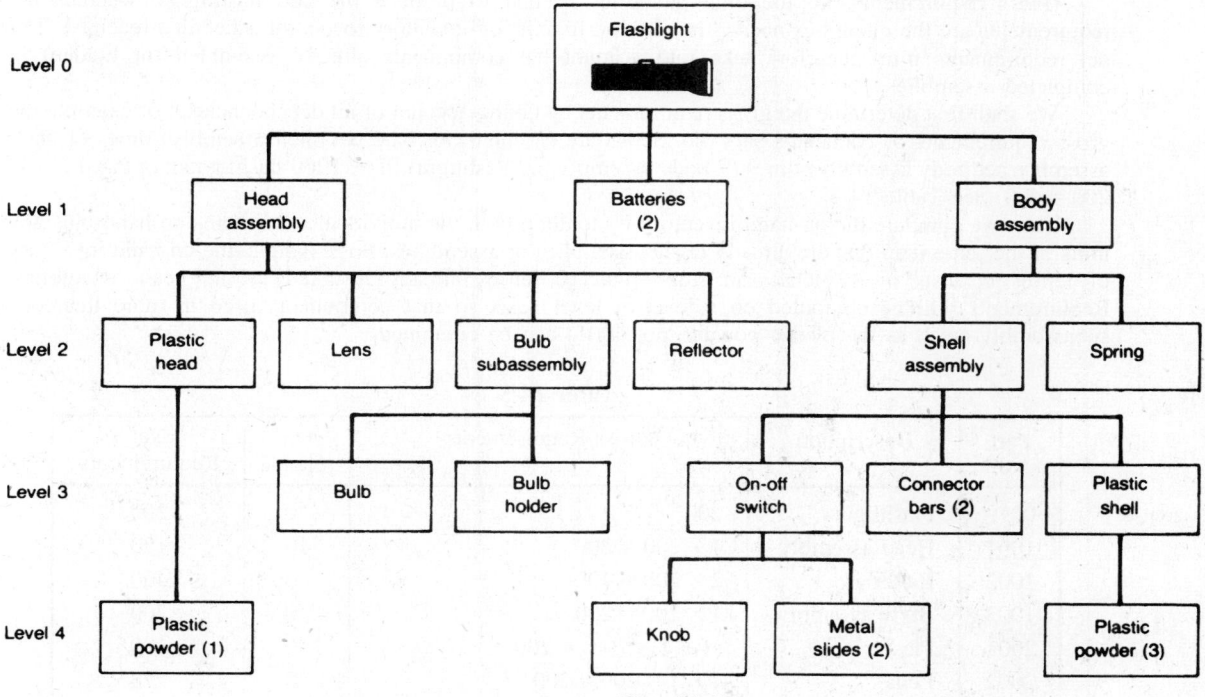

Fig. 14-13

Table 14-4

Bill of Materials			
Item: 0010 Flashlight			Level: 0
Part No.	Description	No.	Level
1001	Head assembly	1	1
2001	Plastic head	1	2
4001	Plastic powder	1	4
2002	Lens	1	2
2003	Bulb assembly	1	2
3001	Bulb	1	3
3002	Bulb holder	1	3
2004	Reflector	1	2
1002	Batteries	2	1
1003	Body assembly	1	1
2005	Shell assembly	1	2
3003	On-off switch	1	3
4002	Knob	1	4
4003	Metal slides	2	4
3004	Connector bars	2	3
3005	Plastic shell	1	3
4001	Plastic powder	3	4
2006	Spring	1	2

Gross requirements are the total quantities needed to produce the 200 flashlights, whereas *net* requirements are the quantities needed in addition to existing inventory levels (or scheduled receipts). The net requirements must therefore take into account the components already assembled (or hidden) in completed assemblies.

We shall first determine the gross requirements by taking account of all dependencies. For example the gross requirements of connector bars (no. 3004) are (2 connector bars per shell assembly) times (1 shell assembly per body assembly) times (1 body assembly per flashlight) times (200 flashlights), or $2 \times 1 \times 1 \times 200 = 400$. See Table 14-5.

Then we compute the on-hand inventory by totaling both the individual stock items on hand plus any units of the same item that are already in subassemblies or assemblies. For example, the on-hand inventory of lenses consists of 12 lenses in stock plus 10 lenses already installed in the head assemblies. Requirements will be computed on a level-by-level basis so that components used in more than one subassembly (such as the plastic powder, no. 4001) can be combined.

Table 14-5

Part No.	Description	Gross Requirements	On Hand	Net Requirements
0010	Flashlight	200	0	200
1001	Head assembly	$1 \times 200 = 200$	10	190
1002	Batteries	$2 \times 200 = 400$	0	400
1003	Body assembly	$1 \times 200 = 200$	0	200
2001	Plastic head	$(1 \times 1 \times 200) = 200$	10	190
2002	Lens	$1 \times 1 \times 200 = 200$	22	178
2003	Bulb assembly	$1 \times 1 \times 200 = 200$	10	190
2004	Reflector	$1 \times 1 \times 200 = 200$	10	190
2005	Shell assembly	$1 \times 1 \times 200 = 200$	0	200
2006	Spring	$1 \times 1 \times 200 = 200$	50	150
3001	Bulb	$1 \times 1 \times 1 \times 200 = 200$	10	190
3002	Bulb holder	$1 \times 1 \times 1 \times 200 = 200$	10	190
3003	On-off switch	$1 \times 1 \times 1 \times 200 = 200$	15	185
3004	Connector bars	$2 \times 1 \times 1 \times 200 = 400$	0	400
3005	Plastic shell	$1 \times 1 \times 1 \times 200 = 200$	0	200
4001	Plastic powder	$(1 \times 1 \times 1 \times 200) +$ $(3 \times 1 \times 1 \times 1 \times 200) = 800$	10	790
4002	Knob	$1 \times 1 \times 1 \times 1 \times 200 = 200$	15	185
4003	Metal slides	$2 \times 1 \times 1 \times 1 \times 200 = 400$	30	370

14.13 The product structure tree for X is as shown in Fig. 14-14, with the number of units required shown in parentheses. What quantities of E, J, and K are required to complete 500 units of X?

Working from the top (highest level) down, we have:

A: (1)(number of X's) = 1(500) = 500

B: (1)(number of X's) = 1(500) = 500

C: (2)(number of X's) = 2(500) = 1,000

D: (1)(number of A's) = 1(500) = 500

E: (2)(number of A's) = 2(500) = 1,000

F: (2)(number of B's) = 2(500) = 1,000

G: (1)(number of B's) = 1(500) = 500

H: (1)(number of B's) = 1(500) = 500

I: (2)(number of C's) = 2(1,000) = 2,000

J: (4)(number of D's) + (2)(number of G's) = 4(500) + 2(500) = 3,000

K: (1)(number of E's) + (1)(number of I's) = 1(1,000) + 1(2,000) = 3,000

L: (1)(number of E's) + (1)(number of G's) = 1(1,000) + 1(500) = 1,500

Quantities for E, J, and K are 1,000, 3,000, and 3,000, respectively.

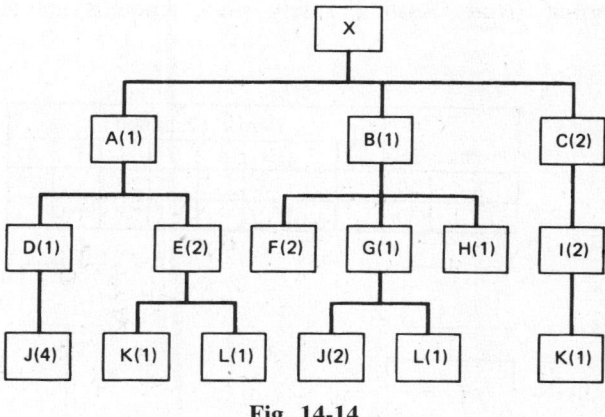

Fig. 14-14

MRP LOGIC

14.14 Complete the material-requirements plan for item X shown in Fig. 14-15. Note that this item has an independent demand that necessitates that a safety stock of 40 units be maintained.

Order quantity = 70 Lead time = 4 weeks Safety stock = 40		Week											
		1	2	3	4	5	6	7	8	9	10	11	12
Gross requirements		20	20	25	20	20	25	20	20	30	25	25	25
Scheduled / Planned receipts			70										
On hand at end of period	65												
Planned-order release													

Fig. 14-15

The material-requirements plan for end item X is shown in Fig. 14-16.

Order quantity = 70 Lead time = 4 weeks Safety stock = 40		Week											
		1	2	3	4	5	6	7	8	9	10	11	12
Gross requirements		20	20	25	20	20	25	20	20	30	25	25	25
Scheduled / Planned receipts			70			70			70			70	
On hand at end of period	65	45	95	70	50	100 ~~30~~	75	55	105 ~~35~~	75	50	95 ~~25~~	70
Planned-order release		70			70			70					

Fig. 14-16

14.15 Clemson Industries produces products X and Y, which have demand, safety stock, and product structure levels as shown in Fig. 14-17. The on-hand inventories are as follows: X = 100, Y = 30, A = 70, B = 0, C = 200, and D = 800. The lot size for A is 250, and the lot size for D is 1,000 (or multiples of these amounts); all the other items are specified on a lot-for-lot (LFL) basis (that is, the quantities are the same as the net requirements). The only scheduled receipts are 250 units of X due in period 2. Determine the order quantities and order release dates for all requirements using an MRP format. (*Note:* Assume safety stock amounts are to be included in the On hand/Available.)

Product	SS	Demand in period							
		1	2	3	4	5	6	7	8
X	50			300			200		250
Y	30							400	

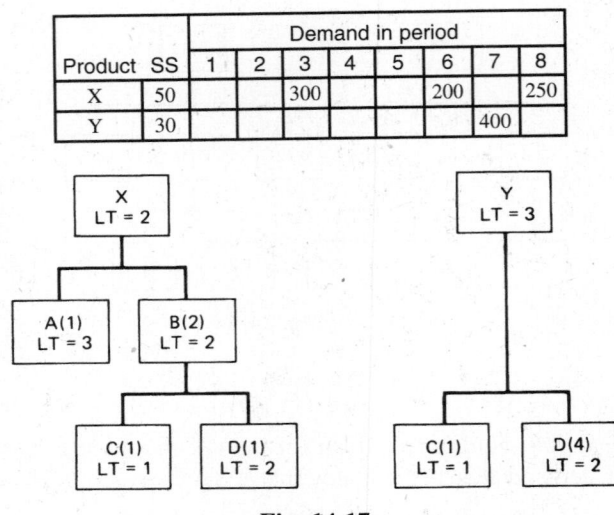

Fig. 14-17

First, establish the codes (lowest level) applicable to each product as shown in Table 14-6. Items C and D appear both at level 1 in product Y and at level 2 in product X, so they are assigned to level 2. Thus their requirements are not netted out until all level 0 and 1 requirements have been netted out.

Table 14-6

Item	Low-Level Code
X	0
Y	0
A	1
B	1
C	2
D	2

Next, set up an MRP format for all items (see Fig. 14-18), and enter the end-item gross requirements for X and Y. They both have low-level codes of 0 and so can be netted out using order quantities that match their requirements (preserving safety stocks, of course). This results in planned-order releases of 200 and 250 units for X (periods 4 and 6) and 400 units of Y (period 4).

Next, explode the planned-order releases for X and Y (that is, multiply them by the quantities required of the level-1 items, A and B). (Note that C and D are not level-1 items.) Projected requirements for A (200 and 250 units) are direct results of the planned-order releases for X. Two units of B are required for each X, so item B's projected requirements in periods 4 and 6 are 400 and 500, respectively. Items A and B are then netted, and the order release dates and amounts are set.

Next, explode the level-2, planned-order releases to the level 3 items. The arrows in Fig. 14-18 show that requirements for C and D come from planned-order releases for both B and Y. End item Y requires 4

Week number		1	2	3	4	5	6	7	8
Item ID: X	Gross requirements			300			200		250
Level code: 0	Scheduled receipts		250						
On hand: 100	On hand/Available	100	350	50	50	50	50	50	50
Lot size: LFL	Net requirements						150		200
LT (wk): 2	Planned-order receipts						200		250
Safety stock: 50	Planned-order releases				200		250		

		1	2	3	4	5	6	7	8
Item ID: Y	Gross requirements						400		
Level code: 0	Scheduled receipts								
On hand: 30	On hand/Available	30	30	30	30	30	30	30	30
Lot size: LFL	Net requirements						400		
LT (wk): 3	Planned-order receipts						400		
Safety stock: 30	Planned-order releases				400				

		1	2	3	4	5	6	7	8
Item ID: A	Gross requirements				200		250		
Level code: 1	Scheduled receipts								
On hand: 70	On hand/Available	70	70	70	120	120	120	120	120
Lot size: 250	Net requirements				130		130		
LT (wk): 3	Planned-order receipts				250		250		
Safety stock:	Planned-order releases	250		250					

		1	2	3	4	5	6	7	8
Item ID: B	Gross requirements				400		500		
Level code: 1	Scheduled receipts								
On hand: 0	On hand/Available	0	0	0	0	0	0	0	0
Lot size: LFL	Net requirements				400		500		
LT (wk): 2	Planned-order receipts				400		500		
Safety stock:	Planned-order releases		400		500				

		1	2	3	4	5	6	7	8
Item ID: C	Gross requirements		400		900				
Level code: 2	Scheduled receipts								
On hand: 200	On hand/Available	200	0	0	0	0	0	0	0
Lot size: LFL	Net requirements		200		900				
LT (wk): 1	Planned-order receipts		200		900				
Safety stock:	Planned-order releases	200		900					

		1	2	3	4	5	6	7	8
Item ID: D	Gross requirements		400		2100				
Level code: 2	Scheduled receipts								
On hand: 800	On hand/Available	800	400	400	300	300	300	300	300
Lot size: 1,000	Net requirements				1700				
LT (wk): 2	Planned-order receipts				2000				
Safety stock:	Planned-order releases		2000						

Fig. 14-18

units of D, so the projected requirements in period 4 are 2,100 units, with 1,600 from Y (that is, 4×400) and 500 from B. Together, they generate a planned-order release for 2,000 units of D in period 2.

LOT SIZING: PART-PERIOD ALGORITHM (PPA)

14.16 The ordering cost to order an item is $225, and carrying cost is $.75 per period. Net requirements per month are as shown in Table 14-7. Use the part-period algorithm to determine the size and timing of orders.

Table 14-7

Month	1	2	3	4	5	6	7	8	9
Requirement	250	150	300	150	100	400	250	200	300

First, express the *ordering cost* in terms of an equivalent number of part periods of carrying cost by dividing the order cost, C_O, by the carrying cost, C_C.

$$\text{PPA order cost} = \text{PPA } C_O = \frac{C_O}{C_C} = \frac{\$225}{\$.75} = 300 \text{ part periods}$$

Next, express the *carrying cost* in terms of part periods by assigning one part-period cost for each time period a unit is held in stock (i.e., weight each unit by the number of periods it is carried).

$$\text{PPA carrying cost} = \text{PPA } C_C = 0/\text{unit if units used during period they arrive}$$
$$= 1/\text{unit for units carried forward 1 period}$$
$$= 2/\text{unit for units carried forward 2 periods, etc.}$$

Next, cumulate requirements until the part-period carrying cost PPA C_C is as close as possible to the part-period ordering cost PPA C_O. Do not divide a period's requirements. Begin with an order to be received in period 1 for the period-1 requirements. Multiply the number of units required times PPA C_O of zero, and add other periods (appropriately weighted).

Cumulate requirements until Σ PPA C_C is closest to 300.

	Month 1	Month 2	Month 3
No. required $\times$ PPA C_c	$250 \times 0 = 0$	$150 \times 1 = 150$	$300 \times 2 = 600$
Cumulative total	0	150	750

The value nearest 300 is 150 (see month 2). The first order will include requirements for months 1 and 2 only (400 units).

Next, continue by reassigning the next (unfilled) order a PPA C_C of zero, and repeat the previous step until the allocation is complete. For convenience we can arrange the results of this procedure in table form, where each additional row identifies another order. See Table 14-8.

Table 14-8

Order No.	Month No. Req'd.	1 250	2 150	3 300	4 150	5 100	6 400	7 250	8 200	9 300
1	No. $\times$ PPA C_C cum. total	250×0 0	150×1 150	300×2 750						
2	No. $\times$ PPA C_C cum. total			300×0 0	150×1 150	100×2 350				
3	No. $\times$ PPA C_C cum. total						400×0 0	250×1 250	200×2 650	
4	No. $\times$ PPA C_C cum. total								200×0 0	300×1 300
Order size		400		550			650		500	

Conclusion: Order 400 units in period 1, 550 in period 3, 650 in period 6, and 500 in period 8.

CRP ACTIVITIES: INFINITE AND FINITE LOADING

14.17 A work center operates 6 days per week on a two-shift-per-day basis (8 hours per shift) and has four machines with the same capability. If the machines are utilized 75 percent of the time at a system efficiency of 90 percent, what is the rated output in standard hours per week?

$$\text{Rated capacity} = \binom{\text{number of}}{\text{machines}}\binom{\text{machine}}{\text{hours}}\binom{\text{percentage of}}{\text{utilization}}\binom{\text{system}}{\text{efficiency}}$$

$$= (4)(8 \times 6 \times 2)(.75)(.90) = 259 \text{ standard hr/week}$$

14.18 The Metric Instrument Company uses an MRP system and plans to adjust capacity when the cumulative deviation exceeds one-half of the forecasted average per week. They have calculated capacity requirements per week for their testing laboratory over the next 8 weeks as shown in the following table.

Week no.	1	2	3	4	5	6	7	8
Hours required	400	380	210	530	420	410	500	350

(*a*) Graph the capacity requirements, showing the average requirement as a dotted line. (*b*) Assume that actual requirements for the first 5 weeks were 390, 460, 280, 510, and 550 units. Compute the cumulative deviation of (actual − planned), and determine whether an adjustment is needed.

(*a*)   See Fig. 14-19. Average requirement = 400 hours per week.

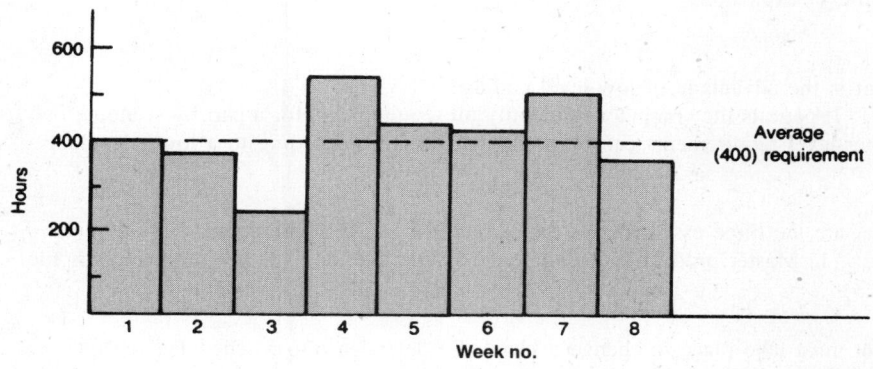

Fig. 14-19

(*b*)   See Table 14-9.

Table 14-9   Report of Cumulative Deviation

Week number	1	2	3	4	5	6	7	8
Planned hours	400	400	400	400	400	400	400	400
Actual hours	390	460	280	510	550			
Cumulative deviation	−10	50	−70	40	190			

No adjustment is needed. The cumulative deviation would have to reach 400 ÷ 2 = 200 hours, but is only 190 hours by the end of the fifth week.

Supplementary Questions and Problems

14.19 (a) What is meant by "time phasing"? (b) What is a "time bucket"?
Ans.   (a) Anticipating requirements and scheduling inventory to arrive when needed. (b) The time interval used for planning (on the MRP planning report).

14.20 (a) Why is the BOM file sometimes called a product structure tree? (b) Suppose a 5-level BOM has one subassembly (AX205) on level 2 and another (BY407) on level 4. Which subassembly has a level code nearest the end item?
Ans.   (a) It is frequently depicted in a treelike diagram to show how all the various materials, parts, and subassemblies are merged to form the finished product. (b) AX205

14.21 Why do MRP programs use single-level BOMs?
Ans.   A series of single-level explosions can typically be processed (on computer) more efficiently than multilevel explosions.

14.22 What is the advantage of low-level coding?
Ans.   It permits the computer to identify all requirements for a part by scanning one level at a time. Then higher-level requirements can be netted out before proceeding to a lower level.

14.23 What are the three essential sources of data for an MRP program?
Ans.   (1) Master production schedule, (2) BOM file, and (3) inventory records file

14.24 What must take place to change a planned-order release to a scheduled receipt?
Ans.   The order must actually be released to a vendor or an internal production shop.

14.25 By using a time-phased plan for component inventories, a firm can reduce average inventory levels from 105 units to 42.5 units. If the average component value is $12 and the reduction applies to 4,000 components, how much of a savings would result? Use inventory carrying costs of 30 percent per year.   *Ans.* $900,000 per year

14.26 Determine the net requirements for the three items shown in Table 14-10.

Table 14-10

	Switches	Microprocessors	Keyboards
Gross requirements	55	14	28
On-hand inventory	18	2	7
Inventory on order (scheduled receipts)	12	12	10

Ans.   25 switches, 0 microprocessors, 11 keyboards

14.27 Given the product structure tree shown in Fig. 14-20 and the inventory shown in Table 14-11, compute the net requirements for A, B, C, D, and E to produce 50 units of X.

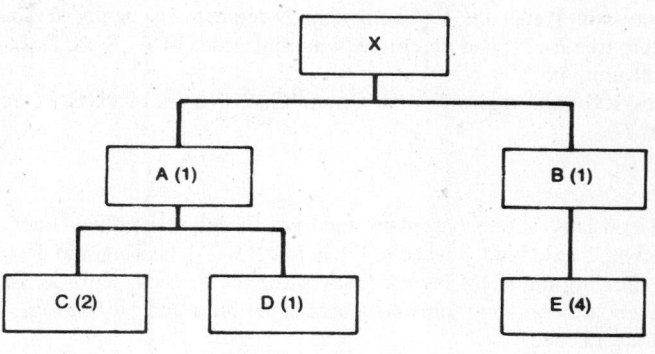

Fig. 14-20

Table 14-11

Component	A	B	C	D	E
Inventory on hand and on order	20	10	15	30	100

Ans.   A = 30, B = 40, C = 45, D = 0, E = 60

14.28 Given the product structure tree shown in Fig. 14-21, compute the net requirements to produce 100 units of subassembly A. No stock is on hand or on order.
   Ans.   400 C's, 200 D's, 400 G's, 400 J's, 800 H's

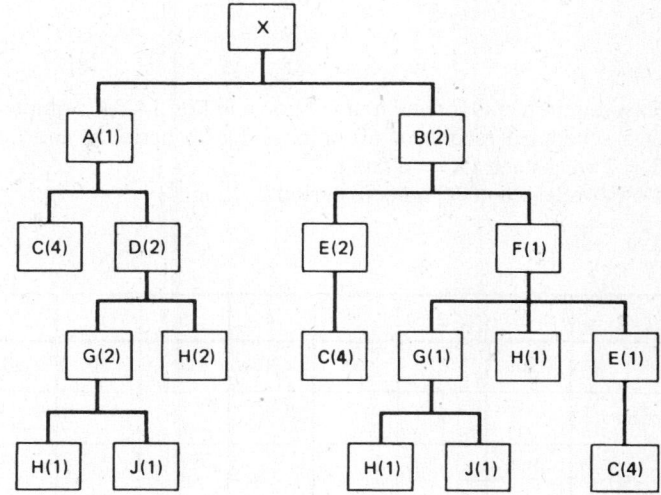

Fig. 14-21

14.29 Given the product structure tree shown in Fig. 14-21 of Prob. 14.28, what net amounts of C are required to produce 200 units of X? The only on-hand inventory is 50 units of subassembly B and 30 units of subassembly F.   *Ans.*   4,880

14.30 End item X is assembled from three major assemblies, A, B, and C. Subassembly A consists of two units of D, two units of E, and one F. To make B, component G and three units of H are needed. Subassembly C requires two units of J and one F. Component D requires two units of J and one unit of K. Construct a product structure tree for X; and determine what quantities of A, B, C, D, E, F, G, H, J, and K are required to produce 100 units of X?

Ans.   Requires 100 units each of A, B, C, and G; 200 units of D, E, F, and K; 300 units of H; and 600 units of J.

14.31   A skateboard consists of one baseplate and two wheel assemblies. Each wheel assembly comprises 1 mounting bracket, 1 axle, and 2 wheels. Each wheel has 1 bearing and 1 steel shell. (*a*) Draw the product structure tree showing the BOM levels. (*b*) Assume the firm has an order for 300 skateboards and has 200 completed skateboards on hand, plus 40 wheel assemblies and 50 bearings. How many *more* bearings (net requirements) are needed?

Ans.   (*a*) Skateboard = level 0; baseplate(1) and wheel assembly(2) = level 1; mounting bracket(1), axle(1), and wheels(2) = level 2; bearing(1) and shell(1) = level 3 (*b*) 270 bearings

14.32   Complete the MRP format shown in Fig. 14-22. How many units are on hand at the end of period 8?
Ans.   85 units

	Week	1	2	3	4	5	6	7	8
Item ID:	Gross requirements	40	85	10	60	130	110	50	170
Level code:	Scheduled receipts								
On hand: 140	On hand/Available								
Lot size: 200	Net requirements								
LT (wk): 3	Planned-order receipts								
Safety stock:	Planned-order releases								

Fig. 14-22

14.33   Given the forecast requirements for end item Y shown in Fig. 14-23,, complete the material-requirements plan. Note that a scheduled receipt of 60 units is due in period 2 and a safety stock of 25 is to be maintained. LT = 2 weeks and OQ = 60 units

Ans.   Planned-order releases of 60 units in period 2, 5, and 8

Week		1	2	3	4	5	6	7	8	9	10
Gross requirements		20	20	20	30	20	20	20	25	20	35
Scheduled / Planned receipts			60								
On hand at end of period	50										
Planned-order release											

Fig. 14-23

14.34   A master scheduler would like to determine whether an order for 200 units of product A can be supplied in period 8. See Fig. 14-24. No stock of any components is on hand or on order, and all order sizes are lot-for-lot. Determine the amount and date of the planned-order releases for all components.

Ans.   400 B in period 5; 1,600 C in period 4 and 400 in period 6; 400 D in period 1; 1,600 E in period 2 and 400 in period 4; 1,600 F in period 3 and 400 in period 5

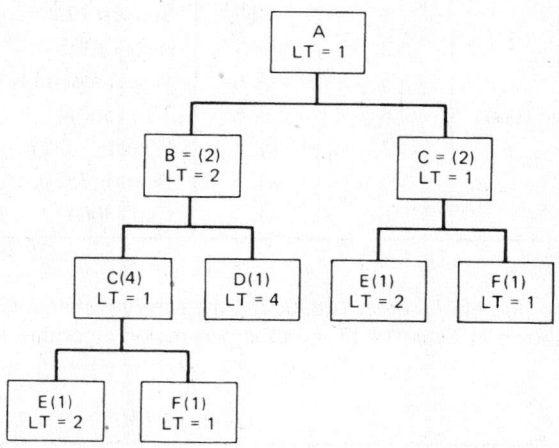

Fig. 14-24

14.35   Industrial Supply Co. produces a maintenance and repair parts cart (MRP cart) for use in warehouses. The cart design and product structure are shown in Fig. 14-25. The firm has 2 axles (number 2005) and 1 wheel assembly (number 2006) in stock. They have an order for 3 carts in period 10. Use an MRP format with lot-for-lot ordering (i.e., order the exact number of units required), and determine the order size and order-release period for all components.   *Ans.*   See Table 14-12.

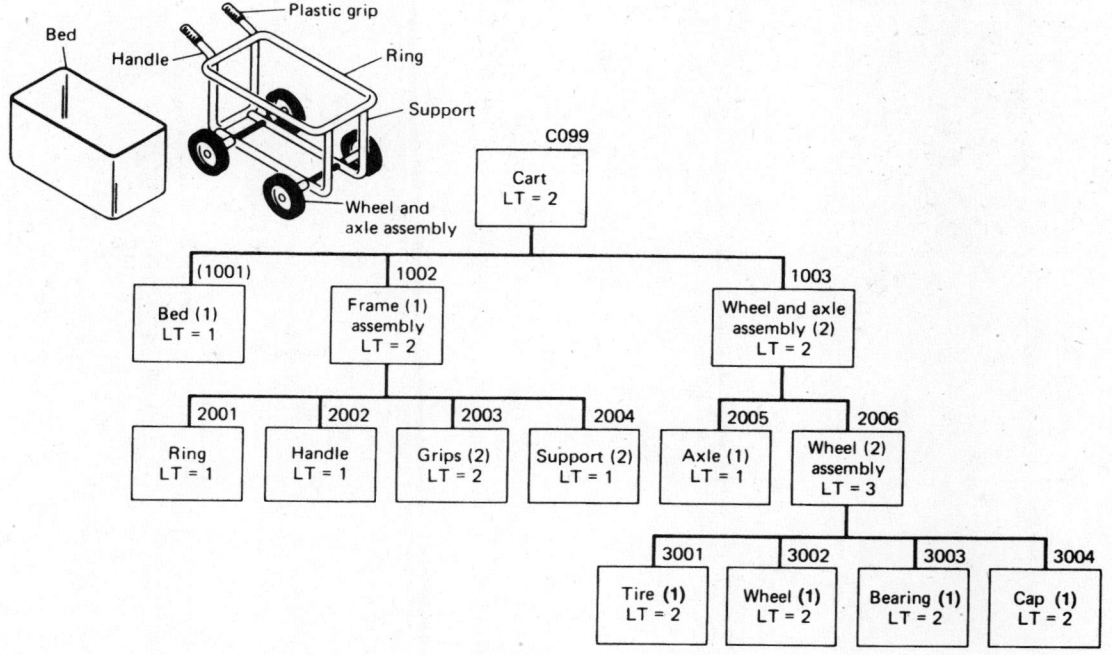

Fig. 14-25

Table 14-12

Item	Quantity	Release Date	Item	Quantity	Release Date
Cart (C099)	3	wk 8	Support (2004)	6	wk 5
Bed (1001)	3	wk 7	Axle (2005)	4	wk 5
Frame (1002)	3	wk 6	Wheel assembly (2006)	11	wk 3
W & A assembly (1003)	6	wk 6	Tire (3001)	11	wk 1
Ring (2001)	3	wk 5	Wheel (3002)	11	wk 1
Handle (2002)	3	wk 5	Bearing (3003)	11	wk 1
Grip (2003)	6	wk 4	Cap (3004)	11	wk 1

14.36 The cost for placing an item on order is $90, and the carrying charge is $.45 per month. Net requirements per month are as shown in Table 14-13. Use the part-period algorithm to determine the size and timing of orders.

Table 14-13

Month	1	2	3	4	5	6	7	8
Requirement	180	150	250	450	125	50	200	210

Ans.   Order 330 units in period 1, 250 in period 3, 625 in period 4, and 410 in period 7

14.37 An office furniture manufacturer has a work center with 3 metal presses that are each operated 7 1/2 hours per shift on a 3-shift-per-day, 6-day-per-week basis. The presses are allocated to furniture production 80 percent of the time, with the remainder reserved for special-order jobs. If the machine efficiency is 95 percent, what is the rated output for furniture production in standard hours per week?
Ans.   308 hours

Chapter 15

Operations Scheduling and Control
Assignment Linear Programming

OVERVIEW OF OPERATIONS SCHEDULING AND CONTROL FUNCTIONS

Production activity takes place in a wide range of (*a*) manufacturing and (*b*) service systems. Although many of the concepts presented in this chapter were originally developed within the context of manufacturing systems, they now have wide applicability in service systems.

The material and capacity functions described in the previous chapter (e.g., MRP and CRP) are primarily concerned with (priority and capacity) *planning* activities. Production, or operations scheduling and control, is concerned with *implementing* the plans, i.e., the detailed scheduling of jobs, assignment of workloads to machines (and people), and the actual flow of work through the system.

Question: Systems control has been defined to include measurement, feedback, comparison with standards, and correction activities. In what sense are operations scheduling and control activities really "control" activities?

Operations scheduling and control activities follow planning and satisfy the basic definition of "control." They involve (1) the *measured* release/monitoring of work in the system, (2) *feedback* of performance data, (3) *comparison* of results to the material and capacity plan objectives, and (4) the implementation of *corrections* when needed.

Figure 15-1 identifies the major scheduling and control functions. *Order release, sequencing, and status control* activities help ensure that production activities follow the (MRP or other) plan by assigning and monitoring the priorities of orders in the system. *Loading, lead time, and input/output controls* are designed to ensure that work centers provide the amount of labor and equipment capacity (time) that is necessary—and was planned—to do the scheduled work.

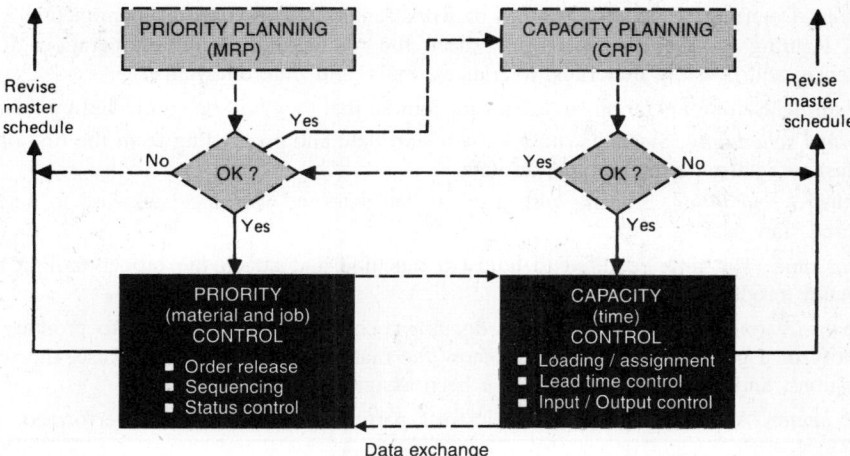

Fig. 15-1 Production activity control system functions

301

Fig. 15-2   Production Activity Control Terms*

1. *Control* (as related to the type of system):

   Flow. Control of *continuous operations* by setting common production rates for all items; feeding work into the system at a specified rate, and monitoring the rate.

   Order. Control of *intermittent operations* by monitoring the progress of each individual order through successive operations in its production cycle.

2. *Control* (as related to jobs and time):

   Priority. Control over the status of jobs and work activities by specifying the order in which materials or jobs are assigned to work centers.

   Capacity. Control over the labor and machine *time* used for jobs and work activities by planning and monitoring the time requirements of key work centers.

3. *Critical ratio.* A dynamic scheduling technique. Priority index numbers are calculated for ranking jobs according to which are in most urgent need of work time so that orders can be shipped on schedule.

4. *Dispatching.* Selecting and sequencing jobs to be run at individual work centers and actually authorizing or assigning the work to be done. The dispatch list is the primary means of priority control.

5. *Expediting.* Finding discrepancies between planned and actual work output, and correcting them by attempting to speed up the processing in less than the normal lead time.

6. *Input control.* Control over the work being sent to a supplying facility, whether this is the shop itself or an outside vendor.

7. *Lead time.* The period between the decision to release an order and the completion of the first units. Includes wait, move, set-up, queue, and run time.

8. *Line of balance.* A charting technique that uses lead times and assembly sequencing to compare planned component completions with actual component completions.

9. *Loading.* Assigning hours of work to work centers in accordance with the available capacity of the work centers.

   Finite capacity. Rescheduling work into other periods if insufficient capacity exists in the required time period.

   Infinite capacity. Assigning work to the given time period whether or not sufficient capacity exists.

10. *Output control.* Dispatching, expediting and any other follow-up necessary to get scheduled work from a work center or vendor.

11. *Priority decision rules.* Rules used by a dispatcher to determine the sequence in which jobs will be done.

12. *Routing.* Determining which machines or work centers will be used to manufacture a particular item. Routing is specified on a route sheet; the route sheet identifies operations to perform, sequence, and possibly materials, tolerances, tools, and time allowances.

13. *Scheduling.* Setting operation start dates for jobs so that they will be completed by their due date.

    Forward scheduling. Starting with a known start date and proceeding from the first operation to the last to determine the completion date.

    Backward scheduling. Starting with a given due date and working backward to determine the required start date.

14. *Set-up time.* The time required to adjust a machine and attach the proper tooling to make a particular product.

15. *Shop order (manufacturing order).* A document conveying the authority to produce a specific quantity of a given item. It may also show the materials and machines to use, the sequence of operations, and the due dates that have been assigned by the scheduler.

16. *Work center.* An area or workstation where a particular type of work is performed.

*The definitions are taken largely from the *American Production and Inventory Control Society (APICS) Dictionary* (modified and/or condensed).

SCHEDULING AND CONTROL FUNCTIONS, DATA, AND OBJECTIVES

Question: What are the six key functions of the operations scheduling and control (per Fig. 15-1)?

(1) *Release orders* to the system in accordance with the priority plan.

(2) *Assign jobs* to specific work centers (including machine *loading*, or shop *loading*).

(3) Provide *sequencing priorities* to specify the order in which jobs are to be processed.

(4) Control the *manufacturing lead time* by *tracking and expediting* jobs if necessary.

(5) Monitor the priority *status of jobs* via summary, scrap, rework, and other reports.

(6) Monitor the capacity *status of facilities via input/output* reports of workload versus capacity.

Functions. Numerous techniques (e.g., graphic, charting, computer algorithms) have been developed to assist in accomplishing the six functions described above—some of which will be reviewed in detail later in this chapter.

Many of the six functions incorporate ancillary activities that are not always clearly and definitively classified into one of the categories listed. For example, the machine or shop loading activities referenced in (2) are sometimes referred to as *short-run capacity planning*, and the input/output controls of (6) as *short-run capacity control*. The sequencing activities mentioned in (3) may be assumed to include the *dispatching function*, which is the actual authorization for the work to begin. The sequencing, tracking, expediting, and status control activities (3, 4, 5, and 6) are frequently associated with the term *shop floor controls*, which is sometimes extended to include (1 and 2). When taken together, the six functions—or major portions of them—are also referred to as *production activity controls*.

Question: What are production activity controls?

Production Activity Controls (*PAC*) are the priority and capacity management techniques used to schedule and control production operations so as to ensure they follow the material (MRP) and capacity (CRP) plans. Figure 15-2 provides some definitions of PAC and other related terms.

Data Requirements. Operations scheduling and control represents a transition from the broader-based planning phase of production to the initiation and control of more detailed day-to-day operations—often referred to as shop floor controls. To accomplish this, the system requires accurate information on (1) the current status of jobs (e.g., what orders are in process and where), (2) what upcoming jobs are available, (3) the adequacy of materials and capacities, (4) equipment and labor utilization, and (5) job progress and efficiency. In addition, the database must contain information on current inventory levels, lot sizes, lead times, work-center capacities, set-up and run times, scrap rates, due dates, etc.

Scheduling Objectives. Finally, the term scheduling is such a central and encompassing activity that it is sometimes used in a general way that assumes the inclusion of other aspects of production activity control.

Question: (*a*) Explain the use of the term "scheduling." (*b*) What are the objectives of scheduling?

(*a*) *Scheduling* refers to the setting of operation start dates so that jobs will be completed by their due date. [It is frequently associated with the sequencing and dispatching functions listed in (3) above.] In a broader sense, scheduling establishes the timing of productive activities that use the organization's human and equipment resources to serve its customers.

(*b*) The *objectives of scheduling* are to optimize the goals of (1) providing the best customer service possible (e.g., inventory, delivery) while at the same time (2) making the most efficient use of the people, equipment, and facilities available to the organization.

SCHEDULING IN HIGH-, INTERMEDIATE-, AND LOW-VOLUME SYSTEMS

Scheduling activities are highly dependent upon the type of underlying system and the volume of output delivered by the system.

Question: How do scheduling activities differ in (a) high-, (b) intermediate-, and (c) low-volume systems?

(a) *High-volume (flow) systems* typically utilize specialized equipment that routes work on a *continuous* basis through the same fixed path of operations, often at a rapid rate. The problems of order release, dispatching, and monitoring are less complex than in low-volume, make-to-order systems. However, material flows must be well coordinated, inventories carefully controlled, and extra care exercised to avoid equipment breakdowns, material shortages, etc., that result in production-line downtime.

(b) *Intermediate-volume (flow and batch) systems* utilize a mixture of equipment and similar processes to produce an *intermittent* flow of similar products on the same facilities. The sequencing of jobs and production-run lengths are of significant concern to schedulers, as they attempt to balance the costs of changeover time against those of inventory accumulations.

(c) *Low-volume (batch or single job) systems* utilize general-purpose equipment that must route orders individually through a unique combination of work centers. The variable work-flow paths and processing times generate queues, work-in-process inventories, and capacity utilization concerns that can require more day-to-day attention than in the high- or intermediate-volume systems.

Figure 15-3 summarizes some relevant characteristics of high-, intermediate-, and low-volume systems. The classifications can overlap, however; some intermittent operations are much like job shops, whereas some low-volume operations are done in batches. And job shops often exist within continuous systems. (*Note:* The table also includes comparative data for projects.)

	High Volume	Intermediate Volume	Low Volume	
Type of production system	Continuous (flow operations)	Intermittent (flow and batch operations)	Job Shop (batch or single jobs)	Project (single jobs)
Key characteristics	• Specialized equipment • Same sequence of operations unless guided by microprocessors and robots	• Mixture of equipment • Similar sequence for each batch	• General-purpose equipment • Unique sequence for each job	• Mixture of equipment • Unique sequence and location for each job
Design concerns	• Line balancing • Changeover time and cost	• Line and worker-machine balance • Changeover time and cost	• Worker-machine balance • Capacity utilization	• Allocating resources to minimize time and cost
Operational concerns	• Material shortages • Equipment breakdowns • Quality problems • Product mix and volume	• Material and equipment problems • Set-up costs and run lengths • Inventory accumulations (run-out times)	• Job sequencing • Work-center loading • Work flow and work in process	• Meeting time schedule • Meeting budgeted costs • Resource utilization

Fig. 15-3   Characteristics of scheduling systems

Scheduling Strategies. Scheduling strategies differ among firms and range from very detailed scheduling to no scheduling at all. A cumulative schedule of total workload is useful for long-range planning of approximate capacity needs. However, detailed scheduling of specific jobs on specific equipment at times far in the future is often impractical—because of inevitable changes.

For continuous systems, detailed schedules (production rates) can often be firmed us as the master schedule is implemented. *For job shop operations*, schedules may be planned on the basis of estimated labor and equipment (standard hour) requirements per week at key work centers. When detailed scheduling is desirable, capacity is sometimes allocated to specific jobs as late as a week, or a few days, before the actual work is to be performed. (A goal of agile manufacturing activities is to enhance the system's ability to respond very quickly to such customer needs.) However, detailed scheduling is not always used; some firms simply rely on priority decision rules such as first come, first served.

ORDER RELEASE

Order release converts a need from a planned-order status to a real order in the shop or with a vendor by assigning it either a shop order or purchase order number. Well-designed scheduling and control systems release work at a reasonable rate that keeps unnecessary backlogs from the production floor. (Releasing all available jobs as soon as they are received from customers is a common cause of increased manufacturing lead times and excess work in process, WIP.)

Question: How does the order release function create a scheduled receipt?

Figure 15-4 illustrates the process. As the shop day and current calendar day coincide, the planned-order release takes place. The order quantity is deleted (from the MRP planned-order release row), and a shop order for that amount is added to the dispatch list, along with a start and due date priority. The order quantity is then reentered (on the MRP form) as a scheduled receipt on the listed due date.

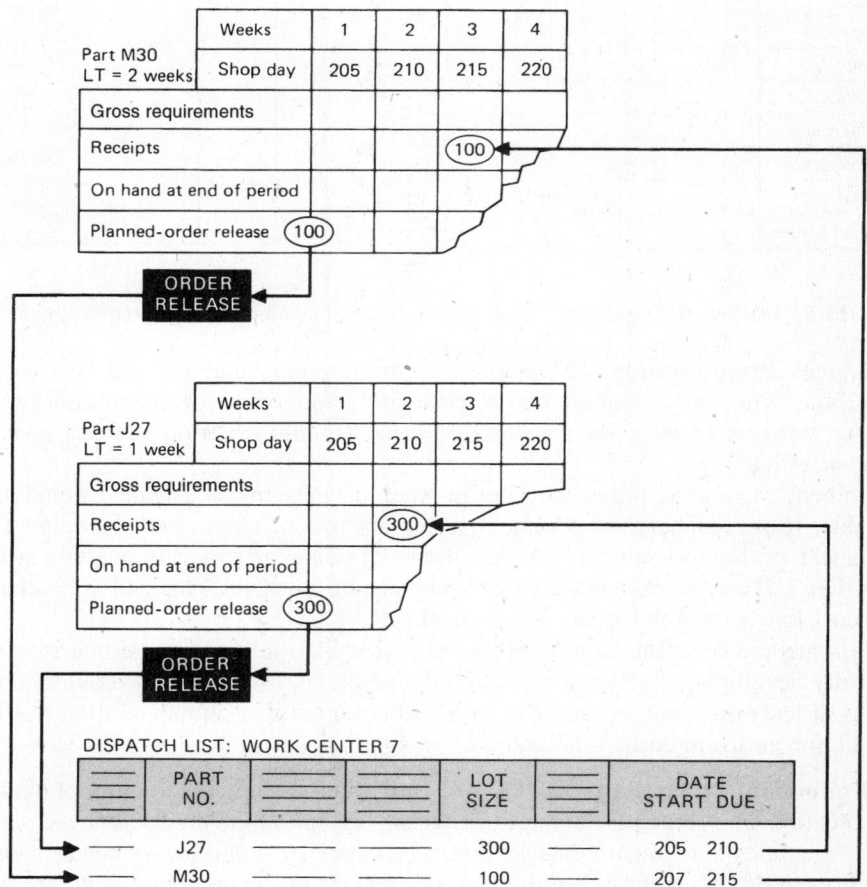

Fig. 15-4   Relationship between planned-order releases and dispatch list

LOADING AND ASSIGNMENT

Loading assigns jobs to specific work centers (WCs) in an effort to complete the work within the planned time and cost constraints. In addition to allowing adequate time *for* operations, planners must allow for time *before*, *between*, and *after* operations. When planning the work flow through consecutive work centers, planners frequently load only selected (gateway and bottleneck) work centers. Loading procedures must also allow for maintenance, changeover, and unplanned events.

Gantt Charts.   When jobs can be done in any of several different work centers, various criteria are employed to help in the assignment of work. Some charting methods, originally developed by Henry Gantt in the early 1900s, continue to be among the most widely used tools in use today, except they are now done on computer.

Question: What are Gantt charts?

Gantt charts are visual (bar chart) aids used to depict the sequencing, load on facilities, or progress associated with work effort over a well-defined time period. Figure 15-5 illustrates a general form of scheduling chart (*a*) and a load chart (*b*). In the load chart, the work assigned to specific work centers is identified by the shop order numbers (SO #). Time periods X-ed out are unavailable (e.g., in maintenance).

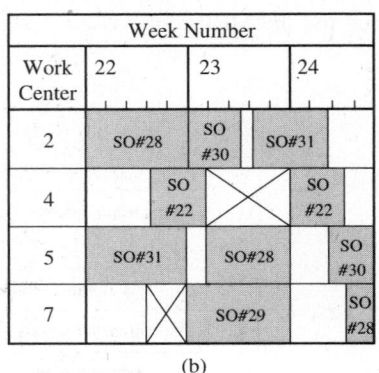

Fig. 15-5   (*a*) Scheduling chart   (*b*) Load chart ⊠ Unavailable delay (maintenance, etc.)

Assignment Linear Programming.   Mathematical programming methods are also useful techniques for loading facilities. The assignment method is particularly applicable for assigning jobs to machines or work centers, or workers to jobs, on the basis of some (single) criterion such as cost, performance, quantity, time, or efficiency.

The assignment method requires that the number of jobs to be assigned equal the number of workers available. If the numbers are not equal, a dummy row or column is added and assigned a zero criterion coefficient (to help identify the job to eliminate) or a high criterion coefficient (to be sure the job never gets done). If any worker-machine assignments are infeasible, the cell is blocked out or given an exorbitant cost that will prohibit any assignment.

The solution method for minimization problems involves forming a square matrix of criteria values and systematically developing zeros in the cells until a zero exists for each row-column combination. As soon as there is at least one zero in each row and each column, an optimal solution has been obtained. See Prob. 15.11 for an example of this method.

Optimized Production Technology (OPT).   A (proprietary) finite loading tool that has received considerable attention in operations literature is known as *optimized production technology*, or OPT. OPT consists of modules that contain data on products, customer orders, work center capacities, etc., as well as algorithms to do the actual scheduling. A key feature of the computer software is the ability of the program to simulate the load on the system, identify bottleneck (as well as other) operations, and develop alternative production schedules.

SEQUENCING AND DISPATCHING

Sequencing activities are most closely identified with detailed scheduling, for they specify the order in which jobs are to be processed at the various work centers. The form that conveys this order and translates the schedule into action is the dispatch list, as depicted in Fig. 15-4.

Question: (*a*) What is dispatching, and (*b*) what is the function of the dispatch list?

(*a*)   *Dispatching* is the schedule-generating procedure that involves selecting and sequencing jobs to be run at specific work centers and authorizing the work to commence.

(*b*)   The *function of the dispatch list* is to implement the schedule in a manner that retains any order priorities assigned at the (MRP) planning phase.

Forward versus Backward Scheduling.   Forward scheduling is akin to releasing all available jobs to the shop in that it begins as soon as requirements are known. This immediate release can result in early completion of a job at a possible cost of more work in process and in higher inventory carrying costs than necessary. Backward (or *set back*) scheduling uses the same lead-time offset logic as MRP. Components are delivered ''when needed'' rather than ''as soon as possible.''

Example 15.1   A job due at the end of period 12 requires a 2-period lead time for material acquisition, 1 period of run time for operation 1, 2 periods for operation 2, and 1 period for final assembly. Allow 1 period of transit time prior to each operation. Illustrate the completion schedule under (*a*) forward and (*b*) backward scheduling approaches.

Forward scheduling (*a*) is shown above the time period line, and backward scheduling (*b*) below the line in Fig. 15-6.

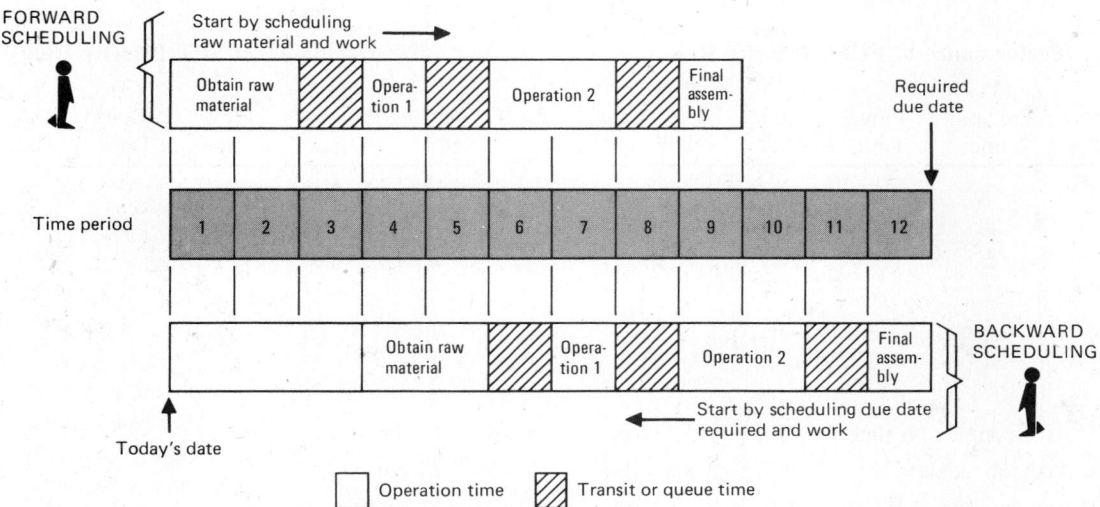

Fig. 15-6   Forward and backward scheduling

Priority Decision Rules.   Job shops typically have many jobs waiting to be processed. *Priority decision rules* are simplified guidelines (heuristics) for determining the sequence in which jobs will be done. The simplest rules assign jobs on the basis of a single criterion such as first come, first served (FCFS); earliest due date (EDD); least slack, or due date less processing time (LS); shortest processing time (SPT); longest processing time (LPT); and preferred customer order (PCO). Most single-criterion rules are static in that they do not incorporate an updating feature.

Example 15.2   Shown below are the due dates (number of days until due) and process time remaining (number of days) for five jobs that were assigned a letter as they arrived. Sequence the jobs by priority rules: (*a*) FCFS, (*b*) EDD, (*c*) LS, (*d*) SPT, and (*e*) LPT.

Job	Due Date (days)	Process Time (days)
A	8	7
B	3	4
C	7	5
D	9	2
E	6	6

See Table 15-1. The numerical amounts included in parentheses are for reference only.

Table 15-1

	FCFS	EDD	LS	SPT	LPT
1st	A	B(3)	B(−1)	D(2)	A(7)
2d	B	E(6)	E(0)	B(4)	E(6)
3d	C	C(7)	A(1)	C(5)	C(5)
4th	D	A(8)	C(2)	E(6)	B(4)
5th	E	D(9)	D(7)	A(7)	D(2)

Example 15.3 Compare the effectiveness of the FCFS and SPT rules in terms of the (a) average completion time, (b) average job lateness, and (c) average number of jobs at the work center.

Performance of FCFS Priority Rule

Job Seq.	(1) Process Time	(2) Flow Time	(3) Due Date	(2)−(3) Days Late (0 if neg.)
A	7	7	8	0
B	4	11	3	8
C	5	16	7	9
D	2	18	9	9
E	6	24	6	18
	24	76		44

Performance of SPT Priority Rule

Job Seq.	(1) Process Time	(2) Flow Time	(3) Due Date	(2)−(3) Days Late (0 if neg.)
D	2	2	9	0
B	4	6	3	3
C	5	11	7	6
E	6	17	6	11
A	7	24	8	16
	24	60		36

(a) Ave. completion time: $76/5 = 15.2$ days $60/5 = 12.0$ days

(b) Ave. job lateness: $44/5 = 8.8$ days $36/5 = 7.2$ days

(c) Ave. no. jobs at WC: $76/24 = 3.2$ jobs $60/24 = 2.5$ jobs

Johnson's rule yields a minimum processing time for sequencing *n* jobs through two machines or work centers where the same processing sequence must be followed by all jobs. Jobs with the shortest processing times are placed early if that processing time is on the first machine and placed late if that processing time is on the second machine. This procedure maximizes the concurrent operating time of both work centers.

Example 15.4 Wonderloaf Bakery has orders for five specialty jobs (A, B, C, D, and E) that must be processed sequentially through two work centers (baking and decoration). The amount of time (in hours) required for the jobs is shown in the table below. Determine the schedule sequence that minimizes the total elapsed time for the five jobs, and present it in the form of a Gantt chart.

	Time Required for Job (hours)				
	A	B	C	D	E
WC 1 (baking)	5	4	8	7	6
WC 2 (decoration)	3	9	2	4	10

Johnson's rule says to identify the shortest processing time. If it is at the first work center, place the (entire) job as early as possible. If it is at the second work center, place the job as late as possible. Eliminate that job from further consideration, and apply the decision rule to the remaining jobs. Break any ties between jobs by sequencing the job on the first work center earliest and that on the second work center latest. Jobs having the same time at both work centers can be assigned at either end of the available sequence. See Fig. 15-7.

(a) The shortest time is for job C in work center 2 (2 hours). Place job C as late as possible.

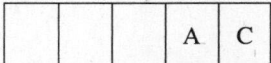

(b) The next shortest time is for job A in work center 2. Place job A as late as possible.

(c) The next shortest time is a tie between jobs B and D. Sequence the job on the first work center (job B) as early as possible.

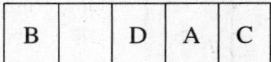

(d) The next shortest time is for job D in work center 2. Place the job as late as possible.

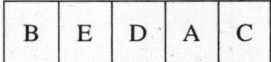

(e) Place job E in the remaining opening.

| B | E | D | A | C |

The sequential times are as follows: (32 hr total)

| Work center 1 | 4 | 6 | 7 | 5 | 8 |
| Work center 2 | 9 | 10 | 4 | 3 | 2 |

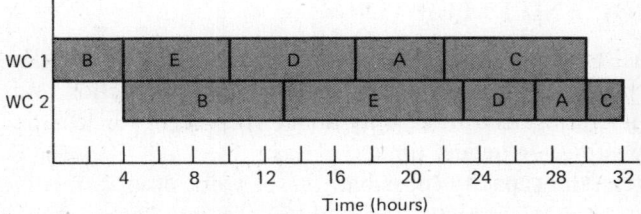

Fig. 15-7 Sequence using Johnson's rule

Critical ratio (CR) is a dynamic dispatching rule that yields a priority index number that expresses the time remaining/work remaining ratio. It can be constantly updated (often daily) with a computer to provide close control. If the CR is less than 1.0, the job is behind schedule; if the CR = 1.0, the job is on schedule; and if the CR is larger than 1.0, the job has some slack.

$$CR = \frac{\text{time remaining}}{\text{work remaining}} = \frac{TR}{WR} \qquad (15.1)$$

where   TR = date due − date now = DD − DN

Example 15.5 Today is day 22 on the production-control calendar, and four jobs are on order as shown. Determine the critical ratio for each job, and assign priority ranks.

Job	Date Due	Workdays Remaining
A	28	8
B	26	2
C	24	2
D	30	12

See Table 15-2.

Table 15-2   Computation of Critical Ratio

Job	Time Remaining (DD − DN) = TR	Work Remaining (WR)	$CR = \dfrac{TR}{WR}$	Priority
A	28 − 22 = 6	8	.75	2
B	26 − 22 = 4	2	2.00	4
C	24 − 22 = 2	2	1.00	3
D	30 − 22 = 8	12	.67	1

With the critical ratio, jobs would be assigned in the order of D, A, C, and B. Job B is the only one with some slack. Jobs A and D have critical ratios of less than 1, meaning that the orders will not be shipped on time unless they are expedited. Job C, with an index of 1, is the only job on schedule.

LEAD-TIME CONTROL AND EXPEDITING

Total *lead time* consists of the time that orders wait in unreleased backlog, plus the manufacturing lead time (the interval between when an order is released to production and when it is available for shipment). Of the manufacturing lead time, only about 10 percent (or less) is usually working, or run, time; the bulk of it is waiting, or queue, time.

For most work centers, the capacity (possible *rate* of work flow) is relatively fixed, much like flow through a funnel. If work is released to the facility at a rate faster than its output rate, both the manufacturing lead times and the work in process will increase. But output will not increase. The only result is a mounting inventory, a worsening shortage of space, lost orders, unnecessary expediting, and increased inventory carrying costs. Furthermore, priority techniques, such as red (urgent) tags, and expediting do not solve capacity problems.

Example 15.6   A farm machinery manufacturer has an output rate of 320 hours per week and has measured the load on his shop as follows:

Unreleased shop orders	640 hr
Work in process:	
Current requirements	960
Long-term orders	320
Total	1.920 hr

(a)   Find the manufacturing lead time, and (b) comment on the inclusion of long-term orders in the lead time.

(a) $$\text{Lead time} = \frac{\text{work in process}}{\text{rate of output}} = \frac{960 \text{ hr} + 320 \text{ hr}}{320 \text{ hr/week}} = 4 \text{ weeks}$$

(b)   Once the long-term orders are released to the shop, they become part of the work in process, and so they are included as "released backlog." If they contain only deferred requirements and are not required to keep the shop loaded, they should not have been released to the shop.

Question: (a) What is expediting, and (b) why are expediters necessary?

(a) *Expediting* is the process of speeding up the progress of an individual order. (b) Theoretically, expediters should not be necessary if systems always functioned as planned. In reality, machines do break down and materials are sometimes late or defective. If the affected job is to be delivered on schedule, someone must give it special handling to ensure it moves through the facility on schedule. The expediter's challenge is to "push" the critical job without causing unnecessary problems for other jobs.

STATUS CONTROL

Feedback, in the form of status and exception reports, completes the link that makes production-control activities a closed loop. Status reports measure the progress of orders and reveal problems such as internal slippages, defects, and lot-size errors. These reports enable shop planners to be in close communication with everyone involved with an order, from purchasing and inventory control personnel to supervisors on the shop floor.

Question: Identify some of the commonly used status reports.

Included are (1) anticipated delay reports, (2) shortage lists, (3) scrap reports, (4) rework reports, and (5) various order status reports. These tell the production progress, number (and percentage) of orders completed on time, reasons for downtime, etc. One of the most useful reports is the input/output report illustrated in the next section.

INPUT/OUTPUT CONTROLS

Input/output controls are one of the most effective capacity-control devices for intermittent production systems. Output reports compare the actual hours of work delivered by a work center with the planned hours. If the cumulative deviation of actual minus planned hours exceeds some preset standard (such as 1 week's average), corrective action in the form of overtime, subcontracting, or revision of the master schedule may be called for.

Example 15.7   Shop 42 has an average capacity capability of 200 hours per week of work. Actual (standard) hours delivered during weeks 9, 10, 11, and 12 were 180, 210, 170, and 160, respectively. (a) Formulate an output control report showing the cumulative deviation. (b) The maximum allowable cumulative deviation limit is 1 week's average. Is corrective action called for?

(a)   See Table 15-3.

Table 15-3   Output Control Report: Shop 42

Hours	Week Number				
	9	10	11	12	13
Actual hours	180	210	170	160	
Planned hours	200	200	200	200	200
Cumulative deviation (actual − planned)	−20	−10	−40	−80	

(*b*)   By the end of week 12, the cumulative deviation (−80) does not exceed the planned average (200), so no corrective action is yet required.

Output control alone does little to control lead times, however. The input to a facility must be equal to or less than the output. Thus combination input/output control reports are a more effective control device. (See Prob. 15.18.) They are structured in different ways, but the result can be interpreted as follows: If the actual minus the planned value is negative, it signifies that there was less input (or output) than planned. If it is positive, it signifies that there was more input (or output) than planned.

SCHEDULING IN SERVICE SYSTEMS

Scheduling of services (e.g., transportation, medical, professional) is sometimes more of a problem than the scheduling of manufacturing activities for two major reasons:

(1)   Services cannot be stored (inventoried) to meet demand peaks at later times.

(2)   Demand may be quite variable, with peaks more pronounced than for goods.

For example, unused airline seats cannot be "saved" for later use. And income tax deadlines create *yearly* peak demands for accountants, just as our three-meal-a-day convention creates *hourly* peaks for restaurants.

Question: What goals underlie the scheduling of service facilities?

Two common goals are (1) a high-level customer service and (2) a highly efficient use of the service system capacity. This involves a trade-off (facilities are limited but services cannot be preproduced).

Although some service facilities can adjust to meet peak demands (e.g., using part-time workers), the capacity of most service facilities (e.g., airlines, hospitals, theaters, hotels) is relatively fixed. For this reason, service organizations frequently attempt to shift demand by a pricing strategy (e.g., weekend air travel) or to control demand by their scheduling policies.

Question: What methods are used to adapt customer demands to the limited capacity of service facilities?

(1)   *Appointment systems.* Appointments control customer arrival times and the number of customers in the system at one time (e.g., doctors' offices and auto repair shops).

(2)   *Reservation systems.* Reservations are used when customers occupy some facility (e.g., a hotel room, airline seat, rental car). They provide the lead time and assurance of demand that enables the organization to realize a high utilization of its capacity.

(3)   *Pricing incentives and backlogs.* Incentives and backlogs are imprecise ways of shifting or adjusting services demand, but they are widely used. For example, discount fares are commonly used to smooth out airline demands, and backlogs are used in appliance installation and repair.

Questions and Solved Problems

SCHEDULING IN HIGH-, INTERMEDIATE-, AND LOW-VOLUME SYSTEMS

15.1 How is flow control accomplished in continuous flow systems?

In flow systems, the goods (or services) follow a common path, established by the design of the system. Flow control involves measurement of the rate of movement past control points and comparison of this actual data to capacities and standard rates of flow. If actual flows are significantly different from standards, the reason for this can be assessed and acted upon.

15.2 What is a shop floor control system?

A *shop floor control system* is a production control system that utilizes empirical data from the shop floor along with data in a database to maintain and communicate status information on shop orders and work centers for purposes of (1) assigning priorities, (2) monitoring quantities, (3) communicating status, (4) providing output data for capacity control and (5) accounting purposes, and (6) providing measures of efficiency and utilization of the firm's productive resources.

15.3 How can effective shop floor controls improve customer service?

An effective shop floor control system can result in (*a*) lower cost due to reduced rework and scrap, (*b*) improved (reduced) lead times, (*c*) fewer parts shortages, and (*d*) a higher percentage of products delivered on schedule. Reliable operations also increase customer confidence.

15.4 (*Run-out times in intermediate-volume systems*)   A firm that produces athletic supplies on an intermittent production system has the production and inventory characteristics shown in Table 15-4 as of June 30. Rank the items in terms of the urgency of scheduling replenishment inventories.

Table 15-4

Item	Inventory on Hand	Released Orders (and WIP)	Demand Rate
A (soccer balls)	20 cases	30 cases	12 cases/wk
B (volleyballs)	110 cases	—	32 cases/wk
C (basketballs)	70 cases	30 cases	40 cases/wk

One method of scheduling the production of produce-to-stock inventory is to assign the highest priority to items with the lowest *run-out time*.

$$\text{Run-out time} = \frac{\text{inventory on hand} + \text{orders in process}}{\text{demand rate}} \qquad (15.2)$$

$$A = \frac{20 + 30}{12} = 4.2 \text{ weeks} \qquad B = \frac{110}{32} = 3.4 \text{ weeks} \qquad C = \frac{70 + 30}{40} = 2.5 \text{ weeks}$$

The scheduling sequence should be basketballs first, then volleyballs, and then soccer balls.

15.5 How does dispatching in continuous (flow) systems differ from dispatching in intermittent (order) systems?

Dispatching in continuous systems can be less complex when it is concerned primarily with releasing authority to produce and (perhaps) with specification of options. In intermittent systems each order must be routed, prioritized, and individually controlled.

GANTT CHARTS

15.6   Explain the symbols used in a Gantt progress chart.

See Fig. 15-8. The review date is the end of day 5/10. Job A is 2 days behind schedule and has 1 day of work remaining. Job B was not scheduled to start until job A was finished and has not yet been started. Job C has progressed through a scheduled maintenance activity and is on (or slightly ahead of) schedule.

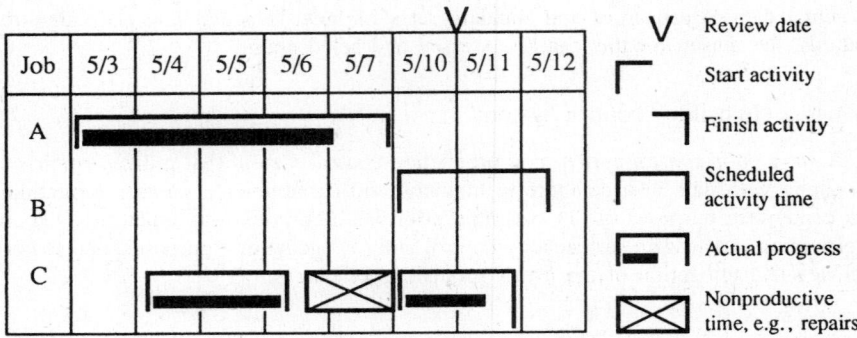

Fig. 15-8   Gantt progress chart

15.7   A foundary producing rod castings experienced the work progress shown below as of July 7. Using standard symbols, formulate a Gantt chart depicting the progress on this job.

Activity	Scheduled Start Date	Scheduled End Date	Actual End Date	Notes
Cast rod	June 28	July 5	July 6	Maintenance scheduled for July 1 took extra day
Form wires	June 29	July 7	Open	Is one-half day behind schedule
Fabricate end caps	July 5	July 7	Open	Started on July 7
Assemble heater	July 11	July 13	Open	Assembly area unavailable on July 8

See Fig. 15-9.

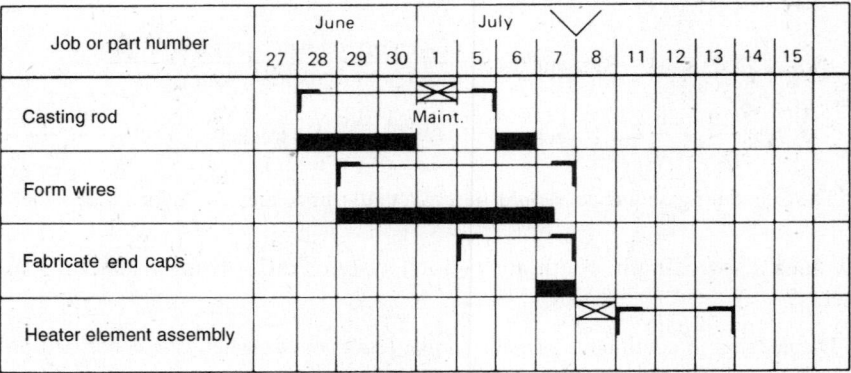

Fig. 15-9   Progress chart

15.8   The hours shown in Table 15-5 are required to complete six jobs that are routed through four work centers. The hours available at the work centers are 40 hours at number 4, 32 at number 5, 36 at number 8, and 30 at number 12. Develop a load chart for assigning the jobs to the work centers.

Table 15-5

	Hours Required at Work Center			
Job Number	#4	#5	#8	#12
A21	4	2	7	4
A22	7	—	—	8
A23	4	10	12	3
B14	—	5	6	5
B15	2	4	5	—
B16	8	1	6	7

Begin with the total time available at each work center, and subtract the time for each job to obtain a cumulative total of unused time; for example, $40 - 4 = 36$, $36 - 7 = 29$. See Table 15-6 and Fig. 15-10.

Table 15-6

Job Number	Work Center 4 hours		Work Center 5 hours		Work Center 8 hours		Work Center 12 hours	
	Required	Available	Required	Available	Required	Available	Required	Available
A21	4	40	2	32	7	36	4	30
A22	7	36	—	30	—	29	8	26
A23	4	29	10	30	12	29	3	18
B14	—	25	5	20	6	17	5	15
B15	2	25	4	15	5	11	—	10
B16	8	23	1	11	6	6	7	10
Unused:		15	Unused:	10	Unused:	0	Unused:	3

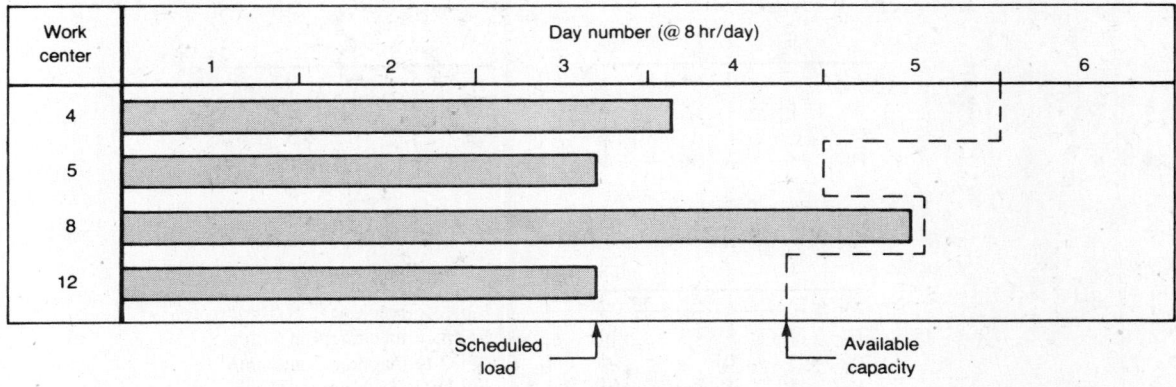

Fig. 15-10

ASSIGNMENT LINEAR PROGRAMMING

15.9 A scheduler has four jobs that can be done on any of four machines with respective times (minutes) as shown in Table 15-7. Determine the allocation of jobs to machines that will result in minimum time.

Table 15-7

Job	Machine			
	1	2	3	4
A	5	6	8	7
B	10	12	11	7
C	10	8	13	6
D	8	7	4	3

The solution method involves five steps.

Step 1. Subtract the smallest number in each row from all others in the row, and enter the results in the form of a new matrix.

Step 2. Using the new matrix, subtract the smallest number in each column from all others in the column, again forming a new matrix.

Step 3. Check to see if there is a zero for each row and column, and draw the minimum number of lines necessary to cover all zero in the matrix.

Step 4. If the number of lines required is less than the number of rows, modify the matrix again by adding the smallest uncovered number to all values at line intersections and subtracting it from each uncovered number, including itself. Leave the other (lined-out) numbers unchanged.

Step 5. Check the matrix again via zero-covering lines, and continue with the modification (step 3) until the optimal assignment is obtained.

The five steps result in the following (Fig. 15-11):

1. *Row subtraction*

	1	2	3	4
A	0	1	3	2
B	3	5	4	0
C	4	2	7	0
D	5	4	1	0

2. *Column subtraction*

	1	2	3	4
A	0	0	2	2
B	3	4	3	0
C	4	1	6	0
D	5	3	0	0

3. *Cover all zeros*

	1	2	3	4
A	0	0	2	1
B	3	4	3	0
C	4	1	6	0
D	5	3	0	0

4. *Modify matrix*

	1	2	3	4
A	0	0	2	3
B	2	3	2	0
C	3	0	5	0
D	5	3	0	1

5. *Cover zeros again*

	1	2	3	4
A	0	0	2	3
B	2	3	2	0
C	3	0	5	0
D	5	3	0	1

Optimum assignments

Job A to machine 1 at 5 min
Job B to machine 4 at 7 min
Job C to machine 2 at 8 min
Job D to machine 3 at 4 min

Fig. 15-11

Note that the final allocation (square boxes in step 5 on page 316) should begin with those jobs that are limited to one machine (B and D), for once they are assigned, this may constrain the assignment of the remaining jobs (A and C).

15.10 How can the assignment method be used to maximize profits rather than minimize costs?

Profits are converted to relative costs. First, subtract every number in a matrix from the largest number in the matrix. Then, follow the same procedure as in minimization problems.

15.11 A bank operations manager has five tellers (T) whom she must assign to customer services of checking accounts (C), foreign exchange (F), notes (N), and savings (S) accounts. Three tellers are not yet qualified for foreign exchange, and one teller cannot handle notes. Work-sampling studies have shown that, working under constant queues, the tellers can handle the number of customers per hour shown in Table 15-8. Assuming the manager wishes to serve as many customers as possible, what assignments should be made? The extra teller will be assigned a data processing task.

Table 15-8

	Customers Served per Hour			
	Checking	Foreign	Notes	Savings
T1	60	X	30	50
T2	70	60	40	50
T3	30	X	10	30
T4	40	X	X	60
T5	40	70	50	80

The number of workers does not balance the number of assignments, so we must add an extra column (call it D = data processing) and assign it a low priority (zero service) so that it will absorb the poorest teller. See Table 15-9.

Table 15-9

	C	F	N	S	D
T1	60	X	30	50	0
T2	70	60	40	50	0
T3	30	X	10	30	0
T4	40	X	X	60	0
T5	40	70	50	80	0

Since this is a maximization problem, we must first convert the matrix values to relative costs by subtracting all values from the largest number, 80, as shown in Table 15-10.

Table 15-10

	C	F	N	S	D
1	20	X	50	30	80
2	10	20	40	30	80
3	50	X	70	50	80
4	40	X	X	20	80
5	40	10	30	0	80

We then follow the same steps as we did in solving Prob. 15.9. See Fig. 15-12.

1. *Row subtraction*

	C	F	N	S	D
1	0	X	30	10	60
2	0	10	30	20	70
3	0	X	20	0	30
4	20	X	X	0	60
5	40	10	30	0	80

2. *Column subtraction*

	C	F	N	S	D
1	0	X	10	10	30
2	0	0	10	20	40
3	0	X	0	0	0
4	20	X	X	0	30
5	40	0	10	0	50

3. *Cover zeros*

	C	F	N	S	D
1	0	X	10	10	30
2	0	0	10	20	40
3	0	X	0	0	0
4	20	X	X	0	30
5	40	0	10	0	50

4. *Modify matrix*

	C	F	N	S	D
1	0	X	0	10	20
2	0	0	0	20	30
3	10	X	0	10	0
4	20	X	X	0	20
5	40	0	0	0	40

5. *Cover zeros again*

	C	F	N	S	D
1	[0]	X	0	10	20
2	0	[0]	0	20	30
3	10	X	0	10	[0]
4	20	X	X	[0]	20
5	40	0	[0]	0	40

Fig. 15-12

Step 5 requires five lines to cover all zeros, so a solution has been reached. Note that the maximum number of lines will never exceed the number of rows. Assigning tellers and tasks that have only one choice first, we assign T4 to S and T3 to D. We can then assign T1 to C, T2 to F, and T5 to N. See Table 15-11.

Alternatively, we could assign T2 to C, T5 to F, and T1 to N and have an equally optimal assignment as revealed by the total number of customers served per hour. See Table 15-12.

Table 15-11

Solution	
T1 at checking	= 60
T2 at foreign	= 60
T3 at data processing	= 0
T4 at savings	= 60
T5 at notes	= 50
Cust/hr	230

Table 15-12

Alternative Solution	
T1 at notes	= 30
T2 at checking	= 70
T3 at data processing	= 0
T4 at savings	= 60
T5 at foreign	= 70
Cust/hr	230

SEQUENCING: PRIORITY DECISION RULES

15.12 The times required to process four jobs received in order (A, B, C, D) are as shown in Table 15-13. Determine the processing sequence that would result from using the priority rules (*a*) FCFS,

(*b*) EDD, (*c*) SPT, and (*d*) critical ratio.

Table 15-13

Job	Job Time (days)	Due Date (days)
A	8	6
B	11	18
C	4	24
D	12	14

Job	(*a*) FCFS Order	(*b*) EDD Due Date	Order	(*c*) SPT Job Time	Order	(*d*) CR TR/WR	Order
A	1	6	1	8	2	.75	1
B	2	18	3	11	3	1.64	3
C	3	24	4	4	1	6.00	4
D	4	16	2	12	4	1.17	2

15.13 Describe, in words, the meaning of the performance measures: (*a*) average completion time, (*b*) average job lateness, and (*c*) average number of jobs at the work center.

(*a*) *Average completion time* is the average number of days a job spends *in the system*— with all released jobs now in the system. It is the sum of the flow times for all the jobs divided by the number of jobs.

(*b*) *Average job lateness* is the total of the days late of all the jobs divided by the number of jobs. (The days late for each individual job is the difference between flow time and due date.)

(*c*) *Average number of jobs at the work center* is the average number at the work center from the current time until all the listed jobs are finished. It is the sum of the flow times for all the jobs currently in the system (i.e., job days) divided by the total required process time for those jobs.

15.14 Find the (*a*) average completion time, (*b*) average job lateness, and (*c*) average number of jobs at the WC for the FCFS and CR priority decision rules of Prob. 15.12.

Performance of FCFS Priority Rule

Job Seq.	(1) Process Time	(2) Flow Time	(3) Due Date	(2)–(3) Days Late (0 if neg.)
A	8	8	6	2
B	11	21	18	4
C	4	25	24	1
D	12	37	14	23
	35	91		30

Performance of CR Priority Rule

Job Seq.	(1) Process Time	(2) Flow Time	(3) Due Date	(2)–(3) Days Late (0 if neg.)
A	8	8	6	2
D	12	20	14	6
B	11	31	18	13
C	4	35	24	11
	35	94		32

(*a*)   Ave. completion time: 91/4 = 22.8 days   94/4 = 23.5 days

(*b*)   Ave. job lateness: 30/4 = 7.5 days   32/4 = 8.0 days

(*c*)   Ave. no. jobs at WC: 91/35 = 2.6 jobs   94/35 = 2.7 jobs

15.15 Six jobs are released to a shop. Rank them according to the critical ratio (1 = highest).

	A	B	C	D	E	F
Time remaining (days)	20	4	15	45	27	6
Work remaining (days)	18	2 ·	5	50	27	12

	A	B	C	D	E	F
CR = TR ÷ WR	1.1	2.0	3.0	.9	1.0	.5
Priority rank	(4)	(5)	(6)	(2)	(3)	(1)

15.16 (*Johnson's rule with three WCs*)   A job shop has eight shop orders that must be processed sequentially through three work centers. Each job must be finished in the same sequence in which it was started. Times (in hours) required at the various work centers are as shown in Table 15-14. Use Johnson's rule to develop the job sequence that will minimize the completion time over all shop orders.

Table 15-14

Job no.	A	B	C	D	E	F	G	H
WC #1 time	4	8	5	9	3	4	9	6
WC #2 time	6	4	7	1	4	2	5	2
WC #3 time	8	7	9	7	9	8	9	7

This is a special (3 WC) case of Johnson's rule, which may be used if the largest of the times at the middle work station (WC #2) is less than or equal to the smallest time required at one or both of the other two WCs. The (modified) procedure is to add the job times of WC #1 + WC #2 and of WC #2 + WC #3. Then use these combined times to solve the problem in the standard two-station approach as seen in Table 15-15.

Table 15-15

Job no.	A	B	C	D	E	F	G	H
WC #1 + WC #2	10	12	12	10	7	6	14	8
WC #2 + WC #3	14	11	16	8	13	10	14	9

Now, the times are scanned, and the lowest value (6) is with the first combination (WC #1 + WC #2), so we place job F as early as possible. The next lowest (7) is also with the first combination, so job E goes second. Jobs D and H are tied for next, with D going as late as possible (last), and H going third. Continuing on, we have the sequence:

1st	2nd	3rd	4th	5th	6th	7th	8th
F	E	H	A	C	G	B	D

INPUT/OUTPUT CONTROL

15.17 Why is input control an essential part of shop floor controls?

Input control is important as a means of controlling lead times and work in process (WIP). If input exceeds output, a backlog develops and manufacturing lead times lengthen.

15.18 An electronics shop has an average output capacity of 240 hours per week. However, it also has a (released) backlog of 80 hours of work. The supervisor plans to keep the incoming work at 240 hours per week over the next 5 weeks but work off the backlog by using equal amounts of overtime during the next 4 weeks. Actual input and output turned out to be the values shown below. (a) Complete the input/output report, (b) find the backlog at the end of week 16, and (c) analyze your findings.

Week number	12	13	14	15	16
Input: actual hours	280	230	200	235	260
Output: actual hours	250	270	265	240	230

(a) To work off the 80 hours of backlog over 4 weeks will require 80 hours ÷ 4 weeks = 20 hours of overtime per week. Added to the regular-time capacity of 240 hours per week, this results in a planned output of 260 hours per week. (Backlog should be down to 80 − 20 = 60 hours by end of week 1, 40 by week 2).

Table 15-16   Input/Output Report for Electronics Shop

		Week number	12	13	14	15	16
Input	(1)	Actual input (hr)	280	230	200	235	260
	(2)	Planned input (std. hr)	240	240	240	240	240
	(3)	Deviation (actual − plan)	40	−10	−40	−5	20
	(4)	Cumulative deviation	40	30	−10	−15	5
Output	(5)	Actual output (hr)	250	270	265	240	230
	(6)	Planned output (std. hr)	260	260	260	260	240
	(7)	Deviation (actual − plan)	−10	10	5	−20	−10
	(8)	Cumulative deviation	−10	0	5	−15	−25
	(9)	Planned backlog	60	40	20	0	0
	(10)	Actual backlog [(9) + (4) − (8)]	110	70	5	0	30

(b) With the planned input and output, the backlog should be zero at the end of week 15. It is zero, as calculated from the cumulative deviation amounts.

(c) Part of the reason the backlog is zero is because the actual input to the shop was less than planned (i.e., cumulative deviation of −15 in week 15). Note that although the planned output was for 260 hours in week 15, the backlog was already down to 5 hours at the end of week 14, and there was only enough work to keep it busy for the 240 regular hours in week 15. Actual input jumped up in week 16, but the shop was not geared to the higher output (actual = 230 hours), so the backlog went back up to 30.

15.19 Work center 4 has an average capacity of 260 hours a week. It had a released backlog of 160 hours and an unreleased backlog of 180 hours as of the beginning of week 21. The planner scheduled to work off the backlog over the next 5 weeks by scheduling a 10 percent reduction in planned input, along with 6 hours of overtime each week. Actual input and output hours for the 5 weeks were as shown in Table 15-17. Depict this situation in an input/output report for WC4, and determine the released backlog at the end of week 25.

Table 15-17

Week number	21	22	23	24	25	26
Actual input	230	240	235	250	220	
Actual output	265	260	270	280	280	

Planned reductions were to come from:

(1)   Reduced input: 5 wk @ 26 hr/wk = 130 hr

(2)   Increased output: 5 wk @ 6 hr/wk = 30 hr

160 hr

Table 15-18   Input/Output Report WC4

	Week number	21	22	23	24	25	26
Input	Actual hours	230	240	235	250	220	
	Planned hours	234	234	234	234	234	260
	Cumulative deviation (actual − planned)	− 4	+ 2	+ 3	+ 19	+ 5	
Output	Actual hours	265	260	270	280	280	
	Planned hours	266	266	266	266	266	260
	Cumulative deviation (actual − planned)	− 1	− 7	− 3	+ 11	+ 25	

The initial released backlog (as of week 20) was 160 hours. Actual reductions came from (see Table 15-18):

(1)   Reduced input: $130 - 5 = 125$

(2)   Increased output: $30 + 25 = \underline{\ 55}$

Total reductions $= 180$

Net effect as of the end of week 25 was

$$160 \text{ hr} - 180 \text{ hr} = -20 \text{ hr}$$

(*Analysis:* Five more hours than planned of input were released to the shop, so the input reduction was only 125 hours, but 30 more hours than planned of output were delivered. Therefore, all the released backlog is removed. In addition, 20 hours of the original 180 hours of unreleased backlog has been worked off.)

Supplementary Questions and Problems

15.20 What are the major activities associated with (*a*) priority control and (*b*) capacity control?

Ans. (*a*) Order release, sequencing, and status control; (*b*) loading/assignment, lead-time control, input/output control

15.21 Distinguish between (*a*) a Gantt progress chart and (*b*) a Gantt load chart.

Ans. (*a*) A progress chart depicts activities on a sequential time scale that clearly specifies which activities must be completed prior to other activities. (*b*) A load chart shows the scheduled workload, maintenance, and idle time on key machines, along with the accumulated backlog.

15.22 A piece of mining equipment requires the manufacturing times shown in Table 15-19. Each of the activities must be done sequentially, except that the steel fabrication can begin 2 weeks after purchasing begins, and the hydraulics and electrical activities can be done concurrently. Construct a Gantt scheduling chart for this job.

Table 15-19

Activity	Weeks	Activity	Weeks
1. Engineering	3	5. Electrical	4
2. Purchasing	3	6. Control	1
3. Steel fabrication	1	7. Field test	2
4. Hydraulics	2	8. Packaging	1

Ans. Chart should show the following activities occurring during the weeks noted: engineering (1–3), purchasing (4–6), steel fabrication (6), hydraulics (7–8), electrical (7–10), control (11), field test (12–13), package (14).

15.23 Collins Heating Co. has four central-heating installations to design within an 8-week period (40 hours per week). They also have four capable designers, each of whom has been asked to estimate how long it would take to do each job. The work operations scheduler has compiled the estimates shown in Table 15-20. (*a*) Use assignment linear programming methods to determine how the jobs should be assigned so as to minimize the work time. (*b*) Assuming that the estimates are correct, can the jobs be completed within the 8-week period without planning for overtime? (*c*) Assuming one designer per job and no overtime, could the work be completed in 5 weeks? (*d*) In 3 weeks?

Table 15-20

Designers	Hours to Complete Job			
	1	2	3	4
A	100	140	280	70
B	130	160	200	60
C	80	130	300	90
D	150	110	250	50

Ans. (*a*) Assign A to 4 at 70, B to 3 at 200, C to 1 at 80, and D to 2 at 110 for 460 hours total. (*b*) Time available = (40)(4)(48) = 1,280 hours, so the jobs can be done without overtime. (*c*) Yes, no individual job requires more than 5 weeks. (*d*) No, although the total time available is (40)(4)(3) =

480 hours, there is not sufficient time (200 hours) for drafter B to complete his job in 3 weeks (i.e., $3 \times 40 = 120$ hours available).

15.24 What is the major restriction in applying Johnson's rule?
Ans.   Requires same sequence for both machines

15.25 A market research firm has seven customer orders that must be processed sequentially through two activities: (1) data compilation and (2) analysis. Estimated times (in hours) are as shown.

	A	B	C	D	E	F	G
(1) Data compilation	5	7	2	1	8	3	16
(2) Analysis	4	9	7	2	2	9	5

(*a*) Use Johnson's rule to develop a schedule that will permit all work to be completed in the minimum amount of time. (*b*) What is the total time required to process the seven jobs?
Ans.   (*a*) Sequence is D, C, F, B, G, A, E (*b*) 44 hours

15.26 Use Johnson's rule to determine the sequence that results in the minimum flow time for the seven jobs listed below. All jobs must follow the same sequence of machine first and then polish. Times are in minutes.

<table>
<tr><td></td><td colspan="7">Time Required to Do Job</td></tr>
<tr><td></td><td>A</td><td>B</td><td>C</td><td>D</td><td>E</td><td>F</td><td>G</td></tr>
<tr><td>Machine</td><td>10</td><td>6</td><td>5</td><td>4</td><td>6</td><td>9</td><td>7</td></tr>
<tr><td>Polish</td><td>2</td><td>3</td><td>12</td><td>5</td><td>9</td><td>11</td><td>6</td></tr>
</table>

(*a*) What is the optimal sequence of jobs? (*b*) What is the minimum time flow to finish these seven jobs?   *Ans.*   (*a*) Sequence is D, C, E, F, G, B, A; (*b*) 52 minutes

15.27 A wood pattern shop has five shop orders that must be processed through six work centers during the coming week. The capacities of the work centers (in hours) are number $9 = 40$, number $10 = 20$, number $11 = 20$, number $12 = 20$, number $13 = 20$, number $14 = 30$. See Table 15-21. (*a*) Is the shop capacity sufficient to complete all jobs? (*b*) Assume that the scheduling guidelines require 4 hours of move time *between* work centers (not included above). Rank the shop orders according to the longest processing (and transit) time.

Table 15-21

Shop Order	Hours Required at Work Center					
	9	10	11	12	13	14
A	4	3	—	—	7	5
B	6	9	13	—	3	4
C	12	—	7	10	5	7
D	6	4	—	—	—	8
E	11	2	—	9	8	4

Ans. (*a*) Total shop load is 147 hours versus a capacity of 150 hours. However, work center 13 is overloaded, so it may restrict shop capacity. The balance is reasonably close, however. (*b*) With 4 hours of move time, the processing time per order is A = 31, B = 51, C = 57, D = 26, and E = 50. Therefore, the rank according to LPT is C, B, E, A, D.

15.28 The orders shown in Table 15-22 were received in a job shop where scheduling is done by priority decision rules.

Table 15-22

Job Number	Shop Calendar Date Received	Due	Production Days Required
870	317	368	20
871	319	374	30
872	320	354	10
873	326	373	25
874	333	346	15

In what sequence would the jobs be ranked according to the following decision rules: (*a*) earliest due date, (*b*) shortest processing time, (*c*) least slack, (*d*) first come, first served?
Ans. (*a*) 874, 872, 870, 873, 871 (*b*) 872, 874, 870, 873, 871 (*c*) 874, 873, 872, 871, 870, (*d*) 870, 871, 872, 873, 874

15.29 A defense contractor in Chicago has six different jobs in process with delivery requirements as shown in Table 15-23. Today is day 60, and the contractor uses a critical-ratio scheduling technique. Rank the jobs according to priority, with the first being the highest.

Table 15-23

	A	B	C	D	E	F
Promised (date due)	60	72	67	72	65	70
Days of work remaining	2	2	5	4	5	6

Ans. Priority 1st = A, 2nd = E, 3rd = C, 4th = F, 5th = D, 6th = B

15.30 A manufacturing coordinator has the shop orders shown below due to be shipped a week (5 working days) from now. Sequence the jobs according to priority as established by (*a*) least slack and (*b*) critical ratio.

Shop order number	427	430	432	433	435	436
Number of days of work remaining	2	4	7	6	5	3

Ans. (*a*) The least-slack sequence is 432, 433, 435, 430, 436, 427.
(*b*) The CR sequence is 432, 433, 435, 430, 436, 427.
Note: When the time remaining is constant across all jobs, the least slack and critical ratio result in the same priority.

15.31 Indiana Equipment Co. has a small shop with an output capacity of 280 hours per week. The production controller has been asked to provide a delivery estimate for a sales representative in Tampa. She generally allows 3 weeks for order release to obtain materials and tooling. The current lead-time situation is as shown.

Unreleased backlog	1,400 hours
Active orders (except current)	540 hours
Current orders	300 hours

Assuming the controller follows a first come, first served policy, what is the best delivery time (in weeks) she can offer?

Ans.   During the ninth week. (*Note:* There are already 8 weeks of work (total), and the order release can be accomplished while the job is in unreleased backlog.)

15.32 The Automated Billing Equipment Co. produces electric meter reading equipment that automatically transmits meter readings to a billing center at preset times. Production controllers in the factory have estimated the standard hours of capacity requirements at an assembly workstation over the next quarter as shown. The company uses an MRP system with output controls designed to signal corrective action when the cumulative deviation exceeds one-half the weekly average as computed from the forecast.

Week number	1	2	3	4	5	6	7	8	9	10	11	12
Estimated hours	370	340	290	350	360	410	320	350	330	340	380	360

Suppose the company plans to produce at a steady rate equal to the average estimated requirement. (*a*) Formulate an estimated-requirements graph showing the average requirement as a dotted line. (*b*) Assume that actual standard hours delivered over the first 8 weeks are 360, 310, 340, 280, 360, 300, 270, 370. Construct an output-control chart, and determine the cumulative deviation. (*c*) Is corrective action warranted? If so, when?

Ans.   (*a*) Average requirement = 350 hours (*b*) cumulative deviation at end of week 8 = −210 hours
(*c*) Yes, during (or at end of) week 7, when the cumulative deviation exceeds |175| for the first time.

15.33 The following input/output report describes the status of jobs at work center 7. The unreleased backlog at the end of period 10 is 120 hours. (*a*) Compute the cumulative deviations for the charts above. (*b*) Has the backlog been worked off (eliminated) by the end of week 16? Explain. (*c*) Is the output less than or greater than planned? Why?

	Period number	11	12	13	14	15	16
Input	Actual hours	245	220	310	280	300	285
	Planned hours	240	240	300	300	300	300
	Cumulative deviation						
Output	Actual hours	295	295	310	270	305	285
	Planned hours	300	300	300	300	300	300
	Cumulative deviation						

Ans.   (*a*) Cumulative deviation of input in period 16 is −40 and for output is also −40. (*b*) Yes, 1,640 hours were input and 1,760 hours were output, so 120 hours of backlog were worked off. (*c*) Output was 40 hours less than planned, because input was 40 hours less than planned. It appears that work center 7 did not have enough work to do.

Chapter 16

Operations Analysis and Maintenance
Calculus, Learning Curves, Queuing, and Simulation

OPERATIONS ANALYSIS AND CONTROL

Operations change over time. Product life cycles expire, and opportunities for new goods and services must be constantly evaluated. As facilities become worn or obsolete, choices must be made with respect to maintaining existing processes or incorporating updated technologies. Analysis and maintenance are essential to the continuity and effectiveness of competitive operations.

Question: Define operations analysis.

Operations analysis is the use of analytical and quantitative methods to systematically collect and study data relating to the productivity of operations over time.

Question: Explain the terms (*a*) analytical methods, (*b*) data, (*c*) productivity, and (*d*) time, as used in the definition of operations analysis.

(*a*)   *Analytical methods.* The analytical methods used in operations range from deterministic techniques such as calculus to highly stochastic approaches, such as computer simulations. Our coverage here is limited to brief applications of calculus, learning curves, queuing, statistics, and simulation.

(*b*)   *Data.* Any systematic analysis of operations implies an underlying body of knowledge and an availability of relevant data. The principles of economics, organizational management, and operations form the knowledge base that gives direction to the analysis of operations.

(*c*)   *Productivity.* As the ratio of the value of outputs to the cost of inputs, productivity is a relative measure of the effectiveness of a transformation process. Productivity measures (e.g., units shipped per employee) should be meaningful and measurable.

(*d*)   *Time.* The focus on time is a common and unifying concept underlying the analysis of ongoing operations. Time is a consistent measured span of existence that can be viewed profitably (if used effectively) or costly (when used unwisely). The techniques reviewed in this chapter are ways of expressing an interface of operations with time.

Question: In what ways are calculus, learning curves, queuing, and simulation associated with the concept of time?

Calculus is a means of describing the rate of change of activities in time. *Learning curves* depict the improvement in output over time. *Queuing theory* enables us to analyze waiting lines and service times. Finally, *simulation* is a technique for moving a system through time in an accelerated manner so that the time-dependent characteristics can be studied in isolation.

Operations analysis involves abstracting a problem from the overall (*macro*) environment, breaking it down into its component (*micro*) parts, deriving a solution, and implementing it in the total (*macro*) system. *Operations control* follows to ensure that the redesigned system (or solution) conforms to the organizational goals.

ANALYSIS OF GOODS VERSUS SERVICES SYSTEMS

Although physical goods are the tangible result of production, services generate a significant portion of the GDP in many nations. Services are less tangible products and convey value directly to consumers as they are produced. Production and consumption usually occur simultaneously, and no inventory accrues. Service locations tend to be decentralized to meet a highly variable demand with an output that is often customized (e.g., restaurants and tax returns).

Question: What differentiates equipment-based from people-based services?

Equipment-based services (e.g., utility, phone, transit systems) tend to be the most predictable and measurable—and most subject to analysis and control. *People-based services* (e.g., entertainment, legal, and investment advisory services) are more dependent on the individual skill or knowledge levels of the provider and have a more variable quality of output.

Question: How do services add value (or fail to add value) to the assets of a nation?

Some services are ''paper'' transactions that simply redistribute wealth rather than create it. For example, many real estate transactions only change the ownership of goods, and lawyers use lawsuits to help settle disputes over ownership. On the other hand, a real estate transaction that facilitates the location of a new plant or a legal contract that facilitates its construction would be adding value.

Figure 16-1 illustrates the focus of goods-producing facilities on tangible *materials* and the focus of services on *customers*, where the timing of delivery, procedures, and environmental conditions assume more importance. With many services (e.g., hospital) customers become an integral part of the productive environment—though this is not always the case (e.g., auto repair).

Goods	Primary Area of Concern	Services
• Influenced largely by *raw materials*, *labor* supply, and *inventory* considerations	Location and layout	• Influenced strongly by location and convenience of *customers*
• Interface *with machines* • May require technical skill • Motivation is important	Human resource inputs	• Interface *with customers* • Requires more interpersonal skills • Training is important
• Number of *units*	Forecasting	• Number of *customers*
• Availability and timing of *materials* that are built into the product	Inventory management	• Availability and timing of *supplies consumed* by customer
• Gross quantities and types of *products* produced • Specific *end items* to be produced	Aggregate planning and master scheduling	• Gross quantities and types of *customers* served • Specific *types* of customers to be served
• Flow of *materials* and scheduling of *facility* time	Material and capacity planning	• Flow of *customers* and scheduling of *personnel* time
• Priority rules applied to *materials* and *jobs* • Input/output control of *hours* (and *units*) *produced*	Production activity control	• Priority rules applied to *customers* • Input/output control of *hours* (and *customers*) *served*
• Quality inherent in stored product	Quality control	• Quality in the service and process (e.g., time, environment)
• Preventive and repair activities on equipment and product	Maintenance	• Care and attention to individual performing the service

Fig. 16-1 Operations analysis and control in goods versus services systems

APPLICATIONS OF CALCULUS

Analysis of operations often necessitates that data be expressed and evaluated using equations or summary statistics. *Elementary mathematics* may be sufficient to describe production volumes, costs, and revenues that are constant or linearly related. For example, if output is constant at 8 units per hour, the volume y resulting from x hours of production would be $y = 8x$. A graph of volume y versus time x would be a line segment with a slope of 8, where the slope tells the change in one variable (Δy) resulting from a change in another variable (Δx). The slope value of 8 signifies that the output y increases by 8 units for each additional hour x used. If $x = 40$, then $y = 8(40) = 320$ units.

But not all business computations are that simple. For example, the production rate may improve slightly with every unit produced, so that output is not a linear function of time but rather an exponential function of the number of units already produced (i.e., there is a learning or improvement effect). In this case, the output volume is a curvilinear function of the time to produce the first unit, and the slope of the relationship is not constant but variable.

Calculus is a useful deterministic technique for finding the slope or length of curved functions and volumes resulting from nonlinear relationships among the variables. With nonlinear functions, the slope of the curve, or rate of change, in y relative to x varies from one point to another, depending upon the value of x. At any given point x it is equal to the slope of the line tangent to the curve at that point.

Example 16.1   Total costs for product A are linear at $10 + 5x$ and are curvilinear for product B at $10 + 2x^2$, where x is the number of units produced. How would you determine the rate of change in costs as volume increases from 2 units to 4 units for (*a*) product A and (*b*) product B?

(*a*)   The rate of change in costs is measured by the slope of the cost function shown in Fig. 16-2. Because product A has a straight-line (linear) cost function, its slope ($\Delta y/\Delta x$) is constant.

$$\text{Slope} = \frac{\text{change in } y}{\text{change in } x} = \frac{\Delta y}{\Delta x} = \frac{y_2 - y_1}{x_2 - x_1} = \frac{\$30 - \$20}{4 - 2} = \frac{\$10}{2} = \$5/\text{unit}$$

(*b*)   Product B has a second-order cost function (i.e., has an x^2 value), and its slope is not constant but depends upon the value assigned to x. The rate of change is equal to the slope of the line tangent to the curve at the corresponding value of x as seen in Fig. 16-2*b*.

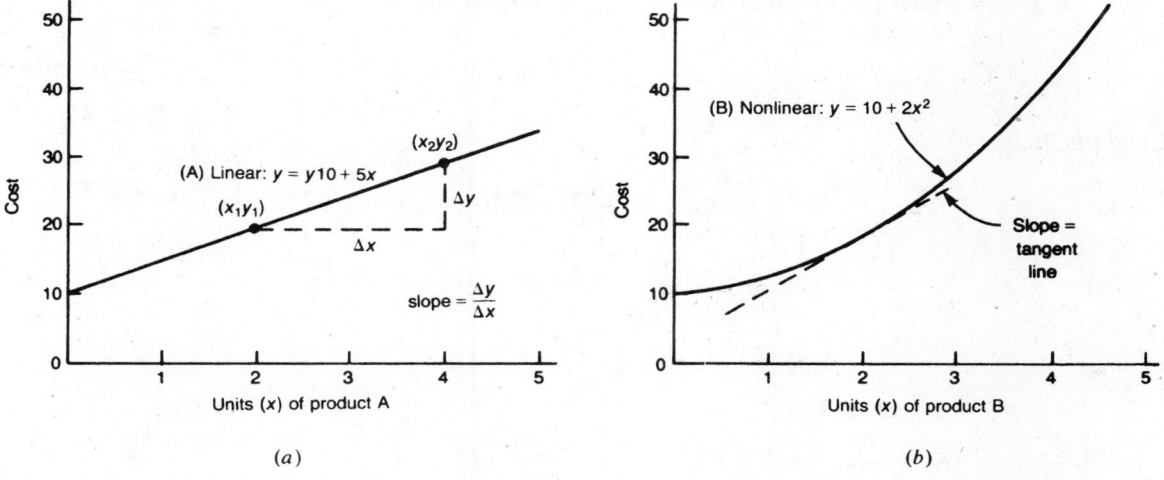

(*a*)   (*b*)

Fig. 16-2

The slope of a function at any point may be determined by finding the first *derivative* of the function. The *derivative* is the calculus expression of the rate at which the function is changing, and is the slope of the tangent to the curve at some point whose abscissa is the given value of x. It is also referred to as the instantaneous rate of change of the curve at that tangent. The function must, of course, be continuous (no breaks) and smooth (no sudden change in direction), but many production functions satisfy these conditions.

Insofar as points of zero slope often identify maximum or minimum values of a function, differential calculus is an important optimizing technique. (See Fig. 16-3.) The local maximum (or minimum) values can be determined by (1) finding the first derivative, (2) setting it equal to zero, (3) solving for the value of x that satisfies the zero derivative equation, and (4) substituting this volume of x back into the original equation.

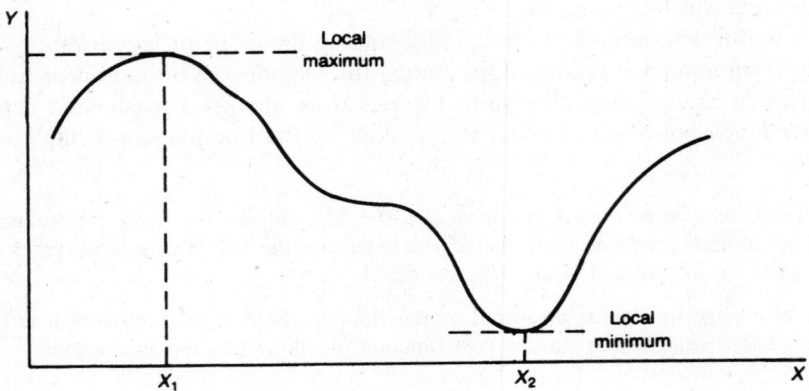

Fig. 16-3 Maximum and minimum points

RULES FOR DIFFERENTIATION

Think of finding the derivative of a function (differentiating) as the problem of finding the rate of change in the vertical dimension (Δy) relative to the change in the horizontal dimension (Δx) at a given point. The standard symbol used to denote the derivative of y with respect to x is dy/dx [or y' or $f'(x)$]. If we let c and n represent constants, we can express the derivatives of various functions as follows:

1. If $y = cx^n$, then the first derivative of y with respect to x is:

$$\frac{dy}{dx} = ncx^{n-1} \qquad (16.1)$$

Example 16.2

(a) $\qquad\qquad y = 9x^2 \qquad \dfrac{dy}{dx} = (2)(9)x^1 = 18x$

(b) $\qquad\qquad y = 5x^{-3} \qquad \dfrac{dy}{dx} = (-3)(5)x^{-4} = -15x^{-4} = \dfrac{-15}{x^4}$

(c) $\qquad\qquad y = -6x^{1/3} \qquad \dfrac{dy}{dx} = \left(\dfrac{1}{3}\right)(-6)x^{-2/3} = -2x^{-2/3} = \dfrac{-2}{x^{2/3}}$

(d) $\qquad\qquad y = \dfrac{1}{4}x^5 \qquad \dfrac{dy}{dx} = (5)\left(\dfrac{1}{4}\right)x^4 = \dfrac{5x^4}{4}$

2. The derivative of a constant is zero, and the derivative of a constant times a variable is the constant.

Example 16.3

(a) $$y = 20 \qquad \frac{dy}{dx} = 0$$

(b) $$y = 7x \qquad \frac{dy}{dx} = 7$$

3. The derivative of the sum or difference of two functions is the sum or difference of the individual derivatives. If we let $g(x)$ be one function, and $h(x)$ another, then if $y = g(x) + h(x)$

$$\frac{dy}{dx} = g'(x) + h'(x) \qquad (16.2)$$

Example 16.4

$$y = 3x^4 + 5x^{-2} \qquad \frac{dy}{dx} = 12x^3 - 10x^{-3}$$

$$y = 2x - 3x^2 \qquad \frac{dy}{dx} = 2 - 6x$$

$$y = 7x^3 - 2x^2 + 3x + 4 \qquad \frac{dy}{dx} = 21x^2 - 4x + 3$$

4. The derivative of a composite function $y = [f(x)]^n$ is:

$$\frac{dy}{dx} = n[f(x)]^{n-1}f'(x) \qquad (16.3)$$

Example 16.5

$$y = (x^2 + 1)^2 \qquad \frac{dy}{dx} = 2(x^2 + 1)(2x)$$

$$= 4x(x^2 + 1) = 4x^3 + 4x$$

$$y = (7 - 4x)^3 \qquad \frac{dy}{dx} = 3(7 - 4x)^2(-4)$$

$$= (-12)(7 - 4x)^2 = -192x^2 + 672x - 588$$

5. The derivative of a product of two functions is the first times the derivative of the second, plus the second times the derivative of the first. If $y = [g(x)][h(x)]$, then

$$\frac{dy}{dx} = g(x)h'(x) + g'(x)h(x) \qquad (16.4)$$

Example 16.6

$$y = (x^3 + 4x)(2x - 3) \qquad \frac{dy}{dx} = (x^3 + 4x)(2) + (2x - 3)(3x^2 + 4)$$

$$= 2x^3 + 8x + 6x^3 + 8x - 9x^2 - 12$$
$$= 8x^3 - 9x^2 + 16x - 12$$

Derivatives of additional functions, such as the quotient of two functions and the natural logarithm of a function, can be found in any good calculus text.

FINDING MAXIMUM AND MINIMUM VALUES

As suggested in Fig. 16-3, the point of zero slope on a function may be either a local maximum or local minimum. (*Note:* The point of zero slope is not necessarily a "global" optimum unless all points on the function are known to lie within these extremes.) Once the point of interest is identified, the type of optimum (i.e., whether a maximum or minimum) can be determined by finding the second derivative of the function at the point in question. The second derivative d^2y/dx^2, or y'', or $f''(x)$ is simply the derivative of the first derivative and represents the rate of change in the slope of the tangent line.

If the value of the second derivative is negative, the point is a local maximum. If the value of the second derivative is positive, the point is a local minimum. In the event that the second derivative is zero, evaluate y' at points close to $x = 0$. If the sign of the first derivative changes from positive on the left of the point to negative on the right, then the point is a maximum; otherwise, it is a minimum.

Example 16.7   A cost curve is defined by the function $y = x^2 - 10x + 30$, where $y =$ cost in dollars and $x =$ hours of labor. Find the point of zero slope, and indicate whether it is a maximum or minimum.

See Fig. 16-4.

$$\frac{dy}{dx} = 2x - 10$$

The point of zero slope is where

$$2x - 10 = 0$$
$$2x = 10$$
$$x = 5$$

$$\frac{d^2y}{dx^2} = 2 \text{ (a positive value)}$$

Thus, $x = 5$ is a minimum cost point.

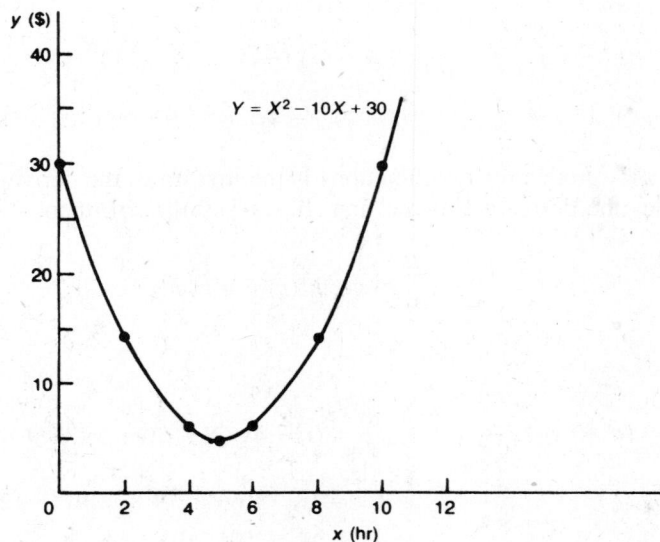

Fig. 16-4

LEARNING CURVE EFFECTS

Learning or *improvement curve effects* are the reductions in time per unit to perform specified activities. As the number of repetitions of doing a task increases, improvement results from the development of individual skills, plus other factors such as better organization of work, improved methods, and enhanced work environment. Learning curve information is useful for planning and scheduling work, budgeting costs, negotiating price and delivery of purchased items, and pricing the firm's own products.

The degree of improvement depends upon the task being performed, but is normally expressed in terms of the percentage of time it takes to complete the unit that represents a *doubling* of output. For example, if an activity followed an 80 percent learning curve and required 100 hours for the first unit, the second would take 80 hours, the fourth 64 hours, and the eighth 51.2 hours. See Fig. 16-5.

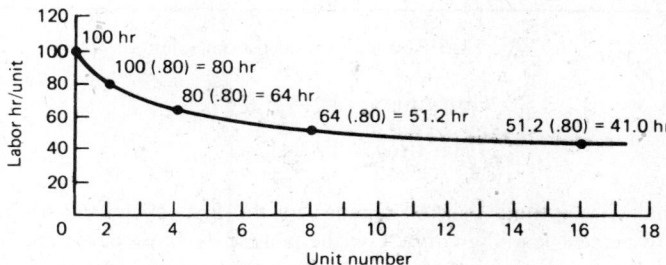

Fig. 16-5 80 percent learning curve

Mathematically, the number of direct labor hours required to produce the nth unit of a product, Y_n, is exponentially related to the time to produce the first unit, Y_1, by the expression

$$Y_n = Y_1 n^b \qquad (16.5)$$

where Y_n = time to produce nth unit

Y_1 = time to produce first unit

n = unit number

$$b = \frac{\log \text{ learning } \%}{\log 2}, \text{ or } \frac{\ln \text{ learning } \%}{\ln 2}$$

Example 16.8 Production of a certain type of television series program follows an 80 percent learning curve and requires 100 hours to complete the first unit. Estimate the time required for the fourth unit of the series.

$$Y_n = Y_1 n^b$$

where $Y_1 = 100$

$n = 4$

$$b = \frac{\log .80}{\log 2} = -.322$$

$$Y_4 = 100(4)^{-.322} = \frac{100}{(4)^{.322}} = 64 \text{ hr}$$

In Eq. 16.5 the exponent (b) is the ratio of two logarithms. Note that either base-10 logarithms or natural logarithms can be used, on condition that the numerator and denominator both have the same base. The learning curve function appears as a straight line on log-log paper.

Although Eq. 16.5 can be used to calculate the individual unit time for any learning percentage, tables are available that simplify this calculation for selected learning curve percentages. Appendix H contains learning curve coefficients for 70 percent, 80 percent, 85 percent, and 90 percent. It also contains cumulative values that enable one to calculate the total time required to complete the specified number of units. Figure 16-6 shows the first few rows of Appendix H.

To use Appendix H, go down the column to the unit number of interest and across (to the right) to the appropriate learning percentage. The time to produce the desired unit Y_n or cumulative number of units Y_N is then the time for the first or base unit, Y_B, multiplied by the appropriate learning curve coefficient, L_{unit} or L_{total}.

Unit Number	70 Percent		80 Percent		85 Percent		90 Percent	
	L_{unit}	L_{total}	L_{unit}	L_{total}	L_{unit}	L_{total}	L_{unit}	L_{total}
1	1.000	1.000	1.000	1.000	1.000	1.000	1.000	1.000
2	.700	1.700	.800	1.800	.850	1.850	.900	1.900
3	.568	2.268	.702	2.502	.773	2.623	.846	2.746
4	.490	2.758	.640	3.142	.723	3.345	.810	3.556

Fig. 16-6 Learning curve coefficients

$$\text{Unit time:} \quad Y_n = Y_B(L_{unit}) \tag{16.6}$$

$$\text{Total time:} \quad Y_N = Y_B(L_{total}) \tag{16.7}$$

Example 16.9 A maintenance activity took 60 minutes for the first repair, and historical records suggest that these activities follow a 70 percent learning curve. Use the table of learning curve coefficients to estimate (a) the time required to complete the fourth such repair and (b) the total time required to complete all four repairs.

$$\text{Unit time:} \quad Y_n = Y_B(L_{unit}) = 60(.490) = 29.4 \text{ min}$$

$$\text{Total time:} \quad Y_N = Y_B(L_{total}) = 60(2.758) = 165.5 \text{ min} = 2.75 \text{ hr}$$

ESTIMATING THE LEARNING PERCENTAGE

Although learning curves apply primarily to labor-intensive activities, they can also reflect productivity improvements from other sources (e.g., fixtures, work-flow redesign, layout, work team benefits, etc.). Care must be taken in estimating the learning percentages, which may come from historical observations of the job under study or from experience with similar products. Learning rates in the 70 to 90 percent range are commonly observed.

Example 16.10 A change in work methods has resulted in a new assembly job, which must be done several times per shift. A capable worker took 50 minutes to do the job the first time and 45 minutes the second time. Estimate (a) the appropriate learning percentage and (b) how much time it should take the worker to do the twentieth assembly.

(a)   This is very limited data, but insofar as the second time (of 45 minutes) does represent a doubling of output over the first time (or 50 minutes), we can use these two values for the estimate.

$$\text{Estimated learning percentage} = 45 \text{ min} \div 50 \text{ min} = .90$$

(b) $$Y_n = Y_B(L_{unit}) = 50(.634) = 31.7 \text{ min}$$

When more extensive historical data are available the (more representative) learning rate can be determined by first solving for the value of the exponent, b, and then multiplying it by $\ln 2$ to find the natural logarithm of the learning percentage. The learning percentage is then e raised to this power.

$$b = \frac{\ln(Y_n/Y_1)}{\ln n} \tag{16.8}$$

Since b is the ratio of $(\ln \text{ learning percentage}) \div \ln 2$, the ln learning percentage is:

$$\ln \text{ learning } \% = b \ln 2$$

The learning percentage is then the natural logarithm base ($e = 2.718$) raised to this exponent:

$$\text{Learning } \% = e^{(b \ln 2)} \qquad (16.9)$$

Example 16.11 An activity required 100 hours to complete the first unit, but only 41.0 hours for unit 16. Find the learning percentage. (*Note:* From Fig. 16-5, this should work out to 80 percent.)

$$b = \frac{\ln(Y_n/Y_1)}{\ln n} = \frac{\ln(41.0/100)}{\ln 16} = \frac{\ln .41}{\ln 16} = \frac{-.8916}{2.7726} = -.3216$$

$$\text{Learning } \% = e^{(b \ln 2)} = e^{(-.3216 \ln 2)} = e^{(-.2229)} = .80$$

QUEUING MODELS

Queuing theory is a quantitative (mathematical) approach to the analysis of systems that involve waiting lines, or queues. Examples range from supermarket checkout counters to banking activities and manufacturing jobs awaiting processing. The waiting lines may form even though the system (facility) has enough capacity, *on the average*, to handle the demand. This is because the arrival times and service times for the customers (jobs) are random and variable.

The *objective of queuing analysis* is to evaluate the service and the costs of a facility so as to maximize its usefulness. This often results in minimizing the total costs associated with the idle time of facilities or services versus the waiting time costs of employees or customers. Numerous computer software programs are available for queuing analysis. Calculations typically seek to estimate:

(1) System utilization ($\% \, U$) or average usage rate of capacity

(2) Mean number of customers in the queue N_q or in the system N_s

(3) Mean time customers spend in the queue T_q or in the system T_s

(4) Related idle facility and waiting time costs

Figure 16-7 illustrates the structure of four variations of queuing systems. The simplest of these is a single-channel, single-phase system. Multiple-channel, single-phase systems, such as those found at banks and toll-road pay stations, have more than one service facility. Multiple-phase systems incorporate two or more service activities and are more difficult to analyze mathematically. Simulation is often the most feasible technique for analysis of multiple-phase systems.

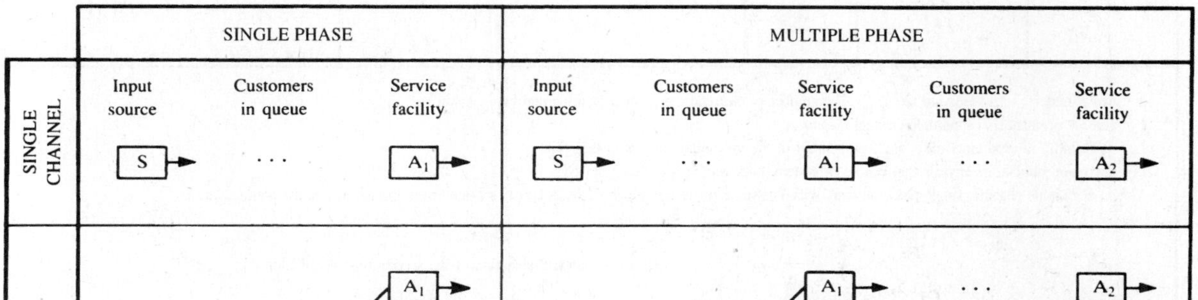

Fig. 16-7 Types of queuing systems

Table 16-1 Equations for Queuing Model Computations*

	Model 1	Model 2	Model 3	Model 4
Channel:	Single	Single	Single	Multiple
Phase:	Single	Single	Single	Single
Arrival rate:	Poisson	Poisson	Poisson	Poisson
Service rate:	Poisson	Constant	Poisson	Poisson
Queue length:	Unlimited	Unlimited	Limited	Unlimited
Designation:	M/M/1	M/D/1	M/M/1/Q	M/M/C
T_s = mean time in system	$\dfrac{1}{\mu - \lambda}$	$T_q + \dfrac{1}{\mu}$	$T_q + \dfrac{1}{\mu}$	$T_q + \dfrac{1}{\mu}$
N_s = mean number in system	$\dfrac{\lambda}{\mu - \lambda}$	$N_q + \dfrac{\lambda}{\mu}$	$\dfrac{\lambda}{\mu}\left\{ \dfrac{1 - (Q+1)\left(\frac{\lambda}{\mu}\right)^Q + Q\left(\frac{\lambda}{\mu}\right)^{Q+1}}{\left(1 - \frac{\lambda}{\mu}\right)\left[1 - \left(\frac{\lambda}{\mu}\right)^{Q+1}\right]} \right\}$	$N_q + \dfrac{\lambda}{\mu}$
T_q = mean waiting time in queue	$\dfrac{\lambda}{\mu(\mu - \lambda)}$	$\dfrac{\lambda}{2\mu(\mu - \lambda)}$	$\dfrac{N_q}{\lambda(1 - P_Q)}$	$P_0\left[\dfrac{1}{\mu C(C!)\left(1 - \frac{\lambda}{\mu C}\right)^2}\left(\frac{\lambda}{\mu}\right)^C \right] = \dfrac{N_q}{\lambda}$
N_q = mean number in queue	$\dfrac{\lambda^2}{\mu(\mu - \lambda)}$	$\dfrac{\lambda^2}{2\mu(\mu - \lambda)}$	$\left(\dfrac{\lambda}{\mu}\right)^2\left\{ \dfrac{1 - Q\left(\frac{\lambda}{\mu}\right)^{Q-1} + (Q-1)\left(\frac{\lambda}{\mu}\right)^Q}{\left(1 - \frac{\lambda}{\mu}\right)\left[1 - \left(\frac{\lambda}{\mu}\right)^Q\right]} \right\}$	$P_0\left[\dfrac{\lambda\mu\left(\frac{\lambda}{\mu}\right)^C}{(C-1)!(C\mu - \lambda)^2} \right]$
P_n = probability of n units in system	$\left(1 - \dfrac{\lambda}{\mu}\right)\left(\dfrac{\lambda}{\mu}\right)^n$ where $n = 0, 1, \ldots$ $\lambda < \mu$	Not available	$\left[\dfrac{1 - \frac{\lambda}{\mu}}{1 - \left(\frac{\lambda}{\mu}\right)^{Q+1}} \right]\left(\dfrac{\lambda}{\mu}\right)^n$ where $n = 0, 1, \ldots Q$	$P_0 \dfrac{\left(\frac{\lambda}{\mu}\right)^n}{n!}$ for $0 \le n \le C$ $P_0 \dfrac{\left(\frac{\lambda}{\mu}\right)^n}{C! C^{n-c}}$ for $n \ge C$ where $P_0 = \dfrac{1}{\sum\limits_{n=0}^{C-1} \dfrac{\left(\frac{\lambda}{\mu}\right)^n}{n!} + \dfrac{\left(\frac{\lambda}{\mu}\right)^C}{C!}\left[1 - \dfrac{\lambda}{\mu(C)}\right]^{-1}}$ where $\lambda < C\mu$

λ = arrival rate μ = service rate n = number of customers in system (waiting and being served)

C = number of channels in multiple-channel system

P_0 = probability of zero units P_Q = probability of the maximum number in the system

Q = maximum number of arrivals that can be in system, both waiting and being served

Note: For single-channel, single-phase systems with Poisson arrivals and general (nonspecified) service times, the number in the waiting line is

$$N_q = \frac{(\lambda\sigma)^2 + \left(\frac{\lambda}{\mu}\right)^2}{2\left(1 - \frac{\lambda}{\mu}\right)} \qquad \text{where } \sigma = \text{standard deviation of the service time distribution}$$

*Kostas and Dervitsiotis, *Operations Management*, McGraw-Hill, 1981; and Allen, *Probability, Statistics, and Queuing Theory*, Academic Press, 1978.

As depicted in Fig. 16-7, the most relevant characteristics of queuing systems are:

(1) *Input source.* This may be finite or infinite, and it generates customer arrivals, which are assumed to follow a Poisson distribution rate (λ units per period) unless specified otherwise.

(2) *Customers.* They form in a queue length that can theoretically vary from zero to infinity, unless the model used assumes a limited queue length. Customers are allocated to service facilities according to a dispatching rule called the *queue discipline*. A first in, first out discipline is assumed unless otherwise stated.

(3) *Service rate.* The service rate (μ) must be greater than the arrival rate (λ) or the queue can become infinite. *Service rate* (units serviced per period) is also Poisson distributed.

Table 16-1 lists some equations useful for selected queuing problems.

SINGLE-CHANNEL, SINGLE-PHASE SYSTEMS

Computations for single-channel, single-phase queuing models are relatively straightforward. The problem in Example 16.12 assumes an infinite number of customers, unlimited waiting-line length, Poisson arrivals, and negative exponential service times. The queue discipline is first in, first out, with no defections or balking from the waiting line.

Example 16.12   An equivalent service facility has Poisson arrival and service rates and operates on a first come, first served queue discipline. Requests for service average λ = three per day. The facility can service an average of μ = six machines per day. Find the:

(a)   Utilization factor ($\% U$) of the service facility

(b)   Mean time T_s in the system

(c)   Mean number N_s in the system

(d)   Mean waiting time T_q in the queue

(e)   Probability P of finding $n = 2$ machines in the system

(f)   Expected number N_q in the queue

(g)   Percentage of time the service facility is idle (percentage I)

(a)   Utilization factor:

$$\% U = \frac{\text{mean arrival rate}}{\text{mean service rate}} = \frac{\lambda}{\mu} \tag{16.10}$$

$$= \frac{3}{6} = 50\%$$

(b)   Mean time in the system:

$$T_s = \frac{1}{\text{mean service rate} - \text{mean arrival rate}} = \frac{1}{\mu - \lambda} \tag{16.11}$$

$$= \frac{1}{6 - 3} = \frac{1}{3} \text{ day}$$

(c)   Mean number in the system:

$$N_s = (\text{mean time in system})(\text{mean arrival rate}) \tag{16.12}$$

$$= \left(\frac{1}{\mu - \lambda} \right) \lambda = \frac{\lambda}{\mu - \lambda}$$

$$= \frac{3}{6 - 3} = 1 \text{ machine}$$

(d) Mean waiting time:

$$T_q = \text{mean time in system} - \text{service time} \qquad (16.13)$$

$$= \frac{1}{\mu - \lambda} - \frac{1}{\mu} = \frac{\lambda}{\mu(\mu - \lambda)}$$

$$= \frac{1}{6 - 3} - \frac{1}{6} = \frac{1}{6} \text{ day}$$

(e) Probability of $n = 2$ machines in the system:

$$P_n = \text{(probability of no others)(probability of two)} \qquad (16.14)$$

$$= \left(1 - \frac{\lambda}{\mu}\right)\left(\frac{\lambda}{\mu}\right)^n$$

$$= \left(1 - \frac{3}{6}\right)\left(\frac{3}{6}\right)^2 = .125$$

(f) Mean number in the queue:

$$N_q = \text{(mean number in system)} - \text{(mean number being served)} \qquad (16.15)$$

$$= \frac{\lambda}{\mu - \lambda} = \frac{\lambda}{\mu} = \frac{\lambda^2}{\mu(\mu - \lambda)}$$

$$= \frac{3^2}{6(6 - 3)} = \frac{1}{2} \text{ machine}$$

(g) Percentage of idle time:

$$\% I = \text{total} - \text{percentage utilization} = 100 - \% U \qquad (16.16)$$

$$= 100\% - 50\% = 50\%$$

Service Times. Although equations are frequently stated in terms of service rates (i.e., units per time period), data are frequently collected in terms of service times (i.e., time per unit). Poisson service rates have negative exponential service times, which offer a strong probability of short service times but allow for an occasional task that far exceeds the average time. Service rates are the reciprocal of service times.

Example 16.13 A facility that required 60 minutes to service 5 units has negative exponentially distributed service times. What is the service rate?

Per unit service time is $(60 \text{ min}) \div (5 \text{ units}) = 12 \text{ min/unit}$

Service rate is the reciprocal

$$\lambda = \frac{1}{\text{service time}} = \frac{1}{12 \text{ min/unit}} = \frac{.083 \text{ unit}}{\text{min}} = \frac{5.0 \text{ units}}{\text{hr}}$$

Constant Service Rates. When service rates are constant, as in machine-paced systems, the mean waiting time $T_{q(c)}$ and number in the queue $N_{q(c)}$ are reduced by half. (See model 2 in Table 16-1.)

Example 16.14 Metropolitan Collection Co. (MCC) garbage trucks currently wait an average of 6 minutes each trip before being able to dump their load. MCC is considering hauling to a different collection center at an extra cost of $8 per trip for each truck. The new center can process the loads at a constant rate of 30 units per hour. Arrivals at the new center will be Poisson-distributed, with an average rate of 24 loads per hour. The system is a single-channel, single-phase system with unlimited queue length. If waiting time for the trucks is valued at $200 per hour, how much of a savings per hour would result?

The mean waiting time at the new center is estimated as:

$$T_{q(c)} = \frac{\lambda}{2\mu(\mu - \lambda)} = \frac{24}{2(30)(30 - 24)} = \frac{1}{15} \text{ hr}$$

Current waiting cost/trip:
$$\left(\frac{6 \text{ min}}{\text{trip}}\right)\left(\frac{\text{hr}}{60 \text{ min}}\right)\left(\frac{\$200}{\text{hr}}\right) = \$20.00$$

Less: New waiting cost/trip:
$$\left(\frac{1}{15}\frac{\text{hr}}{\text{trip}}\right)\left(\frac{\$200}{\text{hr}}\right) = -13.33$$

$$\text{Savings} = \$6.67/\text{trip}$$

The extra cost of \$8/trip exceeds the savings of \$6.67/trip in waiting time; the change is not worthwhile.

MULTIPLE-CHANNEL, SINGLE-PHASE SYSTEMS

Multiple-channel, single-phase systems (model 4 of Table 16-1) are commonly encountered in banks, post offices, and numerous retail activities where a single line of customers may be served at any one of two or more workstations. In addition to the assumptions of Poisson arrivals, Poisson service rates (i.e., negative exponential service times), and unlimited queue length, we also assume that the servers work independently of each other but at the same average service rate μ.

For multiple-channel, single-phase systems, where C = number of channels or servers, the average utilization of capacity is $\lambda/C\mu$. The calculations (model 4) can be simplified by use of Table 16-2, which gives average queue lengths N_q for various ratios of r, the ratio of the arrival rate to the service rate:

$$r = \frac{\lambda}{\mu} \tag{16.17}$$

Once r is computed, and the number of channels C specified, N_q can be taken from the table and other values (T_q, T_s, N_s, etc.) computed more easily.

Example 16.15   A new post office is being designed with six service counters. During peak hours, customers are expected to arrive at a (Poisson-distributed) rate averaging 4 per minute. Service time is negative exponential, with some customers taking only a few seconds and others taking several minutes; the mean is 1 minute and 12 seconds. If all six service counters are staffed, (*a*) what is the average utilization of capacity, (*b*) how many customers, on average, will be in the waiting line, (*c*) what is their average waiting time, and (*d*) what is their average time in the post office?

(*a*)
$$C = 6 \text{ counters (service channels)}$$
$$\lambda = 4 \text{ cust/min}$$

$$\mu = \frac{60 \text{ sec/min}}{72 \text{ sec/cust}} = .833 \text{ cust/min}$$

$$\% \text{ Utilization} = \frac{\lambda}{C\mu} = \frac{4}{(6)(.833)} = 80\%$$

(*b*) N_q can be taken from Table 16-2 where $r = \dfrac{\lambda}{\mu} = \dfrac{4}{.833} = 4.80$

Thus, $N_q = 2.07$ customers.

(*c*)
$$T_q = \frac{N_q}{\lambda} = \frac{2.07 \text{ cust}}{4 \text{ cust/min}} = .518 \text{ min} = 31 \text{ sec}$$

(*d*)
$$T_s = T_q + \frac{1}{\mu} = \frac{.518 \text{ min}}{\text{cust}} + \frac{1}{.833 \text{ cust/min}} = 1.72 \text{ min}$$

Table 16-2*

Average Queue Lengths (Nq) for C = 1 to 11 Channels for Common Ratios of r = λ/μ
All values assume Poisson arrivals, negative exponential service times

$r = \frac{\lambda}{\mu}$	1	2	3	4	5	6	7
.10	.0111						
.20	.0500	.0020					
.30	.1285	.0069					
.40	.2666	.0166					
.50	.5000	.0333	.0030				
.60	.9000	.0593	.0061				
.70	1.6333	.0976	.0112				
.80	3.2000	.1523	.0189				
.90	8.1000	.2285	.0300	.0041			
1.0		.3333	.0454	.0067			
1.2		.6748	.0904	.0158			
1.4		1.3449	.1778	.0324	.0059		
1.6		2.8444	.3128	.0604	.0121		
1.8		7.6734	.5320	.1051	.0227	.0047	
2.0			.8888	.1739	.0398	.0090	
2.2			1.4907	.2770	.0659	.0158	
2.4			2.1261	.4305	.1047	.0266	.0065
2.6			4.9322	.6581	.1609	.0426	.0110
2.8			12.2724	1.0000	.2411	.0659	.0180
3.0				1.5282	.3541	.0991	.0282
3.2				2.3856	.5128	.1452	.0427
3.4				3.9060	.7365	.2085	.0631
3.6				7.0893	1.0550	.2947	.0912
3.8				16.9366	1.5184	.4114	.1292
4.0					2.2164	.5694	.1801
4.2					3.3269	.7837	.2475
4.4					5.2675	1.0777	.3364
4.6					9.2885	1.4867	.4532
4.8					21.6384	2.0708	.6071

$r = \frac{\lambda}{\mu}$	6	7	8	9	10	11
3.0	.0991	.0282	.0077			
3.2	.1452	.0427	.0122			
3.4	.2085	.0631	.0189			
3.6	.2947	.0912	.0283	.0084		
3.8	.4114	.1292	.0412	.0127		
4.0	.5694	.1801	.0590	.0189		
4.2	.7837	.2475	.0827	.0273	.0087	
4.4	1.0777	.3364	.1142	.0389	.0128	
4.6	1.4867	.4532	.1555	.0541	.0184	
4.8	2.0708	.6071	.2092	.0742	.0260	
5.0	2.9375	.8102	.2786	.1006	.0361	.0125
5.2	4.3004	1.0804	.3680	.1345	.0492	.0175
5.4	6.6609	1.4441	.5871	.1779	.0663	.0243
5.6	11.5178	1.9436	.6313	.2330	.0883	.0330
5.8	26.3726	2.6481	.8225	.3032	.1164	.0443
6.0		3.6828	1.0707	.3918	.1518	.0590
6.2		5.2979	1.3967	.5037	.1964	.0775
6.4		8.0768	1.8040	.6454	.2524	.1008
6.6		13.7692	2.4198	.8247	.3222	.1302
6.8		31.1270	3.2441	1.0533	.4090	.1666
7.0			4.4471	1.3471	.5172	.2119
7.2			6.3135	1.7288	.6521	.2677
7.4			9.5102	2.2324	.8203	.3364
7.6			16.0379	2.9113	1.0310	.4211
7.8			35.8956	3.8558	1.2972	.5250
8.0				5.2264	1.6364	.6530
8.2				7.3441	2.0736	.8109
8.4				10.9592	2.6470	1.0060
8.6				18.3223	3.4160	1.2484
8.8				40.6824	4.4806	1.5524

* Elwood S. Buffa, *Modern Production/Operations Management*, 6th ed. John Wiley & Sons, New York, 1980, pp. 644–645. Reprinted by permission of John Wiley & Sons, Inc.

MAINTENANCE OBJECTIVES

Question: Define maintenance.

Maintenance is any activity designed to keep equipment or other assets in the condition that will best support organizational goals.

Although maintenance is sometimes trivialized to an objective of minimizing long-run maintenance costs, maintenance encompasses a broad range of activities that affect the long-run viability of an organization. Included are concerns of safety, reliability, employment stability, and even economic survival. One of the major issues firms must face is achieving the optimal balance between preventive and breakdown maintenance.

Question: Distinguish between preventive and breakdown maintenance.

(a)   *Preventive maintenance* (*PM*) is the routine inspection and service activities designed to detect potential failure conditions and make minor adjustments or repairs that will help prevent major operating problems.

(b)   *Breakdown maintenance* is the repair, often of an emergency nature and at a cost premium, of facilities and equipment that have been used until they fail to operate.

The reliability demanded by MRP and JIT systems, coupled with a commitment to total quality, has placed increased emphasis on preventive maintenance programs. Effective programs necessitate an accurate and accessible records system, trained personnel, regular inspections, and service.

PROBABILITY MODELS FOR BREAKDOWNS

Example 16.16   The housewares plant of a chemical company has 15 identical molding machines that produce a variety of molded products that generate a profit of $100 per machine per day. The machines fail according to a Poisson distribution with an average of 2.2 machines down each day.

(a)   What is the chance of having exactly three machines down on a given day?

(b)   What is the expected amount of lost profit per day due to this Poisson failure rate of 2.2 per day?

(a)   Since failures follow the Poisson distribution, the probability of X machines failing on any given day is:

$$P(X) = \frac{\lambda^x e^{-\lambda}}{x!}$$

where X = number of machines broken down = 3
   λ = mean failure rate = 2.2/day
   e = 2.718

$$P(X = 3) = \frac{(2.2)^3 e^{-2.2}}{3!} = .1966 = 20 \text{ percent chance}$$

(Note that the values may be calculated or taken from Appendix D as $P(X = 3) = P(X \leq 3) - P(X \leq 2) =$
.819 - .623 = .196)

(b)   The expected loss per day is:

$$E(X) = X \cdot P(X)$$

where x = amount of loss = $100/machine-day
   $P(X)$ = mean value of distribution = 2.2 machines/day

Therefore, $E(X) = 100(2.2) = \$220/\text{day}$

EXPECTED-VALUE MODEL FOR ESTIMATING BREAKDOWN COST

Expected-value and simulation techniques are useful for estimating breakdown costs. The expected-value model requires data on the frequency and cost of past breakdowns.

Example 16.17   Worldwide Travel Services (WTS) has experienced the number of breakdowns per month in its automated reservations processing system over the past 2 years shown below.

Number of breakdowns	0	1	2	3	4
Number of months this occurred	2	8	10	3	1

Each breakdown costs the firm an average of $280. For a cost of $150 per month, WTS could have engaged a data processing firm to perform preventive maintenance, which is guaranteed to limit the breakdowns to an *average* of one per month. (If the breakdowns exceed this limit, the firm will process WTS data free of charge.) Which maintenance arrangement is preferable from a cost standpoint, the current breakdown policy or a preventive maintenance contract arrangement?

By converting the frequencies into a probability distribution and determining the expected cost per month of breakdowns, we have the information shown in Table 16-3.

Table 16-3

Number of Breakdowns X	Frequency in Months $f(X)$	Frequency in Percentages $P(X)$	Expected Value $X \cdot P(X)$
0	2	.083	.0
1	8	.333	.333
2	10	.417	.834
3	3	.125	.375
4	1	.042	.168
	24		1.710

$$\text{Expected breakdown cost per month} = \left(\frac{1.71 \text{ breakdowns}}{\text{month}}\right)\left(\frac{\$280}{\text{breakdown}}\right) = \frac{\$479}{\text{month}}$$

Preventive maintenance cost per month: Since the data processing firm guarantees to limit the cost to an "average" of one breakdown per month and the expected number (1.710) is greater than 1, we may assume that WTS will, in the long run, always incur the cost of one breakdown per month.

Average cost of breakdown/month	$280
Maintenance contract cost/month	150
Total	$430

Preventive maintenance advantage = $479 − $430 = $49/month

SIMULATION MODEL FOR ESTIMATING BREAKDOWN COST

Computer simulations of breakdown and repair time values can also be used to estimate breakdown costs and help reach decisions on the appropriate crew size. The following (limited) example illustrates the methodology; a larger (computer) simulation would be recommended.

Example 16.18 A management analyst is attempting to study the total cost of the present maintenance policy for machinery in a decentralized section of a shoe manufacturing plant in Boston. The analyst has collected historical data and simulated breakdowns of machinery over a 16-hour period as shown in Table 16-4.

Table 16-4

Request for Repair (arrival time)	Total Repair Time Required (worker-hours)
0100	1.0
0730	3.0
0800	.5
1150	2.0
1220	.5
	7.0

The firm has two maintenance technicians and charges their time (working or idle) at $34 per hour each. The downtime cost of the machines, from lost production, is estimated at $360 per hour. (*a*) Determine the simulated service maintenance cost. (*b*) Determine the simulated breakdown maintenance cost. (*c*) Determine the simulated total maintenance cost. (*d*) Would another technician be justified?

(*a*)   Simulated service maintenance cost:

$$\text{Service cost} = (2 \text{ technicians})(\$34/\text{hr})(16 \text{ hr}) = \$1,088$$

(*b*)   Simulated breakdown maintenance cost (note that we assume that two technicians are twice as effective as one and reduce the downtime accordingly):

Table 16-5

(1)	(2) Repair Time Required (2 Technicians)		(3)	(4)	(5)	(6)
Request Arrival Time	Hr	Min	Repair Time Begins	Repair Time Ends	Machine Downtime(hr) (2 Technicians)	Machine Downtime(hr) (3 Technicians)
0100	.50	30	0100	0130	.50	.33
0730	1.50	90	0730	0900	1.50	1.00
0800	.25	15	0900	0915	1.25	.67
1150	1.00	60	1150	1250	1.00	.67
1220	.25	15	1250	1350	.75	.33
	3.50				5.00	3.00

The machine downtime is shown in Table 16-5, in hours, in column 5, as the decimal difference between the request arrival time (1) and the ending repair time (4). Note that on the 0800 breakdown the technicians were not available until 0900, when they finished the earlier job.

$$\text{Breakdown cost} = (\$360/\text{hr})(5 \text{ hr}) = \$1,800$$

(*c*)   Simulated total maintenance cost:

$$\text{Total cost} = \text{service} + \text{breakdown} = \$1,088 + \$1,800 = \$2,888/\text{period}$$

(*d*)   The machine downtime hours for three technicians would have to be calculated in the same way as was done for two. The calculations are not included, but the final result is shown in column 6.

$$\text{Service maintenance cost} \quad = (3)(\$34)(16) = \$1,632$$
$$\text{Breakdown maintenance cost} = (\$360)(3 \text{ hr}) = \underline{1,080}$$
$$\text{Total} \qquad\qquad\qquad\qquad\qquad\qquad \$2,712$$

There appears to be an advantage to adding a third technician.

Questions and Solved Problems

RULES FOR DIFFERENTIATION

16.1   Find the first derivative of the functions:

$$(a) \quad y = 10x^4 \qquad (b) \quad M = 3Q^2 + 2Q \qquad (c) \quad P = -\frac{8}{x^{1/2}}$$

(a)
$$\frac{dy}{dx} = 40x^3$$

(b)
$$\frac{dM}{dQ} = 6Q + 2$$

(c)   Rewrite equation to: $P = -8x^{-1/2}$

$$\frac{dP}{dx} = \left(-\frac{1}{2}\right)(-8)x^{-1/2 - 2/2} = 4x^{-3/2} = \frac{4}{x^{3/2}} = \frac{4}{\sqrt{x^3}}$$

16.2   Find the first derivative of the functions:

$$(a) \quad y = x^4 - 3x^2 + 7x + 12 \qquad (b) \quad y = (3 + 4x)^3 \qquad (c) \quad y = (4x + 7)(x^2 - 1)$$

(a)
$$\frac{dy}{dx} = 4x^3 - 6x + 7$$

(b)
$$\frac{dy}{dx} = 3(3 + 4x)^2(0 + 4) = 12(3 + 4x)^2$$

(c)
$$\frac{dy}{dx} = (4x + 7)(2x) + (x^2 - 1)(4) = 12x^2 + 14x - 4$$

16.3   A firm makes \$8 revenue on each unit sold, so the total revenue function is $R(x) = 8x$. Find and explain the marginal revenue.

Marginal revenue is the instantaneous rate of change in total revenue, or derivative of the total revenue function.

$$\frac{dR}{dx} = R'(x) = 8$$

Thus the slope of the total revenue function, or average rate of change (marginal revenue), is \$8.

16.4   A cost curve is defined by the function $y = x^3/3 - 2x^2 + 3x$ where $y = $ cost (in dollars) and $x = $ pounds of a specified raw material. (*a*) Graph the function. (*b*) Find the point of zero slope. (*c*) Indicate whether the zero slope points are local maximums or minimums.

(*a*)   See Fig. 16-8:

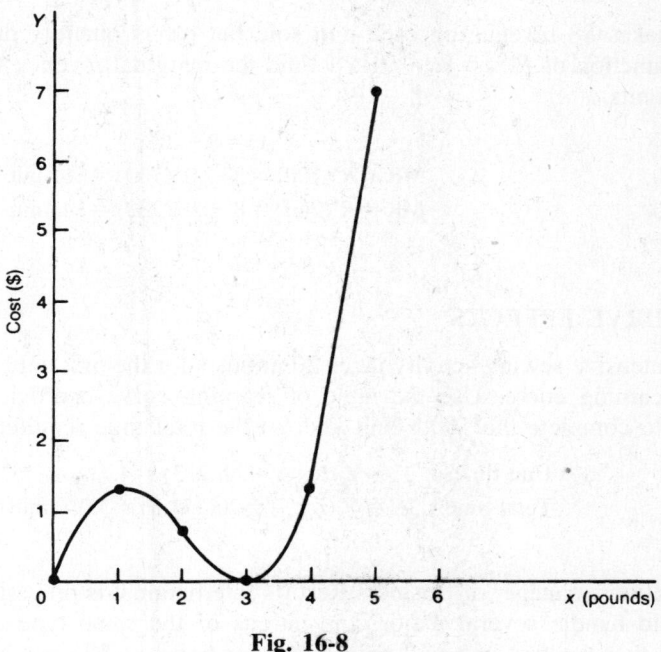

Fig. 16-8

(b)   This is a cubic equation.

$$\frac{dy}{dx} = x^2 - 4x + 3$$

Therefore, the points of zero slope are where

$$x^2 - 4x + 3 = 0$$

Factoring we have

$$(x - 1)(x - 3) = 0$$

Setting each of these terms equal to zero we have

$$x - 1 = 0 \qquad x - 3 = 0$$

The equation has two zero derivative points (roots) that are at $x = 1$ and $x = 3$, where the y values are 4/3 and 0, respectively.

(c)   The second derivative is:

$$\frac{d^2y}{dx^2} = 2x - 4$$

We can substitute the x values of 1 and 3 into the second derivative equation to determine its sign.

At $x = 1$, $$\frac{d^2y}{dx^2} = 2(1) - 4 = 2 - 4 = -2$$

A negative sign means the point $y = 4/3$ is a maximum.

At $x = 3$, $$\frac{d^2y}{dx^2} = 2(3) - 4 = 6 - 4 = 2$$

A positive sign means the point $y = 0$ is a minimum.

16.5 A firm makes $8 revenue on each unit sold but offers quantity discounts that result in a total revenue function of $R(x) = 8x - .01x^2$. Find the marginal revenue (MR) at sale volumes of 100 and 200 units.

$$R'(x) = 8 - .02x$$

At $x = 100$, $\quad$ MR $= R'(100) = 8 - .02(100) = \$6/$unit

At $x = 200$, $\quad$ MR $= R'(200) = 8 - .02(200) = \$4/$unit

LEARNING CURVE EFFECTS

16.6 A labor-intensive sewing activity takes 20 minutes for the first unit, and production follows an 85 percent learning curve. Use the table of learning curve coefficients to estimate (a) the time required to complete the 500th unit and (b) the total time required to complete all 500 units.

Unit time: $\quad Y_n = Y_B(L_{unit}) = 20(.233) = 4.7$ min

Total time: $\quad Y_N = Y_B(L_{total}) = 20(151.5) = 3,030$ min $= 50.5$ hr

16.7 The operations manager of Business Resorts International is preparing a budget for the labor cost required to handle several major conventions of the same type during the coming year. She estimated that the first one will take 300 labor hours at $25 per hour. Assuming that the work follows a 78 percent learning curve, what is the firm's estimated labor cost for the eighth convention?

Appendix H does not have values for a 78 percent learning rate, so we must either approximate it (e.g., by interpolation) or use the expression in Eq. 16.5.

$$Y_n = Y_1 n^b \quad \text{where} \quad b = \frac{\ln \text{ learning } \%}{\ln 2} = \frac{\ln .78}{\ln 2} = -.3585$$
$$= (300)8^{-.3585}$$
$$= 142.4 \text{ hr}$$

Estimated cost $= (142.4 \text{ hr})(\$25/\text{hr}) = \$3,560$

16.8 (*Using historical data*) Northern Appliance Co. has a new facility for producing home freezers. The firm has gone through a preliminary manufacturing period and believes it is experiencing an 88 percent learning curve. The 200th unit has required 1.40 labor hours for an assembly activity. Estimate the comparable time for (a) the 100th unit, (b) the 500th unit, and (c) the 5,000th unit.

(a) The 100th unit has already been completed. We could estimate its time from Eq. 16.5 if we had the time for unit 1, which we do not have. However, knowing the time for unit 200 ($Y_n = 1.40$ hours), we can easily work backward, using Eq. 16.5, to find the time for unit 1.

$$Y_n = Y_1 n^b \quad \text{where} \quad b = \frac{\ln \text{ learning } \%}{\ln 2} = \frac{\ln .88}{\ln 2} = -.1844$$

$$1.40 = (Y_1)200^{-(.1844)}$$

$$= \frac{Y_1}{200^{(.1844)}} = \frac{Y_1}{2.6568}$$

$$Y_1 = (1.40)(2.6568) = 3.7196 \text{ hr}$$

Using $Y_1 = 3.7196$ hours, we can now compute the time for unit 100.

$$Y_{100} = Y_1 n^b \qquad \text{where} \quad b = -.1844$$
$$= (3.7196)100^{(-.1844)}$$
$$= (3.7196)(.4277) = 1.59 \text{ hr}$$

(b) $$Y_{500} = (3.7196)500^{(-.1844)} = 1.18 \text{ hr}$$

(c) $$Y_{5000} = (3.7196)5000^{(-.1844)} = .77 \text{ hr}$$

QUEUING AND MAINTENANCE APPLICATIONS

16.9 Patients arrive at a medical clinic with an arrival rate that is Poisson distributed with a mean of 6 per hour. Treatment (service) time averages 8 minutes and can be approximated by the negative exponential distribution. Find (a) the mean waiting time, (b) the mean number in the queue, and (c) the percentage of idle time.

$$\lambda = \text{arrival rate} = 6/\text{hr}$$

$$\mu = \text{service rate} = \left(\frac{\text{unit}}{8 \text{ min}}\right)\left(\frac{60 \text{ min}}{\text{hr}}\right) = 7.5/\text{hr}$$

(a) $$T_q = \frac{\lambda}{\mu(\mu - \lambda)} = \frac{6}{7.5(7.5 - 6)} = .53 \text{ hr} = 32 \text{ min}$$

(b) $$N_q = \frac{\lambda^2}{\mu(\mu - \lambda)} = \frac{6^2}{7.5(7.5 - 6)} = 3.20 \text{ units}$$

(c) $$\% I = 100 - \% U = 100 - \frac{\lambda}{\mu} = 100 - \frac{6}{7.5} = 20\%$$

16.10 A service desk at an office of charity in a large city is staffed by one social worker who receives an average of three people (customers) an hour seeking some kind of aid (food, housing, etc.). The social worker talks with them and assists them directly or guides them to other facilities in the city. This takes an average of 15 minutes. Assume that arrivals are Poisson distributed, that service time has an exponential distribution, and that the queue discipline is first come, first served. Find (a) the probability of zero customers, (b) the probability of two customers in the office at one time, (c) the average number of customers in the system, (d) the average time a customer spends in the system, including service time, and (e) the average number of customers waiting to be served.

Remember that $\lambda = \text{arrival rate} = 3$ per hour and $\mu = \text{service rate} = (60 \text{ min}/15 \text{ min per cust.}) = 4$ per hour.

(a) $$P_0 = 1 - \frac{\lambda}{\mu} = 1 - \frac{3}{4} = .25$$

(b) $$P_2 = \left(1 - \frac{\lambda}{\mu}\right)\left(\frac{\lambda}{\mu}\right)^n = \left(1 - \frac{3}{4}\right)\left(\frac{3}{4}\right)^2 = .14$$

(c) $$N_s = \frac{\lambda}{\mu - \lambda} = \frac{3}{4 - 3} = 3 \text{ customers}$$

(d) $$T_s = \frac{1}{\mu - \lambda} = \frac{1}{4 - 3} = 1 \text{ hr}$$

(e) $$N_q = \frac{\lambda^2}{\mu(\mu - \lambda)} = \frac{3^2}{4(4 - 3)} = \frac{9}{4} = 2.25 \text{ customers}$$

16.11 Use the data from Prob. 16.10 to find the number of people waiting for service if the service time is normally distributed with a mean of 15 minutes and a standard deviation of .025 minutes.

From Table 16-1 for nonspecified service times:

$$N_q = \frac{(\lambda\sigma)^2 + \left(\frac{\lambda}{\mu}\right)^2}{2\left(1 - \frac{\lambda}{\mu}\right)} = \frac{(3 \times .025)^2 + \left(\frac{3}{4}\right)^2}{2\left(1 - \frac{3}{4}\right)} = 1.14 \text{ customers}$$

Note that the assumption of the normal distribution substantially increases the queue length.

16.12 Use the data from Prob. 16.10, except assume that the number of persons in the system is limited to five (model 3 of Table 16-1). Find (a) the probability of two customers in the office at one time, (b) the probability of five customers in the office at one time, (c) the average number of people in the system, (d) the average number in the queue, (e) the average waiting time, and (f) the average time in the system.

(a)
$$P_2 = \left[\frac{1 - \frac{\lambda}{\mu}}{1 - \left(\frac{\lambda}{\mu}\right)^{Q+1}}\right]\left(\frac{\lambda}{\mu}\right)^n = \left[\frac{1 - \frac{3}{4}}{1 - \left(\frac{3}{4}\right)^{5+1}}\right]\left(\frac{3}{4}\right)^2 = .17$$

(b)
$$P_5 = \left[\frac{1 - \frac{3}{4}}{1 - \left(\frac{3}{4}\right)^{5+1}}\right]\left(\frac{3}{4}\right)^5 = .072$$

(c)
$$N_s = \frac{\lambda}{\mu}\left\{\frac{1 - (Q+1)\left(\frac{\lambda}{\mu}\right)^Q + Q\left(\frac{\lambda}{\mu}\right)^{Q+1}}{\left(1 - \frac{\lambda}{\mu}\right)\left[1 - \left(\frac{\lambda}{\mu}\right)^{Q+1}\right]}\right\} = \left(\frac{3}{4}\right)\left\{\frac{1 - (5+1)\left(\frac{3}{4}\right)^5 + 5\left(\frac{3}{4}\right)^{5+1}}{\left(1 - \frac{3}{4}\right)\left[1 - \left(\frac{3}{4}\right)^{5+1}\right]}\right\}$$

$$= 1.68 \text{ persons}$$

(d)
$$N_q = \left(\frac{\lambda}{\mu}\right)^2\left\{\frac{1 - Q\left(\frac{\lambda}{\mu}\right)^{Q-1} + (Q-1)\left(\frac{\lambda}{\mu}\right)^Q}{\left(1 - \frac{\lambda}{\mu}\right)\left[1 - \left(\frac{\lambda}{\mu}\right)^Q\right]}\right\} = \left(\frac{3}{4}\right)^2\left\{\frac{1 - 5\left(\frac{3}{4}\right)^{5-1} + (5-1)\left(\frac{3}{4}\right)^5}{\left(1 - \frac{3}{4}\right)\left[1 - \left(\frac{3}{4}\right)^5\right]}\right\}$$

$$= 1.08 \text{ persons}$$

(e)
$$T_q = \frac{N_q}{\lambda[1 - P_Q]} = \frac{1.08}{3[1 - .072]} = .388 \text{ hr} = 23.3 \text{ min}$$

(f)
$$T_s = T_q + \frac{1}{\mu} = .388 + \left(\frac{1}{4}\right) = .638 \text{ hr} = 38.3 \text{ min}$$

16.13 The time required to replace a filter on any of 500 industrial mixers can be considered a constant at 15 minutes per filter. Maintenance records show that the failure rate of filters is distributed according to a Poisson distribution, with a mean of 2 per hour. (a) Find the average number of mixers waiting for a filter replacement. (b) Find the average waiting time of a mixer for repair.

(a)
$$N_{q(c)} = \frac{\lambda^2}{2\mu(\mu - \lambda)}$$

where λ = arrival rate = 2/hr

μ = service rate = (filter/15 min)(60 min/hr) = 4/hr

$$N_{q(c)} = \frac{2^2}{2(4)(4-2)} = .25 \text{ mixer}$$

(b) $$T_{q(c)} = \frac{\lambda}{2\mu(\mu - \lambda)} = \frac{2}{2(4)(4-2)} = .125 \text{ hr} = 7.50 \text{ min}$$

16.14 Rent-A-Dent Ltd. receives an average of 15 requests per day for older-model cars. It can fill 20 such requests per day. However, if fewer than 3 cars are rented, the company loses money as follows: If only 2 cars are rented, the loss equals $220 per day; if only 1 car is rented, the loss equals $260 per day; if no cars are rented, the loss equals $290 per day. The losses are, of course, offset by gains from renting 3 or more cars. Considering the *losses only*, what is the expected value of the loss per day? Assume that there are Poisson arrivals and service rates, and that there is unlimited line length with no defects from the queue.

$$P(n = 2) = \left(1 - \frac{\lambda}{\mu}\right)\left(\frac{\lambda}{\mu}\right)^2 = \left(1 - \frac{15}{20}\right)\left(\frac{15}{20}\right)^2 = .141$$

$$P(n = 1) = \left(1 - \frac{15}{20}\right)(.75)^1 = \qquad\qquad .188$$

$$P(n = 0) = \left(1 - \frac{15}{20}\right)(.75)^0 = (.25)(1) = \qquad \frac{.250}{.579}$$

Expected loss $= \Sigma[XP(X)] = \$220(.141) + \$260(.188) + \$290(.250) = \$152.40/\text{day}$

Supplementary Questions and Problems

16.15 Find the first derivative of the functions (a) $y = 4x^3$ (b) $y = x^3 - 10x^2$ (c) $y = x^2 - 120x + 80$ (d) $y = 50x$ *Ans.*   (a) $12x^2$ (b) $3x^2 - 20x$ (c) $2x - 120$ (d) 50

16.16 Find the first derivative of the functions (a) $y = (x - 2)^2$ (b) $y = 5x(x^3 - 4)$ (c) $y = 5/x^2$
Ans.   (a) $2x - 4$ (b) $20x^3 - 20$ (c) $-10/x^3$.

16.17 At what value of Q does the total cost function TC $= 4Q^2 - 240Q + 75$ have a slope of zero?
Ans.   Solving for first derivative and setting it $= 0$ gives $Q = 30$

16.18 Find the points of zero slope for the function $y = 4x + 16/x$, and indicate whether they are maximums or minimums.
Ans.   At $x = 2$, the y value of 16 is a minimum. At $x = -2$, the y value of -16 is a maximum. (Curve is asymptotic to y axis.)

16.19 Find the points of zero slope for the function $y = x - 4/x^2$, graph it, and indicate whether they are maximums or minimums.   *Ans.*   At $x = -2$ is a maximum ($y = -3$)

16.20 A firm's revenue function is $R(x) = 8x - .002x^2$, and its cost function is $C(x) = 85 + .20x$. Find (a) the marginal revenue and (b) the marginal cost.   *Ans.*   (a) $8 - .004x$ (b) $.20$.

16.21 Given the following cost and revenue functions: $C(x) = 2x^3 - 20x^2 + 10x + 50$, $R(x) = 200x - 2x^2$. Find the (*a*) profit function, (*b*) the marginal profit function, (*c*) the profit at 1, 4, and 10 units of production, and (*d*) the marginal profit at 1, 4, and 10 units.
   Ans.   (*a*) $P(x) = -2x^3 + 18x^2 + 190x - 50$ (*b*) $P'(x) = -6x^2 + 36x + 190$ (*c*) 156, 870, 1,650 (*d*) 220, 238, -50

16.22 The maintenance cost for a shop is defined by a fixed component of \$12,000 per month, plus a variable component of $\$5x^2 - \$300x$, where x is the number of hours of preventive maintenance per month. (*a*) What is the maintenance cost when 20 hours of preventive maintenance are used? (*b*) What is that cost when 35 hours of preventive maintenance are used? (*c*) What amount of preventive maintenance will minimize the total monthly cost?   *Ans.*   (*a*) \$8,000 (*b*) \$7,625 (*c*) 30 hours for a cost of \$7,500

16.23 A manufacturer of radar assemblies has received a contract for 32 units and has produced the first one in 100 hours. If the activity follows a 90 percent learning curve, how long will it take to produce (*a*) the second unit, (*b*) the fourth unit, (*c*) the last unit?   *Ans.*   (*a*) 90 hours (*b*) 81 hours (*c*) 59 hours

16.24 Cranston Manufacturing Co. must determine whether to purchase new equipment for \$18,500 to assist on a contract operation that has been following a 90 percent learning curve. The company has just completed the fourth unit, which took 30 days at a direct labor cost of \$950 per day, and the contract calls for a total of eight units. The new equipment is expected to increase the time of the fifth unit to 31 days, but would generate other improvements that would put the operation on a 70 percent learning curve. Is the new installation economically justified?   *Ans.*   Yes, savings of \$24,945 do outweigh the \$18,500 cost.

16.25 Rocket Control Inc., a Long Beach firm, does control-panel wiring for solid-fueled rocket engines. The firm is currently preparing delivery estimates for a government contract for 80 panels. The first unit is expected to take 200 worker-hours, and the firm usually experiences an 84 percent learning curve for this type of work. (*a*) What average time per unit can be expected for the first three units? (*b*) How many worker-hours should be scheduled for the 40th unit?   *Ans.*   (*a*) 173 hours per unit (*b*) 79.1 hr

16.26 Airframe International engineers have developed a new design for a wing section. Assembling the first unit required 5,000 direct labor hours at \$30 per hour. This type of work follows a 90 percent learning curve, and the firm has orders for 50 such wing sections. (*a*) How many direct labor hours can be expected for the tenth unit? (*b*) Estimate the direct labor cost (to the nearest \$100) for the third unit.
   Ans.   (*a*) 3,524 hours (*b*) \$127,000.

16.27 Given Poisson arrivals of 15 per hour, a Poisson service rate of 18 per hour, and a single-channel, single-phase queue with unlimited line, find (*a*) T_s (*b*) N_s (*c*) T_q (*d*) N_q.
   Ans.   (*a*) .33 hour (*b*) 5 units (*c*) .28 hour (*d*) 4.17 units

16.28 An automatic car wash uses a chain drive that requires a constant time of 5 minutes per car to wash and clean a car. The demand for service follows a Poisson distribution with a mean of $\lambda =$ four per hour. (*a*) Express the service time as a service rate. (*b*) What is the mean waiting time (in minutes) for service?   *Ans.*   (*a*) .20 car per minute (*b*) 1.26 minutes

16.29 A large television service firm has Poisson arrival rates and negative exponential service times, and serves customers 24 hours per day on a first come, first served basis. If the firm receives service orders at a mean rate of 30 per day and has the personnel and facilities to handle up to 35 per day, (*a*) how many sets, on the average, will it have in its shop at any one time; (*b*) how many hours, on the average, will a customer have to wait *before the service firm starts work* on his or her set; (*c*) how many hours, on the average, would the customer have to wait before the firm started work on his or her set if the firm had a *constant* service rate of 35 per day?   *Ans.*   (a) 6 sets (b) 4.11 hours (c) 2.06 hours.

16.30   Use the data from Prob. 16.10, except assume that two social workers are on duty (model 4). Otherwise: $\lambda = 3$ per hour, $\mu = 4$ per hour, FCFS. Find (a) P_0, (b) P_2, (c) N_q, (d) N_s, (e) T_q, (f) T_s.
Ans.   (a) .455 (b) .128 (c) .123 customer (d) .873 customer (e) 2.46 minutes (f) 17.46 minutes. *Note:* These values are considerably smaller than the single-channel values of Prob. 16.10.

16.31   Given a multiple-channel, single-phase queuing problem that satisfies the conditions of model 4 (Table 16-1), where $C = 6$, $\lambda = 4$ customers per minute, and $\mu = 45$ seconds per customer. (a) Use the equations of model 4 to compute the average queue length N_q, and (b) compute N_q using the r ratio and Table 16-2.   *Ans.*   Both answers should be the same except for rounding: $N_q \cong .10$

Chapter 17

Project Management
CPM/PERT

This chapter concerns the management of relatively large-scale work activities done on a one-time basis to accomplish a well-defined goal. Examples include constructing a manufacturing plant, designing and producing a new product, and transferring patients and health-care activities from an old to a new hospital. Each of these is a significant undertaking that includes numerous activities, some of which are interdependent and must be performed in a sequential order.

PROJECT PLANNING, SCHEDULING, AND CONTROL

Question: What is a project?

A *project* is a unique set of activities that must be completed to achieve a specific objective within a limited time (and cost) period by utilizing appropriate resources, often at a job site.

Projects are sometimes viewed as subsets of larger *programs* that can involve multiple organizations (e.g., a governmental space exploration program). The projects, in turn, are composed of *activities* that may be further subdivided into *tasks* and even more definitively into *work packages*.

Project management requires attention to the same managerial concerns as other production activities (i.e., planning, scheduling, control). However, the need to coordinate a multitude of diverse tasks to achieve a unique output within a specified time frame distinguishes project management techniques from the continuous, batch, and even job shop managerial techniques described in earlier chapters.

Question: What are some of the primary concerns associated with the planning, scheduling, and control of projects?

See Fig. 17-1, which depicts the progression from project planning to project scheduling and to project control. Major emphasis in this chapter is on the network techniques of CPM and PERT.

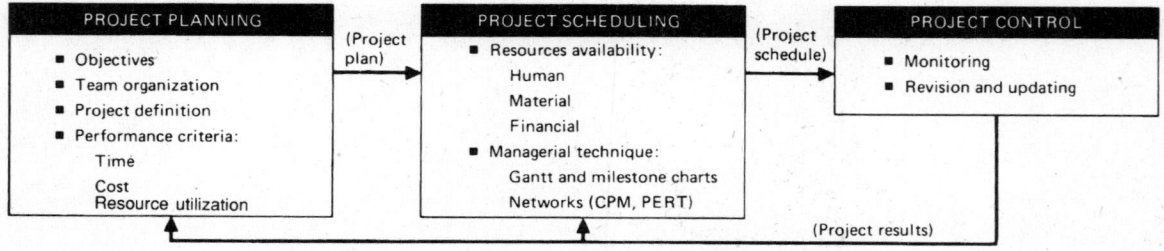

Fig. 17-1   Project management flowchart

Project Planning.   Planning begins with well-defined *objectives*. The *project team* may be drawn from several organizational departments, e.g., engineering, production, marketing, and accounting. *Project definition* involves identifying the controllable and uncontrollable variables involved, and establishing project boundaries. *Performance criteria* should relate to the project objectives, which are often evaluated in terms of time, cost, and resource utilization.

Project Scheduling. Scheduling involves programming the resource requirements and expected progress over the project's time horizon. Figure 17-2 illustrates a chart method of scheduling and tracking the personnel, material, and financial needs of a project. Each chart compares actual with planned levels. Computer graphics can provide whatever level of detail managers deem to be useful.

Techniques for scheduling projects include traditional Gantt (load and progress) charts and network techniques. Gantt charts are easily understood and easily updated (if on computer), but they do not reflect the interrelationships among resources or the precedence relationships among project activities. Network techniques, though more complicated, show precedence relationships and yield valuable trade-off information for improved use of resources.

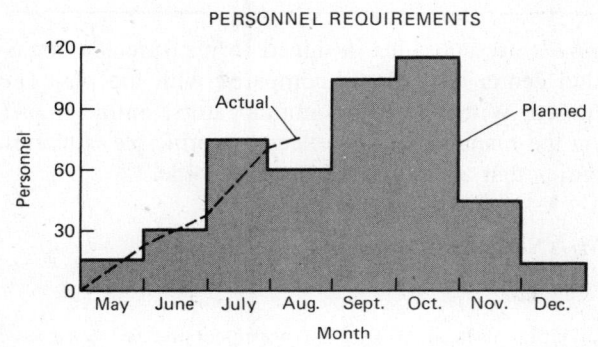

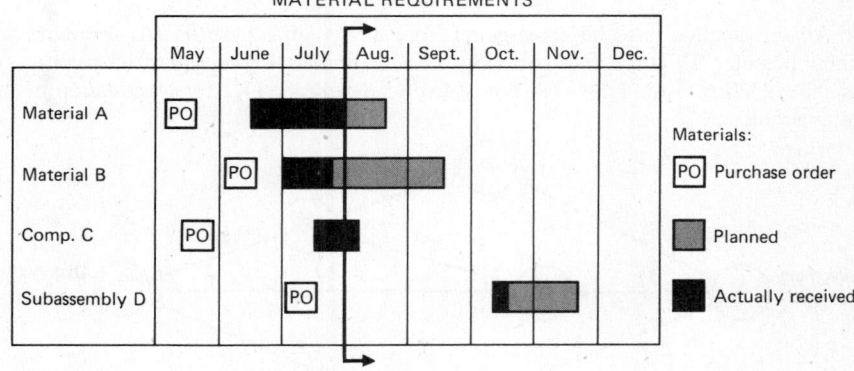

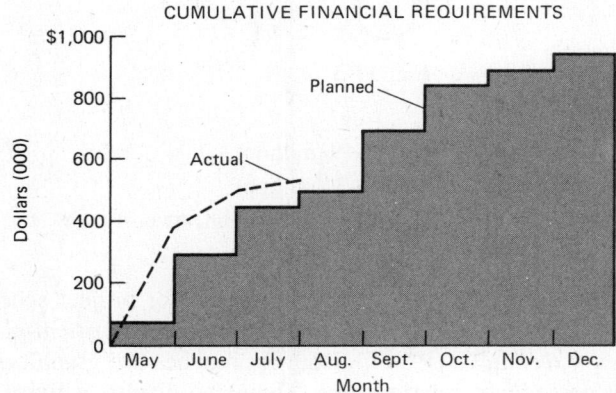

Fig. 17-2   Project resource requirements scheduling

Question: List some advantages of using a network scheduling technique.

1. *Coordinates total project* and all interrelated activities. Shows relationship of each activity to whole project.
2. *Forces logical planning* of all activities. Facilitates work organization and assignment.
3. *Identifies precedence relationships* and activity sequences that are especially critical.
4. *Provides completion time (and/or cost) estimates* and a standard for comparing with actual values.
5. *Facilitates better use of resources* by identifying areas where human, material, or financial resources can be shifted.

Project Control. *Project controls* are activities designed to measure the status of project activities, transmit that data to a command center where it is compared with the plan (i.e., the standard), and initiate corrective action if required. With wireless communicators, online (real time) data analysis is feasible for most projects. Using the management-by-exception principle, attention can then be focused on critical or near-critical activities that are potentially troublesome.

NETWORK MODELS: CPM AND PERT

Question: What is a network diagram?

A *network diagram* is a graph that uses small circles (nodes) connected by arrows or branches (arcs) to represent precedence relationships.

Example 17.1 An oil pipeline is to be constructed from a Wyoming location (A) through some mountainous terrain to a distribution center (F) at the least cost (Fig. 17-3). Alternative routes and construction costs (in millions of dollars) are as shown. What is the least-cost route? (*Note:* See Prob. 17.2 for formulation of this example as a linear-programming problem.)

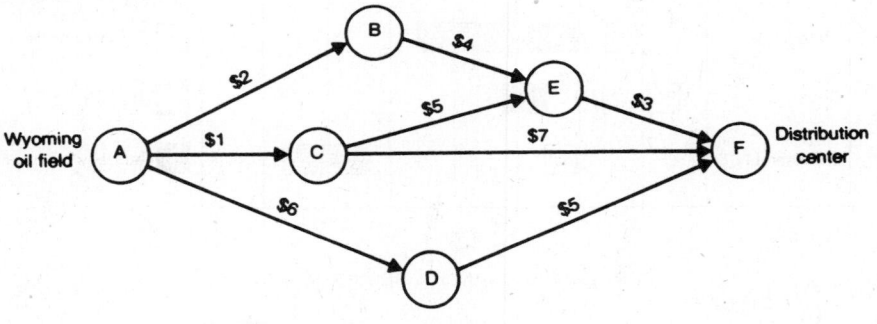

Fig. 17-3

The alternative paths and respective costs are:

$$
\begin{aligned}
\text{Path A–B–E–F} &= \$2 + 4 + 3 = \$ \ 9 \ (\text{million}) \\
\text{A–C–E–F} &= \$1 + 5 + 3 = \$ \ 9 \ (\text{million}) \\
\text{A–C–F} \quad &= \$1 + 7 \quad = \$ \ 8 \ (\text{million}) \leftarrow \text{least-cost path} \\
\text{A–D–F} \quad &= \$6 + 5 \quad = \$11 \ (\text{million})
\end{aligned}
$$

CPM/PERT Historical Development. The network models used for project scheduling today stem largely from developmental efforts in the 1950s. Two systems emerged: *critical-path method* (CPM) and *program evaluation and review technique* (PERT). Both approaches use graphical symbols to depict project activities and show their precedence relationships. They also display activity times and provide some means of estimating project completion times.

Question: Explain the original difference in philosophy between CPM and PERT.

CPM (developed by DuPont and Remington Rand) was designed to improve the scheduling of construction and maintenance activities. Insofar as historical data were available for these types of activities, the CPM model used a single (deterministic) time estimate for each activity. It also included features for analyzing time-cost trade-offs.

PERT (developed by Lockheed and Booz, Allen & Hamilton in conjunction with the U.S. Navy) was designed to coordinate the efforts of over 3,000 suppliers, contractors, and various government agencies involved in a missile project. Insofar as this was a first-time effort for many of the organizations involved, the scheduling system was developed to incorporate uncertain (probabilistic) time estimates for each activity. This feature enabled analysts to develop probabilistic completion time and cost estimates.

AOA and AON Conventions.   Although the CPM and PERT approaches were significantly different in their early stages of development, the beneficial features from each approach have now been largely adapted by the other; there is no longer a clear-cut distinction between the two methodologies. Some analysts use the CPM/PERT designation to refer to either or both methods.

One difference that remains today involves the graphic method of describing activity relationships. Two methods in common use are the activity-on-node diagrams (more closely associated with CPM) and the activity-on-arrow diagrams (associated with PERT).

Question: Distinguish between the (*a*) activity-on-node and (*b*) activity-on-arrow diagrams.

(*a*)   *Activity-on-node* (*AON*) *diagrams* use circles (or rectangles) to represent project activities and arrows to show the required sequence. AON networks are slightly easier to construct since they avoid the need for certain connecting (dummy) activities. But AON diagrams are less intuitive in that the time span for each activity is depicted at a node, rather than over an interval (as an arrow). Figure 17-4*a* illustrates a network using the AON approach. In the figure, activity A takes 2 days, B takes 5 days, etc.

(*b*)   *Activity-on-arrow* (*AOA*) *diagrams* use arrows (or arcs) to represent activities and nodes (or circles) to represent events. In these models, activities are the component tasks that take time to accomplish. The AOA designation can also signify activity-on-arc. Figure 17-4*b* illustrates a network using the AOA approach. Here, activity 1–2 requires 6 days, 2–5 requires 4 days, etc.

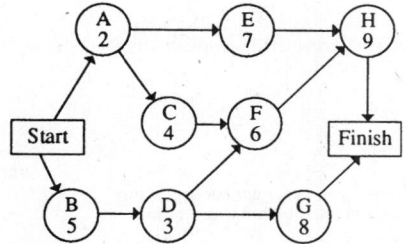

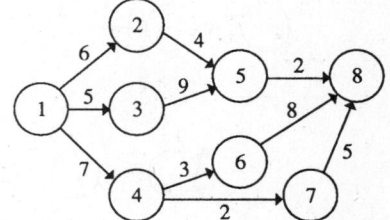

(*a*) Activity-on-node convention
(A, B, . . . , H are activities with
times as shown in days.)

(*b*) Activity-on-arrow convention
(1, 2, . . . , 8 are events; and
1–2, 2–5, etc., are activities
with times as shown in days.)

Fig. 17-4   AON and AOA conventions

Network Symbols.   To illustrate the formulation and use of CPM/PERT networks we shall follow the AOA convention, which considers *activities* to be component tasks that take time and are designated by arrows (→). *Events* are the points in time that indicate some activities have been completed and others may begin. They are the nodes and are designated by circles (○).

Figure 17-5 shows, in a network diagram, the work activities necessary to construct an electrical power plant (the objective). Precedence relationships are indicated by the arrows and circles. For example, the plant design (activity 1–2) must be completed before anything else can take place. Then the selection of the site, vendor, and personnel can take place concurrently. The generator installation (activity 5–7) cannot begin until the site has been prepared (3–5) and the generator has been manufactured (4–5). Note that there are really four paths through the network diagram from event 1 to event 8. The site preparation (3–5) and generator manufacturing (4–5) are on different paths, but since they converge at event 5, either activity could delay the generator installation.

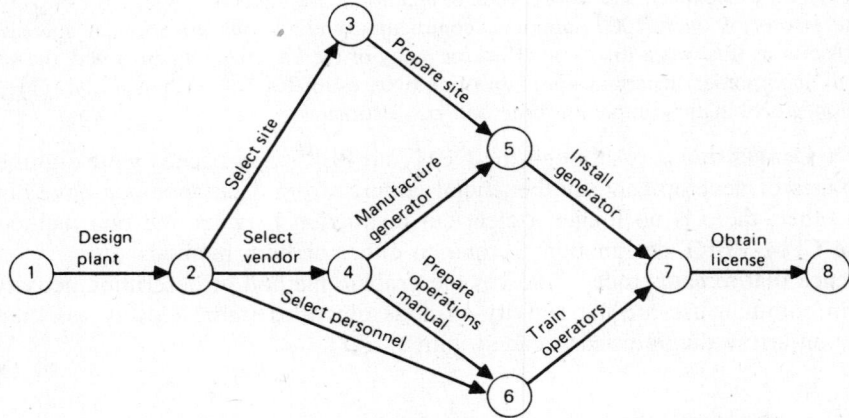

Fig. 17-5 Network diagram for power plant construction

Sometimes precedence relationships are needed even though no time-consuming activities are involved. For example, in Fig. 17-5, suppose the site preparation activity (3–5) cannot begin until the vendor is notified. This means that the vendor selection activity (2–4) must be completed before activity 3–5 can begin. We can indicate this preference requirement by means of a "dummy activity," drawn as a dotted line from event 4 to event 3, that would be assigned a zero time. This dummy activity would then create another unique sequential path (1–2–4–3–5–7–8) through the network.

A summary of common network diagram sequences and arrangements is shown in Fig. 17-6.

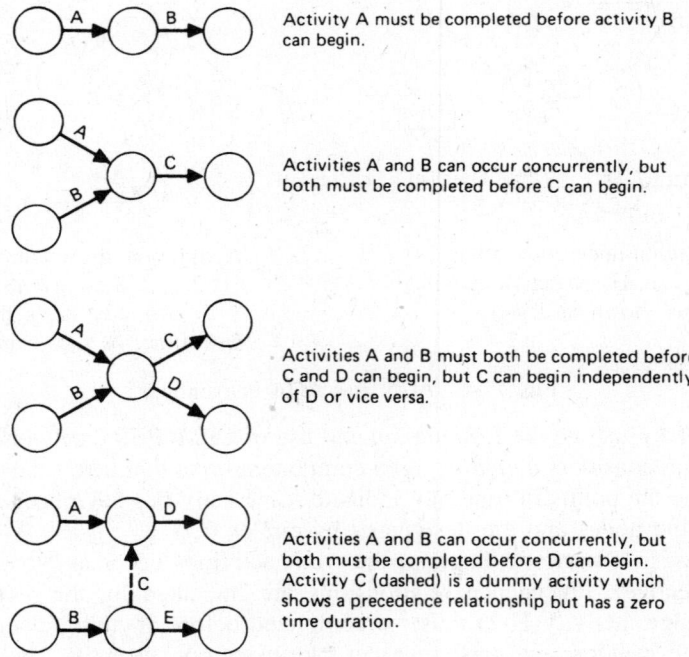

Fig. 17-6 Commonly used network diagram symbols

FINDING THE CRITICAL PATH

A *network path* is a unique sequence of activities from the start to the end of a project. The *critical path* is the path that takes the longest time sequence to complete; it is the minimum time required to complete the project. The steps involved in finding the critical path follow.

Question: List the steps involved in finding the critical path for a project.

> 1. Define the project in terms of activities and events.
> 2. Construct a network diagram showing the precedence relationships.
> 3. Develop a point estimate of each activity time, and show it on the network.
> 4. Compute the time requirement for each path in the network.
> 5. Designate the path with the longest estimated time as the critical path.

The path with the longest time sequence as computed in step 4 is the *critical path*; the activity times of all items on this path are critical to the project completion date. The sum of these activity times is the expected mean time of the critical path (T_E). Other paths will have excess (or slack) time, and the slack associated with any path is simply the difference between T_E and the time for the given path.

Example 17.2   The time estimates for completing the plant construction project of Fig. 17-5 are as shown (in months) on the accompanying network diagram (Fig. 17-7). (a) Determine the critical path. (b) How much slack time is available in the path containing the operations-manual preparation?

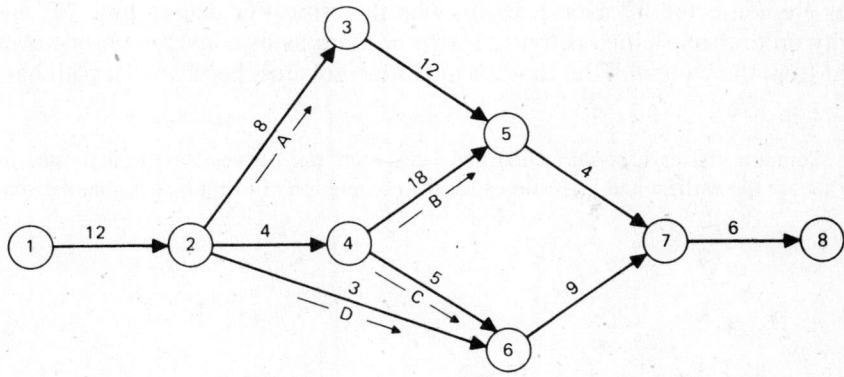

Fig. 17-7

See Table 17-1.

Table 17-1

Path		Times
A:	1–2–3–5–7–8	12 + 8 + 12 + 4 + 6 = 42
B:	1–2–4–5–7–8	12 + 4 + 18 + 4 + 6 = 44
C:	1–2–4–6–7–8	12 + 4 + 5 + 9 + 6 = 36
D:	1–2–6–7–8	12 + 3 + 9 + 6 = 30

(a)   Path B is critical, with a time requirement of 44 months.

(b)   The manual preparation is on path C:

$$\text{Slack} = \text{critical path B} - \text{path C}$$
$$= 44 - 36 = 8 \text{ months}$$

The slack in path C suggests that, other things remaining the same, the manual writing (activity 4–6) could fall behind by 8 months before it would jeopardize the scheduled finish date for the project.

EARLIEST AND LATEST ACTIVITY TIMES

In managing the activities of a project, it is sometimes useful to know how soon or how late an individual activity can be started or finished without affecting the scheduled completion date of the total project. Four symbols are commonly used to designate the earliest and latest activity times.

(1)   ES: the earliest start time for an activity. The assumption is that all predecessor activities are started at their earliest start time.

(2)   EF: the earliest finish time for an activity. The assumption is that the activity starts on its ES and takes its expected time, t. Therefore, $EF = ES + t$.

(3)   LF: the latest finish time for an activity without delaying the project. The assumption is that successive activities take their expected time.

(4)   LS: the latest start time for an activity without delaying the project. $LS = LF - t$.

ES and EF are calculated in a left-to-right sequence (sometimes called a *forward pass*). The ES of an activity is the sum of the times of all preceding activities on that path. Where two paths converge at a node, the longest time path governs.

Latest times are computed in reverse. Begin with the critical or ending time T_E, and subtract each preceding activity up to the specified activity. If two or more paths converge on one event en route, the figure developed from the path with the shortest total time governs, because that path has the least slack.

Example 17.3   Compute the earliest-start (ES) and latest-start (LS) times for the activities in the network of Example 17.2. What are the earliest and latest times for the completion of event 6 such that the schedule will not be delayed?

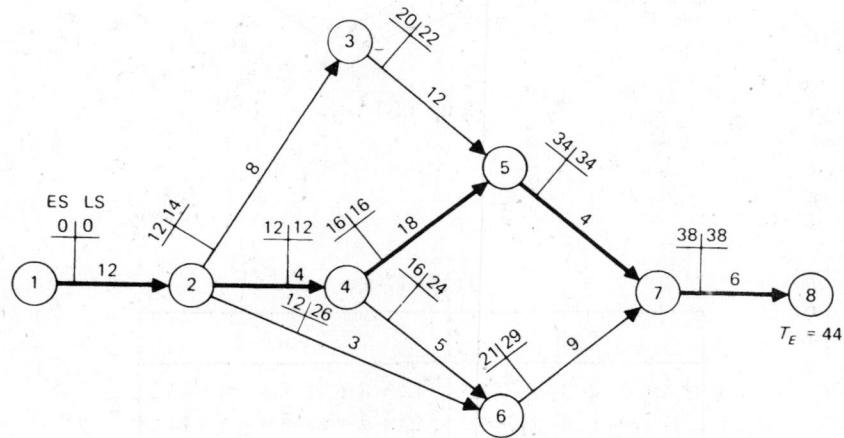

Fig. 17-8

See Fig. 17-8. The ES time (in months) for each activity is shown on the left side of the tee at the beginning of the activity. Activity 1–2 begins at zero, and the other activity times are summed. For example, the ES for activity 6–7 is the maximum of the cumulative times leading to event 6. Thus: via path 1–2–4–6, the time $= 12 + 4 + 5 = 21$. Via path 1–2–6, the time $= 12 + 3 = 15$. Therefore, ES $=$ month 21 because the longest path governs.

The LS time for each activity is on the right side of the tee, and we begin with T_E and work backward. Thus the LS for activity 6–7 is

$$T_E - \text{preceding activity times} = 44 - 6 - 9 = 29$$

Other ES and LS times are shown in Table 17-2, along with slack times. Note that the ES and LS for all activities on the critical path are equal. For activities off the critical path, the LS turns out to be the ES plus the amount of slack in the path (which seems like an easier way to compute it). The table also includes EF and LF times. They are also easily computed, for the EF is simply the ES plus the activity time, and LF is the LS plus the activity time.

Table 17-2

Activity	Time	ES	LS	EF	LF	Slack
1–2	12	0	0	12	12	0
2–3	8	12	14	20	22	2
2–4	4	12	12	16	16	0
2–6	3	12	26	15	29	14
3–5	12	20	22	32	34	2
4–5	18	16	16	34	34	0
4–6	5	16	24	21	29	8
5–7	4	34	34	38	38	0
6–7	9	21	29	30	38	8
7–8	6	38	38	44	44	0

Example 17.4　Compute the slack associated with each activity in Example 17.3.

The slack is shown in the far right column in Table 17-2. The total slack for an activity is the difference between LS and ES (or between LF and EF). Although we associate slack with each activity, it really belongs to the path because once any activity uses up the slack in its path, all activities along that path become critical. Activities along the critical path will always have zero slack if the target date (or planned completion date) for the project is the same as the earliest finish of the last activity.

Free slack is the amount of time an activity can be delayed without delaying the earliest-start time of any succeeding activity.

Example 17.5　Determine the free slack time associated with activities 2–6 and 4–6 in Example 17.2.

The ES of the succeeding activity (6–7) is month 21, so the starting date for activity 2–6 could be delayed 6 months (until month 18) without affecting activity 6–7's ES date. However, if activity 4–6 is delayed any time at all, it will delay the ES of activity 6–7. Thus, activity 2–6 has 6 months of free slack, and activity 4–6 has zero free slack. However, both activities have some total slack because neither is on the critical path.

Activity and path time estimates give project planners a basis for shifting resources to critical-path activities so as to reduce the overall project time.

Example 17.6　The firm in Example 17.2 has determined that by shifting three engineers from writing the manual (activity 4–6) to assisting with manufacturing (activity 4–5), activity 4–5 could be reduced to 15 months, whereas activity 4–6 would be increased to 10 months. What would be the net effect on the schedule?

Path A remains the same, at 42 months
Path B $= 12 + 4 + 15 + 4 + 6 = 41$ months
Path C $= 12 + 4 + 10 + 9 + 6 = 41$ months
Path D remains the same, at 30 months

Path A would become critical, and the new estimated completion time would be 42 months, a 2-month saving over the initial time.

INCORPORATING PROBABILISTIC TIME ESTIMATES

A major contribution of PERT to project scheduling has been the means by which it permits project managers to incorporate uncertainty into a well-structured system of analysis. Using the techniques developed under PERT, managers can accept the unsure time estimates of foremen, suppliers, subcontractors, and other service providers. This information also enables them to develop a central measure of completion time for a project and a measure of dispersion (a standard deviation). Given the mean and standard deviation of the completion time distribution, probabilities of finishing the project in less time or more time than the mean time can be readily determined.

The PERT methodology incorporates uncertainty (and probability) by including three time estimates for each activity rather than one. These estimates are designated as:

- *a: optimistic time*. This is the best time that could be expected if everything went exceptionally well, and it would be achieved only about 1 percent of the time.

- *m: most likely time*. This is the best estimate, or mode expectation.

- *b: pessimistic time*. This is the worst time that could reasonably be expected if everything went wrong, and it would occur only about 1 percent of the time.

The expected mean time (t_e) and variance (σ^2) of each activity are determined as:

$$t_e = \frac{a + 4m + b}{6} \qquad (17.1)$$

$$\sigma^2 = \left(\frac{b - a}{6}\right)^2 \qquad (17.2)$$

where a = optimistic time estimate

   m = most likely time estimate

   b = pessimistic time estimate

Individual activity times are then summed over the respective paths, and the path with the longest time is the critical path. Variances of component activity times along the critical path may also be summed. The ending time distribution is approximately normal with completion time T_E and standard deviation σ

$$T_E = \sum t_e \qquad (17.3)$$

$$\sigma = \sqrt{\sum \sigma_{cp}^2} \qquad (17.4)$$

where σ_{cp}^2 is the variance of individual activities on the critical path.

Given the mean and standard deviation of the ending distribution, the probabilities of various completion times may be calculated using the normal distribution. For example, to determine the probability that a project would exceed time T_x in Fig. 17-9, we would compute

$$Z = \frac{T_x - T_E}{\sigma}$$

then find the probability associated with that Z value from the normal distribution values in Appendix B (or a hand calculator) and subtract it from .5000. The results would represent the shaded area under the curve in Fig. 17-9.

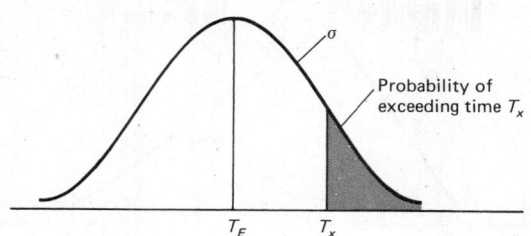

Fig. 17-9   Ending time distribution

Example 17.7   Project planners have sought the judgment of various knowledgeable engineers, foremen, and vendors and have developed the time estimates shown in Table 17-3 for the plant construction project depicted in Fig. 17-5. (*a*) Determine the critical path. (*b*) What is the probability the project will be finished within 4 years? (*c*) What is the probability that it will take more than 55 months?

Table 17-3

Activity		Time Estimates		
Description	Number	a	m	b
Design plant	1–2	10	12	16
Select site	2–3	2	8	36
Select vendor	2–4	1	4	5
Select personnel	2–6	2	3	4
Prepare site	3–5	8	12	20
Manufacture generator	4–5	15	18	30
Prepare manual	4–6	3	5	8
Install generator	5–7	2	4	8
Train operators	6–7	6	9	12
License plant	7–8	4	6	14

Table 17-4

t_e $\dfrac{a + 4m + b}{6}$	σ^2 $\left(\dfrac{b - a}{6}\right)^2$
12.33	1.00
11.67	32.11
3.67	.44
3.00	.11
12.67	4.00
19.50	6.25
5.17	.69
4.33	1.00
9.00	1.00
7.00	2.78

Table 17-5

Path	Times
A: 1–2–3–5–7–8	12.33 + 11.67 + 12.67 + 4.33 + 7.00 = 48.00*
B: 1–2–4–5–7–8	12.33 + 3.67 + 19.50 + 4.33 + 7.00 = 46.83
C: 1–2–4–6–7–8	12.33 + 3.67 + 5.17 + 9.00 + 7.00 = 37.17
D: 1–2–6–7–8	12.33 + 3.00 + 9.00 + 7.00 = 31.33

*Critical path.

(*a*)   Values for t_e and σ^2 for the various activities have been calculated as shown in Table 17-4. The t_e values are entered on the network diagram in Fig. 17-10. The critical path, as determined in Table 17-5, is now A and has been shown by a heavy solid line in the figure.

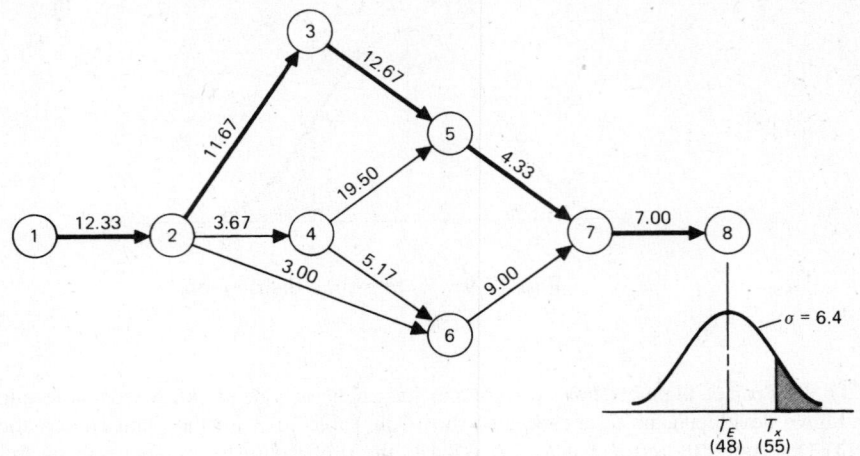

Fig. 17-10   Network diagram showing the critical path and ending time distribution

(b)   The best estimate of completion time is $T_E = 48.0$ months, so there is a 50 percent chance that the project will be finished within the 4-year time period.

(c)   To determine any other completion time probabilities, we must calculate the variances of the distribution of completion times *along the critical path*, sum them, and estimate σ.

$$\sigma = \sqrt{\sum \sigma_{cp}^2} = \sqrt{1.00 + 32.11 + 4.00 + 1.00 + 2.78} = 6.4 \text{ months}$$

$$Z = \frac{T_x - T_E}{\sigma} = \frac{55.0 - 48.0}{6.4} = 1.09$$

$$P(X > T_x) = .5000 - .3621 = .1379$$

Therefore, probability $\cong .14$.

Example 17.8   Although the problem in Example 17.2 and the problem in Example 17.7 had the same mean times for each activity, their critical paths (and critical-path times) differed. Explain why.

The critical paths differed because Example 17.7 incorporates a measure of uncertainty, whereas Example 17.2 does not. For example, the site selection activity (2–3) has a most likely time estimate of 8 months, but a pessimistic time estimate of 36 months, resulting in $t_e = 11.67$ months, in contrast to the 8-month figure used in the CPM calculations.

CRASHING: TIME-COST TRADE-OFFS

An extension of CPM and PERT, referred to as *crashing* a project, focuses attention on the trade-off between time and cost objectives. The *normal* estimate of the time required for each activity (and its associated cost) has already been discussed. The *crash time* estimate is the shortest time that could be achieved if all effort (at any reasonable cost) were made to reduce the activity time. The use of more workers, better equipment, overtime, etc., would generate higher direct costs for individual activities as illustrated in Fig. 17-11. However, shortening the overall time of the project would also reduce certain fixed and overhead expenses of supervision, as well as indirect costs that vary with the length of the project.

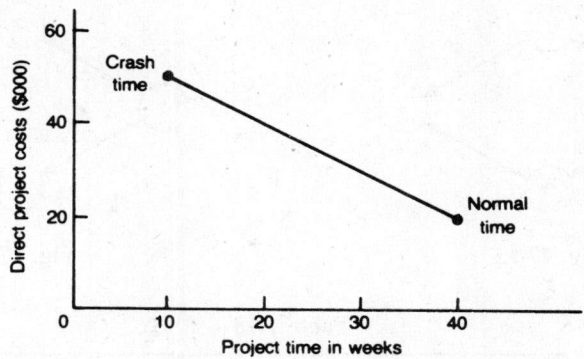

Fig. 17-11   Crash time and crash cost

Time-cost models search for the optimum reductions in time. We seek to shorten the length of a project to the point where the savings in indirect project costs are offset by the increased direct expenses incurred in the individual activities. Following is a simplified illustration.

Example 17.9   A network has four activities with expected times as shown in Fig. 17-12. The minimum feasible times and cost per day to gain reductions in the activity times are shown in Table 17-6. If fixed project costs are $90 per day, what is the lowest cost-time schedule?

Table 17-6

Activity	Minimum Time	Direct Costs of Time Reduction ($)
1–2	2	40 (each day)
1–3	2	35 (first day), 80 (second day)
2–4	4	None possible
3–4	3	45 (first day), 110 (other days)

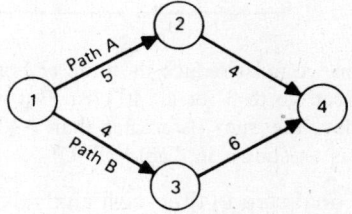

Fig. 17-12

First we must determine the critical path (*) and critical-path time cost (Table 17-7).

Table 17-7

	Path Times	Total Project Cost
Path A	5 + 4 = 9	
Path B	4 + 6 = 10*	10 days × $90/day = $900

Next, we must select the activity that can reduce critical-path time at the least cost. Select activity 1–3 at $35 per day, which is less than the $90 per day fixed cost. Reduce activity 1–3 days as shown in Fig. 17-13. Revise the critical-path time cost (Table 17-8):

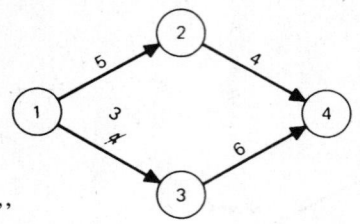

Fig. 17-13

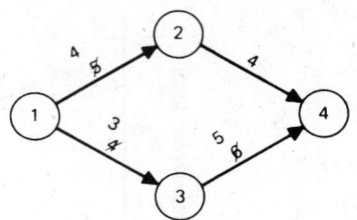

Fig. 17-14

Table 17-8

Revised Path Times	Total Fixed Cost	Savings over Previous Schedule
A: 5 + 4 = 9 B: 3 + 6 = 9	$9 \times \$90 = \$810$	$\$900 - (\$810 + \$35) = \$55$

Both paths are now critical, so we must select an activity on each path. Select activity 1–2 at $40 per day and 3–4 at $45 per day, where $40 + $45 is less than $90. Reduce activity 1–2 to 4 days and 3–4 to 5 days as shown in Fig. 17-14. Revise the critical path time and cost (Table 17-9).

Table 17-9

Revised Path Times	Total Fixed Cost	Savings over Previous Schedule
A: 4 + 4 = 8 B: 3 + 5 = 8	$8 \times \$90 = \$720$	$\$810 - (\$720 + \$40 + \$45) = \$5$

Again we must reduce the time of both paths. Activity 1–2 is a good candidate on path A, for it is still at 4 days and can go to 3 for a $40 cost. But when this cost is combined with the $80 cost for reducing activity 1–3 another day, the sum is greater than $90, so further reduction is not economically justified. The lowest-cost schedule is as shown in Table 17-9.

The final step in time-cost analysis is to compare the *crash times* and the costs associated with them (*crash costs*). A sufficient number of intermediate schedules are computed such that the total of the direct and indirect (fixed) project costs can be plotted.

Example 17.10 Graph the total relevant costs for the previous example, and indicate the optimal time-cost trade-off value.

See Table 17-10 and Fig. 17-15.

Table 17-10

Project Length (days)	Indirect Cost	Activity Reduced	Relevant Direct Cost	Relevant Total Cost
10	$900	None	$ 0	$900
9	810	1–3	$ 0 + $ 35 = 35	845
8	720	1–2 and 3–4	35 + 85 = 120	840
7	630	1–2 and 1–3	120 + 120 = 240	870
6	540	1–2 and 3–4	240 + 150 = 390	930

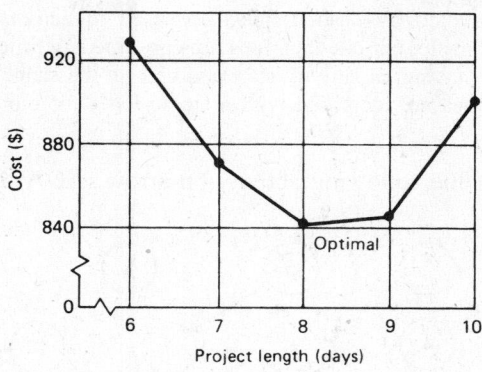

Fig. 17-15

Figure 17-15 is the crash-time diagram for completing the project in 6 to 10 days. The lowest total cost is to complete the project in 8 days at a cost of $840. However, extending it to 9 days adds only $5 to this cost.

Questions and Solved Problems

PROJECT SCHEDULING AND NETWORK FUNDAMENTALS

17.1 How are networks used in applications other than project scheduling?

Network diagrams are frequently used to describe inventory or cash flows, shipping routes, and communication links.

17.2 Are there other techniques for formulating and solving network problems?

Yes. There are many applications of networks, especially in electrical engineering and communications. Some network problems can also be formulated as linear-programming problems. Example 17.1 could be formulated with an objective to minimize the cost of the various links as Min $Z = 2X_{AB} + 4X_{BE} + 3X_{EF} + X_{AC} + 5X_{CE} + 7X_{CF} + 6X_{AD} + 5X_{DF}$. This would be subject to origin, destination, conservation of flow, and non-negativity constraints. The solution (via computer) would then be to use path A–C–F.

17.3 What is the meaning of the following activity-on-node (AON) symbolic relationships?

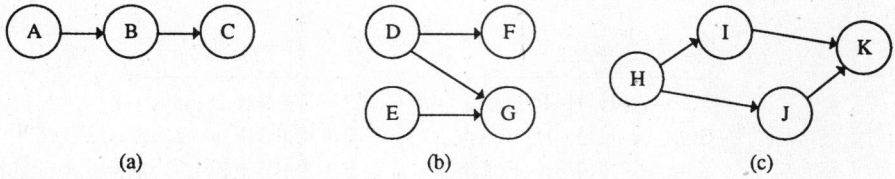

(*a*) Activity A must be completed before B, which must be completed before C.
(*b*) Activity D must be completed before F, and D and E must both be completed before G.
(*c*) Activity H must be completed before I and J, and both I and J must precede K.

17.4 What is the purpose of a ''dummy'' activity?

Dummy activities (denoted by dashed lines) are used in activity-on-arrow diagrams to show an essential precedence relationship, but one that does not consume any time or resources. They help avoid the (unacceptable) situation of having an activity start and end on the same node. With AON networks this is no problem because activities are identified with a single letter (or number).

17.5   What is the meaning of the following activity-on-arrow (AOA) symbolic relationships?

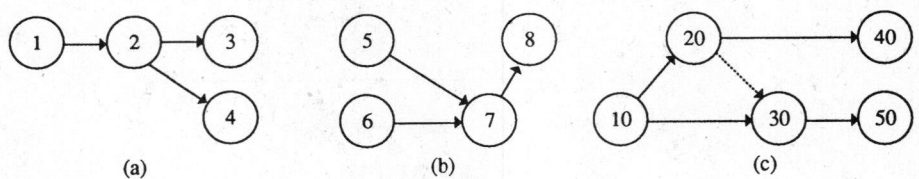

(*a*)   Activity 1–2 must be completed before either 2–3 or 2–4 can be started.

(*b*)   Activities 5–7 and 6–7 must be completed before 7–8 can be started.

(*c*)   Activities 10–20 and 10–30 must both precede activity 30–50. Activity 20–30 is a dummy activity that must be completed before 30–50, but it does not affect 20–40.

17.6   What are two of the major limitations of Gantt charts for use in project management?

Two limitations are (*a*) Gantt charts do not show the interrelationships among activities, which can become quite complex for large projects, and (*b*) they do not reveal the most limiting path of precedence relationships (critical) that should receive the closest attention.

FINDING THE CRITICAL PATH

17.7   Find the critical paths for the two networks illustrated in Fig. 17-4 (repeated below).

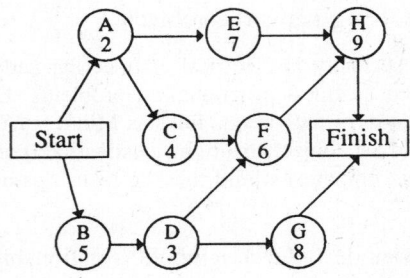

(*a*) Activity-on-node convention

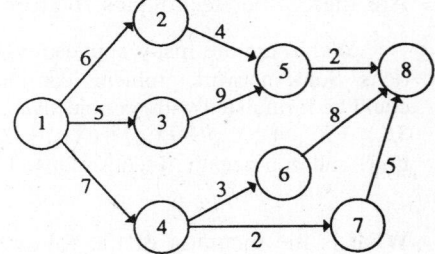

(*b*) Activity-on-arrow convention

(*a*)	Path	Times
	Start–A–E–H–Finish	2 + 7 + 9 = 18 days
	Start–A–C–F–H–Finish	2 + 4 + 6 + 9 = 21 days
	Start–B–D–F–H–Finish	5 + 3 + 6 + 9 = 23 days ← critical path
	Start–B–D–G–Finish	5 + 3 + 8 = 16 days

(*b*)	Path	Times
	1–2–5–8	6 + 4 + 2 = 12 days
	1–3–5–8	5 + 9 + 2 = 16 days
	1–4–6–8	7 + 3 + 8 = 18 days ← critical path
	1–4–7–8	7 + 2 + 5 = 14 days

CPM/PERT APPLICATIONS

17.8 What are some of the benefits from using computer programs for project scheduling?

Many of the project scheduling programs are now quite sophisticated. Computer programs should be considered whenever there are more than a few activities in the network and whenever there will be extensive changes or a need to track and update resource usage figures on a regular basis. Some programs provide excellent charts and extensive graphics of planned versus actual resource usage. Computers are virtually essential for large-scale projects.

17.9 What are some of the problems associated with using CPM/PERT?

(1) Some individuals may resist using it because they are unfamiliar with CPM/PERT, and their lack of knowledge makes them feel insecure.

(2) CPM/PERT forces more detailed planning than some people are willing (or capable) of doing.

(3) It requires the participation of all managers, supervisors, and others, who must make original estimates and update information on a regular basis; otherwise the whole effort is in jeopardy.

(4) Essential activities may be omitted.

(5) Precedence relationships may be incorrectly shown.

(6) Activity time estimates may be substantially off the mark—especially if some supervisors endeavor to include a ''cushion'' to protect themselves from being late.

17.10 The optimistic, most likely, and pessimistic time estimates for an activity are 12, 15, and 24 weeks, respectively. Find the (a) expected mean time and (b) variance for this activity.

(a) $t_e = \dfrac{a + 4m + b}{6} = \dfrac{12 + 4(15) + 24}{6} = 16 \, \text{wk}$

(b) $\sigma^2 = \left(\dfrac{b - a}{6}\right)^2 = \left(\dfrac{24 - 12}{6}\right)^2 = 4 \, \text{wk}$

17.11 The times given in Table 17-11 are the (a) optimistic and (b) pessimistic times associated with the AON network of Fig. 17-4, which has a critical-path time of 23 days (see Prob. 17.7). Find the standard deviation of the ending time distribution.

Table 17-11

Activity	Optimistic Time a	Pessimistic Time b
A	1	5
B	4	6
C	2	8
D	1	5
E	5	8
F	3	8
G	6	11
H	6	13

Table 17-12

$\left(\dfrac{b-a}{6}\right)$	$\left(\dfrac{b-a}{6}\right)^2$	Activities on Critical Path
4/6	16/36	
2/6	4/36	4/36
6/6	36/36	
4/6	16/36	16/36
3/6	9/36	
5/6	25/36	25/36
5/6	25/36	
7/6	49/36	49/36

Table 17-12 gives the variance calculations of activities B, D, F, and H on the critical path.

$$\sigma = \sqrt{\sum \sigma_{\text{cp}}^2} = \sqrt{4/36 + 16/36 + 25/36 + 49/36} = 1.6 \, \text{days}$$

17.12 A small manufacturing firm has developed the following list of activities (Table 17-13) necessary to release a contract for a new plant. Draw the appropriate network diagram.

Table 17-13

Activity Description	Preceding Activity	Activity Time in Weeks		
		Optimistic	Most Likely	Pessimistic
A–B Feasibility study	none	4	6	10
B–C Acquire site	A–B	2	8	24
B–D Prepare plans	A–B	10	12	16
B–F Marketing strategy	A–B	4	5	10
C–D Soil test	B–C	1	2	3
D–E Legal approvals	C–D and B–D	6	8	30
D–F Loan application	C–D and B–D	2	3	4
E–F Evidence approval	D–E	0	0	0
E–G Obtain bids	D–E	6	6	6
F–G Secure financing	D–F and B–F	2	6	12
G–H Release contract	E–G and F–G	2	2	3

See Fig. 17-16. Note the activity E–F (evidencing legal approval) takes no time and is a dummy activity.

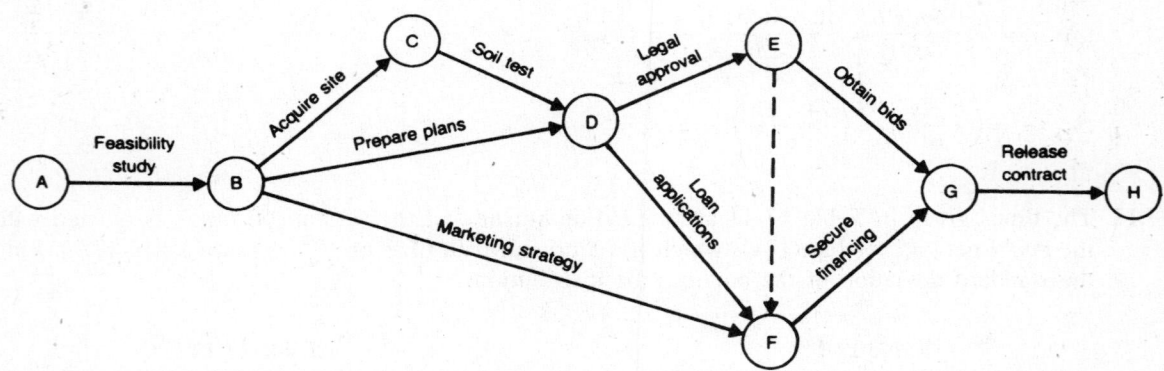

Fig. 17-16

17.13 Compute the expected activity time (t_e) and activity variance (σ^2) for activities A–B and B–C of Prob. 17.12.

A–B:
$$t_e = \frac{a + 4m + B}{6} = \frac{4 + 4(6) + 10}{6} = 6.33 \text{ weeks}$$

$$\sigma^2 = \left(\frac{b - a}{6}\right)^2 = \left(\frac{10 - 4}{6}\right)^2 = \left(\frac{6}{6}\right)^2 = 1.00 \text{ week}$$

B–C:
$$t_e = \frac{a + 4m + b}{6} = \frac{2 + 4(8) + 24}{6} = 9.67 \text{ weeks}$$

$$\sigma^2 = \left(\frac{b - a}{6}\right)^2 = \left(\frac{24 - 2}{6}\right)^2 = \left(\frac{22}{6}\right)^2 = 13.44 \text{ weeks}$$

17.14 Given the data shown in Table 17-14 for a PERT network:

(*a*)   Draw an AOA network diagram, and find the critical path.

(*b*)   What are the parameters of the ending time distribution?

(*c*)   Which activity has the most precise time estimate?

(*d*)   Determine the earliest start, latest start, and slack time for all events in the system.

(*e*)   Each day the project can be shortened is worth $5,000. Should the firm pay $12,500 to reduce activity 3–5 to 2 days?

(*a*)   See Tables 17-15 and 17-16, and Fig. 17-17.

Table 17-14

Preceding Event	Event	Activity Time		
		a	*m*	*b*
1	2	5	6	13
1	3	2	7	12
2	4	1.5	2	2.5
2	5	1	3	5
3	5	4	5	6
3	6	1	1	1
4	7	2	3	10
5	7	4	5	6
6	7	3	5	7

Table 17-15

Activity	$\dfrac{a + 4m + b}{6}$	$\left(\dfrac{b - a}{6}\right)^2$
1–2	7	1.78
1–3	7	2.77
2–4	2	.02
2–5	3	.44
3–5	5	.11
3–6	1	.00
4–7	4	1.78
5–7	5	.11
6–7	5	.44

Table 17-16

Path	Times
A: 1–2–4–7	7 + 2 + 4 = 13
B: 1–2–5–7	7 + 3 + 5 = 15
C: 1–3–5–7	7 + 5 + 5 = 17*
D: 1–3–6–7	7 + 1 + 5 = 13

*Critical path.

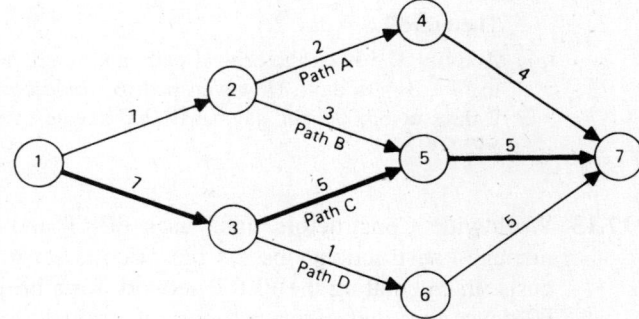

Fig. 17-17

(*b*) $T_E = 17$

$$\sigma_{\text{cp}} = \sqrt{\Sigma\, \sigma_{\text{cp}}^2} = \sqrt{2.77 + .11 + .11} = 1.73$$

(*c*)   The most precise time is for activity 3–6, with a variance of zero.

(*d*)   The ES and LS times for activities on the critical path (path C) are both the same and are simply cumulative totals of the activity times. They are dominating values, for they are maximums in terms of computing ES times (in the forward direction) and minimums in terms of computing LS times (in the reverse direction). See Fig. 17-18. Values for all activities in the network are shown in Table 17-17.

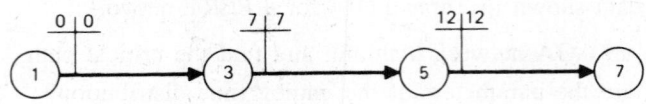

Fig. 17-18

Table 17-17

Activity	Time	ES	LS	Slack
1–2	7	0	2	2
1–3	7	0	0	0
2–4	2	7	11	4
2–5	3	7	9	2
3–5	5	7	7	0
3–6	1	7	11	4
4–7	4	9	13	4
5–7	5	12	12	0
6–7	5	8	12	4

For example, the ES for activity 5–7 is the maximum of

$$\text{Via path B} = 7 + 3 = 10$$
$$\text{Via path C} = 7 + 5 = 12$$

Therefore ES = day 12.   For example, the LS for activity 1–2 is the minimum of

$$\text{Via path A} = 17 - 4 - 2 - 7 = 4$$
$$\text{Via path B} = 17 - 5 - 3 - 7 = 2$$

Therefore LS = day 2.

(e)   Activity 3–5 is on the critical path, and the reduction from 5 to 2 days would reduce the path C time to $17 - 3 = 14$ days. However, path B would become critical at 15 days, so the net reduction would be 2 days at \$5,000 per day = \$10,000 savings versus the \$12,500 cost. The firm should not pay the \$12,500.

17.15   Worldwide Constructors, Inc., uses PERT and expected-value techniques to prepare bids and manage construction jobs. Its bid price is set to give it a 30 percent gross profit over expected costs. In calculating the PERT network for a bridge construction job, T_E was found to be equal to 60 days, and total variance along the critical path was $\sigma^2_{cp} = 36$. See Fig. 17-19. Total expenses for the project are estimated at \$335,000, but if the bridge is not completed within 70 days, there is a penalty of \$50,000. Determine the appropriate bid price.

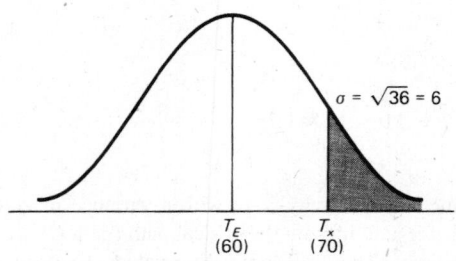

Fig. 17-19

Bid price = expected costs + penalty allowance + profit

where   expected costs = \$335,000

penalty allowance = (amount of penalty)(probability of penalty)

$$Z = \frac{T_x - T_E}{\sigma} = \frac{70 - 60}{6} = 1.67$$

$$P(Z) = .4575$$

$$P(X > T_x) = .5000 - .4525 = .0475$$

Penalty allowance = (\$50,000)(.0475) = \$2,375

Profit = .30(\$335,000 + \$2,375) = \$101,212

Bid price = \$335,000 + \$2,375 + \$101,212 = \$438,587

17.16 *(PERT with learning curve)* An electrical firm has developed a PERT plan for the electrical wiring activity of power plant control panels. It expects that assembly operations will follow a 90 percent learning curve. The project team—composed of workers, electricians, and supervisors— feels that the first assembly will most likely be completed in 14 days but could take as long as 24 days; or if everything went exceptionally well, it would be finished in 10 days. What is the expected assembly time of the fourth unit?

$$t_e = \frac{a + 4m + b}{6} = \frac{10 + 4(14) + 24}{6} = 15$$

From Eq. 16.6 $$Y_n = Y_B(L_{\text{unit}})$$

Then, from Appendix H we have

$$Y_4 = 15(.81) = 12.15 \text{ days}$$

CRASHING: TIME-COST TRADE-OFFS

17.17 The network diagram shown in Fig. 17-20 has the time and direct costs given in Table 17-18. The time-cost trade-offs are cumulative amounts; that is, you can reduce activity 1–2 by 2 weeks, for a total of \$16,000–\$12,000 = \$4,000, or \$2,000 a week.

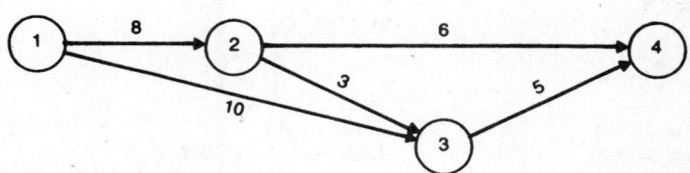

Fig. 17-20

Table 17-18

Activity	Normal Time	Normal Cost	Crash Time	Crash Cost
1–2	8	\$12,000	6	\$13,600
1–3	10	6,000	7	10,500
2–4	6	8,000	4	10,000
2–3	3	14,000	2	17,000
3–4	5	7,000	2	9,700
		\$47,000		

(a)   Compute the total direct costs for completing the project in 16, 15, 14, 13, 12, or 11 weeks.

(b)   The indirect project costs are shown in Table 17-19. Graph the total project cost (direct and indirect), and determine the least-cost project completion time.

Table 17-19

Project duration (weeks)	16	15	14	13	12	11
Indirect costs (dollars)	23,000	19,100	17,200	14,400	13,700	13,200

(a)   For convenience, set up a table (Table 17-20) to show the incremental cost of reducing each activity, the maximum reduction possible, and a far right column for keeping a tally of the reductions used.

Table 17-20

Activity	Normal		Crash		Incremental Cost of Reduction	Reduction	
	Time	Cost	Time	Cost		Max	Used
1–2	8	$12,000	6	$13,600	$ 800/wk	2	✓✓
1–3	10	6,000	7	10,500	1,500/wk	3	✓
2–4	6	8,000	4	10,000	1,000/wk	2	✓
2–3	3	14,000	2	17,000	3,000/wk	1	
3–4	5	7,000	2	9,700	900/wk	3	✓✓✓

Next, compute the times for each path (A = 14, B = 16, C = 15), and identify the critical path (shown dashed in Fig. 17.21). Begin the time-cost analysis by reducing the time for the activity on the critical

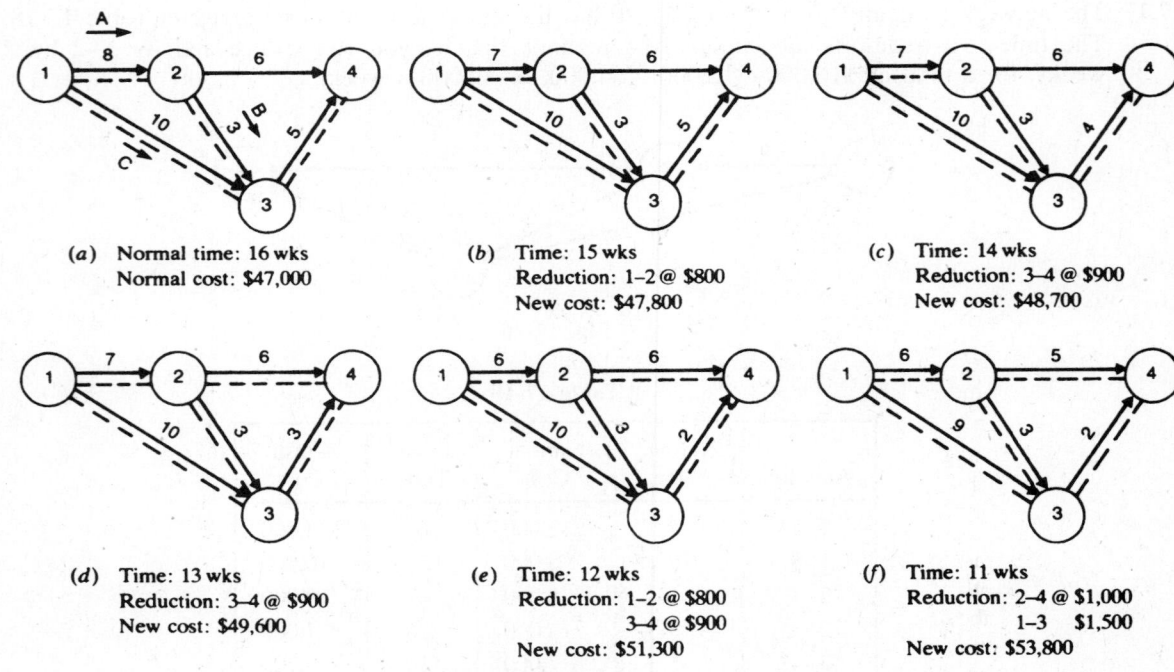

(a)   Normal time: 16 wks
      Normal cost: $47,000

(b)   Time: 15 wks
      Reduction: 1–2 @ $800
      New cost: $47,800

(c)   Time: 14 wks
      Reduction: 3–4 @ $900
      New cost: $48,700

(d)   Time: 13 wks
      Reduction: 3–4 @ $900
      New cost: $49,600

(e)   Time: 12 wks
      Reduction: 1–2 @ $800
              3–4 @ $900
      New cost: $51,300

(f)   Time: 11 wks
      Reduction: 2–4 @ $1,000
              1–3 $1,500
      New cost: $53,800

Fig. 17-21   Crash times and costs

path (or paths) that has the smallest total of all incremental costs (i.e., activity 1–2). Draw a new network diagram, and compute a revised direction cost for each reduced time.

(b)   See Table 17-21 and Fig. 17-22.

Table 17-21

Time (wk)	11	12	13	14	15	16
Direct costs	$53,800	$51,300	$49,600	$48,700	$47,800	$47,000
Indirect costs	13,200	13,700	14,400	17,200	19,100	23,000
Total costs	$67,000	$65,000	$64,000	$65,900	$66,900	$70,000

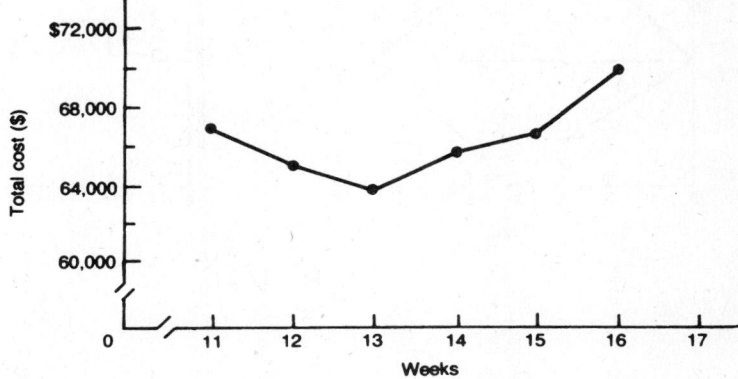

Fig. 17-22

The least-cost project completion time is 13 weeks.

LIMITED RESOURCE ALLOCATION (SUPPLEMENTARY MATERIAL)

17.18   The accompanying schedule-time graph (top two charts of Fig. 17-23) depicts a project with activity times (a through g) as shown on the horizontal axis and a critical path (a, b, e, g) of 7 days. Numbers above the activities represent personnel requirements. Develop an improved personnel balance.

The dashed lines represent slack time and potential relocation zones for activities on the respective paths. Locate maximum and minimum resource requirements, and try to shift activities into slack positions to smooth the demand. The solution is shown in the bottom half of Fig. 17-23 with the revised network and personnel balance as shown. The solution consists simply of shifting activities c and f, which reduces the range of personnel requirements from $20 - 6 = 14$ to $16 - 12 = 4$.

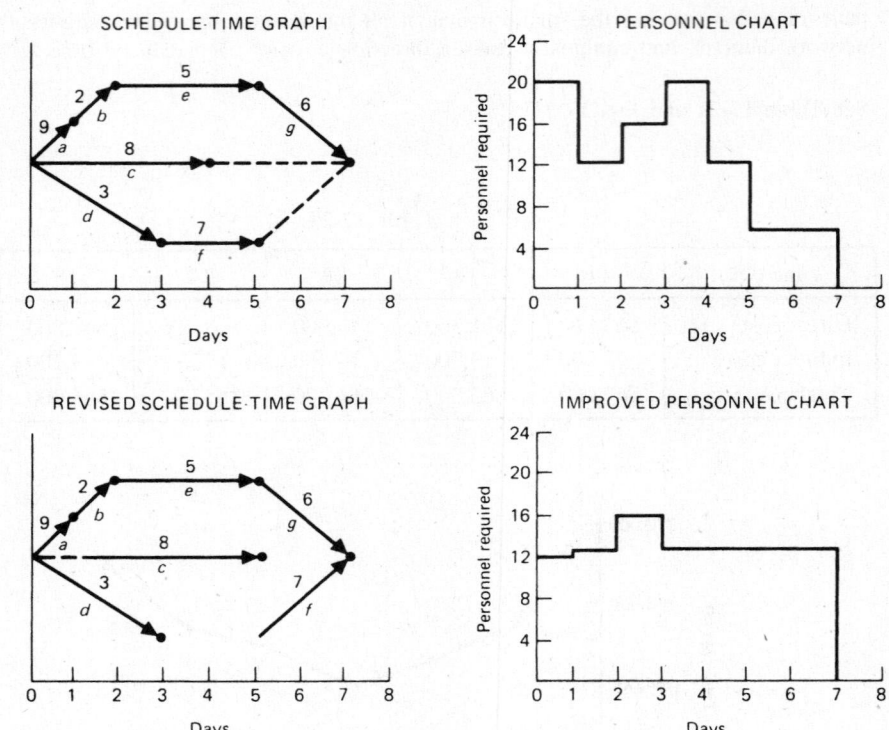

Fig. 17-23

Supplementary Questions and Problems

17.19 A large power transformer is to be transported from a factory (A) to a destination (H). Alternative routes and associated costs ($) are as shown in Fig. 17-24. What is the least-cost route?
Ans. A–D–E–G–H for a cost of $570

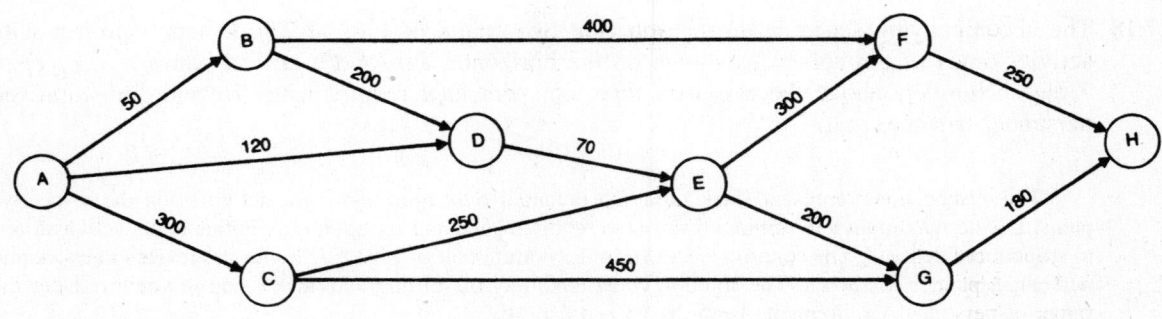

Fig. 17-24

17.20   In Prob. 17.19, assume the values on the network diagram (Fig. 17-24) represent activity times (days) to complete a project. What would be the critical path?   *Ans.*   A–C–E–F–H for 1,100 days

17.21   In developing a CPM/PERT network for a bridge construction project, the construction foreman felt that the optimistic estimate of a concrete pouring activity was 30 days. The project quality-control inspector, citing all the possible delays and rework, said (pessimistically) that it could take as long as 180 days. Both agreed the most likely time was 45 days. Estimate (*a*) the expected activity time t_e and (*b*) the activity variance σ^2.   *Ans.*   (*a*) 65 days (*b*) 625

17.22   The expected completion time of a project is $T_E = 15$ days, and $\sigma^2_{cp} = 4$ days. What is the probability that the project will take 18 or more days to complete?   *Ans.*   .07

17.23   A network has expected times (t_e) in days as shown in Fig. 17-25. The time estimates for activity 6–7 are $a = 1$, $m = 4$, and $b = 7$. For the network, what is the (*a*) expected completion time T_E? (*b*) completion time standard deviation σ_{cp}? (*c*) probability the project will take more than 20.5 days to complete?
Ans. (*a*) 18 days (*b*) 3 (*c*) .20

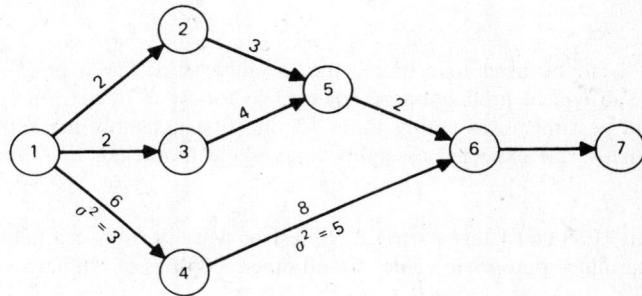

Fig. 17-25

17.24   A microwave relay station construction project is being planned on a network basis with the time estimate data shown in Table 17-22 given in days.

Table 17-22

Activity	a	m	b
1–2	2	3	10
1–3	8	12	20
1–4	10	14	16
2–5	6	10	12
3–5	14	20	26
3–7	3	5	7
4–6	8	12	20
5–7	1	1	1
6–8	6	10	12
7–8	1	3	7

(*a*) Construct an AOA network diagram showing the expected mean time t_e for each activity. (*b*) What is the critical path? (*c*) What is the expected completion time T_E? (*d*) How much slack exists in the path containing event 2? (*e*) What is the latest day event 2 can be completed without delaying the project? (*f*) Find σ_{cp}. (*g*) What is the probability the project will take longer than 41 days to complete?

Ans. (*a*) The network diagram should show activities as arrows and events as circles, beginning with event 1 and ending with 8 (*b*) 1–3–5–7–8 (*c*) 37 days (*d*) 19 days (*e*) day 23 (*f*) 3 days (*g*) .0918

17.25 Given the data in the previous problem, assume that each day of improvement in the completion schedule results in a $3,000 savings (or bonus). For a cost of $2,000 the firm could do any *one* of the following: (*a*) reduce the t_e of activity 3–7 by 3 days; (*b*) reduce the t_e of activity 1–3 by 2 days, or (*c*) reduce the t_e of activities 3–5, 6–8, and 7–8 by 1 day each. Evaluate the alternative choices, and indicate which, if any, is preferable.   *Ans.* (*c*) for a $4,000 savings

17.26 A building contractor company has bid on a job for a water reservoir that must be completed within 34 days ($T_L = 34$) or else the company must pay a $2,000 penalty. If the project is finished within 28 days, the company will get a $1,000 bonus. Expenses associated with the project are estimated to be $30,000. The company has developed a PERT chart of the project and found that $T_E = 31$ days. The variance estimates of the five activities along the critical path are 1.3, 2.2, 2.1, .9, and 2.5 days, respectively. (*a*) What is the probability of obtaining the bonus (accurate to two digits)? (*b*) Assuming that the company wishes to adjust its bid price to allow for the expected bonus or penalty and come out with only a long-run expected profit of $5,000, for what contract price should it be willing to accept the job?   *Ans.* (*a*) .1587 (*b*) $35,160

17.27 A PERT chart is to be used to estimate the assembly time for a new component that is later to be manufactured. Subsequent production is expected to follow a 70 percent learning curve. The optimistic, most likely, and pessimistic assembly times for the first assembly are estimated at 2, 4, and 12 hours, respectively. What is the expected assembly time of the fourth unit?   *Ans.* 2.45 hours

17.28 The earliest-start (ES) and latest-start (LS) times for activity 6–7 of a network diagram are as shown in Fig. 17-26. Determine appropriate values for all other activities of the network, and show them in a similar manner.

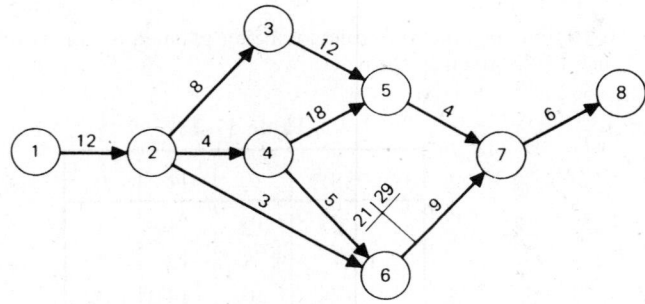

Fig. 17-26

Ans. The chart should show the times found in Table 17-23.

Table 17-23

Activity	1–2	2–3	2–4	2–6	3–5	4–5	4–6	5–7	6–7	7–8
ES	0	12	12	12	20	16	16	34	21	38
LS	0	14	12	26	22	16	24	34	29	38

17.29 *Time Cost.* A network has time (in weeks) and direct costs (in dollars) as shown in Fig. 17-27 and Table 17-24. Crash costs are cumulative totals; i.e., the incremental amount above normal costs can be apportioned equally among the time intervals. (*a*) Compute the total direct costs for finishing the project in 9, 10, 11, 12, or 13 weeks. (*b*) The indirect project costs are shown in Table 17-25. Graph the total project costs (direct and indirect), and determine the least-cost project completion time.

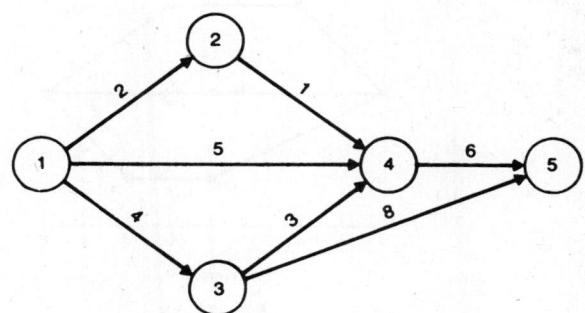

Fig. 17-27

Table 17-24

Activity	Normal Time	Normal Cost	Crash Time	Crash Cost
1–2	2	$ 500	1	$ 800
1–4	5	900	3	1,300
1–3	4	800	3	1,000
2–4	1	400	1	400
3–4	3	1,200	2	1,800
4–5	6	700	4	900
3–5	8	600	4	1,200
		$5,100		

Table 17-25

Project duration	9	10	11	12	13
Indirect costs	$6,000	$6,150	$6,200	$6,500	$7,100

Ans.   (*a*) Direct costs for completion times are as shown in Table 17-26. (*b*) The least-cost project completion time is 11 weeks, at a cost of $5,400 direct + $6,200 indirect = $11,600 total.

Table 17-26

Project duration	9	10	11	12	13
Direct costs	$6,400	$5,650	$5,400	$5,200	$5,100

17.30 *Limited Resource Allocation.* A government naval shipyard has received orders to proceed with a ship construction project and is using CPM. It has developed a schedule-time graph (Fig. 17-28) to show employment requirements over a portion of the project. The numbers above the activities indicate the number of shipfitters required for the respective activities. Develop an improved personnel balance that minimizes the range of the number of shipfitters required.

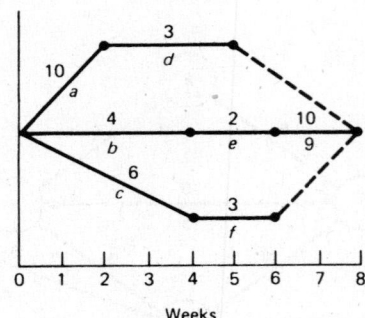

Fig. 17-28

Ans. Use 14 shipfitters during periods 1 and 2, 10 during 3, 13 during 4, 11 during 5 and 6, and 13 during 7 and 8; and so the range is $14 - 10 = 4$.

Chapter 18

Quality Control
Statistical Methods

QUALITY CONTROL AND QUALITY ASSURANCE

Total quality was defined earlier (Chap. 2) as the ability of a good or service to satisfy customer expectations with respect to (1) the product's design, (2) the user's desires, (3) conformance to the design standards, and (4) additional service expectations. Quality control was then associated with a narrower concept that focused upon the third element of this definition.

Question: What is a quality control?

Quality control is the use of (statistical or other) control activities designed to ensure that a good or service meets its proclaimed standards, which may relate to materials, performance, reliability, time, or any quantifiable (objective and measurable) characteristic.

As with any control activity, quality control involves the four elements of measurement, feedback, comparison with a standard, and correction when necessary. Note that the standards can apply to a wide range of measurable criteria.

Importance of Quantifiable Standards.   Industrial quality rests firmly upon standards—not upon who made the product, the price paid for it, or the unspecified preferences of whoever owns or uses it. Expensive products with outstanding features are not necessarily of higher quality than inexpensive products. Products are designed for different markets, and quality is a reflection of how well a specific, or advertised design is achieved.

When quantifiable standards are absent, quality becomes a matter of opinion and is not controllable from a scientific standpoint. Some paintings, entertainment, and personal services fall into this category. Assessments that depend on subjective feelings or substitute measures (such as the name of an artist or one's reputation) belong more to the realm of art than to science. For industrial purposes, the conformance of a good or service to specified standards is the basis for control and the criterion used to determine its acceptability (or unacceptability).

Question: What is quality assurance?

Quality assurance is the system of policies, procedures, and guidelines that establishes and maintains the specified standards of product quality. The system thus includes everyone who is in a position to affect quality, from suppliers through the firm's own production and field service personnel, to customers who use the goods and services.

Question: What are the traditional costs associated with controlling quality?

In the past, quality-control costs have often been classified into (1) inspection and control costs and (2) defective product costs. The quality-control problem has been perceived as one of achieving an optimal level of expenditure on quality-control activities that will minimize the total of the two costs. More recently the costs have been expanded to prevention, appraisal, and failure costs (see Chap. 2). However, some advocates of continuous improvement and zero defect programs argue that defective products are an inherent waste and that advances in quality will virtually always benefit an organization over the long run.

QUALITY MEASURES IN GOODS AND SERVICES

Quality control involves *measurement* of the quality characteristic, *feedback* of the data, *comparison* with specified standards, and *correction* when necessary. Quality control in the production of goods rests heavily upon the measurement of material characteristics. Physical properties, design, and product reliability are key elements. Services often convey intellectual or aesthetic values whose quality is more difficult to measure. Surrogate measures such as service environment or service times are often used to evaluate the quality of services. A comparison of the characteristics of quality in goods and services is found in Fig. 18-1.

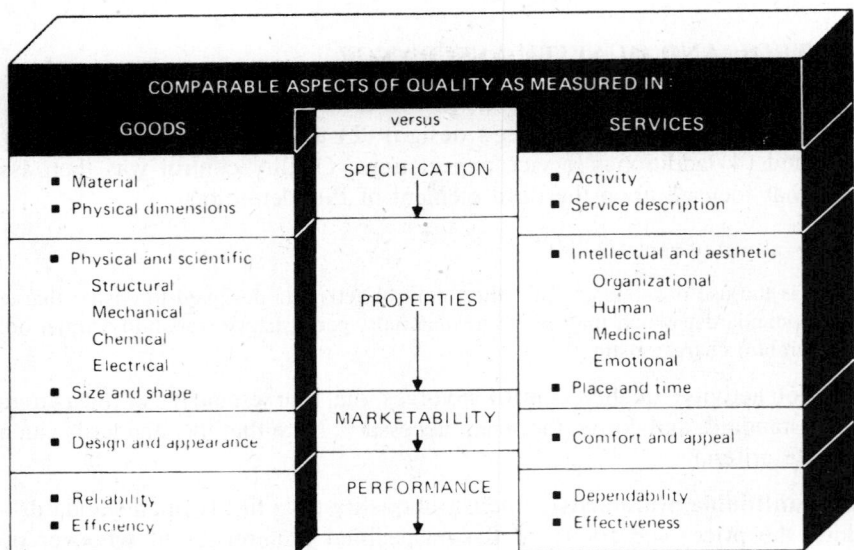

Fig. 18-1   Quality in goods versus services

PROCESS VARIATION AND SAMPLING

Some inherent variation is present in virtually all processes that deliver goods and services. In physical products, variation arises from differences in materials, machines, proficiency of operators, the environment, etc. Thus, one 2×4 piece of lumber is slightly different from another. In services, the skill level of the provider and the service environment cause outputs to vary. For example the times required to process a home loan are likely to vary from one loan officer to the next.

Question: Distinguish between (a) random (common) causes and (b) assignable causes of variation.

(a)   *Random causes of variation* are the inherent factors that are expected as a normal (though often unexplained) aspect of operations. They do not cause a shift in the central tendency of a process.

(b)   *Assignable causes of variation* are identifiable factors that can shift the central tendency of a process or generate some instability. Analysts frequently search for assignable causes of variation.

A process that is in control should be able to accommodate common causes of variation without shifting away from a balanced (or equilibrium) state.

Question: What is process capability?

Process capability is the ability of a process to function within its design parameters with respect to both its central tendency (or target value) and its inherent variability (or tolerance). Process capability is sometimes described by a frequency distribution of the variable under study.

Question: What are tolerances?

Tolerances are the upper and lower limits within which acceptable quality measurements can be expected to lie if the product or process is acceptable. They are often specified by engineering.

Sampling. Variation in output characteristics is a problem because with too much variability, some products may be produced that have characteristics outside the tolerance limits. If all products could be inspected or tested, the defective products might all be found. However, 100 percent inspection of a whole (*population* of N elements) is not always practical, so firms resort to the inspection or testing of a portion (*sample* of n elements) of the output.

Question: Why might sampling be preferred to total inspection of a product or process?

(1) Sampling takes less time and is less costly.

(2) Testing sometimes destroys the product (e.g., fuses, road tests of tires), so testing of all the output may be infeasible.

(3) Sampling can yield highly accurate results at a predictable level of confidence. (Results are sometimes even more accurate than total inspection.)

The major concern in using a sample to assess product or process characteristics is that the sample be truly representative of the entire lot, or population, so that the results observed in the sample can also be inferred to the population. This requires that the sample be randomly selected, e.g., by using a random number table (Appendix A) or computer-generated random numbers.

Question: What constitutes a random sample?

A *random sample* is one wherein each element, or combination of elements, in the population has an equal chance of being selected.

If there were no variability in the population, a sample of only one element would completely reveal the nominal or central value of the variable being controlled. However, with inherent variability in the population, more observations than one are required.

Question: What determines how large a sample is needed to accurately represent a population?

Sample size is a function of the variability of the process and the level of error that is permissible in the result. Small samples are least accurate. For highly variable processes, large samples are needed. However, accuracy does not increase in direct proportion to the sample size. Thus, the reduction in error realized from doubling a sample from 20 to 40 is likely to be much more than that of doubling a (larger) sample of 200 to 400. (See Chap. 10 for sample size calculations.)

Although sample size is not a function of the size of the population—unless the sample constitutes more than 10 percent of the population—some rough guidelines have been used for approximating a sample size. One (imprecise, but simple) expression used by Eastman Kodak* to estimate a sample size (n) from a production lot (N) is $n = \sqrt{2N}$.

Sample size is, however, a function of (1) the standard normal deviate, Z, (2) an estimate of the variability existent in the population, $\sqrt{pq}$ or s, and (3) the level of error that is permissible in the resulting estimates, e. The expressions for sample size (from Chap. 10) are thus:

(*a*) For countable data: (attributes)

$$n = \frac{Z^2 pq}{e^2}$$

(10.8)

(*b*) For measurable data: (variables)

$$n = \left(\frac{Zs}{e}\right)^2$$

(10.1)

---

*Hendrick, Thomas E., and Franklin G. Moore, *Production/Operations Management*, 9th ed., Richard D. Irwin, Homewood, Illinois, 1985.

STATISTICAL METHODS FOR CONTROLLING QUALITY

The collection and analysis of quantitative (measurable) data are an integral part of quality-control activities. This leads directly to the use of statistical analysis, for *statistics* is the body of methods for the collection, analysis, presentation, and interpretation of quantitative data, and for the use of such data.

Question: What is statistical quality control?

Statistical quality control is the use of statistical methods for the purpose of controlling quality.

The statistical methods employed in quality-control work have been classified into three general categories:

    (1)   General-purpose (descriptive and inferential) methods

    (2)   Acceptance sampling plans

    (3)   Process control methods

The general-purpose (generic) methods (1) include a wide range of descriptive check sheets, charts and diagrams, plus traditional inferential methods that rely upon underlying statistical distributions (e.g., normal, binomial, Poisson). Also included are techniques such as regression and analysis of variance. The acceptance sampling plans (2) apply to a wide range of attributes and variables data, and rely heavily upon the statistical theory underlying operating characteristic curves. Although a number of process control methods exist, for many managers, the concept of statistical process control automatically (3) implies the use of control charts (for both attributes and variables data).

GENERAL-PURPOSE STATISTICAL METHODS

Tools of Analysis.   Many organizations have responded to global competition with programs like total quality management. (See Chap. 2.) This has generated a renewed emphasis on the use of both traditional and innovatively new methods of collecting, categorizing, and presenting data for purposes of control. Chapter 2 introduced a few of the relatively descriptive tools used to support continuous improvement. These included (1) the PDCA cycle, (2) lists and diagrams, (3) flowcharts, (4) cause-and-effect diagrams, and (5) benchmarking. Several more statistically based tools could be added to that list.

Question: List some additional statistical tools used for organizing and analyzing quality-related data, along with the major purpose of each.

Technique	Major purpose is to:
1. Database analysis	Answer questions about the purpose and necessity of data; observe any irregularities; and respond to what, when, where, why type of questions
2. Check sheets	Facilitate the recording and analysis of data, and identify problems
3. Histograms	Graphically depict the central tendency and dispersion of observations
4. Flowcharts	Visually depict the inputs/outputs and main steps in a process
5. Run charts	Show the sequence, or progress, of some characteristic over time
6. Pareto analysis	Direct attention to the most critical problem areas, e.g., most defects, under assumption that 80 percent of problems come from 20 percent of causes
7. Scatter diagrams	Show the extent to which one characteristic is correlated with another

Traditional Statistics.   Figure 18-2 describes the significant characteristics of some of the most common statistical distributions used in quality assessment. If the variable of interest in a population is known to follow a normal, Poisson, or other statistical distribution, the probabilities of obtaining specific outcomes from that population can be determined by *deductive logic*, using probability theory. Other applications call for an inference about the quality level of the population on the basis of limited sample evidence. Making this inference is an *inductive* process.

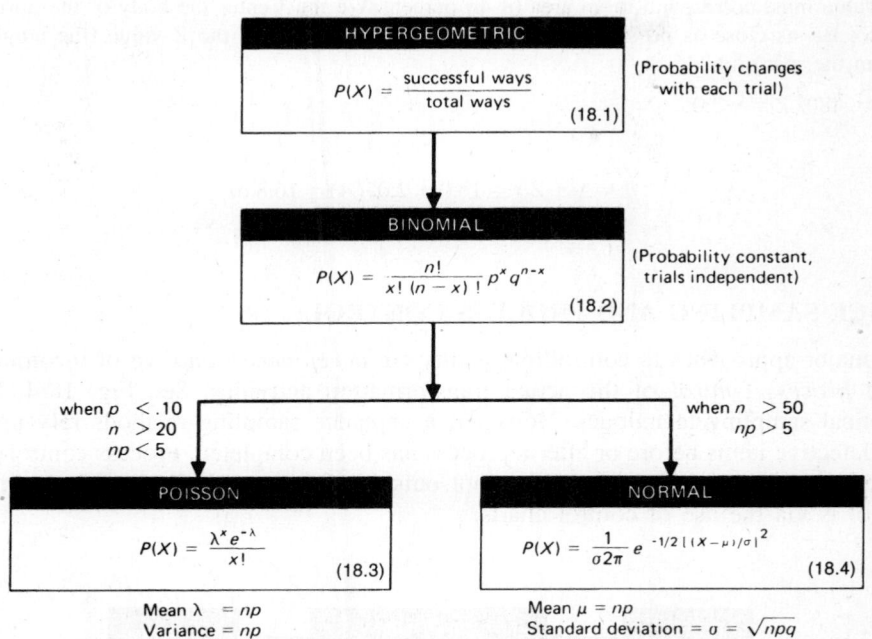

Fig. 18-2 Probability distributions useful in quality control

Example 18.1 (*Binomial distribution*) Fifteen percent of the accounts audited by a CPA firm turn out to have errors that necessitate the payment of additional taxes. What is the chance that exactly 2 accounts taken from a random sample of 10 will owe additional taxes?

This problem calls for the use of deductive logic. The error rate is given as a percentage and can be taken as constant. If we can assume that each account is independent of others, then the binomial distribution applies.

$$P(X = 2 \mid n = 10, \ p = .15) = \frac{n!}{x!(n-x)!} \, p^x q^{n-x} = \frac{10!}{2!8!} \, (.15)^2 (.85)^8 = .2759$$

Note that the solution value could also be obtained more directly from the table of binomial probabilities given in Appendix C.

Example 18.2 (*Normal distribution*) Cans of corn at the Crescent Valley Cannery are filled by a machine that can be set for any desired average weight. The fill is normally distributed with a standard deviation of .4 ounce. If quality standards specify that 98 percent of the cans should contain 16 ounces or more, where should the quality-control supervisor recommend the machine be set on the ounce scale?

The machine average μ must be set high enough so that 48 percent of the cans containing less than the mean still contain 16.0 ounces. Knowing that the fill is normally distributed (Fig. 18-3), we can use the expression for the standard normal deviate.

$$-Z = \frac{X - \mu}{\sigma}$$

$$\therefore \ \mu = X + Z\sigma$$

where X = specified fill = 16.0 oz
   μ = unknown mean setting
   σ = .4 oz
   Z = number of standard deviations from μ to X

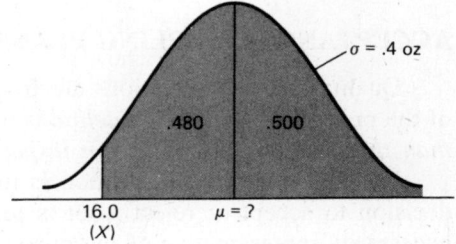

Fig. 18-3

(*Note:* The Z value must correspond to an area of 48 percent. We must enter the body of the normal distribution table—Appendix B—as close as possible to the value of .48 and read off the Z value (the number of standard deviations) from the margin.)

For $P(X) = .480$, $Z = -2.05$.

Therefore $$\mu = X + Z\sigma = 16.0 + 2.05(.4) = 16.8 \text{ oz}$$

ACCEPTANCE SAMPLING AND PROCESS CONTROL

The two major approaches to controlling quality are *acceptance sampling* of incoming or outgoing products, and *process control* of the actual transformation activities. See Fig. 18-4. Both methods involve statistical sampling techniques. However, acceptance sampling methods rely upon estimating the levels of defective items before or after a process has been completed. Process control is more useful during a process to ensure that production is not outside of acceptable limits. The primary means of process control is via the use of control charts.

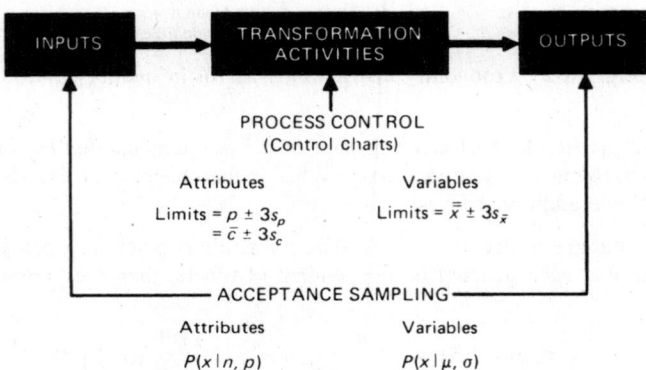

Fig. 18-4 Statistical techniques for controlling quality

The quality characteristic being observed is classified as either an attributes or a variables characteristic. *Attributes characteristics* are either present or not, such as defective or nondefective, or passing a test or failing it. There is no measure of the degree of conformance. For attributes data, a discrete distribution, such as the binomial or Poisson, is used to make inferences about the population characteristic being controlled. *Variables characteristics* are present in varying degrees and are measurable. Examples are dimensions, weights, and times. For variables data, continuous distributions such as the normal are used.

ACCEPTANCE SAMPLING PLANS

Quality-control inspections are frequently made upon receipt of raw materials or upon completion of the product. A general *guideline* is to *inspect whenever the cost of inspection at a given stage is less than the probable loss from not inspecting.*

When 100 percent inspection is uneconomical or infeasible (e.g., if it destroys the product), the decision to accept or reject a lot is made on the basis of sample evidence. But every sample is not necessarily representative of the population. Thus acceptance sampling involves some risk of rejecting good lots and accepting bad lots. The amount of risk can be specified and statistically controlled in terms of a sampling plan.

A *sampling plan* is a decision rule that specifies how large a sample n should be taken and the allowable measurement, number, or percentage c of defectives in the sample. Sampling plans for *attributes* are based upon a qualitative (countable) classification of the item (e.g., percent good or bad), and the probabilities of defectives in the parent population are estimated from discrete distributions such as the binomial and Poisson. Sampling plans for *variables* are based on a quantitative (measurable) characteristic and typically use the normal distribution. We shall let c represent the specified acceptable limit and let X represent the hypothetical or actual percent or number of defectives.

Example 18.3 Illustrate a sampling plan for (*a*) attributes and (*b*) variables.

(*a*) *Attributes plan:* Select a random sample of size $n = 40$, and count the number of defectives X. If $X \le 3$, accept the lot; otherwise, reject it.

(*b*) *Variables plan:* Select a random sample of size $n = 40$, and measure the mean tensile strength, $\overline{X}$. If $\overline{X} \ge 12,000$ psi, accept the lot; otherwise, reject it.

OPERATING CHARACTERISTIC CURVES

An *operating characteristic* curve is a graphic description of a specific sampling plan (n, c combination), which shows the probability that the plan will accept lots of various (possible) quality levels. Figure 18-5*a* illustrates the operating characteristic (OC) curve for an attributes measurement where there is 100 percent inspection of a shipment of N = 100 items. If $\le$ 2 1/2% are defective, the shipment is accepted, and if $>$ 2 1/2% are defective, it is rejected. Assuming accurate inspection, there is no risk of error. If the lot contains either 0, 1, 2 defectives, the probability of acceptance is 1.0, whereas with 3 or more defectives, the $P(\text{accept}) = 0$.

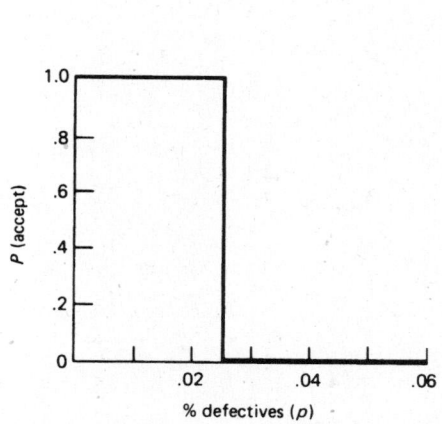

(*a*) OC curve for 100 percent inspection

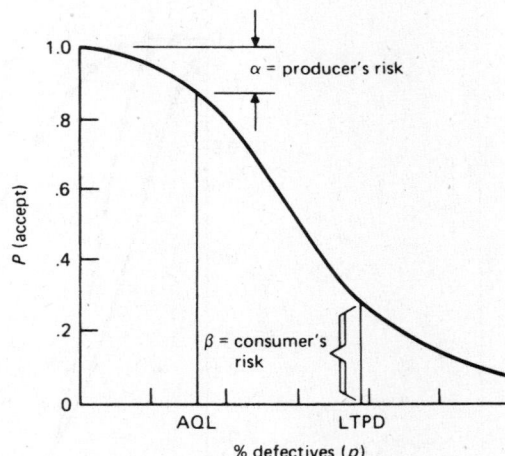

(*b*) OC curve for less than 100 percent inspection

Fig. 18-5 Operating characteristic curves

In Fig. 18-5*b* the shipment is much larger ($N = 1,000$), and the producer and consumer must adopt a sampling plan to reduce inspection costs. Two risks result:

(1) *Producer's risk* is the risk of getting a sample that has a higher proportion of defectives than the lot as a whole and rejecting a good lot. It is designated as the alpha (α) risk. Producers hope to keep this risk low, (e.g., between 1 and 5 percent). If a good lot is rejected, we refer to this as a type I error.

(2)   *Consumer's risk* is the risk of getting a sample that has a lower proportion of defectives than the lot as a whole and accepting a bad lot. It is designated as the beta (β) risk. Consumers want to keep this risk low. If a bad lot is accepted, we refer to this as a type II error.

To derive a sampling plan, the producer and consumer must specify not only the level of the α and β risk but also the lot quality level to which these risks pertain. Thus we must further define *good lot* and *bad lot* in terms of the percent defective in the population.

The *acceptable quality level* (AQL) is the quality level of a good lot. It is the percent defective that can be considered satisfactory as a process average and represents a level of quality that the producer wants accepted with a high probability of acceptance.

The *lot tolerance percent defective* (LTPD) or *limiting quality level* (LQL) is the quality level of a bad lot. It represents a level of quality that the consumer wants accepted with a low probability of acceptance. Lots that have a quality level between the AQL and LTPD are in an indifferent zone between good and bad.

The α risk at the AQL level and β risk at the LTPD level establish two points that largely determine what the sample size n and acceptance number c must be. The appropriate sampling plan (n, c combination) can be found by consulting *tables of* standard plans such as the Dodge and Romig tables, or the U.S. Military Standard MIL-STD-105 tables. Alternatively, a *trial-and-error* procedure can be followed wherein different values of n and c are tried in order to find the combination that yields an OC curve that most closely passes through the two points.

For a small sample, the OC curve is relatively flat, resulting in high risks to both the producer and the consumer. Increasing n makes the OC curve more discriminating, to the point where a 100 percent sample eliminates all risk (Fig. 18-5a). Figure 18-6 shows how increasing the acceptance number from $c \leq 1$ to $c \leq 4$ shifts the risk from producer to consumer. The sampling plan depicted by the dashed line shows the effect of increasing the sample size; it is more discriminating between good (AQL) and bad (LTPD) lots.

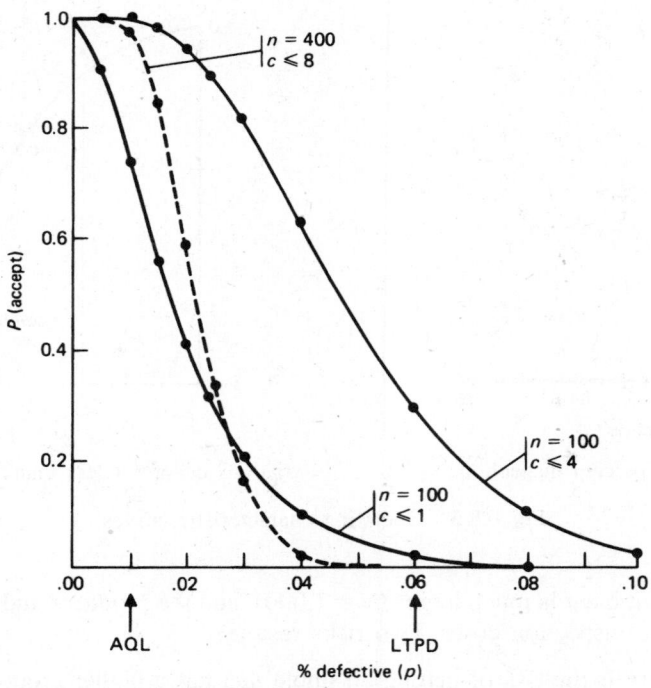

Fig. 18-6   Effect on the OC curve of changing the sample size and the acceptance number

Question: What data points are needed to find a sampling plan that controls the risk to both the producer and the consumer?

    (*a*)   Producers' risk, α, and the AQL level to which it applies, i.e., α at AQL

    (*b*)   Consumers' risk, β, and the LTPD level to which it applies, i.e., β at LTPD

Question: Do the sample size (*n*) and acceptance number (*c*) values completely define an OC curve?

Yes, each *n* and *c* combination results in a different OC curve.

SAMPLING PLANS FOR ATTRIBUTES

Once the α and β risks are set (at the AQL and LTPD points, respectively), a sampling plan (*n*, *c* values) can be determined. If the data are expressed in terms of proportions (e.g., percent defective), then the binomial or Poisson distribution is used to calculate the probabilities of acceptance for the attributes sampling plan.

Example 18.4 A shipment of 1,000 semiconductors is to be inspected on a sampling basis. The producer and consumer have agreed to adopt a plan whereby the α risk is limited to 5 percent at AQL = 1 percent defective, and the β risk is limited to 10 percent at LTPD = 5 percent defective. Construct the OC curve for the sampling plan $n = 100$ and $c \leq 2$, and indicate whether this plan satisfies the requirements.

To construct the OC curve we must determine the probabilities of acceptance of the shipment for various possible values of the true percentage of defectives in the population. Since the shipment is accepted when there are ≤ 2 defectives in the sample, the probabilities we seek are $P(x \leq 2)$, given the alternative values of the population. If we were working with a binomial distribution, we could write this probability as

$$P(X \leq 2 \mid n, p)$$

and obtain the values from a calculator or Appendix C. However, from Fig. 18-2 note that the binomial probabilities of defectives can be approximated by a Poisson distribution here because the sample size (100) is >20. We appear to be working with a small percent defective of $p < .10$, and np looks to be in the neighborhood of 5. Using Appendix D we can obtain the Poisson probabilities as

$$P(X \leq 2 \mid \lambda)$$

where X = number of defectives in sample

        λ = mean of Poisson distribution = np

        p = (alternative) percentage of defectives in population

Thus, for the AQL percentage of $p = .01$, we can find the probability of acceptance of the lot as

$$P(X \leq 2 \mid \lambda)$$

where $\lambda = np = (100)(.01) = 1$. Therefore

$$P(X \leq 2 \mid \lambda = 1) = .92$$

(from Appendix D).

Probabilities for other possible values of the true mean are given in Fig. 18-7*a*, and these values are plotted as an OC curve in Figure 18-7*b*.

Note that this plan ($n = 100$, $c \leq 2$) yields an α risk of .08 and a β risk of .12. Both exceed the respective limits of .05 and .10. Since both risks are exceeded, a larger sample size will be required, and the calculations will have to be repeated.

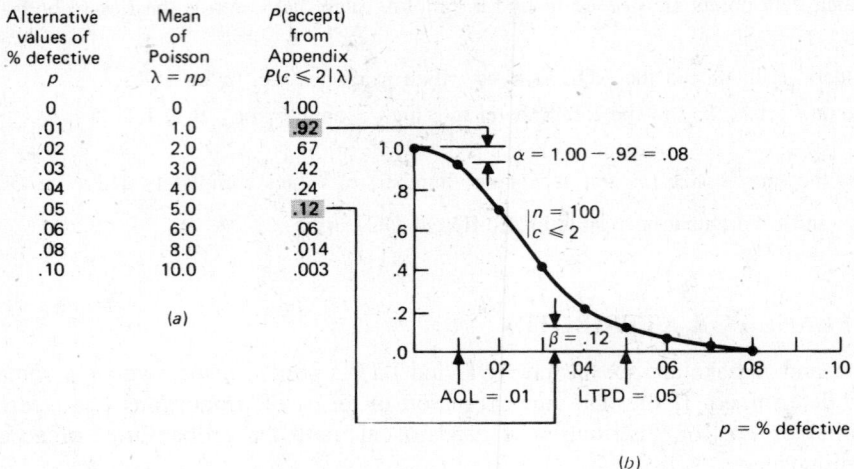

Alternative values of % defective p	Mean of Poisson $\lambda = np$	P(accept) from Appendix $P(c \leq 2 \mid \lambda)$
0	0	1.00
.01	1.0	.92
.02	2.0	.67
.03	3.0	.42
.04	4.0	.24
.05	5.0	.12
.06	6.0	.06
.08	8.0	.014
.10	10.0	.003

(a)

(b)

Fig. 18-7   OC curve values and operating characteristic curve for sampling plan $n = 100$, with $c \leq 2$

SAMPLING PLANS FOR VARIABLES

Sampling plans for variables require measurements of the characteristic being controlled. Measurements yield more information than counts (i.e., measures of dispersion), and the n and c values of sampling plans can be calculated more directly. They capitalize on the fact that means of sufficiently large samples are normally distributed in a sampling distribution that has a mean $= \mu$ and standard error $\sigma_{\bar{x}} = \sigma/\sqrt{n}$.

The first sampling plan example (Example 18.5) assumes that the sample size is predetermined and that only one risk is to be controlled (i.e., either α or β). Use the normal distribution, and solve for the critical limit c. If c is given, solve for n.

Example 18.5   (*Given α risk only: solving for c*)   A metals firm produces titanium castings whose weights are normally distributed with a standard deviation of $\sigma = 8$ pounds. Casting shipments averaging less than 200 pounds are considered poor quality and the firm would like to minimize such shipments. Design a sampling plan for a sample of $n = 25$ that will limit the risk of rejecting lots that average 200 pounds to 5 percent.

The problem situation is described schematically in Fig. 18-8. We assume that the distribution of sample means is approximately normal, with mean $\mu = 200$ and standard error:

$$\sigma_{\bar{x}} = \frac{\sigma}{\sqrt{n}} = \frac{8}{\sqrt{25}} = 1.6 \text{ lb}$$

The limit c is then $c = \mu - Z\sigma_{\bar{x}}$

where Z = value corresponding to area of .450
= 1.64 (from Appendix B).

Thus, $c = 200 - 1.64(1.6) = 197.4$ lb

Plan: Take a random sample of $n = 25$ ingots and determine the mean weight. If $\bar{x} > 197.4$ lb, accept the shipment; otherwise, reject it.

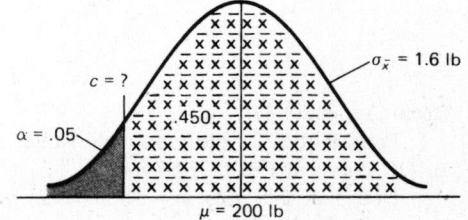

Fig. 18-8

In Example 18.5, with the limit set at 197.4 pounds, the risk of rejecting a lot that really averages 200 pounds (a good lot) is limited to 5 percent. The plan was established wholly on the basis of the α risk and a given sample size. A larger sample size would, of course, be more discriminating. For example, with a sample of $n = 100$ the reject limit could be raised to 198.7 pounds.

If both the producers (α) risk and consumers (β) risk are specified, the required sample size (n) and reject limit (c) can be computed by setting up two equations for c rather than one.

Example 18.6   (*Given α and β risks: solving for n and c*)   A metals firm produces titanium castings whose weights are normally distributed, with a standard deviation of $\sigma = 8.0$ pounds. Casting shipments averaging 200 pounds are of good quality, and those averaging 196 pounds are of poor quality. Design a sampling plan that satisfies the following requirements: (*a*) The probability of rejecting a lot with an average weight of 200 pounds is .05. (*b*) The probability of accepting a lot with an average weight of 196 pounds is .10.

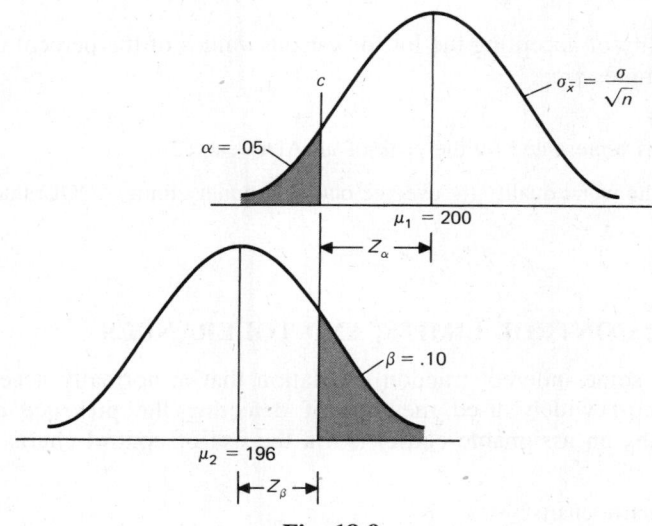

Fig. 18-9

The problem situation is described schematically in Fig. 18-9. The solution procedure is to first set up simultaneous equations defining the reject limit c in terms of Z standard errors. Then solve for n, and substitute it back into either one of the equations to find c. The two equations locating c are

(*a*)   From above:
$$c = \mu_1 - Z_\alpha \frac{\sigma}{\sqrt{n}} = 200 - 1.645 \frac{(8)}{\sqrt{n}}$$

(*b*)   From below:
$$c = \mu_2 + Z_\beta \frac{\sigma}{\sqrt{n}} = 196 + 1.28 \frac{(8)}{\sqrt{n}}$$

Setting the two equations for c equal to each other:

$$200 - 1.645 \frac{(8)}{\sqrt{n}} = 196 + 1.28 \frac{(8)}{\sqrt{n}}$$

$$n = \left(\frac{23.40}{4}\right)^2 = 34$$

Therefore,
$$c = 200 - 1.645 \frac{(8)}{\sqrt{34}} = 197.7 \text{ lb}$$

Plan: Take a random sample of $n = 34$ ingots, and determine the mean weight. If $\bar{x} > 197.7$ pounds, accept the shipment; otherwise, reject it.

AVERAGE OUTGOING QUALITY LEVELS

An *average outgoing quality* (AOQ) curve shows the average expected quality in all outgoing lots after the rejected lots from the sample have been 100 percent inspected and all defects removed. Incoming lots with a small percentage of defects will be passed with a resultant high outgoing quality. Those with a slightly larger proportion of defectives will result in the worst outgoing quality, because lots that have a large proportion of defects will end up undergoing 100 percent inspection, with only the acceptable items being passed. The AOQ curve has the true percentage of defectives in lots undergoing inspection on the *x* axis and the percentage of defectives (P_D) in lots of size N after inspection on the *y* axis.

$$\text{AOQ} = \frac{P_D P_A (N - n)}{N} \qquad\qquad (18.5)$$

where P_A is the probability of accepting the lot for various values of the percent defective (from the OC curve), and n is the sample size.

Question: What point is represented by the peak of an AOQ curve?

The peak represents the worst quality (or average outgoing quality limit, AOQL) that can be expected for the given sampling plan.

CONTROL CHARTS, CONTROL LIMITS, AND TOLERANCES

All processes have some inherent (random) variation that is normally acceptable in a product or process. One of the most widely used methods of detecting the presence of an unacceptable or nonrandom variation (i.e., an assignable cause) is via the use of control charts.

Question: What are control charts?

Control charts are graphic devices used to monitor sample statistics that describe characteristics of a product or process on a regular basis over time. Figure 18-10 illustrates a control chart for monitoring the diameter of plastic frisbees produced on molding machines. Note that the chart has (1) a mean or target value at 12.00 inches, (2) an upper control limit at UCL = 12.10 inches, and (3) a lower control limit at LCL = 11.90 inches. Each plotted point represents the average diameter of a sample of several plastic disks.

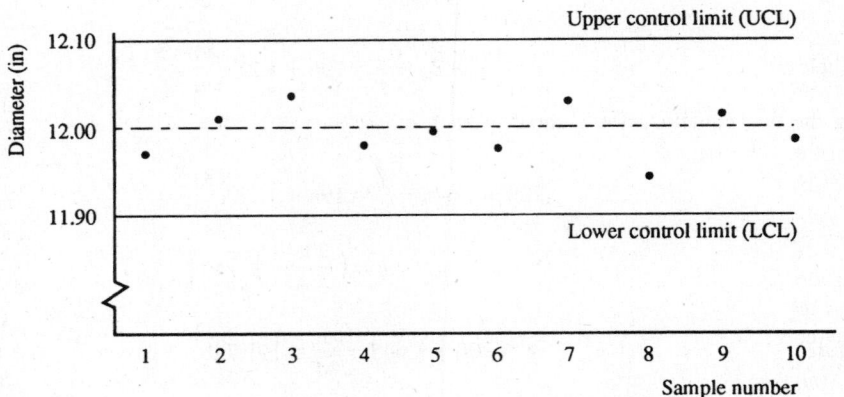

Fig. 18-10   Example control chart

Process Control. When the sample statistics are clustered around the central value within the control limits, a process is said to be "in control." A process may be off target or "out of control" and should be investigated for an assignable cause if the statistic:

(1) Takes on a value outside the control limits

(2) Has a predominance of points on one side of the center line

(3) Exhibits a trend (five points) going consistently toward either limit

(4) Has two points in a row at more than 2/3 the distance to a control limit

(5) Has very erratic behavior or sudden changes in magnitude

Question: Distinguish between variables control charts and attributes control charts.

Variables control charts, such as mean $\overline{X}$ and range R charts, are used to monitor continuous (measurable) data (e.g., the weight or dimensions of a product).

Attributes control charts, such as proportion p and number c charts, are used to monitor discrete (countable) data (e.g., the percentage or number of defects in a product).

Tolerance Limits. Many processes have broad natural tolerance limits within which most individual observations lie. However, as the central limit theorem states, sample means and sample proportions exhibit much less variation than individual values. See Fig. 18-11 where T_{UN} and T_{LN} are the upper and lower natural tolerance limits and UCL and LCL are the upper and lower control limits for means.

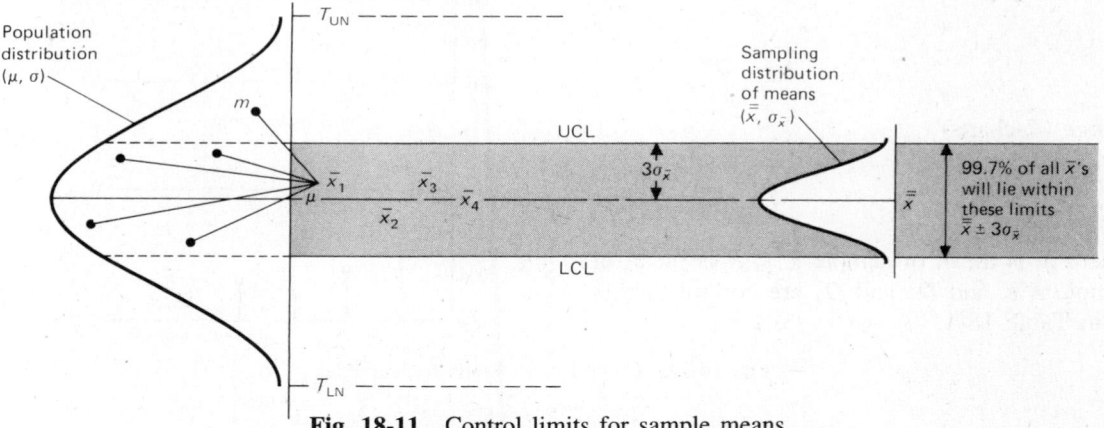

Fig. 18-11   Control limits for sample means

Control limits are the boundaries within which sample statistics can be expected to vary due simply to the randomness of the sample used. They are computed from the relatively tight sampling distributions and are typically set at 3 (or possibly 2) standard errors away from the process average. When a process is "in control," 99.7 percent of the sample averages should be within ±3 standard errors of the chart centerline. If sample averages fall outside the control limits, some assignable cause is probably responsible and corrective action should be taken.

Example 18.7 A control chart is established, with limits of ±2 standard errors, for use in monitoring samples of size $n = 20$. Assume the process is in control. (*a*) Would you expect many *individual values* to lie outside these limits? (*b*) How likely would a *sample mean* fall outside the control limits? (*c*) What kind of error would be committed in erroneously concluding that the process is out of control?

(*a*) Yes, the limits are set to control mean values, not individual values. (*b*) Assuming normality, 95.5 percent of the sample means are within ±2 standard errors, so about 4.5 percent of the means would lie outside. (*c*) Type I. This is concluding the process is out of control when it is not.

PROCESS CONTROL FOR VARIABLES VIA CONTROL CHARTS

Variables control charts, such as mean $\overline{X}$ and range $\overline{R}$ charts, are used to monitor continuous (measurable) data (e.g., the weight or dimensions of a product). Figure 18-12 illustrates and lists some expressions for computing control limits for variables. A control chart of means ($\overline{X}$ chart) reveals variation *among* samples means and is used to signal a shift in the process mean. The range ($\overline{R}$) chart monitors variability *within* samples and is used to signal a change in the spread or dispersion of the data. The range is a common measure of dispersion, and the standard expressions for means (Eqs. 18.6 and 18.7) have been adapted to the use of ranges (Eqs. 18.8 and 18.9) by using standardized conversion factors as given in Table 18-1. These factors also facilitate calculation of range control limits (Eqs. 18.10 and 18.11).

Means ($\overline{X}$-charts)

$$\text{UCL}_{\overline{X}} = \overline{\overline{X}} + Z\sigma_{\overline{X}} \qquad (18.6)$$

$$\text{LCL}_{\overline{X}} = \overline{\overline{X}} - Z\sigma_{\overline{X}} \qquad (18.7)$$

Stating the equations in terms of ranges we have

$$\text{UCL}_{\overline{X}} = \overline{\overline{X}} + A_2\overline{R} \qquad (18.8)$$

$$\text{LCL}_{\overline{X}} = \overline{\overline{X}} - A_2\overline{R} \qquad (18.9)$$

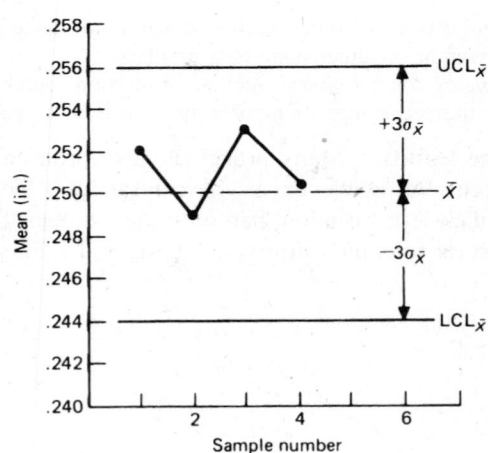

Range ($\overline{R}$-charts)

$$\text{UCL}_R = D_3\overline{R} \qquad (18.10)$$

$$\text{LCL}_R = D_4\overline{R} \qquad (18.11)$$

where $\overline{\overline{X}}$ is mean of sample $\overline{X}$'s, $\overline{R}$ is mean of sample R's, and D_3 and D_4 are control factors from Table 18-1.

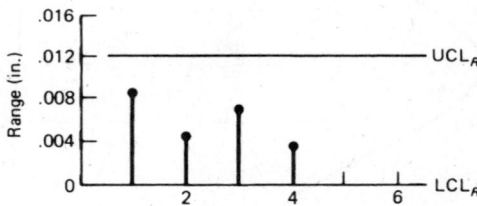

Fig. 18-12   Control chart limits for variables

Question: Outline the procedure for establishing and using control limits.

1.   Select job and quality characteristic to be monitored.
2.   Take 20 to 25 samples of size n, and compute $\overline{X}$ and R for each.
3.   Establish and graph the control limits.
4.   Plot $\overline{X}$ and R points, and look for causes of any points outside limits.
5.   Discard points outside limits with assignable causes, and recalculate the control limits.
6.   Use revised limits, and begin regular sampling activities (assuming the sampling is economically justified).

Example 18.8   (*Variables chart*)   A precision casting process is designed to produce blades having a diameter of $10.000 \pm .025$ centimeters. To establish control limits, 20 samples of $n = 5$ blades are randomly selected from the first 2,000 blades produced as follows (Table 18-2).

**Table 18-1   Factors for Three-Sigma Control Limits
for $\bar{X}$ and R Charts**

Number of Observations in Subgroup n	Factor for $\bar{X}$ Chart A_2	Factors for R Chart LCL D_3	UCL D_4
2	1.88	0	3.27
3	1.02	0	2.57
4	0.73	0	2.28
5	0.58	0	2.11
6	0.48	0	2.00
7	0.42	0.08	1.92
8	0.37	0.14	1.86
9	0.34	0.18	1.82
10	0.31	0.22	1.78
11	0.29	0.26	1.74
12	0.27	0.28	1.72
13	0.25	0.31	1.69
14	0.24	0.33	1.67
15	0.22	0.35	1.65
16	0.21	0.36	1.64
17	0.20	0.38	1.62
18	0.19	0.39	1.61
19	0.19	0.40	1.60
20	0.18	0.41	1.59
25	0.15	0.46	1.54

Source: Adapted (and supplemented) from E. L. Grant and R. S. Leavenworth, *Statistical Quality Control,* 6th ed. (New York: McGraw-Hill Book Company, 1988), p. 670.

Table 18-2

Sample 1	Sample 2	Sample 3	$\cdots$	Sample 20
10.010	10.018	$\cdots$		10.004
9.989	9.992	$\cdots$		9.988
10.019	9.996	$\cdots$		9.990
9.978	10.014	$\cdots$		10.019
10.008	10.005	$\cdots$		9.983
50.004	50.025	$\cdots$		49.984
$\bar{X} = 10.0008$	10.0050	$\cdots$	$\cdots$	9.9968
$R = .041$	.026	$\cdots$	$\cdots$	.0036

The grand mean, $\bar{\bar{X}}$, of the sample means and the mean of the sample ranges, $\bar{R}$, were found to be

$$\bar{\bar{X}} = \frac{\Sigma \bar{X}\text{'s}}{\text{no. samples}} = \frac{10.0008 + 10.0050 + \cdots + 9.9968}{20} = 10.002 \text{ cm}$$

$$\bar{R} = \frac{\Sigma R\text{'s}}{\text{no. samples}} = \frac{.041 + .026 + \cdots + .036}{20} = .032 \text{ cm}$$

(a) Find the control limits for the sample means, and (b) find the control limits for the sample ranges.

(a)   Mean:
$$\text{UCL}_{\bar{X}} = \bar{\bar{X}} + A_2\bar{R} = 10.002 + .577(.032) = 10.020 \text{ cm}$$
$$\text{Center} = \bar{\bar{X}} \qquad\qquad\qquad = 10.002 \text{ cm}$$
$$\text{LCL}_{\bar{X}} = \bar{\bar{X}} - A_2\bar{R} = 10.002 - .577(.032) = 9.984 \text{ cm}$$

(b)   Range:
$$\text{UCL}_R = D_3\bar{R} = (2.114)(.032) = .068 \text{ cm}$$
$$\text{Center} = \bar{R} \qquad\qquad = .032 \text{ cm}$$
$$\text{LCL}_R = D_4\bar{R} = (.000)(.032) \quad = .000 \text{ cm}$$

PROCESS CONTROL FOR ATTRIBUTES VIA CONTROL CHARTS

Attributes control charts, such as proportion p and number c charts, are used to monitor discrete (countable) data (e.g., the percentage or number of defects in a product). Figure 18-13 lists some expressions for computing control limits for attributes. A control chart for proportions (p chart) is based on the binomial distribution (or normal approximation) and is sensitive to a change in the proportion of defectives in a process. The numbers (c chart) is based upon the Poisson distribution, which assumes a small (rare event) probability of a defect occurring. It is especially useful for controlling the defect rate when the number of nondefectives is unavailable, because it uses only the average number of defects.

Proportions (p charts)	Numbers (c charts)
$\text{UCL}_p = p + 3s_p \qquad (18.12)$ $\text{LCL}_p = p - 3s_p \qquad (18.13)$ where $p =$ proportion of defectives in sample $= \dfrac{\text{number of defectives}}{\text{total number of items}}$ $s_p = \sqrt{\dfrac{pq}{n}}$ where $n =$ sample size to be used for monitoring	$\text{UCL}_c = \bar{c} + 3s_c \qquad (18.14)$ $\text{LCL}_c = \bar{c} - 3s_c \qquad (18.15)$ where $\bar{c} =$ average number of defects per unit $= \dfrac{\Sigma c}{N} = \dfrac{\text{total no. of defects/unit in samples}}{\text{no. of samples}}$ $s_c = \sqrt{\bar{c}} \qquad (18.16)$

Fig. 18-13   Control limits for attributes data

Example 18.9   (*Attributes p chart*)   A sportswear firm has set up for automated production of a line of sweaters. Twenty samples of size $n = 50$ are to be withdrawn randomly during the first week of production in order to establish control limits for the process. Defects remain in the shipment but bring less revenue, for they eventually sell as "seconds." The defectives detected in the 20 samples are shown in Table 18-3. Compute the control limits for this process.

Table 18-3   Defective Items in 20 Samples of $n = 50$ Sweaters

Sample no.	1	2	3	4	5	6	7	8	9	10	11	12	13	14	15	16	17	18	19	20
No. defectives	2	3	4	1	0	2	4	1	1	3	0	1	2	1	0	3	7	2	1	2
% defective	.04	.06	.08	.02	.00	.04	.08	.02	.02	.06	.00	.02	.04	.02	.00	.06	.14	.04	.02	.04

Total number of defectives = 40.

$$\text{UCL}_p = p + 3s_p$$

where
$$p = \frac{\text{total of defectives}}{\text{total number of items}} = \frac{40}{50 \times 20} = .040$$

$$s_p = \sqrt{\frac{pq}{n}} = \sqrt{\frac{(.040)(.960)}{50}} = .028$$

$$\text{UCL}_p = .040 + 3(.028) = .124$$
$$\text{LCL}_p = p - 3s_p = .040 - 3(.028) = .000$$

By use of these limits, a preliminary chart is constructed (Fig. 18-14) and the data points plotted.

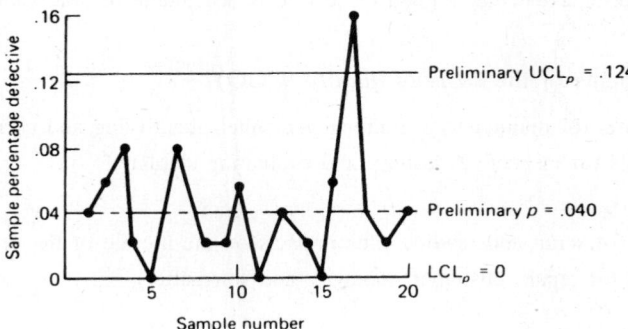

Fig. 18-14

Note that the fraction defective in sample 17 is outside the upper control limit. Suppose the reason for this is investigated, and the cause is found to be that a new machine was phased in at that point before receiving final adjustments from a mechanic. This data point is then discarded, and a new value for p and new control limits are calculated.

$$p = \frac{33}{50 \times 19} = .0347 \qquad s_p = \sqrt{\frac{(.0347)(.9653)}{50}} = .0259$$

$$\text{UCL}_p = .0347 + 3(.0259) = .112 \qquad \text{LCL}_p = .0347 - 3(.0259) = .000$$

None of the remaining sample values fall outside the new limits, so these limits become the standard for controlling the process in the future.

Example 18.10 *(Attributes c chart)* The Metropolitan Transit System uses the number of written passenger complaints per day as a measure of its service quality. For 10 days, the number of complaints received was as shown in Table 18-4. Compute the $3s_c$ control limits.

We seek limits for the number of defects per unit, where defects are written customer complaints and the unit is 1 day. Thus the Poisson distribution applies:

Table 18-4

Day (sample) no.	1	2	3	4	5	6	7	8	9	10	Total
No. of complaints/day	4	8	2	0	3	9	10	0	6	4	46

$$\bar{c} = \text{average number of defects per unit} = \frac{46 \text{ complaints}}{10 \text{ days}} = 4.6 \text{ complaints/day}$$

$$s_c = \sqrt{\bar{c}} = \sqrt{4.6} = 2.14$$
$$\text{UCL}_c = \bar{c} + 3s_c = 4.6 + 3(2.14) = 11.0$$
$$\text{LCL}_c = \bar{c} - 3s_c = 4.6 - 3(2.14) = 0 \qquad \text{(negative values are assigned zero)}$$

The (process) average is 4.6 complaints per day, and control limits are from zero to 11 complaints per day.

Questions and Solved Problems

18.1 Which of the following descriptions of quality is most appropriate for quality-control purposes?

(*a*)   The inherent worth of a product in monetary terms

(*b*)   The degree to which a product conforms to specified standards

(*c*)   The value assigned to the product by the consumer that acquires it

(*d*)   The relationship between the price of the product and its potential lifetime

     As a control activity, quality control necessarily relies upon the existence of standards (as a basis for comparison) to determine whether a good or service is acceptable or not. Therefore (*b*).

18.2 What are components of the *costs of quality* (COQ)?

(1)   Prevention costs (of maintenance, training personnel, identifying and correcting defects)

(2)   Appraisal costs (of inspecting, testing, and evaluating status)

(3)   Failure costs

    (*a*)   Internal (of scrap and rework, plus impacts on the morale of the organization)

    (*b*)   External (of repair, customer goodwill, and warranties)

18.3 Quality is frequently associated with goods, but we also speak of it with respect to services, e.g., quality health care. Is it really feasible to control the quality of a service such as health care?

     It is feasible if a responsible party sets standards (e.g., materials used, reliability, time to administer treatments, etc.), measures performance against those standards, and takes action to ensure they are upheld. However, if there is (political or other) disagreement about what standards should apply, or if performance measurement is sloppy or corrective measures are not taken, then the controls will most likely be ineffective.

PROCESS VARIATION AND SAMPLING

18.4 The philosopher Thomas Aquinas in reasoning back to a first cause (God) argues that effects don't just happen; effects are always caused by something else. If every effect does indeed have a cause, how can one conclude that some causes of variation are "random"?

     Random does not mean "uncaused." It simply means that the cause is not subject to control. The random causes may be due to a myriad of indistinguishable factors, but we lump them all together and refer them as normal (expected) variation.

18.5 Using the Eastman Kodak expression, develop a rough estimate of the size of sample needed to estimate the characteristics of a population of size (*a*) 2,000 and of size (*b*) 20,000. Comment on the difference.

(*a*)   If 2,000: $n = \sqrt{2N} = \sqrt{2(2,000)}$ = 63 observations

(*b*)   If 20,000: $n = \sqrt{2N} = \sqrt{2(20,000)} = 200$ observations

Note that although population (*b*) is 10 times larger than (*a*), the sample size is not 10 times larger.

18.6 If samples are to accurately reflect the characteristics of populations, it seems logical that large populations should require larger samples than smaller populations. How do you account for the fact that "sample size is not a function of the size of the population"?

Sample size is a function of the dispersion of values in the population, not how many values there are. Calculations of n are based upon the central limit theorem (which recognizes that sample means and proportions are normally distributed). As shown in Eqs. 10.8 and 10.1, the sample size is then a function of (1) the standard normal deviate, Z, (2) an estimate of the variability existent in the population, pq or s, and (3) the level of error that is permissible in the resulting estimates, e.

18.7 How large a sample is needed to give a 95 percent confidence estimate that is within ± 3 percent of the proportion defective in a population if the estimated proportion is 12 percent?

This is a proportion (attributes) problem, so we can use Eq. 10.8:

$$n = \frac{Z^2 pq}{e^2} = \frac{(1.96)^2(.12)(.88)}{(.03)^2} = 451 \text{ observations}$$

18.8 How large a sample would be needed to give 98 percent confidence in an estimate of weight in a population of cereal boxes if the standard deviation is estimated to be .8 ounce and an accuracy of $\pm.1$ ounce is desired?

This is a measurable (variables) problem, so we can use Eq. 10.1:

$$n = \left(\frac{Zs}{e}\right)^2 = \left(\frac{(2.33)(.8)}{.1}\right)^2 = 348 \text{ observations}$$

GENERAL-PURPOSE STATISTICAL METHODS

18.9 Why are quality-control activities so closely associated with statistics?

Control activities necessitate the availability of quantitative data, and statistics is the body of methods for the collection and analysis of data. Moreover, in addition to providing description, statistical theory enables analysts to use samples to make inferences about entire populations.

18.10 A QC manager has five mobile phones (A, B, C, D, E) and needs to select phones for three performance tests. In how many ways can phones be selected for three tests if:

(a) Any phone can be used for any or all of the tests, so both duplication and different order of selection count as a different way (*multiple choices*)?

(b) No phone can be used for more than one test, but the order of selection makes a difference (*permutations*)?

(c) No phone can be used for more than one test, and the order of selection of the phones does not count (*combinations*)?

Let $x = 3$ phones chosen from $n = 5$ phones.

(a) Multiple choices

$$N^x = 5^3 = 5 \cdot 5 \cdot 5 = 125 \text{ ways}$$

(b) Permutations

$$P_x^n = \frac{n!}{(n-x)!} = \frac{5!}{(5-3)!} = \frac{5 \cdot 4 \cdot 3 \cdot 2 \cdot 1}{2 \cdot 1} = 60 \text{ ways}$$

(c) Combinations

$$C_x^n = \frac{n!}{x!(n-x)!} = \frac{5!}{3!(5-3)!} = \frac{5 \cdot 4 \cdot 3 \cdot 2 \cdot 1}{3 \cdot 2 \cdot 1 \cdot 2 \cdot 1} = 10 \text{ ways}$$

18.11 A quality-control sample of $n = 100$ items is taken, and the standard deviation is calculated to be .250 inch.

   (a)   Estimate the standard error of the mean.

   (b)   What would be the standard error if the sample size were 1,000 instead of 100?

   (a)
$$s_{\bar{x}} = \frac{s}{\sqrt{n}} = \frac{.250}{\sqrt{100}} = .025 \text{ inch}$$

   (b)
$$s_{\bar{x}} = \frac{.250}{\sqrt{1,000}} = .008 \text{ inch}$$

18.12 Which probability distribution will yield an appropriate answer in a reasonable amount of time if we wish to know the probability of getting 10 or fewer defects in a sample of 400 from a population that is 1 percent defective?

   (a)   hypergeometric   (c)   normal   (e)   Student t

   (b)   binomial   (d)   Poisson

$$n = 400 = {>}20$$
$$np = (400)(0.01) = 4 = {<}5$$
$$p = .01 = {<}.10$$

   Poisson distribution is appropriate

$$P(X \leq 10 \mid \lambda = 4) = .997 \text{ (from Appendix D)}$$

18.13 In an industrial plant, the mean weight of a certain packaged chemical is $\mu = 82.0$ kg and standard deviation is $\sigma = 4.0$ kg. If a sample of $n = 64$ packages is drawn from the population for inspection, find the probability that

   (a)   An *individual package* in the sample will exceed 82.5 kg. (Assume that the population is normally distributed for this part.)

   (b)   The *sample mean* will exceed 82.5 kg.

   (a) $Z = \dfrac{x - \mu}{\sigma}$

$$= \frac{82.5 - 82.0}{4} = .125$$

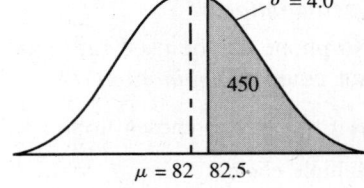

   $P(Z) = .050$ (Appendix B)
   $P(X > 82.5) = .500 - .050 = .450$

   (b) $Z = \dfrac{\bar{x} - \mu}{\sigma_{\bar{x}}}$ where $\sigma_{\bar{x}} = \dfrac{\sigma}{\sqrt{n}} = \dfrac{4}{\sqrt{64}} = .5$
$$Z = \frac{82.5 - 82.0}{.5} = 1.0$$

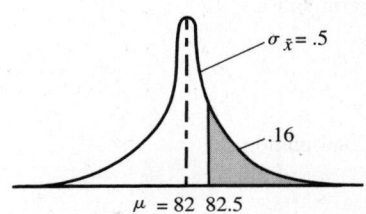

   $P(Z) = .34$ (Appendix B)
   $P(\bar{x} > 82.5) = .50 - .34 = .16$

18.14 (*Hypergeometric distribution*)   A shipment of 20 chips received 6 weeks ago was delivered to an assembly area without inspection upon receipt of delivery. Four of the chips were installed in a space vehicle, and the remainder were mixed with existing inventory. The supplier has just notified the firm that five of the chips were defective.

(*a*)   What is the probability that all four chips installed in the vehicle were good?

(*b*)   What is the probability that there was one defective in the lot of four?

(*a*)
$$P(X = 4 \text{ good}) = \frac{\text{successful ways}}{\text{total ways}}$$

The successful and total number of ways of selecting 4 chips from 20 must be computed recognizing that no chips can be used more than once and that a different order of selection of the same 4 would not change anything. In this case we are concerned with *combinations* of x items chosen from $n = 20$:

$$C_x^n = \frac{n!}{x!(n-x)!}$$

(See Prob. 18.10 for additional distinction between multiple choices, permutations, and combinations.) Note that the lot is known to have contained

- The number of successful ways to select 4 good from the 15 is the combinations of 4 in 15. $\Big\} \ C_4^{15}$

  For each of these ways

15 good
+5 defective
20 total

- The number of successful ways to select 0 defectives from 5 is the combinations of 0 in 5. $\Big\} \ C_0^5$

- The total number of ways to select 4 transistors from 20 is the combination of 4 in 20. $\Big\} \ C_4^{20}$

Thus,
$$P(X = 4) = \frac{C_4^{15} \cdot C_0^5}{C_4^{20}} = \frac{\dfrac{15!}{4!11!} \cdot \dfrac{5!}{0!5!}}{\dfrac{20!}{4!16!}} = .28$$

(*b*)
$$P(X = 3) = \frac{C_3^{15} \cdot C_1^5}{C_4^{20}} = \frac{\dfrac{15!}{3!12!} \cdot \dfrac{5!}{1!4!}}{\dfrac{20!}{4!16!}} = .47$$

18.15 (*Binomial distribution*)   Ten percent of the fire bricks baked in an obsolete oven turn out to be defective in some way. What is the chance that exactly 2 will be defective in a random sample of 10?

The defective rate is given as a percentage and can be taken as constant. If we can assume each brick produced is independent of the previous brick, then the binomial distribution applies.

$$P(X = 2 \mid n = 10, \ p = .10) = \frac{n!}{x!(n-x)!} \, p^x q^{n-x} = \frac{10!}{2!8!} (.10)^2 (.90)^8 = .1937$$

Note that the solution value could also be obtained more directly from the table of binomial probabilities given in Appendix C. From there one could quite easily obtain all the values for the probability distribution of $P(X = 0)$, $P(X = 1)$, $P(X = 2), \ldots, P(X = 10)$ for samples of size $n = 10$ when the proportion in the population is .10 defective.

18.16 (*Poisson approximation*)   A very large shipment of textbooks comes from a publisher who usually supplies about 1 percent with imperfect bindings. What is the probability that among 400 textbooks taken from this shipment, exactly 3 will have imperfect bindings?

This problem could be solved by using the binomial expression for $P(X = 3 \mid n = 400,\ p = .01)$. However, unless one has a calculator that handles exponents, the solution would be tedious. It can be closely approximated by the Poisson distribution since $p < .10$, $n > 20$, and $np < 5$.

$$P(X) = \frac{\lambda^x e^{-\lambda}}{x!}$$

where $\lambda = np = 400(.01) = 4.0$

$x = 3$

$e = 2.718(\text{constant})$

$$P(X = 3 \mid \lambda = 4.0) = \frac{\lambda^x e^{-\lambda}}{x!} = \frac{4^3 e^{-4}}{3!} = \frac{(64)(.018)}{3 \cdot 2} = .195$$

The solution could also be obtained more directly from the table of summed Poisson probabilities given in Appendix D. Go down the λ column to $\lambda = 4$ and then right to the columns where the events are designated as $\leq c$ (rather than $\leq x$). Since we need $c = 3$, we just find the difference between $c \leq 3$ and $c \leq 2$, which is $.433 - .238 = .195$.

18.17 (*Normal approximation*)   In a precious metals manufacturing process, 20 percent of the ingots contain impurities and must be remelted after inspection. If 100 ingots are selected for shipment without inspection, what is the probability 15 or more ingots will contain impurities?

This binomial problem $P(X \geq 15 \mid n = 100,\ p = .20)$ may be solved by using the normal approximation to the binomial, since $n > 50$ and $np = 100(.20) > 5$. (See Fig. 18-15.)

$$\mu = np = 100(.20) = 20$$
$$\sigma = \sqrt{npq} = \sqrt{20(.8)} = 4$$

Since we are using a continuous distribution to estimate discrete probabilities, the appropriate continuous value for ≥ 15 becomes 14.5

$$Z = \frac{X - \mu}{\sigma} = \frac{14.5 - 20}{4} = -1.375$$

$$P(Z) = .415 \text{ (from Appendix B)}$$
$$P(X > 14.5) = .415 + .500 = .915$$

The normal distribution is, of course, useful in its own right for continuous variables, aside from its use as an approximation to the binomial probabilities.

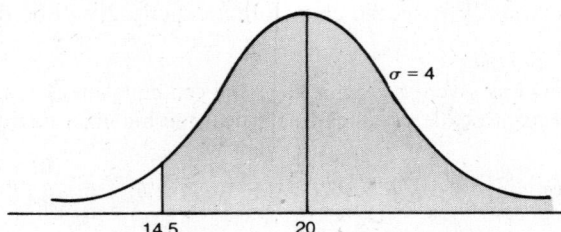

$\sigma = 4$

14.5   20

Fig. 18-15   Normal approximation to binomial

ACCEPTANCE SAMPLING

18.18 What is (*a*) the purpose of acceptance sampling and (*b*) the name of the statistical function that shows the degree to which a sampling plan is able to discriminate between lots of high quality and lots of low quality?

(a) The purpose of acceptance sampling is to decide whether a lot meets predetermined standards.

(b) Operating characteristic (OC) curves show how well a given sampling plan functions to accept good lots or reject poor ones.

18.19 How might one readily distinguish between attributes and variables data?

Attributes are countable (e.g., 196 good, 4 defective), whereas variables are measurable data.

18.20 Name the two risks associated with sampling and testing?

(1) The α, or producers' risk (i.e., that good products will be rejected)

(2) The β, or consumers' risk (i.e., that defective products will be accepted)

SAMPLING PLANS FOR ATTRIBUTES

18.21 An orange grower (producer) and packing plant (consumer) have agreed on a sampling plan that calls for a sample of 150 oranges from a shipment and acceptance if 3 or less are spoiled. (a) Does this plan meet the specifications of limiting the grower's risk of rejecting lots that are as good as 2 percent spoiled to less than or equal to 20 percent, and does it limit the packing plant's risk of accepting shipments that are as bad as 4 percent spoiled to less than or equal to 25 percent? (b) What would be the effect of changing the acceptance value to $c \geq 4$?

(a) This is an attributes plan calling for $\alpha \leq .20$ at AQL = .02 and $\beta \leq .25$ at LTPD = .04. Though we could construct an entire OC curve, it is necessary to check only two points: AQL and LTPD.

p	$\lambda = np$	$P(x \leq 3 \mid \lambda)$		Risk
.02	$(150)(.02) = 3.0$	$P(x \leq 3 \mid \lambda = 3.0) = .647$	so	$\alpha = 1.000 - .647 = .353$
.04	$(150)(.04) = 6.0$	$P(x \leq 3 \mid \lambda = 6.0) = .151$	so	$\beta = .151$

The grower's risk (α) of .35 is larger than .20 and so is unsatisfactory (too high). The packing plant's risk (β) of .15 is less than .25 and so is satisfactory.

(b) With $c \leq 4$: $P(x \leq 4 \mid \lambda = 3.0) = .815$, so $\alpha = 1.000 - .815 = .185$

$P(x \leq 4 \mid \lambda = 6.0) = .285$, so $\beta = .285$

The grower's risk (α) is less than .20 and is satisfactory, but the packing plant's risk (β) is greater than .25 and so is unsatisfactory.

SAMPLING PLANS FOR VARIABLES

18.22 A videogame manufacturer purchases a 4-inch plastic disk from a supplier, where $\sigma = .30$ inch. The manufacturer wishes to design a sampling plan that limits the risk of accepting shipments with an average diameter of ≤ 3.900 inches to .01 and also limits the chance of accepting disks of ≥ 4.100 inches to .01. The sample size should also be large enough to limit the risk of rejecting a lot that really averages 4.000 inches to .10.

(a) Find the appropriate sampling plan (i.e., n and c values).

(b) Construct the OC curve for this sampling plan. What is the probability of accepting a lot that averages (1) 3.920 inches, (2) 3.959 inches, (3) 4.000 inches, (4) 4.041 inches?

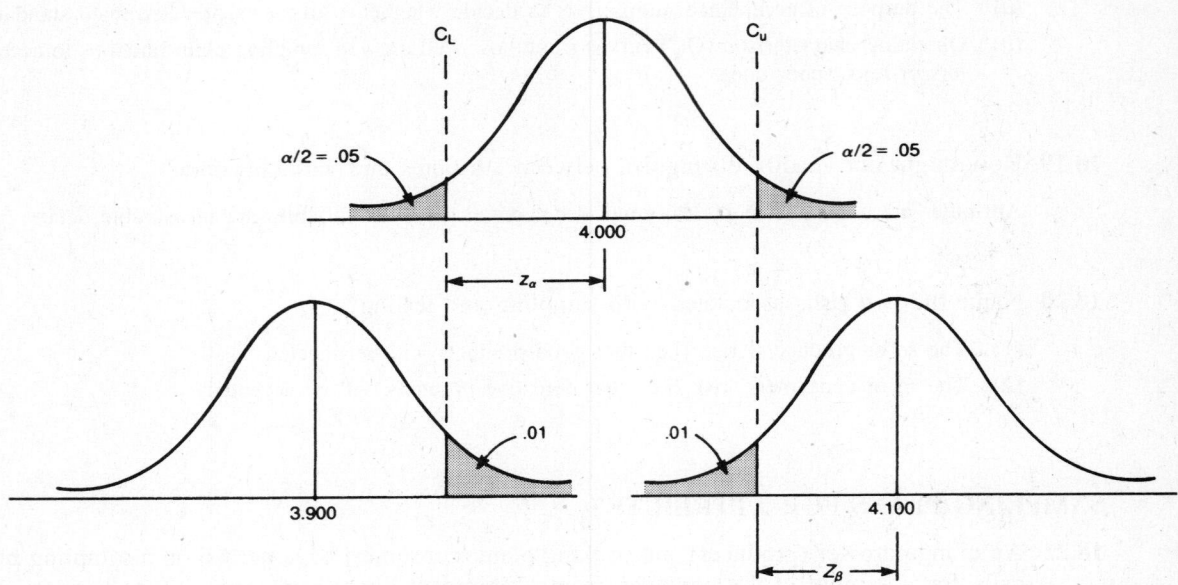

Fig. 18-16

(a)   See Fig. 18-16. For normal distribution area of .450,

$$Z_\alpha = 1.645$$

and for .490 $$Z_\beta = 2.33$$

Establishing equations for the limit C_L we have from below:

$$C_L = 3.900 + Z_\beta \frac{\sigma}{\sqrt{n}} = 3.900 + 2.33 \frac{(.2)}{\sqrt{n}}$$

and from above:

$$C_L = 4.000 - Z_\alpha \left(\frac{\sigma}{\sqrt{n}}\right) = 4.000 - 1.645 \frac{(.2)}{\sqrt{n}}$$

Setting the two equations equal we have

$$3.900 + \frac{.466}{\sqrt{n}} = 4.000 - \frac{.329}{\sqrt{n}}$$

$$\frac{.795}{\sqrt{n}} = .100$$

Thus, $$n = 63.2$$

$$C_L = 3.900 + \frac{2.33(.2)}{\sqrt{63.2}} = 3.959 \text{ inches}$$

Using similar procedures, the upper limit is 4.041 inches.

   The sampling plan is as follows: Take a random sample of $n = 63$ disks, and measure the diameter. If $\bar{x}$ is from 3.959 inches to 4.041 inches, accept the shipment; otherwise, reject it.

(b)   For the OC curve (Fig. 18-17), use the normal distribution to compute P(accept), which is the area between C_L and C_u for selected values of μ. For example, at $\mu = 3.920$

$$Z = \frac{\overline{X} - \mu}{\sigma_{\bar{x}}} = \frac{3.959 - 3.920}{.2/\sqrt{63}} = \frac{.0390}{.0252} = 1.5478$$

Thus, $$P(Z) = .4392$$

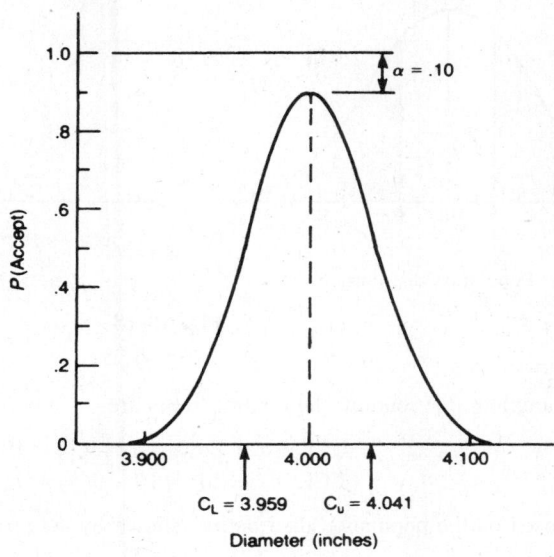

Fig. 18-17

and $\qquad$ $P(\text{accept}) = .5000 - .4392 = .0608$

(1) at $\mu = 3.920$, $P(\text{accept}) = .061$

(2) at $\mu = 3.959$, $P(\text{accept}) = .500$

$\qquad$ at $\mu = 3.980$, $P(\text{accept}) = .785$

(3) at $\mu = 4.000$, $P(\text{accept}) = .900$

(4) at $\mu = 4.041$, $P(\text{accept}) = .500$

AVERAGE OUTGOING QUALITY LEVELS

18.23 An OC curve reveals that lots with a true percentage of defectives of 2 percent have a probability of being accepted of $P_A = .67$. If the sampling plan for lots of $N = 1,000$ called for samples of size $n = 100$, what would be the average outgoing quality (AOQ) level?

$$\text{AOQ} = \frac{P_D P_A (N - n)}{N} = \frac{(.02)(.67)(1,000 - 100)}{1,000} = .012 = 1.2\%$$

In the above example, since the sample size is $n = 100$ and 2 percent of the items in the sample are defective (on the average), then the 2 defective items would be removed from the sample and replaced before the lot was allowed to continue on its way. For lots of $N = 1,000$ items, the number of defects would then be reduced to 2 percent of the 900 uninspected items. This amounts to 18 in 1,000, or 1.8 percent of the items. Since the probability of acceptance of a lot of 2 percent defectives is (from the OC curve) only .67, then the expected value or average outgoing quality level for lots (of 2 percent defectives) is (.67)(1.8 percent) or 1.2 percent defective.

CONTROL CHARTS FOR VARIABLES

18.24 A control chart is established for a normally distributed variable that has $\mu = 10$, and $\sigma = 1$. If control limits for samples of size $n = 16$ are set at ± 3 standard errors, what percent of the individual values in the population would lie outside the control limits?

See Fig. 18-18.

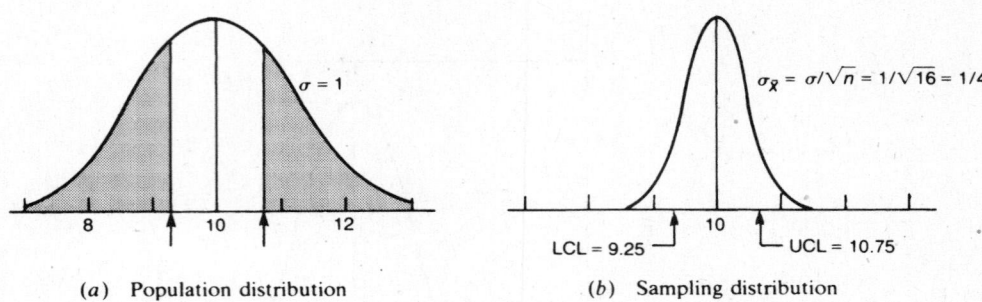

(a) Population distribution   (b) Sampling distribution

Fig. 18-18

From the sampling distribution, the control limits are:

$$\text{UCL} = \bar{\bar{x}} + Z\sigma_{\bar{x}} = 10 + 3(\tfrac{1}{4}) = 10\tfrac{3}{4}$$
$$\text{LCL} = \bar{\bar{x}} - Z\sigma_{\bar{x}} = 10 - 3(\tfrac{1}{4}) = 9\tfrac{1}{4}$$

When superimposed on the population distribution (shown by the arrows), the percentage of area included would be:

Within UCL: $Z = \dfrac{x - \mu}{\sigma} = \dfrac{10.75 - 10.0}{1} = .75$ therefore $P(Z) = .273$

Within LCL: $Z = \dfrac{x - \mu}{\sigma} = \dfrac{9.25 - 10.0}{1} = .75$ therefore $P(Z) = \underline{.273}$
                                                                                        Total   .546

The area outside control limits is $1.000 - .546 = .454$, so approximately 45 percent of the individual observations would lie outside the control limits!

18.25 Belgium Fuel Co. manufacturers uranium pellets to a specified diameter of $.500 \pm .005$ centimeter. In 25 random samples of 9 pellets each, the overall mean of the means $(\bar{\bar{X}})$ and the range $(\bar{R})$ were found to be .501 centimeter and .003 centimeter, respectively. Find the control limits for an $\bar{X}$ and R chart that includes the specified tolerances.

$$\text{UCL}_{\bar{x}} = \bar{\bar{X}} + A_2 \bar{R} = .501 + (.337)(.003) = .502$$
$$\text{LCL}_{\bar{x}} = \bar{\bar{X}} - A_2 \bar{R} = .501 - (.337)(.003) = .500$$
$$\text{UCL}_R = D_3 \bar{R} = (1.816)(.003) = .0054$$
$$\text{LCL}_R = D_4 \bar{R} = (.184)(.003) = .0006$$

CONTROL CHARTS FOR ATTRIBUTES

18.26 A daily-sample of 30 items was taken over a period of 14 days in order to establish attributes control limits. If 21 defectives were found, what should be the LCL_p and UCL_p?

$$p = \frac{\text{no. of defectives}}{\text{total observations}} = \frac{21}{420} = .05$$

$$s_p = \sqrt{\frac{pq}{n}} = \sqrt{\frac{(.05)(.95)}{30}} = .04$$

$$\text{UCL}_p = p + 3s_p = .05 + 3(.04) = .17$$
$$\text{LCL}_p = p - 3s_p = .05 - 3(.04) = 0$$

Supplementary Questions and Problems

18.27　Using the Eastman Kodak expression, approximate the sample size needed to obtain central tendency data from a population of 4,000 items.　　*Ans.*　90 observations

18.28　A random sample of 400 items is drawn from a production process in order to test the hypothesis that the process has 10 percent defectives. Eighty defectives are found. (*a*) What is the theoretical (hypothesized) standard error or proportion (σ_p)? (*b*) What is the estimated standard error of proportion based upon the sample evidence only (s_p)?　　*Ans.*　(*a*) .015 (*b*) .02

18.29　The operations department of a city-owned gas company has a quality service performance standard of no more than four complaints per hour. If the company averages four complaints per hour, what is the probability of 30 minutes passing with no complaints?　　*Ans.*　.135

18.30　If defective components are coming off an assembly line at an average rate of 3.5 per minute, what is the probability that more than 5 defects will arrive in 1 minute?　　*Ans.*　.142

18.31　The manufactured weight of boxes of laundry soap is known to be normally distributed with a mean of 20 pounds and a standard deviation of .4 pound. Approximately what percent of the boxes in a carload shipment could be expected to weigh less than 19.5 pounds if an incoming receipt inspection is made?　　*Ans.*　10.6 percent

18.32　If $\alpha = .04$ and $\beta = .08$ when the true percent defectives for sampling purposes have been correspondingly designated as 2 percent and 5 percent, the AQL and LTPD points would correspond to what values, respectively?　　*Ans.*　.02 and .05

18.33　A producer and consumer have agreed upon the sampling plan $n = 120$, $c \leq 3$. If the AQL is .02 and the LTPD is .06, what are the α and β risks associated with this plan?
Ans.　　Use procedure similar to Prob. 18.21. $\alpha = .221$, $\beta = .072$

18.34　Northeast Paper Co. packages a large volume of tissue under a brand name for a national chain food store. Occasionally the packages are defective because they are from end cuts, the color is bleached, or they are not properly sealed. The paper company and food chain have agreed to adopt a sampling plan so that the risk to Northeast Paper Co. or rejecting lots that are as good as .5 percent defective ($p = .005$) is limited to 2 percent and the risk of the food chain accepting lots as bad as 4 percent defective is no more than 5 percent. (*a*) Construct an OC curve for the sampling plan ($n = 200$, $c \leq 3$). (*b*) Does this plan satisfy the agreed-upon paper company risk? (*c*) Does this plan satisfy the food chain risk?
Ans.　(*a*) Use Poisson (*b*) yes (*c*) yes

18.35　A national bank has established quality standards for its branch banks and allocates a portion of its salary budget on this basis. One measure of service level is the time required to complete all arrangements for opening a checking account. A time of more than 12 minutes is considered *poor service*, and times have a known standard deviation of 4.2 minutes.
　　　Design a variables sampling plan, for a sample of $n = 36$ observations, that will allow the headquarters to sample branch banks so that the risk of rejecting a branch's claim (that it averages 12 minutes or less) is limited to 1 percent (when the true mean time is really 12 minutes).
Ans.　If $\bar{x} \leq 13.63$ minutes, accept.

18.36　The QC supervisor at National Bakery has been asked to direct the receipt inspection of a carload shipment of flour. Each bag is supposed to weigh at least 50 kilograms, and the Chicago Mill has said that the standard deviation is 4 kilograms. Management wishes to limit the risk of rejecting a good lot to 2 percent. On the other hand, if the true mean weight of the bags is only 48 kilograms, management wants to limit the chance of accepting the shipment to 5 percent. (a) Diagram the situation in terms of a sampling distribution showing the α and β risks. (b) How large a sample size is required? (c) What is the critical value c of the sample mean that will satisfy the given conditions?

Ans.   (a) $\alpha = .02$ when $\mu_1 = 50$, $\beta = .05$ when $\mu_2 = 48$ (b) 55 bags (c) 48.88 kg

18.37   A supplier and distributor have agreed upon a contract for a shipment of bags of dog food that are supposed to weigh at least 60.0 pounds ($\sigma = 2.0$ pounds). When the true mean weight of the shipment is 60.0 pounds, the α risk is to be limited to .03; and if the true mean weight is as low as 59.0 pounds, the chance of accepting the lot is to be no more than .04. What sample size is required?

Ans.   Use procedure similar to Example 18.6. $n = 53$

18.38   Variables and attributes charts both monitor the central tendency of a process. What additional monitoring can be done of variables data that cannot be done with attributes data?

Ans.   Can also monitor ranges.

18.39   Suppose the points plotted on a control chart are all within the control limits, but the last three points are on the upper side, about three-quarters of the distance to the UCL. Is the process "in control"?

Ans.   Probably not. [The likelihood of having three observations that far from the centerline (and all on one side of it) is extremely small. The process is probably off its target and should be investigated in search of an assignable cause.]

18.40   In an effort to set up a control chart of a process, samples of size $n = 25$ are taken, and it is determined that $\overline{\overline{X}} = .98$ centimeter and the standard deviation $(s) = .020$ centimeter. Find the control limits for the process.   Ans.   UCL $= .992$ cm, LCL $= .968$ cm

18.41   A sample of $n = 100$ items reveals that 10 percent are defective. What would be the 99.7 percent control limits for the process, if this data were used to establish them?   Ans.   .01 and .19

18.42   The U.S. Department of Testing (USDT) requires that the 100-pound-bag shipments of the Prairie Seed Co. do in fact average 100 pounds or over. Sample data (Table 18-5) from $N = 10$ samples of $n = 6$ bags each showed the following weight deviations from 100 pounds (over, $+$, and under, $-$).

Table 18-5

Sample number	1	2	3	4	5	6	7	8	9	10
	2	3	4	1	6	0	2	−1	1	5
	−1	0	5	2	4	−2	3	−1	2	4
	4	3	6	2	3	2	2	2	0	5
	1	1	2	0	4	−1	2	0	0	2
	0	2	4	4	1	3	4	1	4	3
	2	1	2	2	4	0	6	−1	3	0
Σ	8	10	23	11	22	2	19	0	10	19
Mean $(\overline{X})$	1.33	1.67	3.83	1.83	3.67	.33	3.17	0	1.67	3.17
Range (R)	5	3	4	4	5	5	4	3	4	5

(a) What are the center line ($\overline{\overline{X}}$) and the upper and lower control limits for $\overline{X}$? (b) What are the upper and lower control limits for the range? (Round your calculations to two significant digits beyond the decimal.)

Ans.   (a) Initial calculations yield $\overline{\overline{X}} = 102.07$ and limits of 104.10 pounds and 100.04 pounds, but the mean from sample 8 is outside the limits. Removing sample 8, we have $\overline{\overline{X}} = 102.30$ pounds, and limits of 104.39 and 100.21. (b) Using the revised mean of 102.30, the range limits are 8.68 pounds and 0 pounds.

18.43   A quality-control policy requires setting up control limits on the basis of data from random samples of $n = 100$ per day taken from a 10-day pilot run of a plastics molding activity. A total of 200 defectives were found. (a) What are the UCL_p and LCL_p for the process (in percentage of defectives)? (b) If samples of $n = 100$ continue to be taken, what would be the control limits in numbers of defectives (rather than in percentage)? (*Hint:* This is not c. $\mu = np$ and $\sigma = \sqrt{npq}$.)

Ans.   (a) UCL = .32, LCL = .08 (b) 32 and 8

Appendix A

Random Number Table

27767	43584	85301	88977	29490	69714	94015	64874	32444	48277
13025	14338	54066	15243	47724	66733	74108	88222	88570	74015
80217	36292	98525	24335	24432	24896	62880	87873	95160	59221
10875	62004	90391	61105	57411	06368	11748	12102	80580	41867
54127	57326	26629	19087	24472	88779	17944	05600	60478	03343
60311	42824	37301	42678	45990	43242	66067	42792	95043	52680
49739	71484	92003	98086	76668	73209	54244	91030	45547	70818
78626	51594	16453	94614	39014	97066	30945	57589	31732	57260
66692	13986	99837	00582	81232	44987	69170	37403	86995	90307
44071	28091	07362	97703	76447	42537	08345	88975	35841	85771
59820	96163	78851	16499	87064	13075	73035	41207	74699	09310
25704	91035	26313	77463	55387	72681	47431	43905	31048	56699
22304	90314	78438	66276	18396	73538	43277	58874	11466	16082
17710	59621	15292	76139	59526	52113	53856	30743	08670	84741
25852	58905	55018	56374	35824	71708	30540	27886	61732	75454
46780	56487	75211	10271	36633	68424	17374	52003	70707	70214
59849	96169	87195	46092	26787	60939	59202	11973	02902	33250
47670	07654	30342	40277	11049	72049	83012	09832	25571	77628
94304	71803	73465	09819	58869	35220	09504	96412	90193	79568
08105	59987	21437	36786	49226	77837	98524	97831	65704	09514
64281	61826	18555	64937	64654	25843	41145	42820	14924	39650
66847	70495	32350	02985	01755	14750	48968	38603	70312	05682
72461	33230	21529	53424	72877	17334	39283	04149	90850	64618
21032	91050	13058	16218	06554	07850	73950	79552	24781	89683
95362	67011	06651	16136	57216	39618	49856	99326	40902	05069
49712	97380	10404	55452	09971	59481	37006	22186	72682	07385
58275	61764	97586	54716	61459	21647	87417	17198	21443	41808
89514	11788	68224	23417	46376	25366	94746	49580	01176	28838
15472	50669	48139	36732	26825	05511	12459	91314	80582	71944
12120	86124	51247	44302	87112	21476	14713	71181	13177	55292
95294	00556	70481	06905	21785	41101	49386	54480	23604	23554
66986	34099	74474	20740	47458	64809	06312	88940	15995	69321
80620	51790	11436	38072	40405	68032	60942	00307	11897	92674
55411	85667	77535	99892	71209	92061	92329	98932	78284	46347
95083	06783	28102	57816	85561	29671	77936	63574	31384	51924

Source: Paul G. Hoel, *Elementary Statistics*, 2d ed., John Wiley & Sons, Inc., New York, 1966. Reproduced by permission of the publisher.

Appendix B

Areas under the
Normal Probability Distribution

Values in the table represent the proportion of area under the normal curve between the mean ($\mu = 0$) and a positive value of z.

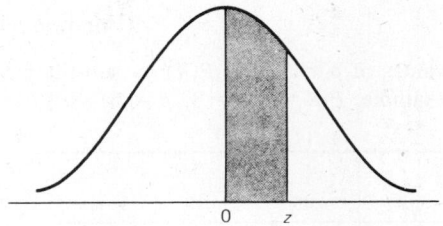

z	.00	.01	.02	.03	.04	.05	.06	.07	.08	.09
.0	.0000	.0040	.0080	.0120	.0160	.0199	.0239	.0279	.0319	.0359
.1	.0398	.0438	.0478	.0517	.0557	.0596	.0636	.0675	.0714	.0753
.2	.0793	.0832	.0871	.0910	.0948	.0987	.1026	.1064	.1103	.1141
.3	.1179	.1217	.1255	.1293	.1331	.1368	.1406	.1443	.1480	.1517
.4	.1554	.1591	.1628	.1664	.1700	.1736	.1772	.1808	.1844	.1879
.5	.1915	.1950	.1985	.2019	.2054	.2088	.2123	.2157	.2190	.2224
.6	.2257	.2291	.2324	.2357	.2389	.2422	.2454	.2486	.2517	.2549
.7	.2580	.2611	.2642	.2673	.2703	.2734	.2764	.2794	.2823	.2852
.8	.2881	.2910	.2939	.2967	.2995	.3023	.3051	.3078	.3106	.3133
.9	.3159	.3186	.3212	.3238	.3264	.3289	.3315	.3340	.3365	.3389
1.0	.3413	.3438	.3461	.3485	.3508	.3531	.3554	.3577	.3599	.3621
1.1	.3643	.3665	.3686	.3708	.3729	.3749	.3770	.3790	.3810	.3830
1.2	.3849	.3869	.3888	.3907	.3925	.3944	.3962	.3980	.3997	.4015
1.3	.4032	.4049	.4066	.4082	.4099	.4115	.4131	.4147	.4162	.4177
1.4	.4192	.4207	.4222	.4236	.4251	.4265	.4279	.4292	.4306	.4319
1.5	.4332	.4345	.4357	.4370	.4382	.4394	.4406	.4418	.4429	.4441
1.6	.4452	.4463	.4474	.4484	.4495	.4505	.4515	.4525	.4535	.4545
1.7	.4554	.4564	.4573	.4582	.4591	.4599	.4608	.4616	.4625	.4633
1.8	.4641	.4649	.4656	.4664	.4671	.4678	.4686	.4693	.4699	.4706
1.9	.4713	.4719	.4726	.4732	.4738	.4744	.4750	.4756	.4761	.4767
2.0	.4772	.4778	.4783	.4788	.4793	.4798	.4803	.4808	.4812	.4817
2.1	.4821	.4826	.4830	.4834	.4838	.4842	.4846	.4850	.4854	.4857
2.2	.4861	.4864	.4868	.4871	.4875	.4878	.4881	.4884	.4887	.4890
2.3	.4893	.4896	.4898	.4901	.4904	.4906	.4909	.4911	.4913	.4916
2.4	.4918	.4920	.4922	.4925	.4927	.4929	.4931	.4932	.4934	.4936
2.5	.4938	.4940	.4941	.4943	.4945	.4946	.4948	.4949	.4951	.4952
2.6	.4953	.4955	.4956	.4957	.4959	.4960	.4961	.4962	.4963	.4964
2.7	.4965	.4966	.4967	.4968	.4969	.4970	.4971	.4972	.4973	.4974
2.8	.4974	.4975	.4976	.4977	.4977	.4978	.4979	.4979	.4980	.4981
2.9	.4981	.4982	.4982	.4983	.4984	.4984	.4985	.4985	.4986	.4986
3.0	.4987	.4987	.4987	.4988	.4988	.4989	.4989	.4989	.4990	.4990

Source: From Paul G. Hoel, *Elementary Statistics*, 2d ed., John Wiley & Sons, Inc., New York, 1966. Reproduced by permission of the publisher.

Binomial Distribution Values

$$P(X|n, p) = \frac{n!}{x!\,(n-x)!}\, p^x q^{n-x}$$

(For population data, replace p with π.)

For values of $p > .5$, find $P(X)$ by substituting $(1 - p)$ for p. Then look up $P(n - X)$.
For example: $P(X = 2 \mid n = 8,\ p = .6) = P[(n - X) \mid n,\ (1 - p)] = P(6 \mid n = 8,\ .4) = .0413$.

						p					
n	X	.05	.10	.15	.20	.25	.30	.35	.40	.45	.50
1	0	.9500	.9000	.8500	.8000	.7500	.7000	.6500	.6000	.5500	.5000
	1	.0500	.1000	.1500	.2000	.2500	.3000	.3500	.4000	.4500	.5000
2	0	.9025	.8100	.7225	.6400	.5625	.4900	.4225	.3600	.3025	.2500
	1	.0950	.1800	.2550	.3200	.3750	.4200	.4550	.4800	.4950	.5000
	2	.0025	.0100	.0225	.0400	.0625	.0900	.1225	.1600	.2025	.2500
3	0	.8574	.7290	.6141	.5120	.4219	.3430	.2746	.2160	.1664	.1250
	1	.1354	.2430	.3251	.3840	.4219	.4410	.4436	.4320	.4084	.3750
	2	.0071	.0270	.0574	.0960	.1406	.1890	.2389	.2880	.3341	.3750
	3	.0001	.0010	.0034	.0080	.0156	.0270	.0429	.0640	.0911	.1250
4	0	.8145	.6561	.5220	.4096	.3164	.2401	.1785	.1296	.0915	.0625
	1	.1715	.2916	.3685	.4096	.4219	.4116	.3845	.3456	.2995	.2500
	2	.0135	.0486	.0975	.1536	.2109	.2646	.3105	.3456	.3675	.3750
	3	.0005	.0036	.0115	.0256	.0469	.0756	.1115	.1536	.2005	.2500
	4	.0000	.0001	.0005	.0016	.0039	.0081	.0150	.0256	.0410	.0625
5	0	.7738	.5905	.4437	.3277	.2373	.1681	.1160	.0778	.0503	.0312
	1	.2036	.3280	.3915	.4096	.3955	.3602	.3124	.2592	.2059	.1562
	2	.0214	.0729	.1382	.2048	.2637	.3087	.3364	.3456	.3369	.3125
	3	.0011	.0081	.0244	.0512	.0879	.1323	.1811	.2304	.2757	.3125
	4	.0000	.0004	.0022	.0064	.0146	.0284	.0488	.0768	.1128	.1562
	5	.0000	.0000	.0001	.0003	.0010	.0024	.0053	.0102	.0185	.0312
6	0	.7351	.5314	.3771	.2621	.1780	.1176	.0754	.0467	.0277	.0156
	1	.2321	.3543	.3993	.3932	.3560	.3025	.2437	.1866	.1359	.0938
	2	.0305	.0984	.1762	.2458	.2966	.3241	.3280	.3110	.2780	.2344
	3	.0021	.0146	.0415	.0819	.1318	.1852	.2355	.2765	.3032	.3125
	4	.0001	.0012	.0055	.0154	.0330	.0595	.0951	.1382	.1861	.2344
	5	.0000	.0001	.0004	.0015	.0044	.0102	.0205	.0369	.0609	.0938
	6	.0000	.0000	.0000	.0001	.0002	.0007	.0018	.0041	.0083	.0156
7	0	.6983	.4783	.3206	.2097	.1335	.0824	.0490	.0280	.0152	.0078
	1	.2573	.3720	.3960	.3670	.3115	.2471	.1848	.1306	.0872	.0547
	2	.0406	.1240	.2097	.2753	.3115	.3177	.2985	.2613	.2140	.1641
	3	.0036	.0230	.0617	.1147	.1730	.2269	.2679	.2903	.2918	.2734
	4	.0002	.0026	.0109	.0287	.0577	.0972	.1442	.1935	.2388	.2734
	5	.0000	.0002	.0012	.0043	.0115	.0250	.0466	.0774	.1172	.1641
	6	.0000	.0000	.0001	.0004	.0013	.0036	.0084	.0172	.0320	.0547
	7	.0000	.0000	.0000	.0000	.0001	.0002	.0006	.0016	.0037	.0078
8	0	.6634	.4305	.2725	.1678	.1002	.0576	.0319	.0168	.0084	.0039
	1	.2793	.3826	.3847	.3355	.2670	.1977	.1373	.0896	.0548	.0312
	2	.0515	.1488	.2376	.2936	.3115	.2065	.2587	.2090	.1569	.1094
	3	.0054	.0331	.0839	.1468	.2076	.2541	.2786	.2787	.2568	.2188
	4	.0004	.0046	.0185	.0459	.0865	.1361	.1875	.2322	.2627	.2734
	5	.0000	.0004	.0026	.0092	.0231	.0467	.0808	.1239	.1719	.2188
	6	.0000	.0000	.0002	.0011	.0038	.0100	.0217	.0403	.0703	.1094
	7	.0000	.0000	.0000	.0001	.0004	.0012	.0033	.0079	.0164	.0312
	8	.0000	.0000	.0000	.0000	.0000	.0001	.0002	.0007	.0017	.0039

n	X	p .05	.10	.15	.20	.25	.30	.35	.40	.45	.50
9	0	.6302	.3874	.2316	.1342	.0751	.0404	.0207	.0101	.0046	.0020
	1	.2985	.3874	.3679	.3020	.2253	.1556	.1004	.0605	.0339	.0176
	2	.0629	.1722	.2597	.3020	.3003	.2668	.2162	.1612	.1110	.0703
	3	.0077	.0446	.1069	.1762	.2336	.2668	.2716	.2508	.2119	.1641
	4	.0006	.0074	.0283	.0661	.1168	.1715	.2194	.2508	.2600	.2461
	5	0000	.0008	.0050	.0165	.0389	.0735	.1181	.1672	.2128	.2461
	6	.0000	.0001	.0006	.0028	.0087	.0210	.0424	.0743	.1160	.1641
	7	.0000	.0000	.0000	.0003	.0012	.0039	.0098	.0212	.0407	.0703
	8	.0000	.0000	.0000	.0000	.0001	.0004	.0013	.0035	.0083	.0176
	9	.0000	.0000	.0000	.0000	.0000	.0000	.0001	.0003	.0008	.0020
10	0	.5987	.3487	.1969	.1074	.0563	.0282	.0135	.0060	.0025	.0010
	1	.3151	.3874	.3474	.2684	.1877	.1211	.0725	.0403	.0207	.0098
	2	.0746	.1937	.2759	.3020	.2816	.2335	.1757	.1209	.0763	.0439
	3	.0105	.0574	.1298	.2013	.2503	.2668	.2522	.2150	.1665	.1172
	4	.0010	.0112	.0401	.0881	.1460	.2001	.2377	.2508	.2384	.2051
	5	.0001	.0015	.0085	.0264	.0584	.1029	.1536	.2007	.2340	.2461
	6	.0000	.0001	.0012	.0055	.0162	.0368	.0689	.1115	.1596	.2051
	7	.0000	.0000	.0001	.0008	.0031	.0090	.0212	.0425	.0746	.1172
	8	.0000	.0000	.0000	.0001	.0004	.0014	.0043	.0106	.0229	.0439
	9	.0000	.0000	.0000	.0000	.0000	.0001	.0005	.0016	.0042	.0098
	10	.0000	.0000	.0000	.0000	.0000	.0000	.0000	.0001	.0003	.0010

n	x	p .05	.10	.15	.20	.25	.30	.35	.40	.45	.50
20	0	.3585	.1216	.0388	.0115	.0032	.0008	.0002	.0000	.0000	.0000
	1	.3774	.2702	.1368	.0576	.0211	.0068	.0020	.0005	.0001	.0000
	2	.1887	.2852	.2293	.1369	.0669	.0278	.0100	.0031	.0008	.0002
	3	.0596	.1901	.2428	.2054	.1339	.0716	.0323	.0123	.0040	.0011
	4	.0133	.0898	.1821	.2182	.1897	.1304	.0738	.0350	.0139	.0046
	5	.0022	.0319	.1028	.1746	.2023	.1789	.1272	.0746	.0365	.0148
	6	.0003	.0089	.0454	.1091	.1686	.1916	.7712	.1244	.0746	.0370
	7	.0000	.0020	.0160	.0545	.1124	.1643	.1844	.1659	.1221	.0739
	8	.0000	.0004	.0046	.0222	.0609	.1144	.1614	.1797	.1623	.1201
	9	.0000	.0001	.0011	.0074	.0271	.0654	.1158	.1597	.1771	.1602
	10	.0000	.0000	.0002	.0020	.0099	.0308	.0686	.1171	.1593	.1762
	11	.0000	.0000	.0000	.0005	.0030	.0120	.0336	.0710	.1185	.1602
	12	.0000	.0000	.0000	.0001	.0008	.0039	.0136	.0355	.0727	.1201
	13	.0000	.0000	.0000	.0000	.0002	.0010	.0045	.0146	.0366	.0739
	14	.0000	.0000	.0000	.0000	.0000	.0002	.0012	.0049	.0150	.0370
	15	.0000	.0000	.0000	.0000	.0000	.0000	.0003	.0013	.0049	.0148
	16	.0000	.0000	.0000	.0000	.0000	.0000	.0000	.0003	.0013	.0046
	17	.0000	.0000	.0000	.0000	.0000	.0000	.0000	.0000	.0002	.0011
	18	.0000	.0000	.0000	.0000	.0000	.0000	.0000	.0000	.0000	.0002
	19	.0000	.0000	.0000	.0000	.0000	.0000	.0000	.0000	.0000	.0000
	20	.0000	.0000	.0000	.0000	.0000	.0000	.0000	.0000	.0000	.0000

Source: National Bureau of Standards, *Tables of the Binomial Probability Distribution*, Applied Mathematics Series, U.S. Department of Commerce, 1950.

Appendix D

Poisson Distribution Values

$$P(X \le c \mid \lambda) = \sum_{0}^{c} \frac{\lambda^x e^{-\lambda}}{x!}$$

The table shows 1,000 times the probability of c or less occurrences of an event that has an average number of occurrences of λ.

Values of c

λ	0	1	2	3	4	5	6	7	8	9	10
.02	980	1000									
.04	961	999	1000								
.06	942	998	1000								
.08	923	997	1000								
.10	905	995	1000								
.15	861	990	999	1000							
.20	819	982	999	1000							
.25	779	974	998	1000							
.30	741	963	996	1000							
.35	705	951	994	1000							
.40	670	938	992	999	1000						
.45	638	925	989	999	1000						
.50	607	910	986	998	1000						
.55	577	894	982	998	1000						
.60	549	878	977	997	1000						
.65	522	861	972	996	999	1000					
.70	497	844	966	994	999	1000					
.75	472	827	959	993	999	1000					
.80	449	809	953	991	999	1000					
.85	427	791	945	989	998	1000					
.90	407	772	937	987	998	1000					
.95	387	754	929	984	997	1000					
1.00	368	736	920	981	996	999	1000				
1.1	333	699	900	974	995	999	1000				
1.2	301	663	879	966	992	998	1000				
1.3	273	627	857	957	989	998	1000				
1.4	247	592	833	946	986	997	999	1000			
1.5	223	558	809	934	981	996	999	1000			
1.6	202	525	783	921	976	994	999	1000			
1.7	183	493	757	907	970	992	998	1000			
1.8	165	463	731	891	964	990	997	999	1000		
1.9	150	434	704	875	956	987	997	999	1000		
2.0	135	406	677	857	947	983	995	999	1000		

Source: Adapted from E. L. Grant, *Statistical Quality Control,* McGraw-Hill Book Company, New York, 1964. Reproduced by permission of the publisher.

Values of c

λ	0	1	2	3	4	5	6	7	8	9	10	11	12	13	14	15	16	17	18	19	20	21	22
2.2	111	355	623	819	928	975	993	998	1000														
2.4	091	308	570	779	904	964	988	997	999	1000													
2.6	074	267	518	736	877	951	983	995	999	1000													
2.8	061	231	469	692	848	935	976	992	998	999	1000												
3.0	050	199	423	647	815	916	966	988	996	999	1000												
3.2	041	171	380	603	781	895	955	983	994	998	1000												
3.4	033	147	340	558	744	871	942	977	992	997	999	1000											
3.6	027	126	303	515	706	844	927	969	988	996	999	1000											
3.8	022	107	269	473	668	816	909	960	984	994	998	999	1000										
4.0	018	092	238	433	629	785	889	949	979	992	997	999	1000										
4.2	015	078	210	395	590	753	867	936	972	989	996	999	1000										
4.4	012	066	185	359	551	720	844	921	964	985	994	998	999	1000									
4.6	010	056	163	326	513	686	818	905	955	980	992	997	999	1000									
4.8	008	048	143	294	476	651	791	887	944	975	990	996	999	1000									
5.0	007	040	125	265	440	616	762	867	932	968	986	995	998	999	1000								
5.2	006	034	109	238	406	581	732	845	918	960	982	993	997	999	1000								
5.4	005	029	095	213	373	546	702	822	903	951	977	990	996	999	999	1000							
5.6	004	024	082	191	342	512	670	797	886	941	972	988	995	998	999	1000							
5.8	003	021	072	170	313	478	638	771	867	929	965	984	993	997	999	1000							
6.0	002	017	062	151	285	446	606	744	847	916	957	980	991	996	999	999	1000						
6.2	002	015	054	134	259	414	574	716	826	902	949	975	989	995	998	999	1000						
6.4	002	012	046	119	235	384	542	687	803	886	939	969	986	994	997	999	1000						
6.6	001	010	040	105	213	355	511	658	780	869	927	963	982	992	997	999	1000						
6.8	001	009	034	093	192	327	480	628	755	850	915	955	978	990	996	998	999	1000					
7.0	001	007	030	082	173	301	450	599	729	830	901	947	973	987	994	998	999	1000					
7.2	001	006	025	072	156	276	420	569	703	810	887	937	967	984	993	997	999	999	1000				
7.4	001	005	022	063	140	253	392	539	676	788	871	926	961	980	991	996	998	999	1000				
7.6	001	004	019	055	125	231	365	510	648	765	854	915	954	976	989	995	998	999	1000				
7.8	000	004	016	048	112	210	338	481	620	741	835	902	945	971	986	993	997	999	1000				
8.0	000	003	014	042	100	191	313	453	593	717	816	888	936	966	983	992	996	998	999	1000			
8.5	000	002	009	030	074	150	256	386	523	653	763	849	909	949	973	986	993	997	999	1000			
9.0	000	001	006	021	055	116	207	324	456	587	706	803	876	926	959	978	989	995	998	999	1000		
9.5	000	001	004	015	040	089	165	269	392	522	645	752	836	898	940	967	982	991	996	998	999	1000	
10.0	000	000	003	010	029	067	130	220	333	458	583	697	792	864	917	951	973	986	993	997	998	999	1000

Present-Value Factors for Future Single Payments

Periods until Payment	1%	2%	4%	6%	8%	10%	12%	14%	15%	16%	18%	20%	22%	24%	25%	26%	28%	30%	35%	40%
1	.990	.980	.962	.943	.926	.909	.893	.877	.870	.862	.847	.833	.820	.806	.800	.794	.781	.769	.741	.714
2	.980	.961	.925	.890	.857	.826	.797	.769	.756	.743	.718	.694	.672	.650	.640	.630	.610	.592	.549	.510
3	.971	.942	.889	.840	.794	.751	.712	.675	.658	.641	.609	.579	.551	.524	.512	.500	.477	.455	.406	.364
4	.961	.924	.855	.792	.735	.683	.636	.592	.572	.552	.516	.482	.451	.423	.410	.397	.373	.350	.301	.260
5	.951	.906	.822	.747	.681	.621	.567	.519	.497	.476	.437	.402	.370	.341	.328	.315	.291	.269	.223	.186
6	.942	.888	.790	.705	.630	.564	.507	.456	.432	.410	.370	.335	.303	.275	.262	.250	.227	.207	.165	.133
7	.933	.871	.760	.665	.583	.513	.452	.400	.376	.354	.314	.279	.249	.222	.210	.198	.178	.159	.122	.095
8	.923	.853	.731	.627	.540	.467	.404	.351	.327	.305	.266	.233	.204	.179	.168	.157	.139	.123	.091	.068
9	.914	.837	.703	.592	.500	.424	.361	.308	.284	.263	.225	.194	.167	.144	.134	.125	.108	.094	.067	.048
10	.905	.820	.676	.558	.463	.386	.322	.270	.247	.227	.191	.162	.137	.116	.107	.099	.085	.073	.050	.035
11	.896	.804	.650	.527	.429	.350	.287	.237	.215	.195	.162	.135	.112	.094	.086	.079	.066	.056	.037	.025
12	.887	.788	.625	.497	.397	.319	.257	.208	.187	.168	.137	.112	.092	.076	.069	.062	.052	.043	.027	.018
13	.879	.773	.601	.469	.368	.290	.229	.182	.163	.145	.116	.093	.075	.061	.055	.050	.040	.033	.020	.013
14	.870	.758	.577	.442	.340	.263	.205	.160	.141	.125	.099	.078	.062	.049	.044	.039	.032	.025	.015	.009
15	.861	.743	.555	.417	.315	.239	.183	.140	.123	.108	.084	.065	.051	.040	.035	.031	.025	.020	.011	.006
16	.853	.728	.534	.394	.292	.218	.163	.123	.107	.093	.071	.054	.042	.032	.028	.025	.019	.015	.008	.005
17	.844	.714	.513	.371	.270	.198	.146	.108	.093	.080	.060	.045	.034	.026	.023	.020	.015	.012	.006	.003
18	.836	.700	.494	.350	.250	.180	.130	.095	.081	.069	.051	.038	.028	.021	.018	.016	.012	.009	.005	.002
19	.828	.686	.475	.331	.232	.164	.116	.083	.070	.060	.043	.031	.023	.017	.014	.012	.009	.007	.003	.002
20	.820	.673	.456	.312	.215	.149	.104	.073	.061	.051	.037	.026	.019	.014	.012	.010	.007	.005	.002	.001
21	.811	.660	.439	.294	.199	.135	.093	.064	.053	.044	.031	.022	.015	.011	.009	.008	.006	.004	.002	.001
22	.803	.647	.422	.278	.184	.123	.083	.056	.046	.038	.026	.018	.013	.009	.007	.006	.004	.003	.001	.001
23	.795	.634	.406	.262	.170	.112	.074	.049	.040	.033	.022	.015	.010	.007	.006	.005	.003	.002	.001	
24	.788	.622	.390	.247	.158	.102	.066	.043	.035	.028	.019	.013	.008	.006	.005	.004	.003	.002	.001	
25	.780	.610	.375	.233	.146	.092	.059	.038	.030	.024	.016	.010	.007	.005	.004	.003	.002	.001	.001	
26	.772	.598	.361	.220	.135	.084	.053	.033	.026	.021	.014	.009	.006	.004	.003	.002	.002	.001		
27	.764	.586	.347	.207	.125	.076	.047	.029	.023	.018	.011	.007	.005	.003	.002	.002	.001	.001		
28	.757	.574	.333	.196	.116	.069	.042	.026	.020	.016	.010	.006	.004	.002	.002	.002	.001	.001		
29	.749	.563	.321.	.185	.107	.063	.037	.022	.017	.014	.008	.005	.003	.002	.002	.001	.001	.001		
30	.742	.552	.308	.174	.099	.057	.033	.020	.015	.012	.007	.004	.003	.002	.001	.001	.001			

Appendix F

Present-Value Factors for Annuities

Years (N)	1%	2%	4%	6%	8%	10%	12%	14%	15%	16%	18%	20%	22%	24%	25%	26%	28%	30%	35%	40%
1	.990	.980	.962	.943	.926	.909	.893	.877	.870	.862	.847	.833	.820	.806	.800	.794	.781	.769	.741	.714
2	1.970	1.942	1.886	1.833	1.783	1.736	1.690	1.647	1.626	1.605	1.566	1.528	1.492	1.457	1.440	1.424	1.392	1.361	1.289	1.224
3	2.941	2.884	2.775	2.673	2.577	2.487	2.402	2.322	2.283	2.246	2.174	2.106	2.042	1.981	1.952	1.923	1.868	1.816	1.696	1.580
4	3.902	3.808	3.630	3.465	3.312	3.170	3.037	2.914	2.855	2.798	2.690	2.589	2.494	2.404	2.362	2.320	2.241	2.166	1.997	1.849
5	4.853	4.713	4.452	4.212	3.993	3.791	3.605	3.433	3.352	3.274	3.127	2.991	2.864	2.745	2.689	2.635	2.532	2.436	2.220	2.035
6	5.795	5.601	5.242	4.917	4.623	4.355	4.111	3.889	3.784	3.685	3.498	3.326	3.167	3.020	2.951	2.885	2.759	2.643	2.385	2.168
7	6.728	6.472	6.002	5.582	5.206	4.868	4.564	4.288	4.160	4.039	3.812	3.605	3.416	3.242	3.161	3.083	2.937	2.802	2.508	2.263
8	7.652	7.325	6.733	6.210	5.747	5.335	4.968	4.639	4.487	4.344	4.078	3.837	3.619	3.421	3.329	3.241	3.076	2.925	2.598	2.331
9	8.566	8.162	7.435	6.802	6.247	5.759	5.328	4.946	4.772	4.607	4.303	4.031	3.786	3.566	3.463	3.366	3.184	3.019	2.665	2.379
10	9.471	8.983	8.111	7.360	6.710	6.145	5.650	5.216	5.019	4.833	4.494	4.192	3.923	3.682	3.571	3.465	3.269	3.092	2.715	2.414
11	10.368	9.787	8.760	7.887	7.139	6.495	5.937	5.453	5.234	5.029	4.656	4.327	4.035	3.776	3.656	3.544	3.335	3.147	2.752	2.438
12	11.255	10.575	9.385	8.384	7.536	6.814	6.194	5.660	5.421	5.197	4.793	4.439	4.127	3.851	3.725	3.606	3.387	3.190	2.779	2.456
13	12.134	11.343	9.986	8.853	7.904	7.103	6.424	5.842	5.583	5.342	4.910	4.533	4.203	3.912	3.780	3.656	3.427	3.223	2.799	2.468
14	13.004	12.106	10.563	9.295	8.244	7.367	6.628	6.002	5.724	5.468	5.008	4.611	4.265	3.962	3.824	3.695	3.459	3.249	2.814	2.477
15	13.865	12.849	11.118	9.712	8.559	7.606	6.811	6.142	5.847	5.575	5.092	4.675	4.315	4.001	3.859	3.726	3.483	3.268	2.825	2.484
16	14.718	13.578	11.652	10.106	8.851	7.824	6.974	6.265	5.954	5.669	5.162	4.730	4.357	4.033	3.887	3.751	3.503	3.283	2.834	2.489
17	15.562	14.292	12.166	10.477	9.122	8.022	7.120	6.373	6.047	5.749	5.222	4.775	4.391	4.059	3.910	3.771	3.518	3.295	2.840	2.492
18	16.398	14.992	12.659	10.828	9.372	8.201	7.250	6.467	6.128	5.818	5.273	4.812	4.419	4.080	3.928	3.786	3.529	3.304	2.844	2.494
19	17.226	15.678	13.134	11.158	9.604	8.365	7.366	6.550	6.198	5.877	5.316	4.844	4.442	4.097	3.942	3.799	3.539	3.311	2.846	2.496
20	18.046	16.351	13.590	11.470	9.818	8.514	7.469	6.623	6.259	5.929	5.353	4.870	4.460	4.110	3.954	3.808	3.546	3.316	2.850	2.497
21	18.857	17.011	14.029	11.764	10.017	8.649	7.562	6.687	6.312	5.973	5.384	4.891	4.476	4.121	3.963	3.816	3.551	3.320	2.852	2.498
22	19.660	17.658	14.451	12.042	10.201	8.772	7.645	6.743	6.359	6.011	5.410	4.909	4.488	4.130	3.970	3.822	3.556	3.323	2.853	2.498
23	20.456	18.292	14.857	12.303	10.371	8.883	7.718	6.792	6.399	6.044	5.432	4.925	4.499	4.137	3.976	3.827	3.559	3.325	2.854	2.499
24	21.243	18.914	15.247	12.550	10.529	8.985	7.784	6.835	6.434	6.073	5.451	4.937	4.507	4.143	3.981	3.831	3.562	3.327	2.855	2.499
25	22.023	19.523	15.622	12.783	10.675	9.077	7.843	6.873	6.464	6.097	5.467	4.948	4.514	4.147	3.985	3.834	3.564	3.329	2.856	2.499
26	22.795	20.121	15.983	13.003	10.810	9.161	7.896	6.906	6.491	6.118	5.480	4.956	4.520	4.151	3.988	3.837	3.566	3.330	2.856	2.500
27	23.560	20.707	16.330	13.211	10.935	9.237	7.943	6.935	6.514	6.136	5.492	4.964	4.524	4.154	3.990	3.839	3.567	3.331	2.856	2.500
28	24.316	21.281	16.663	13.406	11.051	9.307	7.984	6.961	6.534	6.152	5.502	4.970	4.528	4.157	3.992	3.840	3.568	3.331	2.857	2.500
29	25.066	21.844	16.984	13.591	11.158	9.370	8.022	6.983	6.551	6.166	5.510	4.975	4.531	4.159	3.994	3.841	3.569	3.332	2.857	2.500
30	25.808	22.396	17.292	13.765	11.258	9.427	8.055	7.003	6.566	6.177	5.517	4.979	4.534	4.160	3.995	3.842	3.569	3.332	2.857	2.500

Appendix G

Equations and Factors for 10 Percent Interest

n	To Find F, Given P: $(1+i)^n$ $(F\|P)_{10}^n$	To Find P, Given F: $\dfrac{1}{(1+i)^n}$ $(P\|F)_{10}^n$	To Find A, Given F: $\dfrac{i}{(1+i)^n - 1}$ $(A\|F)_{10}^n$	To Find A Given P: $\dfrac{i(1+i)^n}{(1+i)^n - 1}$ $(A\|P)_{10}^n$	To find F, Given A: $\dfrac{(1+i)^n - 1}{i}$ $(F\|A)_{10}^n$	To Find P, Given A: $\dfrac{(1+i)^n - 1}{i(1+i)^n}$ $(P\|A)_{10}^n$
1	1.100	.9091	1.00000	1.10000	1.000	.909
2	1.210	.8264	.47619	.57619	2.100	1.736
3	1.331	.7513	.30211	.40211	3.310	2.487
4	1.464	.6830	.21547	.31547	4.641	3.170
5	1.611	.6209	.16380	.26380	6.105	3.791
6	1.772	.5645	.12961	.22961	7.716	4.355
7	1.949	.5132	.10541	.20541	9.487	4.868
8	2.144	.4665	.08744	.18744	11.436	5.335
9	2.358	.4241	.07364	.17364	13.579	5.759
10	2.594	.3855	.06275	.16275	15.937	6.144
11	2.853	.3505	.05396	.15396	18.531	6.495
12	3.138	.3186	.04676	.14676	21.384	6.814
13	3.452	.2897	.04078	.14078	24.523	7.103
14	3.797	.2633	.03575	.13575	27.975	7.367
15	4.177	.2394	.03147	.13147	31.772	7.606
16	4.595	.2176	.02782	.12782	35.950	7.824
17	5.054	.1978	.02466	.12466	40.545	8.022
18	5.560	.1799	.02193	.12193	45.599	8.201
19	6.116	.1635	.01955	.11955	51.159	8.363
20	6.727	.1486	.01746	.11746	57.275	8.514
21	7.400	.1351	.01562	.11562	64.002	8.649
22	8.140	.1228	.01401	.11401	71.403	8.772
23	8.954	.1117	.01257	.11257	79.543	8.883
24	9.850	.1015	.01130	.11130	88.497	8.985
25	10.835	.0923	.01017	.11017	98.347	9.077
26	11.918	.0839	.00916	.10916	109.182	9.161
27	13.110	.0763	.00826	.10826	121.100	9.237
28	14.421	.0693	.00745	.10745	134.210	9.307
29	15.863	.0630	.00673	.10673	148.631	9.370
30	17.449	.0573	.00608	.10608	164.494	9.427
31	19.194	.0521	.00550	.10550	181.943	9.479
32	21.114	.0474	.00497	.10497	201.138	9.526
33	23.225	.0431	.00450	.10450	222.252	9.569
34	25.548	.0391	.00407	.10407	245.477	9.609
35	28.102	.0356	.00369	.10369	271.024	9.644
40	45.259	.0221	.00226	.10226	442.593	9.779
45	72.890	.0137	.00139	.10139	718.905	9.863
50	117.391	.0085	.00086	.10086	1163.909	9.915
55	189.059	.0053	.00053	.10053	1880.591	9.947
60	304.482	.0033	.00033	.10033	3034.816	9.967
65	409.371	.0020	.00020	.10020	4893.707	9.980
70	789.747	.0013	.00013	.10013	7887.470	9.987
75	1,271.895	.0008	.00008	.10008	12708.954	9.992
80	2,048.400	.0005	.00005	.10005	20474.002	9.995
85	3,298.969	.0003	.00003	.10003	32979.690	9.997
90	5,313.023	.0002	.00002	.10002	53120.226	9.998
95	8,556.676	.0001	.00001	.10001	85556.760	9.999
100	13,780.612	.0001	.00001	.10001	137796.123	9.999

Learning Curve Coefficients

Unit Number	70%		80%		85%		90%	
	Unit Time L_{unit}	Total Time L_{total}	Unit Time L_{unit}	Total Time L_{total}	Unit Time L_{unit}	Total Time L_{total}	Unit Time L_{unit}	Total Time L_{total}
1	1.000	1.000	1.000	1.000	1.000	1.000	1.000	1.000
2	.700	1.700	.800	1.800	.850	1.850	.900	1.900
3	.568	2.268	.702	2.502	.773	2.623	.846	2.746
4	.490	2.758	.640	3.142	.723	3.345	.810	3.556
5	.437	3.195	.596	3.738	.686	4.031	.783	4.339
6	.398	3.593	.562	4.299	.657	4.688	.762	5.101
7	.367	3.960	.534	4.834	.634	5.322	.744	5.845
8	.343	4.303	.512	5.346	.614	5.936	.729	6.574
9	.323	4.626	.493	5.839	.597	6.533	.716	7.290
10	.306	4.932	.477	6.315	.583	7.116	.705	7.994
11	.291	5.223	.462	6.777	.570	7.686	.695	8.689
12	.278	5.501	.449	7.227	.558	8.244	.685	9.374
13	.267	5.769	.438	7.665	.548	8.792	.677	10.052
14	.257	6.026	.428	8.092	.539	9.331	.670	10.721
15	.248	6.274	.418	8.511	.530	9.861	.663	11.384
16	.240	6.514	.410	8.920	.522	10.383	.656	12.040
17	.233	6.747	.402	9.322	.515	10.898	.650	12.690
18	.226	6.973	.394	9.716	.508	11.405	.644	13.334
19	.220	7.192	.388	10.104	.501	11.907	.639	13.974
20	.214	7.407	.381	10.485	.495	12.402	.634	14.608
25	.191	8.40	.355	12.31	.470	14.80	.613	17.71
30	.174	9.30	.335	14.02	.450	17.07	.596	20.73
35	.160	10.13	.318	15.64	.434	19.29	.583	23.67
40	.150	10.90	.305	17.19	.421	21.43	.571	26.54
45	.141	11.62	.294	18.68	.410	23.50	.561	29.37
50	.134	12.31	.284	20.12	.400	25.51	.552	32.14
60	.122	13.57	.268	22.89	.383	29.41	.537	37.57
70	.112	14.74	.255	25.47	.369	33.17	.524	42.87
80	.105	15.82	.244	27.96	.358	36.80	.514	48.05
90	.099	16.83	.235	30.35	.348	40.32	.505	53.14
100	.094	17.79	.277	32.65	.340	43.75	.497	58.14
200	.066	25.48	.182	52.72	.289	74.79	.447	105.0
300	.053	31.34	.159	69.66	.263	102.2	.420	148.2
400	.046	36.26	.145	84.85	.245	127.6	.402	189.3
500	.041	40.58	.135	98.85	.233	151.5	.389	228.8
1,000	.029	57.40	.108	158.7	.198	257.9	.350	412.2
2,000	.020	80.96	.087	254.4	.168	438.9	.315	742.3
3,000	.016	98.90	.076	335.2	.153	598.9	.296	1047

Portions adapted (and modified) from Richard B. Chase and Nicholas J. Aquilano, *Production and Operations Management*, 6th Edition, Irwin, 1992, p. 584; and from William J. Stevenson, *Production/Operations Management*, 4th edition, Irwin, 1993, p. 409.

Appendix I

Normally Distributed Random Numbers

	1	2	3	4	5	6	7	8
1	.34	−.25	−.97	−.62	.37	−1.89	−.79	−87
2	−1.09	1.13	.99	.72	−.82	.46	−.41	.35
3	−1.87	.35	−.56	−.53	.91	− 48	1.31	.95
4	1.57	.75	1.20	2.29	.02	.67	−.41	.35
5	2.09	−1.54	1.02	−1.06	.65	−2.05	.73	−1.06
6	.37	.64	1.26	−.39	−.25	.53	.29	−.14
7	.03	−.71	1.08	.53	.28	.37	.27	−1.06
8	1.42	−.41	−.60	.75	−1.02	.91	2.11	.35
9	−.26	.99	−1.09	3.29	−.62	1.23	−1.36	.79
10	.93	.29	−.46	.63	1.84	−.36	.46	−1.00

Index